CRYSTAL GROWTH AND EVALUATION OF SILICON
for VLSI and ULSI

CRYSTAL GROWTH AND EVALUATION OF SILICON
for VLSI and ULSI

Golla Eranna

CRC Press
Taylor & Francis Group
Boca Raton London New York

CRC Press is an imprint of the
Taylor & Francis Group, an **informa** business

A CHAPMAN & HALL BOOK

Ingot cover image: Wacker Chemie AG

CRC Press
Taylor & Francis Group
6000 Broken Sound Parkway NW, Suite 300
Boca Raton, FL 33487-2742

First issued in paperback 2016

© 2015 by Taylor & Francis Group, LLC
CRC Press is an imprint of Taylor & Francis Group, an Informa business

No claim to original U.S. Government works

Version Date: 20150225

ISBN 13: 978-1-138-03419-8 (pbk)
ISBN 13: 978-1-4822-3281-3 (hbk)

Library of Congress Cataloging-in-Publication Data

Eranna, G.
    Crystal growth and evaluation of silicon for VLSI and ULSI / author, Golla Eranna.
      pages cm
    Includes bibliographical references and index.
    ISBN 978-1-4822-3281-3 (hardback : alk. paper) -- ISBN 1-4822-3281-2 (hardback : alk. paper) 1. Integrated circuits--Ultra large scale integration--Materials. 2. Silicon crystals--Testing. 3. Crystal growth. 4. Semiconductor wafers--Materials. I. Title.

TK7871.15.S55E73 2015
621.39'50284--dc23
                                           2015006228

Visit the Taylor & Francis Web site at
http://www.taylorandfrancis.com

and the CRC Press Web site at
http://www.crcpress.com

In Memory of My Parents

G. Sanjivappa

G. Govindamma

Contents

Preface xiii
About the Author xvii

1. Introduction 1
1.1   Silicon: The Semiconductor 2
1.2   Why Single Crystals 2
1.3   Revolution in Integrated Circuit Fabrication Technology
      and the Art of Device Miniaturization 4
1.4   Use of Silicon as a Semiconductor 6
1.5   Silicon Devices for Boolean Applications 11
1.6   Integration of Silicon Devices and the Art of Circuit
      Miniaturization 12
1.7   MOS and CMOS Devices for Digital Applications 18
1.8   LSI, VLSI, and ULSI Circuits and Applications 18
1.9   Silicon for MEMS Applications 20
1.10 Summary 23
References 23

2. Silicon: The Key Material for Integrated Circuit Fabrication Technology 27
2.1   Introduction 27
2.2   Preparation of Raw Silicon Material 28
2.3   Metallurgical-Grade Silicon 29
2.4   Purification of Metallurgical-Grade Silicon 31
2.5   Ultra-High Pure Silicon for Electronics Applications 37
2.6   Polycrystalline Silicon Feed for Crystal Growth 37
2.7   Summary 41
References 41

3. Importance of Single Crystals for Integrated Circuit Fabrication 45
3.1   Introduction 45
3.2   Crystal Structures 45
      3.2.1   Different Crystal Structures in Nature 47
      3.2.2   Cubic Structures 47
3.3   Diamond Crystal Structure 47
      3.3.1   Silicon Crystal Structure 47
      3.3.2   Silicon Crystals and Atomic Packing Factors 48
3.4   Crystal Order and Perfection 48
3.5   Crystal Orientations and Planes 50
3.6   Influence of Dopants and Impurities in Silicon Crystals 54
3.7   Summary 58
References 58

4. Different Techniques for Growing Single-Crystal Silicon 59
  4.1   Introduction 59
  4.2   Bridgman Crystal Growth Technique... 60
  4.3   Czochralski Crystal Growth/Pulling Technique 60
        4.3.1   Crucible Choice for Molten Silicon.. 64
        4.3.2   Chamber Temperature Profile 72
        4.3.3   Seed Selection for Crystal Pulling... 77
        4.3.4   Environmental and Ambient Control
               in the Crystal Chamber... 82
        4.3.5   Crystal Pull Rate and Seed/Crucible Rotation 84
        4.3.6   Dopant Addition for Growing Doped Crystals 94
               4.3.6.1   Boron 94
               4.3.6.2   Phosphorus... 95
               4.3.6.3   Arsenic.. 96
               4.3.6.4   Gallium 96
               4.3.6.5   Nitrogen.. 96
               4.3.6.6   Antimony... 97
               4.3.6.7   Germanium... 99
        4.3.7   Methods for Continuous Czochralski Crystal Growth......................... 100
        4.3.8   Impurity Segregation Between Liquid and
               Grown Silicon Crystals 102
        4.3.9   Crystal Growth Striations.. 107
        4.3.10 Use of a Magnetic Field in the Czochralski
               Growth Technique 108
        4.3.11 Large-Area Silicon Crystals for VLSI and
               ULSI Applications.. 117
        4.3.12 Post-Growth Thermal Gradient and Crystal Cooling
               after Pull-Out.. 122
  4.4   Float-Zone Crystal Growth Technique... 124
        4.4.1   Seed Selection... 125
        4.4.2   Environment and Chamber Ambient Control 125
        4.4.3   Heating Mechanisms and RF Coil Shape.................................... 125
        4.4.4   Crystal Growth Rate and Seed Rotation 126
        4.4.5   Dopant Distribution in Growing Crystals 128
        4.4.6   Impurity Segregation between Liquid and Grown
               Silicon Crystals.. 130
        4.4.7   Use of Magnetic Fields for Float-Zone Growth 130
        4.4.8   Large Area Silicon Crystals and Limitations
               of Shape and Size 131
        4.4.9   Thermal Gradient and Post-Growth Crystal Cooling 135
  4.5   Zone Refining of Single-Crystal Silicon 135
  4.6   Other Silicon Crystalline Structures and Growth Techniques 136
        4.6.1   Silicon Ribbons... 136
        4.6.2   Silicon Sheets 137
        4.6.3   Silicon Whiskers and Fibers 137
        4.6.4   Silicon in Circular and Spherical Shapes 137
        4.6.5   Silicon Hollow Tubes... 138

      4.6.6 Casting of Polycrystalline Silicon for Photovoltaic
             Applications 138
  4.7 Summary 138
  References 139

5. From Silicon Ingots to Silicon Wafers 157
  5.1 Introduction 157
  5.2 Radial Resistivity Measurements 157
  5.3 Boule Formation, Identification of Crystal Orientation, and Flats 158
  5.4 Ingot Slicing 162
  5.5 Mechanical Lapping of Wafer Slices 164
  5.6 Edge Profiling of Slices 167
  5.7 Chemical Etching and Mechanical Damage Removal 167
  5.8 Chemimechanical Polishing for Planar Wafers 168
  5.9 Surface Roughness and Overall Wafer Topography 170
  5.10 Megasonic Cleaning 170
  5.11 Final Cleaning and Inspection 171
  5.12 Summary 171
  References 172

6. Evaluation of Silicon Wafers 175
  6.1 Introduction 175
  6.2 Acoustic Laser Probing Technique 175
  6.3 Atomic-Force Microscope Studies on Surfaces 178
  6.4 Auger Electron Spectroscopic Studies 178
  6.5 Chemical Staining and Etching Techniques 181
  6.6 Contactless Characterization 184
  6.7 Deep-Level Transient Spectroscopy 185
  6.8 Defect Decoration by Metals 187
  6.9 Electron Beam and High-Energy Electron Diffraction Studies 188
  6.10 Flame Emission Spectrometry 188
  6.11 Four-Point Probe Technique for Resistivity Measurement
        and Mapping 189
  6.12 Fourier Transform Infrared Spectroscopy Measurements
        for Impurity Identification 191
  6.13 Gas Fusion Analysis 195
  6.14 Hall Mobility 195
  6.15 Mass Spectra Analysis 196
  6.16 Minority Carrier Diffusion Length/Lifetime/Surface
        Photovoltage 197
  6.17 Optical Methods for Impurity Evaluation 199
  6.18 Photoluminescence Method for Determining Impurity
        Concentrations 199
  6.19 Gamma-Ray Diffractometry 201
  6.20 Scanning Electron Microscopy for Defect Analysis 201
  6.21 Scanning Optical Microscope 202
  6.22 Secondary Ion Mass Spectrometer for Impurity Distribution 203

6.23   Spreading Resistance and Two-Point Probe Measurement Technique............205
6.24   Stress Measurements...207
6.25   Transmission Electron Microscopy...209
6.26   van der Pauw Resistivity Measurement Technique
        for Irregular-Shaped Wafers...210
6.27   X-ray Technique for Crystal Perfection and Dislocation Density......................210
6.28   Summary...214
References...214

7. **Resistivity and Impurity Concentration Mapping of Silicon Wafers**......225
7.1   Introduction...225
7.2   Electrically Active and Inactive Impurities...228
7.3   Surface Mapping and Concentration Contours.......................................228
7.4   Surface Roughness Mapping on a Complete Wafer.................................229
7.5   Summary...244
References...245

8. **Impurities in Silicon Wafers**..247
8.1   Effect of Intentional and Unintentional Impurities
       and Their Influence on Silicon Devices...247
8.2   Intentional Dopant Impurities in Silicon Wafers.....................................250
       8.2.1   Aluminum..250
       8.2.2   Antimony...250
       8.2.3   Arsenic...251
       8.2.4   Boron..252
       8.2.5   Gallium...253
       8.2.6   Phosphorus...253
8.3   Unintentional Dopant Impurities in Silicon Wafers.................................254
       8.3.1   Carbon..255
       8.3.2   Chromium...259
       8.3.3   Copper..260
       8.3.4   Germanium..261
       8.3.5   Gold..261
       8.3.6   Helium..262
       8.3.7   Hydrogen..262
       8.3.8   Iron...263
       8.3.9   Nickel..265
       8.3.10   Nitrogen..265
       8.3.11   Oxygen..268
       8.3.12   Tin...281
8.4   Other Metallic Impurities...282
8.5   Summary...282
References...283

9. **Defects in Silicon Wafers**...293
9.1   Introduction...293
9.2   Impact of Defects in Silicon Devices and Structures................................294
9.3   Point Defects and Vacancies...298
9.4   Line Defects..304

9.5 Bulk Defects and Voids 306
9.6 Dislocations and Screw Dislocations 310
9.7 Swirl Defects 312
9.8 Stacking Faults...315
9.9 Precipitations 322
9.10 Surface Pits/Crystal-Originated Particles 326
9.11 Grown Vacancies and Defects...329
9.12 Thermal Donors 331
9.13 Slips, Cracks, and Shape Irregularities 332
9.14 Stress, Bowing, and Warpage..334
9.15 Summary..337
References 337

10. Silicon Wafer Preparation for VLSI and ULSI Processing 347
10.1 Introduction 347
10.2 Purity of Chemicals Used for Silicon Processing..................................347
10.3 Degreasing of Silicon Wafers 348
10.4 Removal of Metallic and Other Impurities 348
10.5 Gettering of Metallic Impurities 351
10.6 Denuding of Silicon Wafers..362
10.7 Neutron Irradiation 366
10.8 Argon Annealing of Wafers 366
10.9 Hydrogen Annealing of Wafers..368
10.10 Final Cleaning, Rinsing, and Wafer Drying 371
10.11 Summary..371
References 372

11 Packing of Silicon Wafers..377
11.1 Packing of Fully Processed Blank Silicon Wafers 377
11.2 Storage of Wafers and Control of Particulate Contamination...........388
11.3 Storage of Wafers and Control of Particulate Contamination
     with Process-Bound Wafers 392
11.4 Summary..392
References 393

Index 395

Preface

Silicon, as a single-crystal semiconductor, has brought about a great revolution in the field of electronics and has touched almost all fields of science and technology. Though available abundantly in the form of silica and various other forms, in nature, it is a tough and challenging job to separate it from its chemical compounds because of its reactivity. As a solid, it is chemically inert and stable, but growing it as a single crystal poses many technological issues and one has to cross many hurdles. This book aims in the direction of getting a good-quality single-crystal silicon wafer suitable for very large scale integration (VLSI) and ultra large scale integration (ULSI) to accommodate many millions of transistors on a single block of silicon. With such a thick concentration density of transistors, failure of any single transistor stalls the electronic function of that part of the chip and the remaining chip becomes unusable for the electronic function for which it was designed. Hence, it is necessary to see that each and every silicon atom in the lattice is perfectly placed without any defects. Practically, it is impossible to achieve this perfection but one can try to minimize the deviations and the defects. This book covers the different approaches to select the basic silicon-containing compound, separation of silicon as metallurgical-grade pure silicon, subsequent purification, single-crystal growth, and the defects and evaluation of the deviations within the grown crystals.

The book has 11 chapters and each chapter covers a specific issue of the topic. In the first chapter, silicon has been highlighted as a semiconductor and the chapter discusses the revolution this has brought to the electronics industry. From the material point of view, silicon is abundant and widespread as a compound, and a material mixed with many metals and nonmetals. But it is the affinity for oxygen and the solubility with Group III and Group V compounds that brought the changes in electrical conductivity and gave birth to many new electronic devices. The art of miniaturization and integration of many devices and components on a single chip gave us complicated high-transistor-density circuits.

In Chapter 2, metallurgical-grade silicon extraction from sand is covered. This metallurgical-grade silicon is not qualified for device application, except that one can use it for low-efficiency solar cells. This silicon is purified through chemical reactions and converted into different chlorosilanes and silane. Through chemical vapor deposition (CVD), these compounds are converted to high-purity stable polysilicon, which becomes the feed for the crystal growth processes. At this stage, metallic impurities play a key role in the polysilicon.

In Chapter 3, the importance of single crystals for integrated-circuit fabrication is discussed. Various crystal structures are discussed but emphasis is given to the diamond crystal structure, as silicon materials form sp^3 hybridization and take this structure when it solidifies from a liquid to solid state. Because of its hybridization process, silicon-crystal lattice is not a densely packed structure, and it is only 34% filled space. The rest is an open structure. The presence of impurities disturbs the regular lattice parameters and creates local strains near the lattice points depending upon the position of the impurity atoms (whether they occupy the substitutional or interstitial locations).

Chapter 4 is important for its discussion of the different techniques used to grow the silicon single crystals. The Bridgman, Czochralski (CZ), float zone (FZ), and zone-refining methods are covered to grow single crystals of silicon. The CZ pulling method is the main technique used now to grow single crystals of silicon and large-sized crystals are possible only with this method. Polysilicon feed is used as a raw material and the entire charge is kept inside the quartz crucible, supported by graphite. By using a seed crystal, it is first allowed to touch the molten silicon and slowly pulled up, allowing the molten silicon to solidify according to the seed crystal. In the case of the FZ process, crucible contamination is eliminated and the crystal is grown vertically with the help of a seed and RF-induction heating element to melt the silicon locally. Large-sized crystals are difficult to grow and a limitation arises due to the induction-coil shape and configuration. Zone refining is used to get better crystals, similar to the float zone method. A number of repetitions provides better silicon crystals and will add to the cost. In addition to these techniques, other crystalline structures of silicon, including silicon ribbons, sheets, whiskers, fibers, circular and spherical shapes, and direct casting of polycrystalline silicon for photovoltaic applications, are also covered to some extent.

Chapter 5 covers how the grown single-crystal ingots are sliced into smaller discs and take shape to become a complete silicon wafer for integrated-circuit fabrication. Mechanical lapping and chemical polishing steps are carried out on these wafers to get a finely polished surface. Chemical etching and cleaning steps on these wafers free the silicon wafer surface from contamination.

Chapter 6 covers different methods to evaluate the silicon wafers in order to know the detailed specifications and their suitability for device applications. It is key to know the silicon wafer and its properties before it is used for any application, either for VLSI and ULSI or for MEMS structures. The methods to extract many important physical parameters about these wafers are listed in alphabetical order. These include its electrical, physical, bulk, and compositional parameters. Optical methods are important in determining the material semiconducting property, impurity composition, and intentional and unintentional dopant species present inside the silicon wafers. The four-point probe, spreading resistance, SIMS, Hall mobility, and DLTS methods are useful to know the electrical parameters and to ascertain the suitability of silicon wafers for any specific device application. Chemical etching methods, metal decoration, x-ray, and γ-ray diffractometry provide the details of crystal perfection and defects present inside the silicon lattice.

In Chapter 7, analysis of the silicon wafer in terms of resistivity and impurity concentration mapping is discussed. These two are key parameters for any silicon wafer as they decide threshold voltage (V_T) and also the gate-oxide integrity of MOSFET transistors. By wafer mapping, one can guess the quality of silicon wafer selected for any processing. The four-point probe approach is the only nondestructive method and it is routinely used for wafer resistivity evaluations.

In Chapter 8, the effect of intentional and unintentional impurities is discussed. Details regarding aluminum, antimony, arsenic, boron, gallium, and phosphorus are studied, as these are the intentional impurities used to alter the electrical properties of silicon. However, there are other impurities present in the silicon and they either occupy the interstitial positions or will try to occupy the lattice defects and become a part of the crystal, disturbing the very nature of the crystal properties. With large-sized defects, or at defect cluster points, they precipitate and form a larger defect and provide extra charge-leakage paths between different locations of the integrated circuits. This disturbs the circuit functionality and will lead to various chip reliability issues. We have discussed the unintentional impurities, such as carbon, chromium, copper, germanium, gold, helium, hydrogen, iron, nickel, nitrogen, oxygen, and tin. The composition of these impurities is difficult to guess, as it basically depends on the raw material selected for the preparation of the metallurgical-grade silicon.

Chapter 9 is about the defects found in regular silicon-crystal lattice. This may be an isolated single defect or a group of such defects joining together to form a cluster, or it can be a continuous defect extending to the physical dimensions of the crystal. We have discussed point, void, dislocation, line, screw dislocations, stacking faults, bulk, precipitations, surface pits and crystal-originated particles, grown vacancies, and thermal donors. Silicon crystal slips, cracks, shape irregularities, built-in stress, wafer bowing, and warpage are also discussed here.

Chapter 10 discusses silicon wafer preparation for VLSI and ULSI processing. This includes the purity of chemicals, degreasing of wafers, removal of metallic and other impurities from the silicon surface, the gettering process for metallic impurities, and creation of denuding zones in silicon wafers. Other methods discussed are neutron irradiation and argon and hydrogen annealing of wafers. Finally, the wafers are given a final cleaning and are properly dried for shipment.

In the last chapter, wafer packaging issues are discussed. Fully processed and cleaned silicon wafers need to be shipped with proper and perfect packing boxes, without causing any damage to the wafers, to users at many different locations. A broken wafer is considered useless and may not find a place in the batch of wafers for VLSI and ULSI processing. Initial packages were simple, but with the increase in size, packaging has become a major issue for wafer shipment and handling. New designs have been developed to control particulate contamination and to provide for proper storage.

I am greatly indebted to our CSIR-CEERI director, Dr. Chandra Shekhar, for his continuous support and also for providing the complete infrastructure, network facilities, and access to e-journals. The artwork for this book was done by Rajeev Soni and Jaspreet Kaur. I received a lot of support from A.K. Gupta, B.C. Pathak, A.K. Sharma, J. Bhargava, P. Kothari, T. Mudgal, and P. Someswara Rao at various stages of the manuscript preparation. Interactions with students helped me in refining the content. I thank all of them.

Sincere thanks are due to Taylor & Francis for giving me the opportunity to take up this project and for the total support I received from the team members. I wish to acknowledge the help from Aastha Sharma, Todd Perry, and other publication team members associated with this task.

I would also like to acknowledge the permissions granted by different societies for using their copyright material, particularly John Lewis, associate director of publications of The Electrochemical Society (ECS), IEEE Proceedings IEEE Transactions on Semiconductor Manufacturing, Journal of Crystal Growth and other Elsevier societies, MicroChemicals GmbH (Germany), MERSEN Immeuble La Fayotte (France), and the Copyright Clearance Center (USA) for clearing many of our requests. Open information photographs, reproduced in this book, are directly taken from general Internet websites and every effort was made to identify the owners of those images to get permissions, but this was not fully successful. We give full credit to the unknown owners of those images used in this book. We also declare that neither the author nor the publisher of this book claim any rights to these reproduced images. These pictures are relevant to this book on silicon crystal growth and the author felt it necessary to use these images for a better explanation and a more perfect understanding of this key technology. Thanks to Google Search for this help.

Last but not least, I acknowledge the continuous support I received from my family members. I thank my wife, Syamala Eranna, and my sons, G. Raviteja and G. Keshavaditya, for their support, patience, and also for critically discussing the content of the book. Much of the editing was carried out by them for this task, and at the same time they spared their valuable long weekends to help me to complete this specific assignment within the stipulated time.

G. Eranna, PhD
Central Electronics Engineering Research Institute
Pilani, India

About the Author

G. Eranna, PhD, obtained his master's degree in physics in 1978 from Sri Venkateswara University, Tirupati, India, with a top rank in the field of semiconductor physics. After that, he joined the Indian Institute of Technology (IIT) Madras, India, to work on silicon processing technology for the development of microwave p-i-n diodes and also obtained his PhD in 1985 from the same institute for his work on the process modeling of diffusion in silicon devices. Later, he moved to the IIT Kharagpur Microelectronic Centre and he was associated with it until 1991. Dr. Eranna joined CEERI, Pilani, India, as a scientist and is presently continuing as a senior principal scientist. He became a Professor under AcSIR, The Academy of Scientific and Innovative Research (CSIR, New Delhi), and regularly lectures to the graduate and research students on the topic of VLSI processing technology. He also maintains a full fledged semiconductor device fabrication laboratory, which includes all the key processing tools, such as mask fabrication, oxidation, diffusion, chemistry and chemical cleaning, photolithography, RIE and dry etching, chemical vapor deposition, ion implantation, sputter metallization, and characterization units.

Dr. Eranna's present research interests include, but are not limited to, microelectromechanical systems (MEMS), MEMS-based micro gas sensor devices, integrated circuit fabrication technology, CMOS process standardization, solid-state miniaturized micro gas sensors, and micro heaters for on-chip utilities for gas sensor applications. He has held a number of key administrative positions during his career. He is also a reviewer for many international journals. Dr. Eranna has received the Dr. K.S. Krishnan Memorial Award from IETE, New Delhi, India, for his scientific contributions. He has more than 100 publications to his credit, including both nationally and internationally reputed journals, conference proceedings, technical reports, and patents. His papers are widely cited and referenced. He has authored two books and has also contributed a chapter for another. He has been a visiting scientist at Technical University Ilmenau, Germany. Presently, he is the chief investigator for the micro gas sensor development project for developing VOC micro gas sensor devices. Dr. Eranna is a life member of IETE, New Delhi, India, the Semiconductor Society of India, and the Indo-French Technical Association (IFTA).

1

Introduction

Crystals are the best examples of the order and symmetry in nature, and this systematic arrangement of atoms or molecules in three-dimensional ways, in perfect order, highlights many interesting aspects of science and technology. Very few materials exhibit this type of natural perfection at atomic levels in their arrangement. Growing or arranging crystals in a regular and perfect order by artificial methods shows an insight into the bonding behavior between these atoms or molecules and the way they arrange themselves methodically, while transforming from a liquid state to a solid state.

In the case of crystalline solids, the atoms of the elements are arranged in a periodic format. When this periodicity extends throughout the bulk of solids, it is referred to as a single crystal or monocrystalline solid. Single crystals exhibit this perfect periodicity up to their physical dimensions where they are terminated with what are known as surface defects. At this point, the crystal arrangement discontinues, often giving a specific external physical shape to the solid body and exposing different crystalline planes of interior atomic arrangement. Exposed surfaces are active because of the unsaturated dangling bonds and show adsorption to various atmospheric gases. Exposing such shapes (or planes) is commonly used for diamonds and other gemstones found in jewelry and ornaments.

If the periodicity extends only to a limited range consisting of small crystalline portions, such solid materials are considered polycrystalline, and the part of that crystalline portion is indicated as a single grain. Each grain thus has a physical limitation and is separated with grain boundaries. With this arrangement, the grains generally adhere together in a random direction at the grain boundaries. When there is an absence of any specific periodicity of the arrangement in the solids without any grain boundaries, the result is an amorphous solid. No external planes are visible in such materials, and these solids can take any physical shape.

Out of the many solids that were put to technical use, it is the semiconductors that became popular and found many scientific applications, thus creating a great deal of interest in devices that were made out of a wide variety of semiconducting materials. The revolutionary effect that semiconductors have had starting in 1948 has resulted in new branches of solid-state physics. The advances in semiconductor theory and technology in the early 1950s provided a stimulus to physicists, chemists, and metallurgists in almost all branches of solid-state research. The electronic properties of these semiconductors have been studied more extensively, and many new semiconductors are being studied in terms of concepts similar to those already developed. The usage of these electronic devices brought about many revolutions in almost every branch of science, technology, communication, business, and entertainment, to name a few.

In addition, new electronic devices, such as diodes, transistors, lasers, and integrated circuits (ICs), made from these semiconductors have formed the basis of a new industry and changed the nature of electronic engineering. Device-processing technology, as well as the associated theory of semiconductors, advanced quite quickly as soon as the good quality of single-crystal semiconductors with controllable and predictable material properties became available commercially [1]. In particular, the first single-crystal semiconductor that could be prepared with a high degree of crystalline perfection and purity was germanium, followed by silicon and gallium arsenide. All of the semiconductors whose electronic properties are reasonably well understood are usually available as fairly large single crystals. The restriction to the crystalline state may appear arbitrary, but is practical from the viewpoint of simplicity.

1.1   Silicon: The Semiconductor

Silicon is the most important semiconducting material and is used for some of the previously mentioned applications, and silicon-based devices constitute almost 95% of the world production when compared to other semiconductor devices. Hence, it has become the most important material among electronic materials, and the production of silicon-based devices has dominated other semiconductor industries. This importance has caused many scientists and researchers to focus on silicon material and silicon-based devices. Today, silicon is the most studied material in the periodic table, and many research groups are active in studying silicon's production, preparation, purification, and crystal growth.

During the early days of semiconductor electronics, discrete transistors were typically made from germanium semiconducting material. The lower melting point of germanium made it easy to grow single crystals, but the real challenge was the impurity doping. The only option at that stage was to change the melt, with different dopants as per the requirement, and also the impurity type to get different doping concentrations. However, this trend has changed—today, lighter, cheaper, mechanically stronger, and above all more abundant silicon has replaced germanium in semiconductor device applications. W. G. Pfann [2] has recorded many interesting events in the semiconductor revolution and the device technologies that grew out of it. By the late 1960s, silicon was favored for two main reasons. Silicon has a larger bandgap (1.1 eV) when compared to germanium (0.66 eV), and this provides a device operating temperature above 100°C. The upper limit of operating temperature for germanium devices is 85°C, whereas for silicon devices it is 150°C. The second advantage is its stable oxide (i.e., silicon dioxide). This oxide is a perfect insulator and quite stable chemically with high dielectric strength [3]. Germanium does not offer any such insulating oxides, and oxides formed with germanium are not stable. Oxygen reacts with germanium to form a compound (GeO_2), but is less stable and is water soluble. However, germanium does exhibit better carrier mobility in comparison to silicon, a required property for high-speed device applications. The remarkable combination of high-quality material properties and the high-quality oxide insulator is unique among all the semiconducting materials and is responsible for the astonishing progress in silicon devices over the past half-century.

1.2   Why Single Crystals

The elements carbon, silicon, germanium, and tin belong to Group IV in the periodic table and have four valence electrons. When arranged in the diamond crystal, each atom has four neighbors that are oriented in specific directions. These directions are normally called the tetrahedral bonding directions. Each atom is at the center of a tetrahedron, the corners of which are occupied by the nearest neighbors. In many respects, silicon and germanium behave similarly [4], especially when comparing their properties at temperatures normalized with respect to melting temperatures.

By sharing electrons with their nearest neighbors, it is possible for the atoms to complete their outer shell of eight electrons, which is a general condition to maintain the octet rule. A crystal in which all the bond formations are completed by the shared electrons or those form covalent bonds with neighboring atoms is called a covalent crystal. The covalent bond may be contrasted with the ionic bond, in which the electron is exchanged between atoms, as in the case of ionic crystals such as sodium chloride. Due to this electron sharing or exchange in the covalent crystals, all the atoms are electrically neutral, whereas in the case of ionic crystals, the atoms are either ionized or electrically charged. In the III–V compounds, the bonding is generally neither completely covalent, as with silicon, nor is it completely ionic. The ionic behavior dominates when the combination of II–VI and I–VII compounds is formed.

A salient feature of the crystalline state is the regularity in the arrangement of the atoms in three dimensions. The is an aid in obtaining conclusions about the possible qualitative variety in the electrical behavior of semiconducting materials. The concept of the molecule is not applied in the case of the crystal, as there is no definite size applicable to it. In the case of polycrystalline silicon (polysilicon or poly-Si), small grains of single-crystal silicon are separated by thin grain boundaries. At the grain boundaries, grains of different orientations meet, resulting in an extremely thin amorphous layer. These boundaries create electrical barriers [5], and therefore, have a significant influence on the mechanical properties. Most of the devices are fabricated using single-crystal silicon, whereas heavily doped poly-Si is useful as a gate material in complementary metal-oxide semiconductor (CMOS) devices and in many metal-oxide semiconductor (MOS)-based circuits.

The approach used to study the crystal growth processes is an intimate mixing of experiments, understanding of physical phenomena, and numerical modeling [6]. Traditional crystal-growing techniques, particularly for single crystals of silicon, include the Czochralski (CZ) pulling method and float-zone (FZ) methods. However, these have evolved to a stage where dislocation-free and high-purity crystals are routinely produced for various applications to fabricate silicon-based devices. The needs of the IC industry have been the main driving force for these advances. Many of the advances in IC manufacturing achieved in recent years would not have been possible without parallel advances in silicon-crystal quality studies and growing defect-free engineering. Many techniques evolved to purify high-quality silicon using a variety of source materials [7–17], and large-size crystal growth techniques were developed to obtain perfect single crystals free from defects and unwanted impurities [18–23]. These studies provided enough impetus for advances in the development of technology for silicon-based ICs.

The technology for growing crystals, crystalline, and epitaxial layers has developed over the past 60-plus years into a multibillion-dollar industry that delivers basic components for many major high-end technologies with medical, military, space, and nuclear physics applications. The increased future requirements with respect to the size and quality of crystals necessitate a transition from empirical developments to increasingly scientific approaches [24]. There is a need for specialized crystal growth engineers and scientists for this important area [25,26]. Among the various materials from which single crystals can be grown from the melt, silicon plays a major role, given its extensive use as a substrate in various electronic applications. Hence, silicon has become the basic starting material for a large class of ICs. The silicon produced for the microelectronics industry is, far and away, the purest and most perfect crystalline material manufactured today. Wafers are cut from the grown ingots of single-crystal silicon that have been pulled from a crucible melt of pure molten polycrystalline silicon. Presently, it is grown routinely and in large volumes to meet the needs of many IC fabrication installations. In addition, the silicon crystal remains the largest cost contributor in total wafer manufacturing.

Critical assessment of silicon wafer-specific material parameters and their impact on the final device yields must be correlated and are to be proven so that the material parameters are not unnecessarily specified. For the crystal growth industry to prosper, the wafer suppliers must be able to improve the overall economic situation. As a result, the perfection of the grown crystals and further understanding of defect formation have improved considerably over the last decade.

1.3 Revolution in Integrated Circuit Fabrication Technology and the Art of Device Miniaturization

What makes silicon-based devices and their fabrication technology so interesting is a difficult question to answer, but other semiconductors are unlikely to replace the present silicon-based devices and give us better performance and usage over other materials [27]. Silicon device-processing technology stands tall among others and continues to dominate at present. Silicon's many individual properties, such as electrical, mechanical, chemical, and thermal, seem an excellent and, as a combined set, quite favorable material for electronic devices [28–30]. Table 1.1 [31–34] lists various physical parameters of silicon.

Most of today's products are designed, or rather refined, using silicon devices. The microchip, an IC currently being used in almost all electronic gadgets and equipment, has turned 55 recently. The progression from a single individual transistor in the early 1950s to the more than 100 million transistors on a single chip of the size of a postage stamp available today is an incredible triumph of human ingenuity. In engineering's accumulative string of inventions, this is the biggest success to date. However, because many fundamental laws of physics make it unlikely that we will see many more developments for this material, we are likely approaching the saturation stage. The result is an ultimate scale in silicon—the smallest silicon transistor, the standard brick for all future silicon buildings and its operational limitations.

Scientists have recently claimed to have created the world's smallest transistor from a "quantum dot" of just seven atoms of silicon. This transistor can be used to regulate and control electrical current flow just like a normal transistor, but it represents a key step into a new age of atomic-scale miniaturization suitable for superfast computers. Here, the silicon atoms are manipulated individually and placed precisely to create a working device. This is the latest technological achievement using silicon as a semiconducting material.

Many governments have recognized micro- and nanotechnologies as enabling technologies with exceptional economic potential and have embraced them as centerpieces of their technology policy [35]. By using technology road-mapping, evolution and revolutions of silicon technology have been studied [36,37] with greater care demanding more and more complicated circuits to help the fast-changing scenario by cramming more number of devices. Walsh et al. [38] have analyzed the revolutionary, evolutionary, cumulative, or obsolescent nature of core competencies and capabilities driven by disruptive technologies from a market-driven perspective. They opined that the change in the required competencies has been more cumulative in nature.

TABLE 1.1

Some Important Properties of Silicon at Room Temperature

Parameter	Physical value
Acceptors	Aluminum (Al)
	Boron (B)
	Gallium (Ga)
Atomic density	5.02×10^{22} atoms/cm^3
Atomic number	14
Atomic weight	28.09 g/Mole
Average optical phonon energy	0.063 eV
Boiling point	2355°C
Breakdown field	3.0×10^5 V.cm^{-1} (approx)
Bulk modulus	9.8×10^{11} dynes/cm^2
C11	16.6×10^{11} dynes/cm^2
C12	6.40×10^{11} dynes/cm^2
C44	7.96×10^{11} dynes/cm^2
Color	Blue-grey
Conduction band density of states	2.80×10^{19} cm^{-3}
Crystal structure	Cubic diamond
Density (solid)	2.328 g/cm^3
Density (liquid)	2.530 g/cm^3
Density of states effective mass of electrons $m_e{}^*$	1.08 m_o
Density of states effective mass of holes $m_h{}^*$	0.81 m_o
Dielectric constant	11.7–11.9
Donors	Antimony (Sb)
	Arsenic (As)
	Phosphorus (P)
Effective density of states (valence band) N_v	1.04×10^{19} cm^{-3}
Effective density of states (conduction band) N_c	2.8×10^{19} cm^{-3}
Electron affinity	4.05 eV
Electron drift mobility	1500 cm^2 V^{-1} s^{-1}
Electron saturation velocity v_{sat}	1.0×10^7 cm s^{-1}
Emissivity at high temperature	0.55 (approx)
Energy gap	1.12 eV
Fracture strength	6.0 GPa
Hardness	7.0 Mohs
Heat capacity (solid)	4.78 cal g^{-1} mol^{-1}K^{-1}
Heat capacity (liquid)	6.76 cal g^{-1} mol^{-1}K^{-1}
Hole drift mobility	475 cm^2 V^{-1} s^{-1}
Hole saturation velocity v_{sat}	1.0×10^7 cm s^{-1}
Index of refraction	3.42
Intrinsic carrier concentration	1.25×10^{10} cm^{-3}
Intrinsic Debye length	24 μm
Intrinsic resistivity	2.3×10^5 Ω-cm
Knoop hardness	950–1150 kg/mm^2
Latent heat of fusion	340 cal/g
Lattice constant	5.43095 Å
Lattice electron mobility	1450 cm^2/V s

(Continued)

TABLE 1.1 (*Continued*)

Parameter	Physical value
Lattice hole mobility	500 cm^2/V s
Melting point	1412°C
Minority carrier lifetime	2.5×10^{-3} s
Modulus of elasticity in <100>direction	1.3×10^{11} N/m^2
Modulus of elasticity in <110>direction	1.7×10^{11} N/m^2
Modulus of elasticity in <111>direction	1.9×10^{11} N/m^2
Optical phonon energy	0.063 eV
Optical phonon mean-free path	7.6 nm (for electrons) 5.5 nm (for holes)
Piezoresistive coefficients	n-Si ($\pi_{11} = -102.2$) p-Si ($\pi_{44} = +138.1$)
Poisson's ratio	0.42
Poisson's ratio in <100>direction	0.28
Poisson's ratio in <111>direction	0.262
Relative permittivity ε_r	11.9
Residual stress	0 (none)
Shear modulus	5.2×10^{11} dynes/cm^2
Silicon radius	1.18 Å
Specific heat	0.7 J/g K 0.169 cal/g K
Symbol	Si
Temperature resistivity coefficient	0.0017 (*p*-type)
Tensile strength in <111>direction	3.5×10^8 N/m^2
Thermal conductivity (solid)	1.57 W/cm K
Thermal conductivity (liquid)	4.30 W/cm K
Thermal diffusivity	0.92 cm^2 s^{-1}
Thermal expansion coefficient	$2.33 - 2.6 \times 10^{-6}$ K^{-1}
Torsion modulus	4050 kg mm^{-2}
Valence band density of states	1.04×10^{19} cm^{-3}
Vapor pressure	1.0×10^{-7} Torr (@ 1050°C)
Young's modulus in <100>direction	13.0×10^{11} dynes/cm^2
Young's modulus in <111>direction	19.0×10^{11} dynes/cm^2

Sources: B. El-Kareh, *Fundamentals of Semiconductor Processing Technology*, Kluwer Academic Publishers, Boston, 1995 [31]; J. Vanhellemont and E. Simoen, "Brother silicon, sister germanium," *210 ECE Meeting Cancun*, Mexico, 29 Oct–3 Nov 2006 [32]; S. K. Ghandhi, *VLSI Fabrication Principles: Silicon and Gallium Arsenide*, John Wiley & Sons, Inc., New York, 1994; M. Madou, *Fundamentals of Microfabrication*, Boca Raton, CRC Press, 1997 [33]; W. E. Beadle, J. C. C. Tsai, and R. D. Plummer, *Quick Reference Manual for Silicon Integrated Circuit Technology*, Wiley, 1985 [34].

1.4   Use of Silicon as a Semiconductor

Silicon is the eighth most abundant element in the Earth's crust, making up to 25.7% by mass next to oxygen element. This material is everywhere, but not in the form of a pure element because of its reactive properties, particularly with oxygen. It is most widely distributed in the form of dusts, sands, planetoids, and planets as various forms of silicon dioxide (silica) or silicates. The earth is more precisely constructed by different silica compounds and mixed compositions with wide variety of metals, with the possible exception of the core. Elemental silicon is a nontoxic substance—it is a safe and proven material that lasts nearly forever. When combined with carbon at higher temperatures, silicon forms silicon carbide, which is a stable compound. In the pure form, it is quite stable in any typical environment, although it will form a thin nascent silicon (sub)oxide of the order of 20 Å on the exposed silicon surfaces because of its chemical affinity toward oxygen [39]. This thin oxide layer quickly forms with atmospheric oxygen and covers the pure silicon even in a room-temperature environment. Once formed, this layer saturates under normal ambient conditions.

As per the periodic table, both silicon ($_{14}Si^{28}$) and carbon ($_6C^{12}$) belong to Group IVA, and they have similar electronic configurations in their outermost electron shells. Therefore, it is expected that they exhibit similar physical and chemical properties. Five principal allotropic forms of carbon are known: diamond, graphite, carbyne, lonsdaleite, and fuller-ene. In contrast to carbon, silicon atoms in all compounds, including amorphous silicon, display a valence of four. Valence electrons in these atoms are in sp^3-hybrid state, and until recently, it was thought that no such formations exist where silicon atoms would have sp^2 and/or sp hybridization [40]. However, for semiconductor device applications, we concentrate only on sp^3 hybridization. Other states are not considered in this book due to insufficient information presently available in the literature. Semiconductor properties are not exclusive for the single-crystal materials. Many noncrystalline materials do exhibit semiconductor properties similar to those of crystalline semiconductors. However, silicon is the best understood and most widely used material for IC fabrication.

Silicon in its pure or intrinsic state is a semiconductor and is not a good candidate for electrical conduction. Its bulk electrical resistance lies somewhere between that of a good conductor, like a metal, and that of an insulator. The intrinsic resistivity of silicon is much higher ($2.3 \times 10^5 \, \Omega$ cm) in its purest form. The electrons in the inner shells of the semiconductors do not participate in deciding the electrical conductivity of the materials and have no influence on these properties, either directly or indirectly. Figure 1.1 shows the band diagram of the intrinsic (normal) silicon where there are no charge carriers present for electrical conduction at 0 K. At room temperature, some of the charge carriers get excited, and electrons shift to the conduction band and an equal number of holes in valence band,

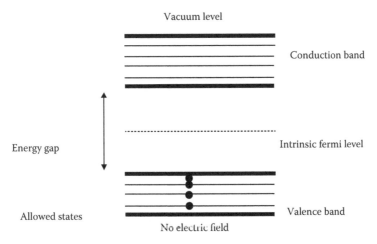

FIGURE 1.1
Band diagram of intrinsic and pure silicon semiconductor.

giving it basic (semi)conducting properties. The conductivity produced by these charge carriers, thermally excited across the energy bandgap, is called intrinsic conductivity. This creation of electrons and holes is a strong function of temperature, and at 0 K, the material shows no electrical conduction and behaves like a perfect insulator.

However, the electrical conductivity of silicon can be changed over several orders of magnitude by introducing impurity atoms into the bulk silicon lattice. These impurity dopants alter the electrical properties by supplying either free electrons or free holes. The presence of these additional electrons and holes drastically alters the conduction properties of silicon as they directly participate in the electrical conduction, along with their intrinsic charge carriers, originally present at room temperature.

Impurity elements present in the regular silicon lattice sites, which use outermost non-bonded (or not paired for covalent bond formation) electrons, are referred to as acceptors since they accept some of the electrons present in the other silicon atoms, leaving vacancies or holes (missing electron sites). This movement of bonded electrons to nonbonded sites permits the charge movement and allows electrical conduction to take place. In this case, the charge movement takes place through the exchange process, and all the lattice atoms participate in the process of electrical conduction. The presence of an extra electron on the acceptor atom changes its state from normal to ionized. The details of electrical conduction are shown in Figure 1.2.

In a similar fashion, other impurity atoms that can donate free (or extra) electrons after completing the octet criteria are referred to as donor atoms. This free electron is not participating in any bond formation and does not interact with the bonded electrons of other lattice atoms, but is held to the impurity site by virtue of its nuclear charge. The binding energy associated with this extra electron is quite small. It can be easily removed by the

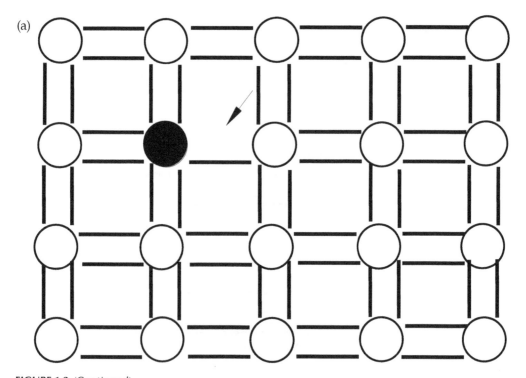

(a)

FIGURE 1.2 *(Continued)*

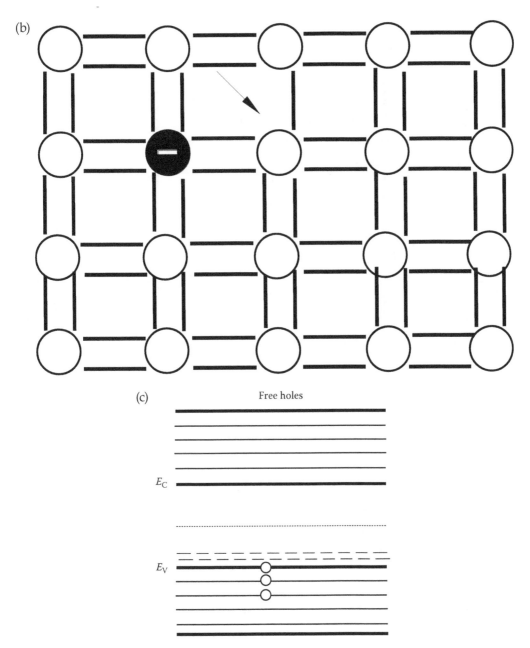

FIGURE 1.2
Silicon doped with trivalent atomic species from Group III of the periodic table element to convert it into *p*-type material. (a) and (b) show the charge movement and (c) shows the band model and free flow of holes used for electrical conduction.

external force and, hence, is responsible for the additional electrical conduction in bulk silicon, which is much more like its normal intrinsic electrical conductivity. Except for the electronic scattering, these free electrons show good mobility. By leaving its extra electron, the donor atoms become ionized due to the nuclear state. The details of electrical conduction due to a free electron is shown in Figure 1.3. Silicon that contains a majority of donor

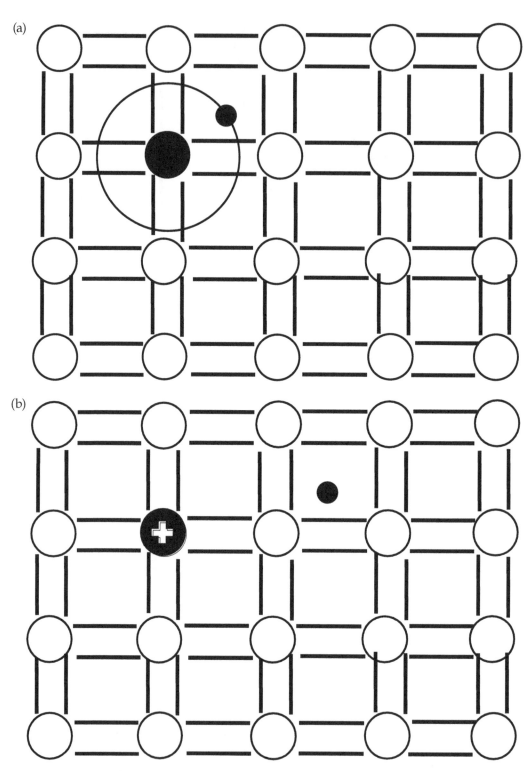

FIGURE 1.3 (*Continued*)

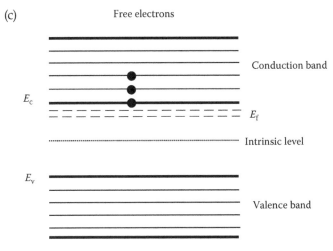

FIGURE 1.3
Silicon doped with pentavalent atomic species from Group V of the periodic table element to convert it into
n-type material. (a) and (b) show the free movement of free electrons and (c) shows the band model and free flow
of electrons for electrical conduction.

TABLE 1.2

Misfit Factors of Typical Impurities in Silicon Lattice

Impurity	Radius (Å)	Misfit Factor
Aluminum (Al)	1.26	0.068
Antimony (Sb)	1.36	0.153
Arsenic (As)	1.18	0.000
Boron (B)	0.88	0.254
Gallium (Ga)	1.26	0.068
Indium (In)	1.44	0.220
Phosphorus (P)	1.10	0.068

Source: B. El-Kareh, *Fundamentals of Semiconductor Processing Technology*, Kluwer Academic Publishers, Boston,
1995 [31].

atoms is known as *n*-type silicon and that which contains a majority of acceptor atoms is
known as *p*-type silicon.

Not all impurity atoms can fit well within the silicon lattice at substitution positions.
Among the popular impurity atoms, only arsenic impurity matches perfectly with the
silicon lattice, whereas other impurities, such as boron, phosphorus, gallium, aluminum,
indium, and antimony, produce strain in the lattice locations. The misfit factors are listed
in Table 1.2 [31]. Because of the limited options, any one of the listed species can be selected
for silicon doping by altering the electrical conduction properties.

1.5 Silicon Devices for Boolean Applications

The processing characteristics and some of the material properties of silicon wafers
depend on its crystal orientation. Traditionally, all the bipolar devices are fabricated in

<111>-oriented silicon wafers, whereas <100> materials are preferred for MOS devices. The silicon crystal <111> planes have the highest density of atoms on the surface, so crystals grow most easily on these planes. Because of this, other silicon properties, such as oxidation, proceed at a higher pace when compared to other crystal planes.

When *n*-type and *p*-type materials are brought together, the region where the silicon changes is called a *p-n* junction. By arranging junctions in certain physical structures, such as *n-p-n*, *p-n-p*, *n*-MOS, *p*-MOS, or *p-n-p-n*, and combining these with other physical structures, various semiconductor devices are constructed. Over the years, silicon semiconductor processing has evolved sophisticated techniques for building these junctions and other structures with special properties.

The electronic properties of metal-oxide semiconductor field-effect transistor (MOSFET) transistors make them useful as an electronic switch to realize many Boolean applications. Initial experiments were carried out using bipolar transistors, but these have been subsequently replaced by MOS transistors. Presently, CMOS circuits are in use. Here, 0 and 1 are explored by using the device's conducting and nonconducting states for this purpose.

1.6   Integration of Silicon Devices and the Art of Circuit Miniaturization

The manufacturing of a silicon chip is a tedious process, involving many steps. For a brief overview, one can refer to Ref. [41]. The process is expensive to maintain and requires elaborate, costly, and sophisticated equipment. A clean room environment, high-purity chemicals, critical gases and gas mixtures, and qualified and trained manpower are some of the key issues. Silicon technology creates devices, wires, and circuits almost concurrently by means of a lithography technique, and this important step will decide the shape, yield, and chip functionality.

The basic starting material used in modern semiconductor industries is a single-crystal wafer, or disk, of silicon, which varies from 75 mm to 300 mm in diameter and is less than 1 mm thick. The fabrication and building up of the entire circuit process is carried out on this single-crystal slab of silicon, which is available in thin, flat, and circular shapes. Typical specifications for different diameter wafers are listed in Table 1.3, for 6-inch (150 mm) wafers in Table 1.4, for 8-inch (200 mm) wafers in Table 1.5, and for 12-inch (300 mm) wafers in Table 1.6 [42]. Depending on the application and the intended device (or circuit), the parameters and specifications do change. The presence of unintended impurities, such as oxygen and carbon, even at the parts per billion (ppb) level, influence the device characteristics and determine the reliability of circuits fabricated in them. Initial defect density, multiplication of these defects during high-temperature thermal cycles, and the total thermal budget of the silicon wafer need to be carefully watched from the beginning to the end of the process cycle. A larger thermal budget may increase the number of secondary defects inside the bulk silicon wafer.

Intrinsic point defects in silicon are an important parameter to evaluate the crystal quality. The properties of these defects in silicon have been studied for more than half a century and are still a major area of research. After an initial focus on the role of vacancies, it was soon realized that self-interstitials also play an important role in many of the physical processes, such as dopant diffusion and defect formation [32]. As a result of a huge research effort, the solubility and diffusivity of intrinsic point defects are quite well known,

TABLE 1.3

Typical Specifications for Large-Size Silicon Wafers Used for Integrated Circuit Fabrication

Diameter	200 ± 0.2 mm
Thickness	725 ± 10 μm
Taper	<10 μm
Global flatness	<3 μm
Local flatness	<1 μm over 20 mm × 20 mm
Bow	<10 μm
Warpage	<10 μm
Orientation	<100> or <111>
Surface finish	10 nm for polished wafer
Defects	Dislocations – 0
	OISF – $\leq 3 \times 10^4/m^2$
	S-pits – $\leq 10^6/m^2$
Oxygen	≤ 2 ppma, <3% radial variation
Carbon	≤ 0.3 ppma
Metals	≤ 0.01 ppba

Sources: Data from Wikipedia, the free encyclopedia; T. F. Shao and F. C. Wang, "Wafer fab manufacturing technology" in *ULSI Technology*, edited by C. Y. Chang and S. M. Sze, McGraw-Hill Companies, Inc., New York, 1996; Lecture notes of Prof. P. K. Chu, University of Honk Kong, Tat Chee Avenue, Kowloon, Hong Kong; S. Franssila, *Introduction to Microfabrication*, John Wiley & Sons, Chichester, 2004 [42], and many other websites available in the literature.

TABLE 1.4

150 mm Diameter Silicon Wafer Specifications for CMOS and MEMS Applications

Crystal/Bulk			
Growth technique			CZ
Diameter			150 ± 0.5 mm
Orientation			100
Orientation tolerance		Degree	±0.2
Off orientation		Degree	0 to 4
Dopant			Boron/phosphorus
Resistivity target range	Prime - boron	Ohm-cm	0.5–50
	Prime - phosphorus	Ohm-cm	1.0–50
Radial resistivity variation	Boron typical	%	<5
	Phosphorus typical	%	<15
Oxygen target range ($\pm$Tol)	Prime - boron	at cm^{-3}	$5 - 7.8 \times 10^{17}$ (±0.5)
	Prime - phosphorus	ASTM F121-83	$6 - 7.5 \times 10^{17}$ (±0.5)
Minority carrier lifetime		μs	>200
Radial oxygen variation	Typical	%	<5
Bulk metal concentration	Fe	at cm^{-3}	$= 5.0 \times 10^{10}$
Bulk carbon concentration	Measured on wafer	at cm^{-3}	$- 2.0 \times 10^{10}$

(*Continued*)

TABLE 1.4 (*Continued*)

Polished Wafer					
Surface metals	Cu, Cr, Fe, Ni	at cm^{-2}	$= 2.5 \times 10^{10}$		
	Al, Zn, K, Na, Ca	at cm^{-2}	$= 5.0 \times 10^{10}$		
LLSs particles	Size	μm	>0.2	>0.16	>0.13
	Pol prime	# per wafer	<15–35	<20–120	<70–600
	Monitor	# per wafer	<15	<20–65	<130–700
Diameter tolerance		mm	±0.2		
Bow and warp	Polished without layer	μm	<20		
Wafer thickness	Standard	μm	675		
Thickness tolerance		μm	±15		
GBIR = TTV		μm	<3.5		
GFLR = TIR		μm	<2.0		
Local flatness	SFQR/ STIRmax	μm	<0.25		
	SFQR/SFPD	μm	<0.18		
	SFQR/ STIRmin	μm	<0.7		
Standard site size		mm^2	25 x 25		
Oxygen-induced stacking fault (OISF)		cm^{-2}	<100		
Dislocations		Etch pits/cm^2	<50		
Slip line		Total wafer	None		
Haze		Total wafer	None		
Other surface defects (pits, dimples, cracks, chips, scratches, organic peels, contamination etc.)		Total wafer	None visible		

Sources: Data from Wikipedia, the free encylopedia; T. F. Shao and F. C. Wang, "Wafer fab manufacturing technology" in *ULSI Technology*, edited by C. Y. Chang and S. M. Sze, McGraw-Hill Companies, Inc., New York, 1996; Lecture notes of Prof. P. K. Chu, University of Honk Kong, Tat Chee Avenue, Kowloon, Hong Kong; S. Franssila, *Introduction to Microfabrication*, John Wiley & Sons, Chichester, 2004 [42], and many other websites available in the literature.

Notes: GBIR, global backside indicated reading; TTV, total thickness variation; TIR, total internal reflection; SFDR, spurious-free dynamic range; STIR, short T1 inversion recovery; SFQR, site flatness quality requirements.

TABLE 1.5

200 mm Diameter Silicon Wafer Specifications for CMOS VLSI Circuits

Crystal/Bulk

Growth technique			CZ		
Diameter			200 ± 0.5 mm		
Orientation			100		
Orientation tolerance		Degree	± 0.2		
Off orientation		Degree	0–4		
Dopant			Boron/phosphorus		
Resistivity target range	Prime - boron	Ohm-cm	0.5–50		
	Prime - phosphorus	Ohm-cm	1.0–50		
Radial resistivity variation	Boron typical	%	<5		
	Phosphorus typical	%	<15		
Oxygen target range ($\pm$Tol)	Prime - boron	at cm^{-3}	5–7.8×10^{17} (±0.5)		
	Prime - phosphorus	ASTM F121-83	6–7.5×10^{17} (±0.5)		
Minority carrier lifetime		μs	>200		
Radial oxygen variation	Typical	%	<5		
Bulk metal concentration	Fe	at cm^{-3}	$= 5.0 \times 10^{10}$		
Bulk carbon concentration	Measured on wafer	at cm^{-3}	$= 2.0 \times 10^{10}$		

Polished Wafer

Surface metals	Cu, Cr, Fe, Ni	at cm^{-2}	$= 2.5 \times 10^{10}$		
	Al, Zn, K, Na, Ca	at cm^{-2}	$= 5.0 \times 10^{10}$		
LLSs particles	Size	μm	>0.2	>0.16	>0.13
	Pol prime	# per wafer	<15–35	<20–120	<70–600
	Monitor	# per wafer	<15	<20–65	<130–700
Diameter tolerance		mm	± 0.2		
Bow and warp	Polished without layer	μm	<20		
Wafer thickness	Standard	μm	725		
Thickness tolerance		μm	± 15		
GBIR = TTV		μm	<3.5		
GFLR = TIR		μm	<2.0		
Local flatness	SFQR/ STIRmax	μm	<0.25		
	SFQR/SFPD	μm	<0.18		
	SFQR/ STIRmin	μm	<0.7		
Standard site size		mm^2	25×25		
Oxygen-induced stacking fault (OISF)		cm^{-2}	<100		

(Continued)

TABLE 1.5 (*Continued*)

Dislocations	Etch pits/cm^2	<50
Slip line	Total wafer	None
Haze	Total wafer	None
Other surface defects (pits, dimples, cracks, chips, scratches, organic peels, contamination, etc.)	Total wafer	None visible

Sources: Data from Wikipedia, the free encyclopedia; T. F. Shao and F. C. Wang, "Wafer fab manufacturing technology" in *ULSI Technology*, edited by C. Y. Chang and S. M. Sze, McGraw-Hill Companies, Inc., New York, 1996; Lecture notes of Prof. P. K. Chu, University of Honk Kong, Tat Chee Avenue, Kowloon, Hong Kong; S. Franssila, *Introduction to Microfabrication*, John Wiley & Sons, Chichester, 2004 [42], and many other websites available in the literature.

Notes: GBIR, global backside indicated reading; TTV, total thickness variation; TIR, total internal reflection; SFDR, spurious-free dynamic range; STIR, short T1 inversion recovery; SFQR, site flatness quality requirements.

TABLE 1.6

300 mm Diameter Silicon Wafer Specifications for CMOS VLSI Circuits

Crystal/Bulk

Diameter			300 ± 0.5 mm		
Orientation			100		
Orientation tolerance		Degree	±0.2		
Off orientation		Degree	0		
Dopant			Boron/phosphorus		
Resistivity target range	Prime - boron	Ohm-cm	0.5–50		
	Prime - phosphorus	Ohm-cm	1.0–50		
Radial resistivity variation	Boron typical	%	<10		
	Phosphorus typical	%	<15		
Oxygen target range (±Tol)	Prime - boron	at cm^{-3}	4.8–7.8 × 10^{17} (±0.5)		
	Prime - phosphorus	ASTM F121-83	4.8–7.8 × 10^{17} (±0.5)		
Minority carrier lifetime		µs	>200		
Radial oxygen variation	Typical	%	<10		
Bulk metal concentration	Fe	at cm^{-3}	= 5.0 × 10^{10}		
Bulk carbon concentration	Measured on wafer	at cm^{-3}	= 2.0 × 10^{10}		

Polished Wafer

Surface metals	Cu, Cr, Fe, Ni	at cm^{-2}	= 5.0 × 10^9		
	Al, Zn, K, Na, Ca	at cm^{-2}	= 2.0 × 10^{10}		
LLSs particles	Size	µm	>0.2	>0.16	>0.12

TABLE 1.6 *(Continued)*

	Polished prime	# per wafer	<30	<40–300	<200–1000
	Monitor	# per wafer	<30	<60	<100
	Monitor	# per wafer	-	-	<50
Diameter tolerance		mm	±0.2		
Bow and warp	Polished without layer	μm	<50		
Wafer thickness	Standard	μm	775		
Thickness tolerance		μm	±25		
GBIR = TTV		μm	<3		
GFLR = TIR		μm	<2.0		
Local flatness	SFQR/STIRmax	μm	<0.2		
	SFQR/SFPD	μm	<0.14		
	SFQR/STIRmin	μm	<0.7		
Standard site size		mm^2	25 × 25		
Oxygen-induced stacking fault (OISF)		cm^{-2}	<100		
Dislocations		Etch pits/cm^2	<50		
Slip line		Total wafer	None		
Haze		Total wafer	None		
Other surface defects (pits, dimples, cracks, chips, scratches, organic peels, contamination etc.)		Total wafer	None visible		

Sources: Data from Wikipedia, the free encyclopedia; T. F. Shao and F. C. Wang, "Wafer fab manufacturing technology" in *ULSI Technology*, edited by C. Y. Chang and S. M. Sze, McGraw-Hill Companies, Inc., New York, 1996; Lecture notes of Prof. P. K. Chu, University of Honk Kong, Tat Chee Avenue, Kowloon, Hong Kong; S. Franssila, *Introduction to Microfabrication*, John Wiley & Sons, Chichester, 2004 [42], and many other websites available in the literature.

Notes: GBIR, global backside indicated reading; TTV, total thickness variation; TIR, total internal reflection; SFDR, spurious-free dynamic range; STIR, short T1 inversion recovery; SFQR, site flatness quality requirements.

although a significant discrepancy still exists between the properties, as determined from defect formation during crystal pulling, equal to high-temperature values, and those determined from metal diffusion experiments carried out to study the crystal defects.

In fact, silicon represents the most important market in bulk crystal growth today, and this has driven an intense activity in numerical modeling since only large companies can afford to use computer simulation-based tools in their research and development (R&D) effort for better quality crystals. The most challenging issues for silicon technology over the coming decade is to find the ideal combination of material properties matching the needs of device manufacturing to support the efforts of customers in order to reach maximum yield. The major driving forces for materials suppliers remain the same as in past years: higher percentage of utilized wafer area, greater throughput and capacity, and lower production costs per wafer. Many research groups are active in this endeavor to get the best quality single crystals possible.

1.7   MOS and CMOS Devices for Digital Applications

Today, most polished CZ wafers are used for MOSFET devices such as dynamic random access memory (DRAM); most epitaxial (epi) wafers (i.e., 2–4 µm thick epi on ~0.7 mm CZ substrates) are used for bipolar ICs, including application-specific integrated circuits (ASICs) and microprocessors. As the industry evolves toward 256 Gbit DRAM and bipolar ICs with 180 million transistors, epitaxial wafers, as well as silicon-on-insulator (SOI) wafers (i.e., 50–200 nm thick silicon, <200 nm SiO_2 on ~0.7 mm thick CZ substrates) will be used increasingly, along with polished CZ wafers [27]. However, in this book, an effort is made to study the nonepitaxial wafers for fabricating very large scale integration (VLSI) and ultra large scale integration (ULSI) circuits. These CZ wafers are further used as a substrate for growing epitaxial layers, and a fresh thin epitaxial layer of silicon is actually used to fabricate ICs.

1.8   LSI, VLSI, and ULSI Circuits and Applications

Iwai and Ohmi [43] explained the tremendous progress that has taken place in the area of silicon IC development, the downsizing of the MOSFET over the last few decades, and how the progress in silicon IC fabrication technologies changed over time. They also covered the wide range of difficulties in further downsizing future circuits. The history of electronic circuits and large scale integration (LSI) is shown in Figure 1.4. Tables 1.7 through 1.9 list the number of transistors on a single silicon chip and the year the achievement was made, the minimum feature size realized on the chip and the corresponding year, and the different devices and the level of circuit complexity over the decades, respectively. Figure 1.5 shows some of the pictures of the process-bound silicon wafers and packaged devices with external contacts.

Accurate control of the defects in single silicon crystals and the wafers is a subject of immense importance in the fabrication of VLSI and ULSI chips. Many aspects of the silicon device industry have changed over the past several years to minimize these defects. Quite a few of the changes have been a result of two simple facts. There has been a huge increase in the costs of both developing large-size wafer silicon products and testing new silicon products by users [44]. Two general problems associated with defectivity in conventional silicon wafers have plagued the industry over the years. One relates to difficulties in specification and the other to the generally complex interaction between the material and the process. Both of these problems are equally important, and each has outward-rippling implications of its own.

In sub-0.25 µm processing, 200 mm wafers meet ever-tightening wafer quality requirements in terms of crystalline defects, heavy metallic impurities, oxygen level, oxygen radial uniformity, minority carrier lifetimes and particles, etc. [31]. However, there must be

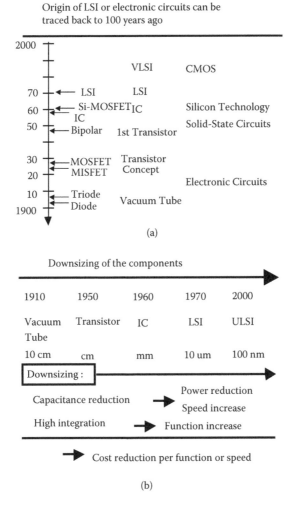

FIGURE 1.4

(a) History of electronic circuits and VLSI and ULSI in twentieth century. (b) Trend in the downsizing of components. (From H. Iwai and S. Ohmi, *Microelectronics Reliability*, **42**, 465–491, 2002 [43].)

fundamental new approaches and solutions to meet quality requirements and reduced costs on 300 mm before an investment into the next diameter change (200 mm) can be realized.

The semiconductor industry has seen quality improvements in silicon crystals grown using the CZ technique. In today's emerging nano-IC manufacturing, using 300 mm wafers meet ever-tightening quality requirements in terms of crystalline defects, heavy metallic impurities (at ppb levels), oxygen concentration levels, radial uniformity of oxygen distribution, minority carrier lifetimes (of both recombination and generation), particles, etc. While production of 300 mm wafers is quite popular, quality requirements will need to be updated and will be met by further developments and refinements in crystal growth technology for larger size wafers, particularly in terms of point and secondary defect control and control of oxygen in grown crystals [18].

The industry's need for continued research in crystal growth for better and higher quality crystals for VLSI substrates and for more and more perfect crystals with zero defects

TABLE 1.7

Single Silicon Chip and the Number of Transistors Accommodated on Them as Recorded at
Different Years

Silicon Chip	No. of Transistors on a Single Chip	Year
First Microprocessor 4004	2,000	1971
8088 Chip	29,000	1978
80286 or 286 Chip	134,000	1982
386 Chip	275,000	1985
Pentium Chip	3,000,000	1993
Pentium-3 Chip	9,500,000	1999
Pentium-4 Chip	55,000,000	2002
Centrino Chip	77,000,000	2003
Pentium-D Chip	230,000,000	2005
Itanium-2 Chip	172,000,000	2006
Intel Core-2 Duo	290,000,000	2006
Intel Core-2 Quad	580,000,000	2007

Sources: Data from Wikipedia, the free encyclopedia; T. F. Shao and F. C. Wang, "Wafer fab manufacturing technology" in *ULSI Technology*, edited by C. Y. Chang and S. M. Sze, McGraw-Hill Companies, Inc., New
York, 1996; Lecture notes of Prof. P. K. Chu, University of Honk Kong, Tat Chee Avenue, Kowloon,
Hong Kong; S. Franssila, *Introduction to Microfabrication*, John Wiley & Sons, Chichester, 2004 [42], and
many other websites available in the literature.

has led many scientific groups to concentrate on the growth techniques. The role of carbon
atom impurity and oxygen precipitations in the crystal lattice are some of the burning
issues explored today [45]. Making these atomic species inactive, if not totally removing
them from the silicon lattice, is a real challenge because of the high chemical affinity of
silicon for these atomic species.

An industry-wide survey of some of the latest 300 mm equipment and materials shows
that the challenges of processing 300 mm wafers are being understood and addressed.
Alternative transistor technologies are advancing into nonplanar and multiple-gate technologies for 32 nm and below design rules. Nearly all practical problems (minority carrier
lifetime, metal precipitation, stacking faults, etc.) associated with metal contaminants have
largely been solved through advances in crystal purity [45]. Intensive studies regarding
control of oxygen and its interactions with intrinsic point defects to provide internal gettering, as opposed to the traditional external gettering layers, have enabled the industry to
migrate to double-side polished substrates.

1.9 Silicon for Microelectromechanical System (MEMS) Applications

In an excellent article by Petersen [46] exposed the excellent mechanical properties of single crystal silicon to use it as a strong material for specific applications. This is in addition
to the perfect electrical properties it originally possesses. This combination has opened a
new field of scientific research coupling both the properties of silicon. Single-crystal silicon
properties are being increasingly employed for many electronic circuits and for VLSI and
ULSI applications—its mechanical properties are equally important to exploit for strong

TABLE 1.8

The Minimum Line Width Achieved at Different Points of Time

Year	Minimum Line Width
1960	30 μm
1964	20 μm
1968	12 μm
1970	8 μm
1973	5 μm
1975	3 μm
1979	2 μm
1983	1.5 μm
1986	1.2 μm
1989	0.8 μm
1992	0.5 μm
1995	0.35 μm
1998	0.25 μm
2000	0.18 μm
2002	0.13 μm
2004	0.09 μm (90 nm)
2006	65 nm
2008	45 nm
2010	32 nm
2012	30 nm

Sources: Data from Wikipedia, the free encyclopedia; T. F. Shao and F. C. Wang, "Wafer fab manufacturing technology" in *ULSI Technology*, edited by C. Y. Chang and S. M. Sze, McGraw-Hill Companies, Inc., New York, 1996; Lecture notes of Prof. P. K. Chu, University of Honk Kong, Tat Chee Avenue, Kowloon, Hong Kong; S. Franssila, *Introduction to Microfabrication*, John Wiley & Sons, Chichester, 2004 [42], and many other websites available in the literature.

TABLE 1.9

Different Devices from Silicon and Circuit Complexity over the Decades

Year	1947	1950	1961	1966	1971	1980	1990	2000
Technology	Transistor invention	Discrete devices	SSI	MSI	LSI	VLSI	ULSI	GSI
Approximate number of transistors per chip	1	1	10	100 to 1,000 (1 K)	1,000 to 20 K	20 K to 1,000 K	1,000 K to 10,000 K	More than 10,000 K
Typical devices	Point contact diode	Junction transistors, diodes	Planar devices, logic gates, flip-flops	Counters, multiplexers, adders	8-bit ROM RAM	16-, 32-bit microprocessors	Special processors	Many more complex circuits

Sources: Data from Wikipedia, the free encyclopedia; T. F. Shao and F. C. Wang, "Wafer fab manufacturing technology" in *ULSI Technology*, edited by C. Y. Chang and S. M. Sze, McGraw-Hill Companies, Inc., New York, 1996; Lecture notes of Prof. P. K. Chu, University of Honk Kong, Tat Chee Avenue, Kowloon, Hong Kong; S. Franssila, *Introduction to Microfabrication*, John Wiley & Sons, Chichester, 2004 [42], and many other websites available in the literature.

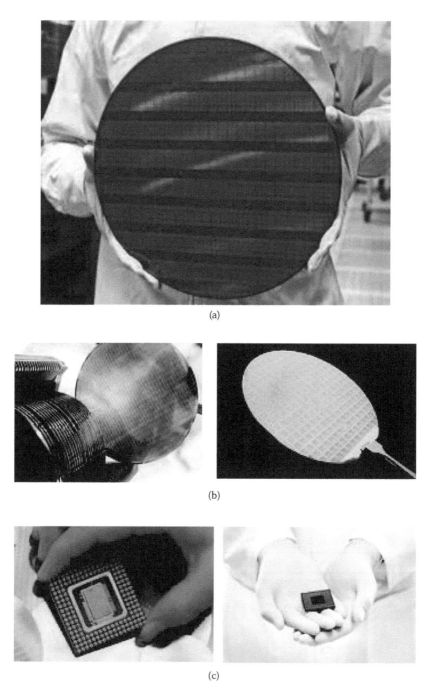

(a)

(b)

(c)

FIGURE 1.5
(a) A fully processed silicon wafer, (b) process-bound silicon wafers, and (c) packaged silicon chip with external contacts. (Information from Wikipedia, the free encyclopedia; T. F. Shao and F. C. Wang, "Wafer fab manufacturing technology" in *ULSI Technology*, edited by C. Y. Chang and S. M. Sze, McGraw-Hill Companies, Inc., New York, 1996; Lecture notes of Prof. P. K. Chu, University of Honk Kong, Tat Chee Avenue, Kowloon, Hong Kong; S. Franssila, *Introduction to Microfabrication*, John Wiley & Sons, Chichester, 2004, and many other websites available in the literature [42].)

and rugged miniaturized mechanical devices. The trends in engineering applications indicate a growing interest in the use of silicon as a mechanical material, with the ultimate goal of developing a broad range of inexpensive, batch-fabricated, high-performance sensors and transducers that can be interfaced with silicon as an electronic material. It is apparent that silicon will continue to be aggressively exploited in a wide variety of mechanical applications complementary to its traditional role as an electronic material. The combination may provide us with miniaturized mechanical devices and components that are unthinkable with any other solid-state material.

1.10 Summary

In this chapter, we explored silicon as a semiconductor and the revolution it brought to the electronics industry. From a material point of view, silicon is abundant and widespread in nature as both a compound and mixed material with many metals and nonmetals. But it is the affinity for oxygen and the solubility with Group III and Group V compounds that brought changes in electrical conductivity and gave birth to many new electronic devices. Despite limitations in charge mobility and the speed at which these devices can be operated, the material flexibility was the reason for silicon's use in Boolean applications. The art of miniaturization and integration of many devices and components gave us complicated VLSI and ULSI circuits. Today, it is difficult to imagine any gadget where silicon chips are not used. Its mechanical properties have opened a new field of research and spawned another revolution in the MEMS category. Integration of both is being explored. Because of these qualities, today, silicon has become the most studied material in the periodic table.

References

1. J. L. Moll, *Physics of Semiconductors*, McGraw-Hill Inc., New York, 1964.
2. W. G. Pfann, "The semiconductor revolution," *Journal of the Electrochemical Society*, **121**, 9C–15C, 1974.
3. G. Fisher, M. R. Seacrist, and R. W. Standley, "Silicon crystal growth and wafer technologies," *Proceedings of the IEEE*, **100**, 1454–1474, 2012.
4. J. Vanhellemont and E. Simoen, "Brother silicon, sister germanium," *Journal of the Electrochemical Society*, **154**, H572–H583, 2007.
5. P. J. French, "Polysilicon: A versatile material for microsystems," *Sensors and Actuators*, **A99**, 3–12, 2002.
6. T. Duffar, "Crystal growth process engineering," *Journal of Optoelectronics and Advanced Materials*, **2**, 432–440, 2000.
7. J. A. Amick, "Purification of rice hulls as a source of solar grade silicon for solar cells," *Journal of the Electrochemical Society*, **129**, 864–866, 1982.
8. L. P. Hunt, J. P. Dismukes, J. A. Amick, A. Schei, and K. Larsen, "Rice hulls as a raw material for producing silicon," *Journal of the Electrochemical Society*, **131**, 1683–1686, 1984.
9. A. Sanjurjo, L. Nanis, K. Sancier, R. Bartlett, and V. Kapur, "Silicon by sodium reduction of silicon tetrafluoride," *Journal of the Electrochemical Society*, **128**, 179–184, 1981.

10. C. H. Lewis, H. C. Kelly, M. B. Giusto, and S. Johnson, "Preparation of high-purity silicon from silane," *Journal of the Electrochemical Society*, **108**, 1114–1118, 1961.
11. A. Yusa, Y. Yatsurugi, and T. Takaishi, "Ultrahigh purification of silane for semiconductor silicon," *Journal of the Electrochemical Society*, **122**, 1700–1705, 1975.
12. W. Lee, W. Yoon, and C. Park, "Purification of metallurgical-grade silicon in fractional melting process," *Journal of Crystal Growth*, **312**, 146–148, 2009.
13. T. L. Chu, G. A. van der Leeden, and H. I. Yoo, "Purification and characterization of metallurgical silicon," *Journal of the Electrochemical Society*, **125**, 661–665, 1978.
14. T. L. Chu and S. S. Chu, "Partial purification of metallurgical silicon by acid extraction," *Journal of the Electrochemical Society*, **130**, 455–457, 1983.
15. K. Abe, T. Matsumoto, S. Maeda, H. Nakanishi, K. Hoshikawa, and K. Terashima, "Oxygen solubility in Si melts: Influence of boron addition," *Journal of Crystal Growth*, **181**, 41–47, 1997.
16. T. Oishi, M. Watanabe, K. Koyama, M. Tanaka, and K. Saegusa, "Process for solar grade silicon production by molten salt electrolysis using aluminum-silicon liquid alloy," *Journal of the Electrochemical Society*, **158**, E93–E99, 2011.
17. J. Cai, X.-t. Luo, G. M. Haarberg, O. E. Kongstein, and S.-I. Wang, "Electrorefining of metallurgical grade silicon in molten $CaCl_2$ based salts," *Journal of the Electrochemical Society*, **159**, D155–D158, 2012.
18. K.-M. Kim, "Materials: Silicon-pulling technology for 2000+," Solid State Technology, **43**, p.69, January 2000, and the references therein.
19. EBSCO Host Trade Publication, "Suppliers' successes with 300 mm tools and materials," Solid State Technology, **44**, p.44, May 2001.
20. W. Sittenthaler, "Wafer suppliers face dilemma in doing much more with less," *WaferNEWS*, **48**, July 2005.
21. E. Dornberger, J. Virbulis, B. Hanna, R. Hoelzl, E. Daub, and W. von Ammon, "Silicon crystals for future requirements of 300 mm wafers," *Journal of Crystal Growth*, **229**, 11–16, 2001.
22. Y. Shiraishi, K. Takano, J. Matsubara, T. Iida, N. Takase, N. Machida, M. Kuramoto, and H. Yamagishi, "Growth of silicon crystal with a diameter of 400 mm and weight of 400 kg," *Journal of Crystal Growth*, **229**, 17–21, 2001.
23. Z. Lu and S. Kimbel, "Growth of 450 mm diameter semiconductor grade silicon crystals," *Journal of Crystal Growth*, **318**, 193–195, 2011.
24. H. J. Scheel, "Historical aspects of crystal growth technology," *Journal of Crystal Growth*, **211**, 1–12, 2000.
25. R. Fornani, "Bulk crystal growth of semiconductors: An overview" in *Comprehensive Semiconductor Science and Technology*, edited by P. Bhattacharya, R. Fornari, and H. Kamimura, Elsevier B.V., 2011.
26. V. V. Voronkov, B. Dai, and M. S. Kulkarni, "Fundamentals and engineering of the Czochralski growth of semiconductor silicon crystals" in *Comprehensive Semiconductor Science and Technology*, edited by P. Bhattacharya, R. Fornari, and H. Kamimura, Elsevier B.V., 2011.
27. H. L. Stormer, "Silicon forever! Really?," *Solid-State Electronics*, **50**, 516–519, 2006.
28. H. R. Philipp and E. A. Taft, "Optical constants of silicon in the region 1 to 10 eV," *Physical Review*, **120**, 37–38, 1960.
29. E. J. Boyd and D. Uttamchandani, "Measurement of the anisotropy of Young's modulus in single-crystal silicon," *Journal of Microelectromechanical Systems*, **21**, 243–249, 2012.
30. M. A. Hopcroft, W. D. Nix, and T. W. Kenny, "What is the Young's modulus of silicon?," *Journal of Microelectromechanical Systems*, **19**, 229–238, 2010.
31. B. El-Kareh, *Fundamentals of Semiconductor Processing Technology*, Kluwer Academic Publishers, Boston, 1995.
32. J. Vanhellemont and E. Simoen, "Brother silicon, sister germanium," *210 ECE Meeting Cancun*, Mexico, 29 Oct–3 Nov 2006.
33. S. K. Ghandhi, *VLSI Fabrication Principles: Silicon and Gallium Arsenide*, John Wiley & Sons, Inc., New York, 1994; M. Madou, Fundamentals of Microfabrication, Boca Raton, CRC Press, 1997.

34. W. E. Beadle, J. C. C. Tsai, and R. D. Plummer, *Quick Reference Manual for Silicon Integrated Circuit Technology*, Wiley, 1985.

35. M. Kautt, S. T. Walsh, and K. Bittner, "Global distribution of micro-nano technology and fabrication centers: A portfolio analysis approach," *Technological Forecasting and Social Change*, **74**, 1697–1717, 2007.

36. R. Phaal, C. J. P. Farrukh, and D. R. Probert, "Technology roadmapping – A planning framework for evolution and revolution," *Technological Forecasting and Social Change*, **71**, 5–26, 2004.

37. S. T. Walsh, "Road mapping a disruptive technology: A case study – The emerging microsystems and top-down nanosystems industry," *Technological Forecasting and Social Change*, **71**, 161–185, 2004.

38. S. T. Walsh, R. L. Boylan, C. McDermott, and A. Paulson, "The semiconductor silicon industry roadmap: Epochs driven by the dynamics between disruptive technologies and core competencies," *Technological Forecasting and Social Change*, **72**, 213–236, 2005.

39. E. A. Taft, "Growth of native oxide on silicon," *Journal of the Electrochemical Society*, **135**, 1022–1023, 1988.

40. A. F. Khokhlov and A. I. Mashin, "On silicon allotropy," *Journal of Optoelectronics and Advanced Materials*, **4**, 523–533, 2002.

41. S. P. Bates, "Silicon wafer processing" *Applied Materials*, Silicon_wafer_processing.pdf, 1–15, Summer, 2000.

42. Data from Wikipedia, the free encyclopedia; T. F. Shao and F. C. Wang, "Wafer fab manufacturing technology" in *ULSI Technology*, edited by C. Y. Chang and S. M. Sze, McGraw-Hill Companies, Inc., New York, 1996; Lecture notes of Prof. P. K. Chu, University of Honk Kong, Tat Chee Avenue, Kowloon, Hong Kong; S. Franssila, *Introduction to Microfabrication*, John Wiley & Sons, Chichester, 2004, and many other websites available in the literature.

43. H. Iwai and S. Ohmi, "Silicon integrated circuit technology from past to future," *Microelectronics Reliability*, **42**, 465–491, 2002.

44. R. Falster, "Defect control in silicon crystal growth and wafer processing," *MEMC Electronic Materials SpA*, Novara 28100, Italy, Defect_Control_in_Silicon_Crystal_Growth.pdf, 1–12, and the references therein.

45. B. L. H. Wilson, "Industry needs in semiconductor crystal growth," *Journal of Crystal Growth*, **79**, 3–11, 1986.

46. K. E. Petersen, "Silicon as a mechanical material," *Proceedings of the IEEE*, **70**, 420–457, 1982.

2

Silicon: The Key Material for Integrated Circuit Fabrication Technology

"You can't see it, but it's everywhere you go."

—Journalist Bridget Booher on silicon

2.1 Introduction

Silicon, the key material for today's integrated circuit fabrication technology, is often referred to as the king among semiconductors due to its usage and the flexibility it offers in the processing of a wide variety of different devices and circuits. Silicon produced for the microelectronics industry is, by far, the purest and most perfect crystalline material manufactured at present for a wide range of applications. Many devices are fabricated routinely using silicon, and in large volumes. The advances in integrated circuit manufacturing achieved in recent years would not have been possible without parallel advances in studying silicon crystal quality and defect engineering evaluations. Essentially all practical problems, such as minority carrier lifetime, metal precipitation, stacking faults, etc. [1], associated with metal contaminants have largely been solved through advances in evaluating the purity of grown crystals.

Silicon belongs to the Group IVA family in the periodic table, along with carbon, germanium, tin, and lead. Its atomic number is 14 and it has an atomic weight of 28.086. Silicon has 24 known isotopes, with mass numbers ranging from 22 to 45. Among them, Si^{28} is the most abundant isotope, with a value of 92.23%, followed by Si^{29}, with a value of 4.67%, and Si^{30}, with a value of 3.1%. These are the most stable isotopes. Silicon melts at 1440°C, and boils at 2355°C. The outer electron configuration of each of these atoms is s^2p^2. Carbon and silicon are nonmetals, white tin and lead are metals, and germanium and gray tin are known semiconductors [2]. The five elements listed in this group have different physical and chemical properties despite being in the same column in the periodic table.

Obtaining high resistivity is generally considered a good quality for silicon wafers, since the intrinsic resistivity of silicon is large—of the order of 2.3×10^5 Ω-cm—and every effort is made to reach to this level. However, electrical resistivity alone does not uniquely specify the purity of silicon. The number of charge carriers measured is a net number of charge carriers available for electrical conduction, which, in the worst situation, may result due to a near compensation of a relatively high concentration of both donor and acceptor species present in the material [3]. Many metallic contamination issues are being explored, and transition-metal contamination is one of the key issues in obtaining silicon wafers that have both high and suitable resistivity.

Silicon is the most abundant element, and we use it every day in the form of various glasses and pottery. Just as carbon is to the living world, silicon is to the nonliving world [2]. It is most widely distributed in the form of sand and various forms of silicon dioxide (silica) or silicates. The combination of silicates with many metallic ions is a common metallurgical issue, but simple silica with minimum metallic composition is a better choice for obtaining metallurgical-grade silicon. Silicon is extracted from silica-rich sands such as quartzite silica (SiO_2) [4]. In nature, silicates typically contain large amounts of unwanted impurities because of the complexity of the Si-O-Si network and its accommodating nature to various metallic ions, either as part of a bridging network lattice or simply as nonbridging any occupying interstitial space available in the solid phase. The initial stages of the manufacturing process are more concerned with large metallic compositions of the starting material and aim for oxides with minimum metallic compositions.

Metallic impurities are known to be harmful in very large scale integration (VLSI) and ultra large scale integration (ULSI) devices for a number of reasons [5]. The presence of metallic precipitates is known to enhance the injection rate of electrons into the SiO_2 layer, and interstitially occupied metallic impurities increase the thermally activated component of the leakage current through *p-n* junctions by providing an extra path for the charge movement. With the continuous decrease in device dimensions, heavily doped junctions are becoming increasingly shallow, and potential metallic precipitation will have an increased effect on junction performance and the yields. The gate oxides are becoming so thin that leakage current is controlled by Fowler-Nordheim tunneling or direct tunneling, which is sensitive to metallic contamination at the Si/SiO_2 interface. Minimizing the metallic contamination or reducing its influence on device performance is an important branch of reliability engineering.

2.2   Preparation of Raw Silicon Material

Silicon was discovered by Jöns Jacob Berzelius at 1824 in Sweden. The origin of the name is the Latin word *silicis* meaning "flint." It was not until 1854 that pure crystalline silicon was produced. This material is silvery grey in color and metallic looking. It has a low density (2.328 g/cm³) and is brittle in nature. Special care is required when handling silicon. Silicon is often compared with carbon, but they differ widely.

Most industries aim to select silica sand that is suitable for producing metallurgical-grade polysilicon, which typically has a purity greater than 98%. This is then further refined to produce solar and semiconductor grades. Polysilicon undergoes various stages of purification, using different methods, to eliminate unwanted impurities. Hence, this metallurgical-grade silicon (MGS) becomes the feedstock for the crystal growth to produce semiconductor-grade, single-crystal silicon that is suitable for device manufacturing.

Raw silicon materials are generally quartzite [6], pelletized granular quartz [7], rice hulls as a source of silicon [8,9], or rice hull ash for pure silica [10]. Crossman and Baker [6] have reported using quartzite as a raw material to obtain MGS from a submerged electrode arc furnace. This is a carbothermic process in which quartzite is reduced in an arc furnace in the presence of carbon. To create upgraded metallurgical silicon, high-purity quartz and carbon are needed. These source materials are limited in availability and are substantially higher in cost than the source materials. Amick et al. [7] suggested an improved arc furnace operating at 100 kW using pelletized granular quartz as a feedstock. They reportedly achieved lowered boron concentration in the silicon, which was evident in the performance of the solar cells fabricated using it. It was further pointed out [8] that rice hulls offer an alternative low-cost, solar-produced source material containing both silicon dioxide and carbon as raw materials. Through simple leaching treatments, the metal ion content of these hulls can be greatly decreased, making them potentially better suitable as a source material.

Hunt et al. [9] pointed out that rice hulls typically contain two basic components that are needed to produce silicon: basic silica and the carbon. The rice hull contains about 15% of fixed carbon. Impurity analyses have indicated that rice hulls from various sources are compositionally similar and that they have low concentrations of aluminum and iron, the major metal impurities in conventional raw materials used to prepare metallurgical silicon. The levels of the other major impurities, such as Ca, K, Mg, and Mn, can be reduced by a factor of about 100 by using hot hydrochloric acid leaching. Burning the rice hulls leaves an ash that consists mainly of silica with about 5% impurity oxides. Kalapathy et al. [10] came out with a simple method for the production of pure silica from rice hull ash. The method is based on alkaline extraction followed by acid precipitation to produce pure silica xerogels with minimal mineral contaminants. The major impurities present in silica, at this stage, were found to be mainly consisting of Na, K, and Ca.

Recently, Oishi et al. [11] proposed a new method for producing solar-grade silicon through molten salt electrolysis using an aluminum-silicon liquid alloy. This process consists of an electrolysis step in which SiO_2 is dissolved in a fluoride melt using an aluminum-silicon liquid alloy cathode. Next, the silicon is precipitated from the liquid alloy. The preliminary electrolysis in a NaF-AlF_3-SiO_2 melt at 1273 K revealed that Si was actually formed by electrolysis when the SiO_2 was reduced by the aluminum. The proposed process has the possibility of producing solar-grade silicon in combination with directional solidification. The silicon's purity was estimated to be acceptable, and as such, the molten salt electrolysis step has no purification effect. Although the contaminants from the molten salt and other materials were not taken into consideration, it indicates the possibility of the process as a new solar-grade silicon production method.

2.3 Metallurgical-Grade Silicon

Today, silicon is primarily extracted by chemically reducing SiO_2 with carbon. A large quantity of MGS is produced in a submerged electrode arc furnace, as shown in Figure 2.1. The furnace is charged with quartzite and carbon in the form of coal, coke, and wood chips. Silica is difficult to decompose into silicon and oxygen and needs high temperatures for this reaction to take place. Very high temperatures on the order of 2000°C are created inside the furnace, and at this temperature, liquid silicon is produced and the oxygen is removed by the presence of carbon, leading to the formation of carbon monoxide (CO) and other silicon compounds such as volatile silicon monoxide (SiO). Several other chemical reactions take place at this temperature. The overall major reactions are as follows:

$$2\ C\ (solid) + SiO_2\ (solid) \rightarrow Si\ (liquid) + SiO\ (gas) + CO\ (gas)$$

$$SiO\ (gas) + 2\ C\ (solid) \rightarrow SiC\ (solid) + CO\ (gas)$$

$$SiC\ (solid) + SiO_2\ (solid) \rightarrow Si\ (liquid) + SiO\ (gas) + CO\ (gas)$$

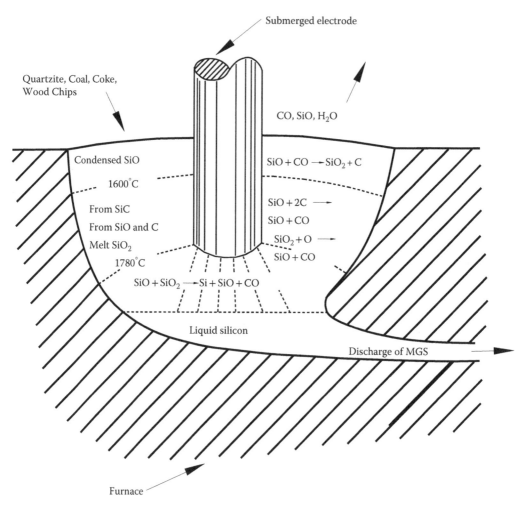

FIGURE 2.1
Schematic of a submerged electrode arc furnace for the production of metallurgical-grade silicon. (From L. D. Crossman and J. A. Baker, "Polysilicon technology" in *Semiconductor Silicon 1977*, Electrochemical Society, Pennington, New Jersey, 1977, p. 18. [6] and C. W. Pearce, "Crystal growth and wafer preparation" in *VLSI Technology*, edited by S. M. Sze, McGraw-Hill, New York, 1988, and the references therein [12].)

Most of these gaseous species escape from the furnace, and liquid silicon is collected from the bottom. The process is power intensive as metallurgical silicon is drawn off and is solidified outside. At this stage, the purity of this silicon is about 98% [6,12] and depends primarily on the purity of the raw material. Major impurities associated with this silicon are typically boron, carbon, and many residual metals, such as aluminum and iron, in the form of donor impurities.

In the case of rice hulls, the system is heated to 500°C until heavy smoke evolution from the decomposition of organic compounds begins to decrease. This process may take up to 30 minutes. The reactor is then heated to 800°C; coking is considered complete only when smoke evolution is no longer visible [9]. During this process, the raw weight of the rice hulls is reduced by one-third. These coked rice hulls are then mixed with water as a lubricant and sucrose as a binder. Sucrose enhances the reactivity of activated carbon during the smelting process. The sucrose content of dry pellets is determined by heating to 300°C under ambient nitrogen. Then, the submerged electrode arc furnace is used in the same way as described in Figure 2.1. The silicon obtained in this method is similar to that obtained using quartzite as a raw material. Although the silicon produced is contaminated with many metallic impurities, the overall purity is as good as the MGS produced using silica sand, and is often considered solar-grade silicon. The main contaminants in this silicon are aluminum, iron, titanium, boron, and phosphorus. The metallic concentrations vary widely at this stage.

In the recently proposed process of Oishi et al. [11] using the molten salt electrolysis method, the presence of impurities is not clear. Because the purification effect of the process on the electrolysis method of $NaF–AlF_3–SiO_2$ is not addresses the presence of specific impurities but tentatively regarded it as zero, and all the impurities in the feed SiO_2 were assumed to be collected in the extracted Al–Si alloy. It is estimated that this silicon has the following impurities: boron, phosphorus, aluminum, iron, and titanium, similar to those reported earlier.

The obtained MGS, with a purity of about 98%, is the most economical form of silicon available for low-efficiency solar cells. The remaining 2% is in the form of impurities and mainly contains aluminum (up to about 0.75% by weight) and iron (up to about 0.5%); however, boron, chromium, copper, magnesium, manganese, nickel, titanium, and vanadium are also present [13]. These concentrations are on the order of several hundred parts per million (ppm). The compositional analysis of Hunt [14] has determined that the main composition of this MGS typically consists of impurity dopants (such as boron and phosphorus), metals (such as aluminum, iron, and others), and nonmetals (such as carbon, oxygen, and silicon carbide). Sigmund [15] has reported the strong presence of magnesium and calcium, which is mainly present on substitutional lattice sites of silicon lattice. The scanning Auger electron spectroscopic studies of Thomas III et al. [16] have reported the presence of vanadium, titanium, nickel, iron, aluminum, and carbon in this MGS as well.

Because of these high metallic impurity contents, metallurgical silicon is unsuitable for device fabrication. Most of these metallic impurities have relatively low segregation coefficients at the melting point of silicon; thus, it is difficult to separate them from the bulk. Zone refining of metallurgical silicon is an option that can provide silicon with a purity sufficient for certain device applications where the requirements are not stringent. It is important to be aware of the specific metallic impurities present when determining the operation with which they will be removed, as well as the operation and reliability of the devices and circuits fabricated using this silicon.

2.4   Purification of Metallurgical-Grade Silicon

Silicon has a high affinity for oxygen, and the reaction is violent. Thus, in all purification processes, it is necessary to avoid the presence of oxygen species. As with organic alkanes, Si–Si bond formation is possible, such as silane (SiH_4), disilane (Si_2H_6), and trisilane (Si_3H_8) compounds. Disilane (Si_2H_6) is more reactive than monosilane and has faster deposition times. The advantage of disilane is probably not significant in regard to semiconductor applications. As a source of amorphous silicon for solar cells, however, disilane provides as much as 10 times faster deposition. Thus, it has been used experimentally in solar cell applications.

All these compounds of silicon with hydrogen are pyrophoric in nature and react with atmospheric oxygen without any ignition, even at room temperature. Otherwise, silicon and many of its compounds are not poisonous and hazardous except as noted here and elsewhere. The only exception is the industrial disease silicosis, which affects industrial workers exposed to silica dust for longer periods. Eckhoff et al. [17] reported that fine silicon powders present a significant explosion hazard, although not as severe as that of fine powders of aluminum metal. They suggested that by taking appropriate precautions, such as preventing electric sparking, strictly controlling the hot work, providing explosion vents, and maintaining good housekeeping, the dust explosion risk in silicon powder producing and handling plants can be reduced to acceptable levels. Other clinical observations are that the fine silicon crystalline dust irritates the skin and eyes on contact. Inhalation will cause irritation to the lungs and mucous membranes. Irritation to the eyes will cause watering and redness. Reddening, scaling, and itching are characteristics of skin inflammation, but few precautions are advised.

There are different ways to purify metallurgical silicon, as shown in Table 2.1. They are broadly classified with three different methods: (1) purification of silicon in liquid (molten) state, (2) solid state (including powder form), and (3) convert to vapor and deposit on hot substrates via the chemical vapor deposition (CVD) route.

Chu et al. [30] suggest a direct purification method by treating metallurgical silicon melt with gaseous reagents, as shown in Figure 2.2. In this process, purification was achieved by treating the melt with chlorine, a chlorine-oxygen mixture, hydrogen chloride, and hydrogen fluoride gases. In this method, halogens and hydrogen halides react with the metallic impurities, convert to volatile halides, and are carried from the reaction chamber by the argon purging gas. Since all the chlorides have negative and relatively large free

TABLE 2.1

Different Methods Used for Purifying Metallurgical-Grade Silicon to Electronic-Grade Silicon Useful for Semiconductor Device Applications

Methods to Purify Metallurgical Silicon	Ref
Chemical decomposition of organosilicon compounds using CH_3SiCl_3, CH_3SiHCl_2, or $(CH_3)_2SiCl_2$	[18]
Chemical leaching technique	[8]
CVD deposition using trichlorosilane route by decomposition method	[19]
Deposition on fines in fluidized bed reactor using silane pyrolysis	[20]
Electrolysis using aluminum-silicon liquid alloy	[11]
Frozen drop method of trichlolosilane	[3]
Electrorefining of metallurgical silicon in molten $CaCl_2$-based salts	[21]
Fractional melting using centrifugal force for separating liquid from cake	[22]
Heterogeneous decomposition of silane in fixed fluidized bed pyrolysis	[23]
Oxidative purification of (tri)chlorosilane silicon source materials	[24]
Partial purification by acid extraction	[25]
Polysilicon deposition on silicon tubes using silane	[26]
Preparation of high-purity silicon from silicon tetrachloride/silane	[27]
Pyrolytic reduction of halosilane vapor and influence of CH_4	[28]
Recrystallization method	[7]
Sodium reduction of silicon tetrafluoride	[29]
Treatment of melt with gaseous reagents	[30]
Ultra-high purification of silane for semiconductor-grade silicon	[31]
Unidirectional solidification method	[30]
Zone refining of metallurgical silicon	[13]

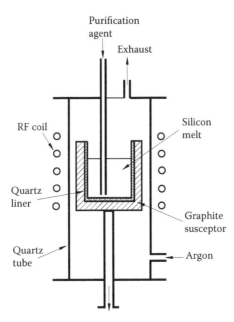

FIGURE 2.2
Schematic diagram of the apparatus for the purification of metallurgical silicon. (From T. L. Chu, G. A. van der Leeden, and H. I. Yoo, *Journal of the Electrochemical Society*, **125**, 661–665, 1978 [30].)

energies of formation and are volatile at the melting point of silicon, this treatment results in removing all possible metals as halides from the molten silicon. Here, the reaction between aluminum and chlorine is thermochemically more favorable than that between silicon and chlorine but less favorable that that between iron and chlorine. The oxygen and hydrogen chloride combination produces volatile $FeCl_3$ and escapes from the molten silicon. In this way, the aluminum and iron impurities are removed from the molten silicon.

Lee et al. [22] came out with a new technique using a fractional melting process of the MGS. This process involves heating an alloy within its liquid-solid region, while simultaneously ejecting liquid from the solid-liquid mixture via centrifugal force. The major impurities, such as iron, titanium, aluminum, and copper, can be separated with this approach. Since these metals have low segregation coefficients in silicon, high purification is possible with this process. It is projected that a further increase in the refining ratio could be realized by either controlling the processing parameters or by reducing the solid fraction. Oishi et al. [11] suggested the electrolysis method using an aluminum-silicon liquid alloy, and Cai et al. [21] suggested the electrolysis of molten silicon based on $CaCl_2$-based salts to purify metallurgical silicon.

Partial purification of pulverized MGS by acid extraction was proposed by Chu and Chu [25]. Silicon may be purified using this technique because of the segregation of metal impurities at the grain boundaries during solidification of the melt. This was achieved by refluxing with acids such as aqua regia, hydrochloric acid, and an equivolume mixture of sulfuric acid and nitric acid. Aqua regia is the most effective at removing metallic impurities. Using this method, more than 80% of the metallic impurities can be removed, but the dissolution of the impurities is slow. The rate of dissolution may be increased by carrying out the acid extraction in a closed system at high temperatures under pressure. It is difficult to identify the endpoint in this case, and working with strong mineral acids at higher temperatures necessitates special precautions and safety issues. With this approach, only a limited number of metallic impurity species can be removed.

The most common approach is to convert the MGS to chlorosilanes and handle these compounds for purification. The MGS is pulverized into a fine powder using mechanical methods, and this powder is treated with anhydrous hydrogen chloride gas. This typically takes place in a fluidized bed reactor. In the presence of a catalyst, this leads to the formation of different chlorosilanes, as shown next:

Si (solid) + HCl (gas) →	SiH_4	silane	b.p.	−112.0°C
	SiH_3Cl	monochlorosilane	b.p.	−22.8°C
	SiH_2Cl_2	dichlorosilane	b.p.	8.0°C
	$SiHCl_3$	trichlorosilane	b.p.	31.8°C
	$SiCl_4$	silicon tetrachloride	b.p.	57.6°C

In the aforementioned chemical reaction, hydrogen is formed as a byproduct. Among the reaction products, silane (SiH_4) and (mono)chlorosilane (SiH_3Cl) are gases at room temperature, whereas dichlorosilane (SiH_2Cl_2) is a liquid/gas with boiling point of about 8°C. Trichlorosilane ($SiHCl_3$) and silicon tetrachloride ($SiCl_4$) are liquids at room temperature with a boiling point of about 31.8°C and 57.6°C, respectively. Purification is achieved through fractional distillation of these compounds. In most cases, trichlorosilane is used for purification purposes; however, other chlorosilanes and silane are also equally popular for this purpose. Most of the metals present in the metallurgical silicon react with HCl and respective metal chlorides are formed. Sirtl et al. [32], Herrick and Woodruff [33], and Narusawa [34,35] presented the silicon depositions in an Si-H-Cl gas phase system by dealing with both homogeneous nucleation and thermofluid effects on the deposition conditions. The major polysilicon processes presently in use are the Siemens process, which uses hydrogen to reduce trichlorosilane; the Union Carbide process, which uses silane decomposition; and the Hemlock Semiconductor process, which uses hydrogen to reduce dichlorosilane [36].

The low boiling point of trichlorosilane is favored for fractional distillation to further purify and extract the liquid. Within this liquid, most of the metals and their chlorides do not show much vapor pressure. In this distillation process, the liquid is purified to the best possible levels, and almost all the metals are retained in the original liquid. This liquid compound undergoes multiple distillation steps, thereby improving the purity up to 99.9999999% ("9N purity") [37]. Ingle and Darnnell [24] have suggested that chemical purification can reduce the number of these repetitive steps and offers a higher level of purification than distillation alone. Furthermore, the introduction of a chemical purification step into a chlorosilane recovery loop can reduce the probability of a potential impurity buildup. The distilled/purified trichlorosilane is further processed to obtain electronic-grade silicon.

The liquid trichlorosilane is commonly converted to solid polysilicon using the Siemens process developed by F. Bischoff [38]. Purified trichlorosilane is used in a chemical vapor deposition process mixed with hydrogen gas as shown in Figure 2.3 [6]. The basic reactor chamber consists of several polycrystalline slim rods placed vertically inside. The purity of the hydrogen gas is important at this point, and any trace gases may contaminate the polysilicon depositions. Residual gases are removed through a separate exit point. Initially, the polycrystalline rods are heated to high temperatures to maintain the required temperature necessary for the chemical reaction to take place. Temperatures between 900°C

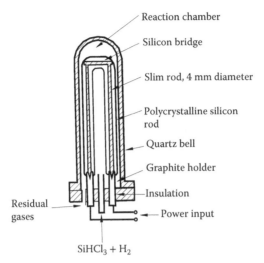

FIGURE 2.3
Schematic of a chemical vapor deposition reactor used for electronic-grade silicon production. (From L. D. Crossman and J. A. Baker, "Polysilicon technology" in *Semiconductor Silicon 1977*, Electrochemical Society, Pennington, New Jersey, 1977, p. 18 [6] and C. W. Pearce, "Crystal growth and wafer preparation" in *VLSI Technology*, edited by S. M. Sze, McGraw-Hill, New York, 1988, and the references therein [12].)

and 1200°C are typical inside the reaction chamber. The heated slim rods serve as the nucleation point for the deposition of polysilicon, and maintaining proper temperature is necessary for the reaction to continue. The chemical reaction is a hydrogen reduction of trichlorosilane as shown:

$$SiHCl_3 \text{ (gas)} + H_2 \text{ (gas)} \rightarrow Si \text{ (solid)} + 3\,HCl \text{ (gas)}$$

Details about the production of electronic-grade silicon using the hydrogen reduction process for chlorosilanes are shown in Figure 2.4. Complete details are discussed elsewhere [40].

Pure silicon slim rods offer very high resistance at low temperatures; thus, increasing the rod temperature to the desired values for the chemical reactions to take place can be an issue. At times, it may be difficult to raise the temperature of these rods to the desired levels, and indirect methods may be necessary. If the required temperature is on the order of 800°C, a blast of air is passed through the reactor enclosure to facilitate the heating of the slim rods [41]. A heating finger coil is another way to raise the required temperature to about 600°C. The heating finger coil is lowered into the reaction space to increase the rod temperature and subsequently withdrawn before the reactions are initiated. Heat-transfer fluids such as trimethylsilyl-endblocked poly-dimethylsiloxane are used to increase the rod temperatures up to 250°C by externally circulating around the reaction chamber [41]. With high temperatures, the resistance of the rods drop and it becomes easy to maintain the desired temperatures inside the reaction chamber electrically. In view of this, initially high voltages are required for the slim rods to reach high temperatures, whereas the actual depositions may continue for longer periods, depending on the desired polysilicon thickness values targeted on these rods. The interior details are shown in Figure 2.5.

Since the depositions are taking place on the surface of the slim rods, maintaining the surface temperature becomes another key point for the chemical reaction to continue with

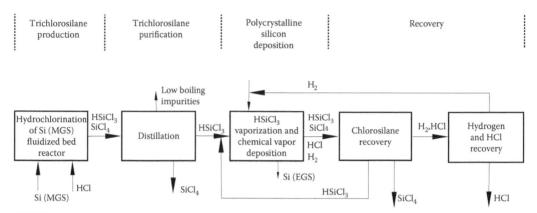

FIGURE 2.4
Schematic diagram of the production of electronic-grade silicon by reducing trichlorosilane with hydrogen. (From C. W. Pearce, "Crystal growth and wafer preparation" in *VLSI Technology*, edited by S. M. Sze, McGraw-Hill, New York, 1988 [12], and the references therein; J. R. McCormick, "Polycrystalline Silicon – 1986," *Semiconductor Silicon 1986*, Electrochem. Soc., Pennington, New Jersey, 1986, p. 43 [39]).

FIGURE 2.5
Thermal conversion of trichlorosilane to polycrystalline silicon on the surface of a heated silicon rod in the Siemens process. (From "Silicon wafers: production, specifications, Si and SiO_2 etching, our portfolio," www.MicroChemicals.eu, MicroChemicals GmbH, Nicolaus-Otto-Strasse 39, Ulm, Germany 89079 [37].)

the desired deposition rates. The surface glow temperature is set by means of an optical pyrometer and held constant for an initial period [38]. The diameter of the rods grows in response to progressive polysilicon depositions, and the glow temperature must be readjusted accordingly. In this way, it is possible to maintain the desired polysilicon deposition rate on the surface. As the rods heat up, the center of the rods will become hotter than the surface. Since the electrical conductivity of silicon is a strong function of temperature, this gradient causes variations in the electric conductivity and in the current density within the rod. For large-diameter, polysilicon-grown rods, the interior temperature may approach the melting point of silicon. Computer models are reported by Li et al. [42] for the rods' growth diameter, applied voltage, power consumption, current distribution, and flow statistics to obtain better results.

2.5   Ultra-High Pure Silicon for Electronics Applications

By using the previously discussed method, polysilicon rods up to 20 cm in diameter and several meters in length can be grown. Figure 2.6 shows one such reactor chamber and the grown polysilicon rods using the Siemens technique. Many production industries use more slim rods for better utilization of the gas feed. This electronic-grade polysilicon will have a concentration purity on the order of 1×10^{14} cm^{-3} [37]. The frozen drop method [3] has been used to assess the concentrations of boron and phosphorus present in the deposited polysilicon. Pratt et al. [19] have reported on the crystallographic texture of the deposited polysilicon and found it to have a sharp <110> fiber texture, with the <110> crystal orientations parallel to the direction of growth, with no preferred orientation about the axis. The texture does not appear to be influenced by the orientation of the slim rod from which the growth initiates. It is further reported that this <110> growth texture has also been found with other polysilicon films grown using similar methods on various other substrates. This electronic-grade polysilicon is the raw material for growing single-crystal silicon.

Other intermediate compounds, such as silicon tetrachloride ($SiCl_4$), dichlorosilane (SiH_2Cl_2), and silane (SiH_4), can be used to grow polysilicon on these slim rods, provided the required temperature necessary for the depositions is maintained. However, an alternative process for the production of electronic-grade silicon—the pyrolysis of silane (SiH_4)—has been gaining in importance. In this process, the reactor, as shown earlier in Figure 2.6, is operated at 900°C and feeds silane gas in place of trichlorosilane. The overall reaction is

$$SiH_4 \text{ (gas)} \rightarrow Si \text{ (solid)} + 2 H_2 \text{ (gas)}$$

In this reaction, no harmful reaction byproducts are generated. Hydrogen is added to the mainstream of silane for better reaction control. This is an advantage over the earlier process based on trichlorosilane gas.

2.6   Polycrystalline Silicon Feed for Crystal Growth

The polycrystalline silicon feed is either extracted in a granular form, or chunks are used as a feed for silicon crystal growth. The purity of this polysilicon is the key issue at this stage. Many purification processes have been explored using silane and chlorosilanes so that the contamination from the metals stays at a minimum level.

Lewis et al. [27] reported a silane purification process wherein thermal decomposition of silane was investigated for preparing high-purity silicon. According to them, chemical

(a)

(b)

(c)

FIGURE 2.6
(a) Basic Siemens reactor. (b) Grown polysilicon rods after a reactor run [4,38]. (c) Multiple slim rods in typical production environment. (From G. Fisher, M.R. Seacrist, and R.W. Standley, Proceedings of the IEEE, **100**, 1454–1474, 2012 [4], http://www.honeywellnow.com/2010/05/03/honeywell-to-help-global-polysilicon-maker-meet-fast-growing-demand-for-solar-panels/, Blaubeuren, Germany [43].)

absorption is considered the most effective means of purifying silane. The silane molecule is considered neutral, and the silicon atom has more than four bonding orbitals. Thus, some of its tetravalent compounds are susceptible to a nucleophilic attack. Yusa et al. [31] used ion-exchanged zeolites to purify silane, primarily sorbing the phosphine (PH_3) component present in the gas. The polysilicon obtained is reported to have higher resistivity values on the order of 10^5 Ω-cm.

Iya et al. [23] used a fixed fluidized bed reactor to heterogeneously decompose silane gas on hot silicon seeds and produce free-flowing particles of silicon at a low cost. The method offered a potential approach to convert high-purity silane into pure, low-cost polysilicon that is directly processable for use as a polysilicon feed for growing single crystals of silicon. Depositions were carried out in the temperature range of 600°C to 900°C. The team recommended that a good operating point for the pyrolysis reactor is 10% silane feed concentration and a bed temperature in the vicinity of 640°C. Under these conditions, the decomposition is primarily heterogeneous and the reaction rates are sufficiently high to completely convert silane to polysilicon.

Hsu et al. [20] experimentally investigated a low-cost, high-throughput method to produce high-purity polysilicon for solar cell applications using fluidized bed pyrolysis of silane. The fine particles formed because of the homogeneous gas phase decomposition, nucleation, and coagulation were found to be clusters of submicron-size nuclei. The size ranged between 1 μm and 10 μm diameters. X-ray diffraction analysis showed that these silicon particles had a crystalline structure. There is an optimum temperature range for this reactor operation, and it was set by the deposition quality. At 550°C, the product adhered loosely and was more granular in nature. As the temperature increased from 600°C to 650°C, the deposit appeared to be more coherent and dense. At 700°C, the deposit was very coherent and dense. Based on the product morphology, the lower limit for operating the fluidized bed reactor in silane pyrolysis is 600°C, and the optimal temperature region is around 700°C. Above 800°C, high levels of fines were formed and operating at this level is not advisable.

This approach became an alternative method to produce high-purity, electronic-grade polysilicon. Both silane and hydrogen gases are injected into the reaction chamber at the bottom, causing a bed of small silicon seed granules to become fluidized. These seed granules provide a place for the surface reactions to take place, and silane decomposes on the surface, causing the seeds to grow. Granular polysilicon is shown in Figure 2.7. Once grown to the desired size, the granules are extracted from the bottom of the reactor while fresh seed particles are fed into the top of the reactor. In this process, these reactors utilize nearly all the silane gas fed into them, providing superior heat and mass transfer characteristics, and the process is more energy efficient. This process is also continuous, while the Siemens process is a batch process. The granules produced this way have more surface area when compared to the silicon produced by the Siemens process, and this may influence the presence of oxygen incorporation adhering on the granular surface when this feed is used for growing single crystals.

Schumacher et al. [44] proposed a new approach to produce solar-cell-grade silicon using bromosilane precursor compounds. A continuous flow reactor process based on the hydrogen reduction of the bromosilanes ($SiBr_4$ and $SiHBr_3$) was proposed, and initial experiments were carried out. Fluidized bed experiments showed both thermal decomposition and hydrogen reduction of $SiHBr_3$ in a reactor to present other attractive closed-loop processes for producing solar-cell-grade polycrystalline silicon.

Figure 2.8 presents the complete flowchart to obtain both solar-grade-silicon and semiconductor-grade silicon [40]. The basic process starts with quartz rock, and MGS

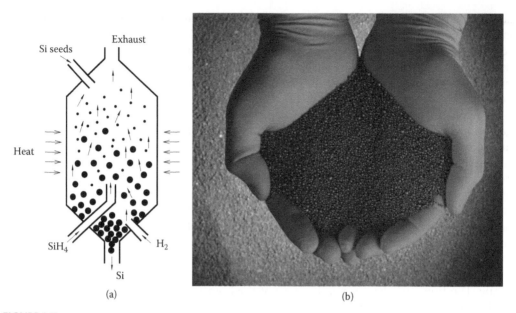

FIGURE 2.7
(a) Granular electronic-grade polysilicon produced in a fluidized bed reactor using silane as a precursor. (b) Produced polysilicon granules. (From G. Fisher, M. R. Seacrist, and R. W. Standley, *Proceedings of the IEEE*, **100**, 1454–1474, 2012 [4].)

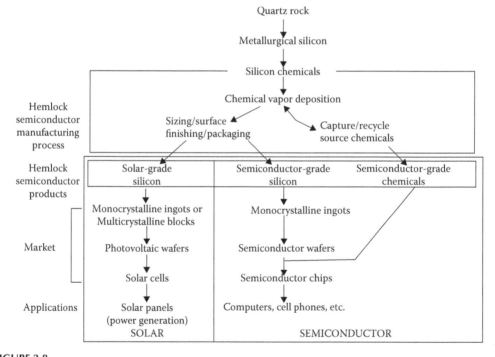

FIGURE 2.8
Complete flowchart for obtaining solar-grade and semiconductor-grade silicon for power generation and for VLSI and ULSI applications. (From http://www.hscpoly.com/content/hsc_prod/default.aspx, Polycrystalline silicon for demanding semiconductor and photovoltaic applications [40].)

is produced that has about 98% purity. This is often used for photovoltaic applications directly. Sometimes, purification is also carried out for this purpose, depending on the point of the application. By using the CVD route, semiconductor-grade silicon is obtained and most of the metallic species are separated. This high-purity polysilicon—now electronic-grade silicon—can be used as a starting feed material for both Czochralski (CZ) and float-zone (FZ) crystal growth processes to grow single-crystal silicon. From the grown ingots, wafers are shaped for the production of semiconductor chips.

2.7 Summary

In this chapter, we have seen how a compound of silicon and oxygen is extracted from sand. Silicon is a stable material, but at higher temperatures, it readily forms oxides and complex mixtures. High-temperature treatment with carbon is used to free the silicon from its oxide compound. In this way, MGS is extracted, but the purity is not sufficient for device application, with the exception of low-efficiency solar cells. MGS is purified through chemical reactions and converted into different chlorosilanes and silane. These compounds are converted to polysilicon solid material through the CVD route, and this polysilicon becomes the feed for the crystal growth process. As a polysilicon, this material is quite stable and can be created in either small chunks or fine granular pieces to feed the crystal growth process. The purity of polysilicon is important at this stage, particularly in terms of metallic impurities.

References

1. R. Falster and V. V. Voronkov, "Intrinsic point defects and their control in silicon crystal growth and wafer processing," *MRS Bulletin*, 28–32, June 2000, and the references therein.
2. J. B. Calvert, "Silicon," http://mysite.du.edu/~jcalvert/phys/silicon.htm, and the references therein.
3. F. H. Horn, "Evaluation of electronic grade silicon and trichlorosilane by a frozen drop method," *Journal of the Electrochemical Society*, **114**, 1307–1311, 1967.
4. G. Fisher, M. R. Seacrist, and R. W. Standley, "Silicon crystal growth and wafer technologies," *Proceedings of the IEEE*, **100**, 1454–1474, 2012, and the references therein.
5. M. Morin, M. Koyanagi, J. M. Friedt, and M. Hirose, "Metal impurity evaluation in silane gas from the qualification of poly-Si layers," *Journal of the Electrochemical Society*, **141**, 274–277, 1994.
6. L. D. Crossman and J. A. Baker, "Polysilicon technology" in *Semiconductor Silicon 1977*, Electrochemical Society, Pennington, New Jersey, 1977, p. 18.
7. J. A. Amick, J. P. Dismukes, R. W. Francis, L.P. Hunt, P.S. Ravishankar, M. Schneider, K. Matthei, R. Sylvain, K. Larsen, and A. Schei, "Improved high-purity arc-furnace silicon for solar cells," *Journal of the Electrochemical Society*, **132**, 339–345, 1985.
8. J. A. Amick, "Purification of rice hulls as a source of solar grade silicon for solar cells," *Journal of the Electrochemical Society*, **129**, 864–866, 1982.
9. L. P. Hunt, J. P. Dismukes, J. A. Amick, A. Schei, and K. Larsen, "Rice hulls as a raw material for producing silicon," *Journal of the Electrochemical Society*, **131**, 1683–1686, 1984.
10. U. Kalapathy, A. Proctor, and J. Shultz, "A simple method for production of pure silica from rice hull ash," *Biosource Technology*, **73**, 257–262, 2000.

11. T. Oishi, M. Watanabe, K. Koyama, M. Tanaka, and K. Saegusa, "Process for solar grade silicon production by molten salt electrolysis using aluminum-silicon liquid alloy," *Journal of the Electrochemical Society*, **158**, E93–E99, 2011.
12. C. W. Pearce, "Crystal growth and wafer preparation" in *VLSI Technology*, edited by S. M. Sze, McGraw-Hill, New York, 1988, and the references therein.
13. T. L. Chu, S. S. Shirley, S. Chu, R. W. Kelm, Jr., and G. W. Wakefield, "Solar cells from zone-refined metallurgical silicon," *Journal of the Electrochemical Society*, **125**, 595–597, 1978.
14. L. P. Hunt, "Compositional analysis of silicon for solar cells," *Journal of the Electrochemical Society*, **131**, 1891–1896, 1984.
15. H. Sigmund, "Solubilities of magnesium and calcium is silicon," *Journal of the Electrochemical Society*, **129**, 2809–2812, 1982.
16. J. H. Thomas III, R. V. D'Aiello, and P. H. Robinson, "A scanning Auger electron spectroscopic study of particulate defects in metallurgical-grade silicon," *Journal of the Electrochemical Society*, **131**, 196–200, 1984.
17. R. K. Eckhoff, S. J. Parker, B. Gruvin, M. Hatcher, and T. Johansson, "Ignitability and explosibility of silicon dust clouds: Influence of dust fineness," *Journal of the Electrochemical Society*, **133**, 2631–2637, 1986.
18. E. Sirtl and H. Seiter, "Vapor-deposited microcrystalline silicon," *Journal of the Electrochemical Society*, **113**, 506–508, 1966.
19. B. Pratt, S. Kulkarni, D. P. Pope, and C. D. Graham, Jr., "Growth texture of polycrystalline silicon prepared by chemical vapor deposition," *Journal of the Electrochemical Society*, **123**, 1760–1762, 1976, and the references therein.
20. G. Hsu, R. Hogle, N. Rohatgi, and A. Morrison, "Fines in fluidized bed silane pyrolysis," *Journal of the Electrochemical Society*, **131**, 660–663, 1984.
21. J. Cai, X.-t. Luo, G. M. Haarberg, O. E. Kongstein, and S.-I. Wang, "Electrorefining of metallurgical grade silicon in molten $CaCl_2$ based salts," *Journal of the Electrochemical Society*, **159**, D155–D158, 2012, and the references therein.
22. W. Lee, W. Yoon, and C. Park, "Purification of metallurgical-grade silicon in fractional melting process," *Journal of Crystal Growth*, **312**, 146–148, 2009.
23. S. K. Iya, R. N. Flagella, and F. S. DiPaolo, "Heterogeneous decomposition of silane in a fixed bed reactor," *Journal of the Electrochemical Society*, **129**, 1531–1535, 1982.
24. W. M. Ingle and R. D. Darnell, "Oxidative purification of chlorosilane silicon source materials," *Journal of the Electrochemical Society*, **132**, 1240–1243, 1985.
25. T. L. Chu and S. S. Chu, "Partial purification of metallurgical silicon by acid extraction," *Journal of the Electrochemical Society*, **130**, 455–457, 1983.
26. D. Cai, L. L. Zheng, Y. Wan, A. V. Hariharan, and M. Chandra, "Numerical and experimental study of polysilicon deposition on silicon tubes," *Journal of Crystal Growth*, **250**, 41–49, 2003.
27. C. H. Lewis, H. C. Kelly, M. B. Giusto, and S. Johnson, "Preparation of high-purity silicon from silane," *Journal of the Electrochemical Society*, **108**, 1114–1118, 1961.
28. L. D. Dyer, "The influence of methane on twin formation during the vapor growth of silicon crystals," *Journal of the Electrochemical Society*, **118**, 957–961, 1971.
29. A. Sanjurjo, L. Nanis, K. Sancier, R. Bartlett, and V. Kapur, "Silicon by sodium reduction of silicon tetrafluoride," *Journal of the Electrochemical Society*, **128**, 179–184, 1981.
30. T. L. Chu, G. A. van der Leeden, and H. I. Yoo, "Purification and characterization of metallurgical silicon," *Journal of the Electrochemical Society*, **125**, 661–665, 1978.
31. A. Yusa, Y. Yatsurugi, and T. Takaishi, "Ultrahigh purification of silane for semiconductor silicon," *Journal of the Electrochemical Society*, **122**, 1700–1705, 1975.
32. E. Sirtl, L. P. Hunt, and D. H. Sawyer, "High temperature reactions in the silicon-hydrogen-chlorine system," *Journal of the Electrochemical Society*, **121**, 919–925, 1974.
33. C. S. Herrick and D. W. Woodruff, "The homogeneous nucleation of condensed silicon in the gaseous Si-H-Cl system," *Journal of the Electrochemical Society*, **131**, 2417–2422, 1984, and the references therein.

34. U. Narusawa, "Si deposition from chlorosilanes: I. Deposition modeling," *Journal of the Electrochemical Society*, **141**, 2072–2077, 1994, and the references therein.

35. U. Narusawa, "Si deposition from chlorosilanes: II. Numerical analysis of thermofluid effects on deposition," *Journal of the Electrochemical Society*, **141**, 2078–2083, 1994.

36. C. L. Yaws, K. Y. Li, and S. M. Chou, "Economics of polysilicon processes" in *JPL Proceedings of the Flat-Plate Solar Array Project Workshop on Low-Cost Polysilicon for Terrestrial Photovoltaic Solar-Cell Applications*, National Aeronautics and Space Database, CDSITC. AEROSPACE, N86-26683, 79–121, 1986.

37. "Silicon wafers: production, specifications, Si and SiO_2 etching, our portfolio," www. MicroChemicals.eu, MicroChemicals GmbH, Nicolaus-Otto-Strasse 39, Ulm, Germany 89079.

38. F. Bischoff, "Apparatus for vapor deposition of silicon," United States Patent Office, 3,335,697 Patented August 15, 1967, Original filed on February 6, 1961, and the references therein.

39. J. R. McCormick, "Polycrystalline Silicon – 1986," *Semiconductor Silicon 1986*, Electrochem. Soc., Pennington, New Jersey, 1986, p. 43.

40. http://www.hscpoly.com/content/hsc_prod/default.aspx. Polycrystalline silicon for demanding semiconductor and photovoltaic applications.

41. A. N. Arvidson, M. H. Greene, and J. R. McCormick, "Process for the production of semiconductor materials," United States Patent Office, 4,724,160 Patented February 9, 1988, originally filed on July 28, 1986, and the references therein.

42. M. Li, A. Mitrasinovic, T. Utigard, G. Plascencia, and A. Warczok, "Silicon rod heat generation and current distribution," *Journal of Crystal Growth*, **312**, 141–145, 2009.

43. http://www.honeywellnow.com/2010/05/03/honeywell-to-help-global-polysilicon-maker-meet-fast-growing-demand-for-solar-panels/Fig.-2_centrothermSiTec24pairCVD-459x620.jpg, Blaubeuren, Germany.

44. J. C. Schumacher, L. Woerner, E. Moore, and C. Newman, "The production of solar cell grade silicon from bromosilanes," http://adsabs.harvard.edu/abs/1979sjcc.rept., Oceanside, CA.

3

Importance of Single Crystals for Integrated Circuit Fabrication

3.1 Introduction

The first step in the manufacturing of integrated circuits (ICs) is the preparation of a perfect single-crystal silicon. In crystalline materials, all atoms (or molecules) are arranged in a systematic periodic fashion, and this periodicity extends throughout the entire crystal in all three dimensions. If the periodicity is limited to shorter distances within the crystal dimensions, then they are referred as polycrystalline materials, and the separation of these polycrystalline areas are defined by physical boundaries. In this arrangement, the crystal consists of relatively small crystals, often referred to as grains, arranged in random directions, but joining at the boundaries. The regions between the grains where the periodicity abruptly changes are called grain boundaries [1]. If the grain sizes are sufficiently small and limited to only to the unit cell dimensions, such solid materials are referred to as amorphous materials and no periodicity is observed.

Solid materials assume a crystalline form because this arrangement of atoms is the minimum energy state. In most solids at room temperature, atoms tend to occupy fixed positions relative to each other. The atoms, however, are in constant vibration about their equilibrium position, depending upon the temperature.

Semiconductor properties are not exclusive to single crystals. Many noncrystalline materials exhibit similar semiconductor properties. However, silicon is the best understood and most widely used material for IC fabrication. Today's microelectronics began with—and are still based on—the purification and crystallization of a single element (i.e., silicon). The silicon wafer used as the starting point in the manufacture of even the most advanced ICs, including very large scale integration (VLSI) and ultra large scale integration (ULSI), is made using materials technology that, in general, has been available since the late 1970s. Refinements have taken place, and today, many scientists are working on this wonderful semiconductor material. They are concentrating on a more precise understanding of the atomic mechanisms in silicon crystals that are needed for small, shallow semiconductor devices to bring out many more capabilities of this material.

3.2 Crystal Structures

A crystal can be described in terms of a periodic lattice, which is used repeatedly to build the entire structure. Each lattice point is attached to and is repeated in three-dimensional space to form the crystal. Crystal planes are best represented by Miller indices, as shown in Figure 3.1. The group of integers (*hkl*) defines a set of equally spaced parallel planes. If the plane crosses the axis on the negative side of the chosen origin, then the corresponding index is negative and is written with a bar over it, such as ($h\bar{k}l$) [1]. For cubic crystals of a single element, the (100), (010), ($\bar{1}$00), (0$\bar{1}$0), and (00$\bar{1}$) planes are indistinguishable, as are the (110), (101) and (011) planes. Many books are available to better understand crystals and crystal-related issues. Here, we touch upon some of the key points related to silicon crystals and their material properties.

Silicon exists with three different forms of microstructures: (1) crystalline, (2) polycrystalline, and (3) amorphous. Polycrystalline (or simply polysilicon) and amorphous silicon are usually deposited as thin films with typical thicknesses not exceeding 5 μm. Single-crystal substrates are commercially available as circular wafers with 50 mm (2-inch), 75 mm (3-inch), 100 mm (4-inch), 150 mm (6-inch), 200 mm (8-inch), 300 mm (12-inch), and 450 mm (18-inch) diameters with varying thicknesses. Larger-diameter (200 mm and above) wafers are preferred for the IC industry, whereas for microelecto-mechanical system (MEMS) applications, most of the wafers are 150 mm (6 inch) and less in diameter.

Single-crystal properties depend on the crystal orientation. This is generally indicated as the anisotropic property in crystals. This can be defined as a difference in a material's physical or mechanical properties when measured along different crystal axes. Metals and alloys tend to be more isotropic, though they can sometimes exhibit significant anisotropic behavior. This is true for the distribution of trace metals when they are in the silicon lattice. Orientation is important in the case of silicon, since many of the crystal's structural and electronic properties are highly anisotropic in nature.

Anisotropic etching techniques, such as deep reactive-ion dry etching, are used in VLSI and ULSI for deep trench isolation and for fabrication processes to create well-defined, microscopic features with high-aspect-ratio structures. These structural features are commonly used in MEMS and microfluidic devices, where the anisotropy of the features is needed to impart the required physical properties to the device. Anisotropic etching could also refer to certain chemical etchants that are etching only in specific crystal orientations.

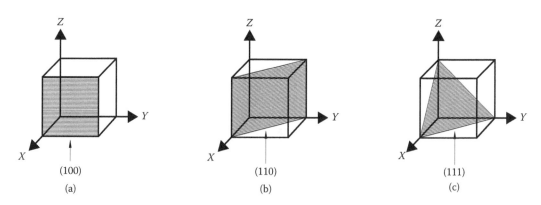

FIGURE 3.1
Miller indices to define different crystal planes of (a) (100), (b) (110) and (c) (111). (From B. El-Kareh, *Fundamentals of Semiconductor Processing Technology*, Kluwer Academic Publishers, Boston, 1995 [1].)

3.2.1 Different Crystal Structures in Nature

The crystal structure of any solid material is highly ordered due to the intrinsic nature of the molecules to form symmetric patterns. Crystal structure and symmetry play a greater role in determining many of its physical properties, such as cleavage planes, electronic band structure, scattering phenomenon, and optical transparency. The unit cell describes the arrangement of the crystal structure. The unit cells are stacked in all three dimensions in terms of lattice parameters (a, b, and c) and the angles (α, β, and γ) between them. Some directions and planes are defined by the symmetry of the crystal system. Almost all crystals have translational symmetry in three directions, but there are crystals with additional symmetry elements, such as rotational symmetry, mirror symmetry, or both. There are seven crystal systems and six crystal families. The crystal systems are triclinic (1), monoclinic (2), orthorhombic (4), rhombohedral (1), tetragonal (2), hexagonal (1), and cubic (3). The number in parentheses indicates the respective Bravais lattices, and there are 14 lattice structures as indicated earlier.

3.2.2 Cubic Structures

The cubic structure is the most common and simplest shape of solid found in nature. There are three main cubic structures: primitive, body centered, and face centered. In cubic crystal systems, there is more than one atom per cubic unit cell. Many metals and simple binary compounds exhibit these structures.

3.3 Diamond Crystal Structure

In the diamond crystal structure, each atom is symmetrically surrounded by four equally spaced atoms and it forms tetrahedral bonds with its four nearest neighbors. This type of bonding is very strong, highly localized, and directional because the distribution of valence electrons around the atom is shifted toward the nearest neighbor. The silicon crystal is, therefore, very hard and has a high melting point. Because of the directionality, the crystal does not assume the closest packing configuration, but allows a large volume per atom.

3.3.1 Silicon Crystal Structure

The silicon crystal has a diamond hexagonal structure and belongs to the cubic crystal system. In these diamond crystals, pairs of valence electrons with opposite spins are shared between four neighboring atoms and form covalent bonds. The $3s$ and $3p$ orbitals of the parent atoms are mixed to form a new set of four equivalent bonding hybridized orbitals, which are directed toward their four nearest neighbors. In this structure, each silicon atom is symmetrically surrounded by four equally spaced atoms; it forms tetrahedral bonds with its four nearest neighbors.

Silicon shares an electron with another silicon atom with a radius of 1.17 Å, but will not make multiple bonds as seen in the case of carbon atom between them. A silicon atom is about 50% larger than a carbon atom, and this could be why it fails to form multiple bonds. Both carbon and silicon find that hybrid sp^3 tetrahedral orbitals are the most stable configuration, with the angle between bonds equal to 109° 28', the tetrahedral angle.

3.3.2 Silicon Crystals and Atomic Packing Factors

The atomic packing factor is the fraction of volume in a crystal structure that is occupied by atoms during their arrangement. It is a dimensionless factor, and its value is always less than unity. For all practical purposes, the packing factor of any crystal is estimated by assuming that the participating atoms are spherical, rigid bodies. No overlapping is considered, and all the participating atoms are considered to be simply touching each other. For a simple and identical type of rigid atoms participating in the crystal, the atomic packing factor is mathematically defined as the ratio of the total spherical volumes of the participating atoms to that of the total volume of the complete unit cell. Mathematically, it is defined as

$$\text{Atomic packing factor (APF)} = \frac{N_{atoms} V_{atom}}{V_{unitcell}}$$

where N_{atoms} is the total number of unit cells in the unit cell, V_{atom} corresponds to the volume of each atom, and $V_{unitcell}$ is the total volume occupied by the unit cell of that crystal structure [2]. For one-component structures, such as simple metals, the best achievable APF is about 0.74, and for multiple components, the value may slightly improve. Table 3.1 lists the different values achievable for several solid crystal structures. When compared to all the crystal structures, only the face-centered cubic and hexagonal close-packed crystals exhibit an APF value of 0.74, indicating that the crystals are 74% packed, with 26% void space in the unit cell. In comparison, silicon has only 0.34 as its APF because of the atomic arrangement in the diamond structure and it is not densely packed. The silicon lattice has a large void space (66%) that is almost double the APF value of 34%. In view of this crystal arrangement, most of the smaller ions and atoms find the space suitable to fit into, leading to many complex silicon compounds.

3.4 Crystal Order and Perfection

A perfect crystal is defined as one that contains no defect in the arrangement of its unit lattice and the atoms involved. The crystal must be free from point, linear, or planar imperfections. Practically, it is difficult to find such crystals in nature. There are a wide variety of crystallographic defects due to interruptions in the arrangement. Point defects occur only at or around a single lattice point, as shown in Figure 3.2a. Here, the point defect is

TABLE 3.1

Different Solid-State Single Crystals and their Atomic Packing Factors

Crystal Structures	APF	Examples
Face-centered cubic crystals (fcc)	0.74	Copper
		Silver
		Gold
Hexagonal close-packed crystals (hcp)	0.74	Zinc
Body-centered cubic crystals (bcc)	0.68	Iron
Simple cubic crystals (sc)	0.52	α-polonium
Diamond cubic crystals	0.34	Carbon
		Silicon

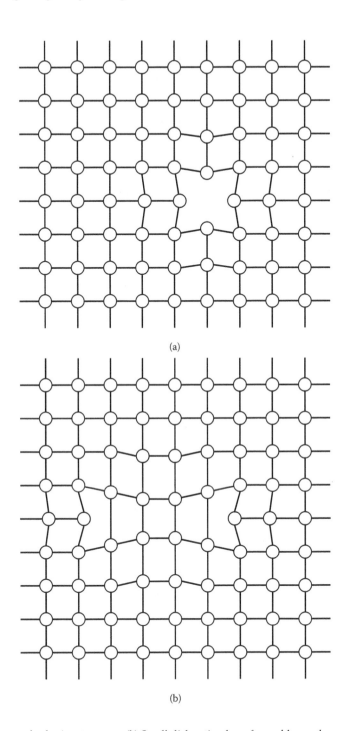

(a)

(b)

FIGURE 3.2
(a) A vacancy in a regular lattice structure. (b) Small dislocation loop formed by agglomeration of vacancies. (From R. Falster and V.V. Voronkov, "Lattice defects in silicon", 123–5.5&5.6.ppt, from Monsanto Electronic Materials Company (MEMC) website, 2006 [3]).

a vacancy, or missing atom. This vacancy is occupied in a perfect crystal. The stability of the surrounding crystal structure assures that the adjacent atom will not simply collapse around the vacancy. As shown in the figure, the presence of a vacancy distorts the lattice network. The vacancy here is active because of the missing coordination bond of the neighboring lattice atoms. This vacancy may attract other free ions present in the interstitial positions to fill that vacant portion of the lattice. Agglomeration of such defects may increase the defect size, as shown in Figure 3.2b, and lead toward the formation of intrinsic stacking faults.

The interstitial defects are atoms that occupy a location in the crystal structure where there is no site for regular arrangement. These positions are generally high-energy configurations. Smaller atoms can occupy these interstitial locations without high energy. Sometimes, two or more atoms share more than one lattice site such that the number of atoms is larger than the number of lattice sites available in the crystal. Self-interstitials are those defects that only contain atoms that are the same as those already present in the lattice sites. Figure 3.3a shows the self-interstitial atom in a lattice; Figure 3.3b shows the agglomeration of self-interstitial atoms in the lattice. Such agglomeration can distort the lattice sites and induce strain, leading to other crystalline defects. This may also lead to extrinsic stacking faults, as shown in Figure 3.3c, leading toward the edge dislocations.

3.5   Crystal Orientations and Planes

Crystals are characterized by a unit cell that repeats in the x, y, and z directions. Crystalline planes and directions are defined using an x, y, and z coordinate system. The [111] direction is defined by a vector with 1-unit components in x, y, and z. Planes are defined by Miller indices—reciprocals of the intercepts of the plane with the x, y, and z axes discussed earlier. It is becoming increasingly important to understand the crystallization processes of silicon on an atomic scale in order to develop high-speed devices and ICs.

The silicon crystal has a diamond hexagonal structure that can be physically constructed from two merged face-centered cubic structures. The atoms share electrons and form tetrahedral covalent bonds in this packing arrangement. Silicon has three primary crystal planes, (100), (110), and (111), and preferentially cleaves along the (111) and (110) planes, as indicated in Figure 3.4a and b.

Silicon is an anisotropic material, which means that the elastic properties depend upon the orientation in the material. The stiffness matrix is a 6×6 tensor and is defined as

$$[c] = \begin{vmatrix} c_{11} & c_{12} & c_{12} & 0 & 0 & 0 \\ c_{12} & c_{11} & c_{12} & 0 & 0 & 0 \\ c_{12} & c_{12} & c_{11} & 0 & 0 & 0 \\ 0 & 0 & 0 & c_{44} & 0 & 0 \\ 0 & 0 & 0 & 0 & c_{44} & 0 \\ 0 & 0 & 0 & 0 & 0 & c_{44} \end{vmatrix}$$

with the following constants:

$$C_{11} = 1.6564 \times 10^{11} \text{ Pa}$$

$$C_{12} = 0.6394 \times 10^{11} \text{ Pa}$$

$$C_{44} = 0.7951 \times 10^{11} \text{ Pa}$$

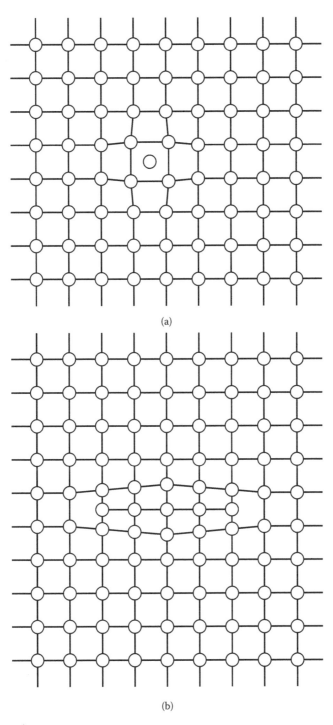

(a)

(b)

FIGURE 3.3 (*Continued*)

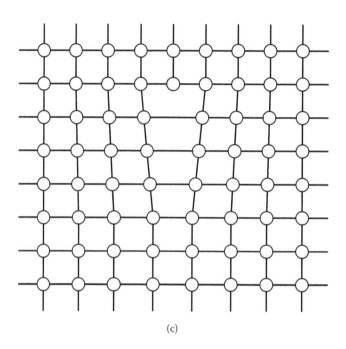

(c)

FIGURE 3.3
(a) Self-interstitial atom inside a regular crystal lattice. (b) Small dislocation loop formed by agglomeration of self-interstitials. (c) Open-loop self-interstitials and edge dislocation inside the lattice. (From R. Falster and V.V. Voronkov, "Lattice defects in silicon", 123–5.5&5.6.ppt, from Monsanto Electronic Materials Company (MEMC) website, 2006 [3]).

Several process parameter and device characteristics are sensitive to wafer orientation. Since the {111} planes have the smallest separation, silicon epitaxy grows faster along a <111> direction than along a <110> or <100> direction. This is also reflected in the chemical etching steps. Since (111) planes have the highest density of atoms, the dissolution of silicon in an etching solution is slowest in the <111> direction [1]. The rate of thermal oxidation and diffusivity of impurities are also dependent on crystal orientation. The oxidation rate of silicon is largest in the <111> direction and smallest in the <100> crystallographic direction. Table 3.2a and b show the angles between crystal planes in decimal and normal angles.

Single-crystal silicon contains only atoms perfectly arranged in a periodic pattern that is assumed to be infinite in three dimensions. In real crystals, however, the atoms terminate at boundaries, known as surface defects, and contain imperfections in the bulk and at the surfaces. These imperfections, also referred as defects, are either inherent to the crystal or they are created during the fabrication process of growth or subsequent device processing. Even the best-grown crystals contain different types of defects. These can drastically alter the electrical and physical properties of the crystal and determine the functionality of the devices. These defects also play an important role in the transport of impurities and charge carriers in the devices.

It is customary to identify the crystal plane direction by using [] notation (for example, [1, 0, –1] direction) and to identify a family of equivalent directions with < > notation (for

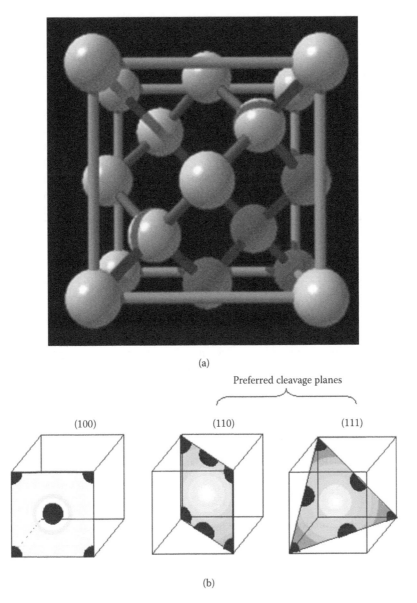

(a)

Preferred cleavage planes

(100)   (110)   (111)

(b)

FIGURE 3.4
(a) Basic silicon diamond structure. (b) Primary crystal planes of single-crystal silicon. The crystals preferably cleave on the (111) and (110) planes. (From A. M. Fitzgerald, "Crack growth phenomena in micro-machined single crystal silicon and design implications for micro electro mechanical systems (MEMS)," Ph.D. thesis, Stanford University, August 2000 [4].)

example, <1, 1, 0> or simply <110> orientation) to understand the distinction between a specific direction and crystal planes. Using () notation identifies a specific plane (for example, crystal plane (113)) and using {} notation indicates a family of equivalent planes, such as {311} planes. A bar above an index is equivalent to a minus sign. Many of these details are explained in Table 3.3 to indicate the equivalent directions for <100>, <110>, and <111> directions.

TABLE 3.2A

Angles between Planes

	100	110	111	010	001	101
100	0°	45°	54.74°	90°	90°	45°
011	90°	60°	35.26°	45°	45°	60°
111	54.74°	35.26°	0°	54.74°	54.74°	35.26°
211	35.26°	30°	19.47°	65.91°	65.91°	30°
311	25.24°	31.48°	29.5°	72.45°	72.45°	31.48°
411	19.47°	33.56°	35.26°	76.37°	76.37°	33.56°
511	15.79°	35.26°	38.94°	78.9°	78.9°	35.26°
611	13.26°	36.59°	41.47°	80.66°	80.66°	36.59°
711	11.42°	37.62°	43.31°	81.95°	81.95°	37.62°

TABLE 3.2B

	100	110	111	010	001	101
100	0°	45°	54° 44′ 24″	90°	90°	45°
011	90°	60°	35° 15′ 36″	45°	45°	60°
111	54° 44′ 24″	35° 15′ 36″	0°	54° 44′ 24″	54° 44′ 24″	35° 15′ 36″
211	35° 15′ 36″	30°	19° 28′ 12″	65° 54′ 36″	65° 54′ 36″	30°
311	25° 14′ 24″	31° 28′ 48″	29° 30′	72° 27′	72° 27′	31° 28′ 48″
411	19° 28′ 12″	33° 33′ 36″	35° 15′ 36″	76° 22′ 12″	76° 22′ 12″	33° 33′ 36″
511	15° 47′ 24″	35° 15′ 36″	38° 56′ 24″	78° 54′	78° 54′	35° 15′ 36″
611	13° 15′ 36″	36° 35′ 24″	41° 28′ 12″	80° 39′ 36″	80° 39′ 36″	36° 35′ 24″
711	11° 25′ 12″	37° 37′ 12″	43° 18′ 36″	81° 57′	81° 57′	37° 37′ 12″

Source: "Crystal planes in semiconductors," http://www.cleanroom.byu.edu/EW_orientation.phtml, maintained by ECEn IMMERSE Web Team. Copyright 1994–2009. Brigham Young University, Provo, Utah, USA [5].

TABLE 3.3

Families of Equivalent Directions and Planes

Type:<100>	Type:<110>	Type:<111>
Equivalent Directions	Equivalent Directions	Equivalent Directions
[100]	[110], [011], [101]	[111]
[010]	[-1-10], [0-1-1], [-10-1]	[-111]
[001]	[-110], [0-11], [-101]	[1-11]
	[1-10], [01-1], [10-1]	[11-1]

Source: "Crystal planes in semiconductors," http//www.cleanroom.byu.edu/EW_orientation.phtml, maintained by ECEn IMMERSE Web Team. Copyright 1994–2009. Brigham Young University, Provo, Utah, USA [5].

3.6 Influence of Dopants and Impurities in Silicon Crystals

There is a limit to the concentration of substitutional impurities that can be incorporated in the silicon lattice sites without seriously disrupting the lattice perfection. This is referred to as the solid solubility limit of the impurity in silicon. The solubility of most of the useful impurities in silicon is relatively small, but it depends on the temperature of the system. The highest concentration of boron in silicon, for example, is less than 1% [1] of the total number of silicon atoms. The solubility value initially increases in response to temperature and then begins to decrease as the crystal melting temperature is approached [6], as shown in Figure 3.5. When the maximum solubility is achieved at a certain temperature, the crystal is said to be saturated with the impurity at that temperature only. If the crystal is cooled to a lower temperature without removing the excess impurity, a supersaturated condition is created. The silicon crystal may return to the saturated condition by precipitating the excess impurities present above the solubility limit corresponding to that temperature. In most cases, such as a high-temperature diffusion step, one rarely reaches the melting point of silicon for impurity incorporation. It is safe to predetermine the temperature and the impurity combination as per the requirement to avoid precipitation issues.

In a perfect crystal, each silicon atom is situated within a tetrahedron and is equidistant from its four neighbors. In a hard sphere model, it can be assigned a tetrahedral radius r_0. Substitutional impurities can also be assigned a hard sphere radius, $r_1 = r_0 (1 \pm \Delta)$, where the ratio Δ/r_0 describes the mismatch in size between the impurity and the host silicon atom, and is called the misfit factor. The misfit factor is indicative of the strain in the lattice caused by introducing impurities into the crystal. Misfit factors of typical dopants in silicon indicate that only arsenic impurity has a perfect fit with the silicon lattice. Figure 3.6a through d explains the stress in the silicon lattice due to the presence of impurities when

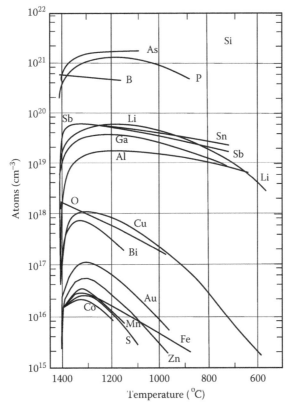

FIGURE 3.5
Solid solubility of typical impurities in silicon. (From F. A. Trumbore, "Solid solubilities of impurity elements in germanium and silicon," *Bell System Technical Journal*, **39**, pp. 205–233, 1960 [6].)

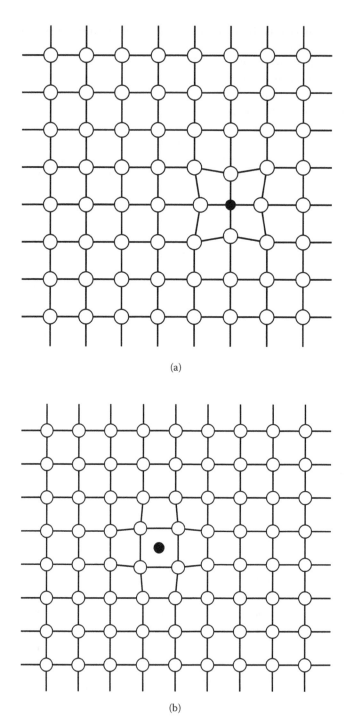

(a)

(b)

FIGURE 3.6 *(Continued)*

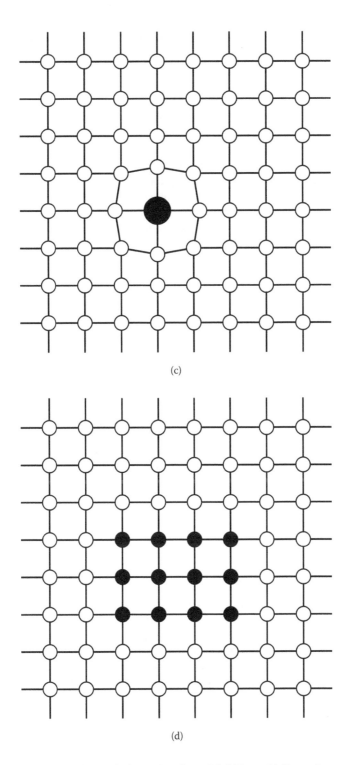

(c)

(d)

FIGURE 3.6

(a) Substitutional impurity atom widening the lattice (tensile strain). (b) Interstitial impurity atom. (c) Substitutional impurity atom comprising the lattice. (d) Precipitate of substitutional impurity atoms. (From R. Falster and V.V. Voronkov, "Lattice defects in silicon", 123–5.5&5.6.ppt, from Monsanto Electronic Materials Company (MEMC) website, 2006 [3].

the substitutional impurity is smaller compared to the host silicon atom, and similarly when it is larger than the size of the host atom. As mentioned, arsenic impurity in silicon is an exception. Misfit factors are shown in Table 1.2 (in Chapter 1).

The impurity atom at the substitutional position in the silicon lattice may not fit the physical space exactly. This problem is mainly seen with the boron impurity atom, which is physically small. Figure 3.6a shows the lattice distortion in such cases. If the same atom is present at the interstitial position, similar distortion is felt, but with an opposite type of lattice distortion, as shown in Figure 3.6b. If the impurity atom is larger than the available space, it will disturb the lattice structure. Figure 3.6c shows such a situation. The presence of large impurities in a single location leads to what is known as precipitation, as shown in Figure 3.6d. This precipitation may be due to different dopant species that were intentionally added or the agglomeration of different metallic species present as unintentional impurities in the lattice sites. These precipitations may lead to leakage current in the fabricated devices. Later chapters discuss the dopant species present in the silicon lattice in more detail.

3.7 Summary

In this chapter, we have seen the importance of single crystals for IC fabrication. Various crystal structures among the solids were discussed, but the diamond crystal structure was emphasized, as this silicon material forms sp^3 hybridization and takes this structure when it solidifies from a liquid. In this structure, one can visualize the crystal from different planes and orientations. Because of its hybridization process, the silicon crystal lattice is not a densely packed structure. It is only 34% filled; the rest is an open structure. The presence of impurities disturbs the regular lattice parameters and creates local strains near the lattice points, depending upon the position of the impurity.

References

1. B. El-Kareh, *Fundamentals of Semiconductor Processing Technology*, Kluwer Academic Publishers, Boston, 1995.
2. http://en.wikipedia.org/w/index.p1hp?title = Atomic_packing_factor&oldid = 570681494, and the references therein.
3. R. Falster and V.V. Voronkov, "Lattice defects in silicon", 123–5.5&5.6.ppt, from Monsanto Electronic Materials Company (MEMC) website, 2006, and the references therein.
4. A. M. Fitzgerald, "Crack growth phenomena in micro-machined single crystal silicon and design implications for micro electro mechanical systems (MEMS)," Ph. D. thesis, Stanford University, August 2000.
5. "Crystal planes in semiconductors," http://www.cleanroom.byu.edu/EW_orientation.phtml, maintained by ECEn IMMERSE Web Team. Copyright 1994–2009. Brigham Young University, Provo, Utah, USA.
6. F. A. Trumbore, "Solid solubilities of impurity elements in germanium and silicon," *Bell System Technical Journal*, **39**, 205–233, 1960.

4

Different Techniques for Growing Single-Crystal Silicon

4.1 Introduction

The silicon industry depends heavily on a ready supply of good-quality silicon wafers for processing integrated circuits. Typical specifications depend on various physical parameters, such as wafer diameter, thickness, degree of flatness, mechanical defects, crystallographic orientation, crystal defect and dislocation density, and polishing. Electrical parameters include dopant type, dopant impurity, resistivity range, and the level of unintentional impurities, in particular, oxygen, carbon, and iron. Other parameters will depend upon the specific application.

Today, very large scale integration (VLSI) and ultra large scale integration (ULSI) circuits require totally defect-free silicon wafers. This is an impossible goal to meet; however, efforts are underway to grow silicon crystals that are as defect-free as possible. The two main methods used to grow silicon crystals today are the Czochralski (CZ) method and float-zone (FZ) method. There are others, but these two provide an ability to grow large-diameter wafers suitable for integrated circuit fabrication. CZ is the most common method for integrated circuit fabrication because of its capability to produce large-diameter crystals that can then be cut to smaller die. For any given wafer diameter (d, in mm) and the target integrated circuit die size (S, in mm²), there is an exact number of integral die pieces that can be sliced out of the wafer. The gross die per wafer (DPW) can be estimated roughly by the following expression:

$$\mathrm{DPW} = d\pi\left(\frac{d}{4S} - \frac{1}{\sqrt{2S}}\right)$$

The gross die count does not take into account the die defect loss, various alignment markings, and test sites on the wafer; in addition, edge pieces contribute to the loss. Increasing the size of the wafer provides more die pieces [1]. For example, if a microprocessor of 15 mm × 15 mm die size is fabricated in an 8-inch (200 mm) diameter silicon wafer, 88 fully processed chips can be obtained. The same effort on a 12-inch (300 mm) wafer will produce 232 die pieces.

4.2 Bridgman Crystal Growth Technique

The Bridgman technique is the oldest technique used to grow single crystals from the melt. In this method, the crucible (typically a circular vacuum-sealed tube) containing the molten material is translated in a furnace along a definite temperature gradient with a high-temperature zone and a low-temperature zone. It is the directional solidification by translating a melt from the hot zone to the cold zone that changes the state of the material from liquid to solid. The Bridgman technique is implemented mostly in a vertical system configuration. Initially, the liquid is allowed to solidify in a capillary portion and then is subsequently moved to condense at a slow rate. This capillary condensation forms a self-seeding process that guides the rest of the molten material. Self-seeding is a random process, and it is difficult to predict the correct orientation in which the crystal will grow. This method is not popular for silicon semiconductors, but is very popular for other compound materials. Since no seed is required here, almost any stable compound can be grown using this approach.

Initial experiments with silicon using the Bridgman method studied the effective segregation coefficients of different dopant species. Ravishankar et al. [2] have reported the effective segregation coefficient for boron dopant to be 0.786 (± 0.036) for CZ-grown crystals and 0.803 (± 0.036) for Bridgman-grown crystals. Similar results are discussed for other impurity species. In this method, single crystals of a small size are grown. Larger size single crystals are difficult to get using this technique.

4.3 Czochralski Crystal Growth/Pulling Technique

Silicon single crystals produced by the CZ process provides a majority of silicon substrates for the fabrication of microelectronic devices. Almost 80% to 90% of the single-crystal silicon used for integrated circuit fabrication is prepared using this method. The ability to grow large-diameter crystals made this method popular and economical because of the large number of die one can get from a single processed wafer. However, the process has its limitations—in particular, the crucible that holds the molten silicon and the heating elements used to maintain the temperatures inside the growth chamber. Oxygen and carbon are the two main contaminants in the growing silicon crystal. Oxygen originates from the quartz crucible holding the silicon melt, and carbon originates from the graphite heating elements and from the quartz crucible mechanical holder (i.e., graphite again). The CZ crystal growth process in which a crystalline ingot is continuously pulled from the silicon melt in a quartz crucible accounts for most of the single-crystal silicon produced today. The process can be viewed as a complex phenomenon in the sense that a complex flow of the melt due to thermal and forced convection plays a significant part in growth rate fluctuations. Figure 4.1 provides a conceptual view of this growth technique, including the various parts: seed, quartz crucible, graphite heater, graphite supporter, crucible support, and the growing crystal. The hot and cold regions demonstrate the different temperature zones inside the chamber.

CZ crystal growth is a dynamic process and involves a continuous growth of a crystal from a hot melt placed in a quartz crucible heated by an electric heater. Electronic-grade silicon pieces, as shown in Figure 4.2, are collected and placed in a quartz crucible, along

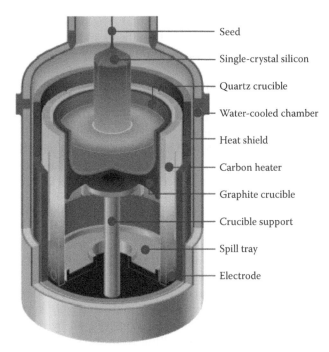

FIGURE 4.1
Basic Czochralski crystal growing apparatus. (From www.egg.or.jp/MSIL/english/index-e.html [1].)

FIGURE 4.2
Final polycrystalline silicon chunks for loading into the crystal growth furnace. (From G. Fisher, M. R. Seacrist, and R. W. Standley, *Proceedings of the IEEE*, **100**, 1454–1474, 2012, and the references therein [3].)

with the dopant to be added to the growing crystal. The complete charge is heated to a temperature well above the melting point of the silicon and is allowed to mix uniformly with the added impurity dopant. Once uniformity has been reached, the entire melt temperature is allowed to stabilize at just above the silicon melting point, approximately 1417°C, in preparation for the next step.

The entire crystal growth assembly is placed in a water-cooled chamber and collectively termed a hot zone, as shown schematically in Figure 4.3a. Practical limitations imposed by the system dynamics require a conical section of crystal, known as the crown, to be grown in the beginning, which is followed by the growth of a cylindrical portion known as the crystal body, or simply the body. The end of the process is marked by the growth of another long conical section, termed the endcone, as indicated in Figure 4.3b. The tip of the crown is normally called the seed end of the crystal, and the tip of the endcone is called the opposite end. Substrate wafers for microelectronic devices are manufactured from the crystal body, while the crown and the endcone are either recycled or discarded. The crown will have a large number of crystal defects, whereas the endcone will have many unintentional impurities. Therefore, the microdefect distribution in the crystal body is of critical significance [4], and only this part of the crystal is useful for integrated circuit fabrication.

Numerical calculation was performed on the fluid flow and mass transfer in large-scale CZ growth of silicon by Chung et al. [5]. The fluid motion in a CZ system is driven by the crystal and crucible rotation (CR), the buoyancy force caused by the density differences, and the shear stress at the free surface resulting from the surface tension gradient. The solidification interface is assumed to be flat. In the real process, however, the solidification interface may be convex or concave to the melt. The meniscus effects in the vicinity of the trijunction point are generally neglected, and the crystal radius is kept constant during

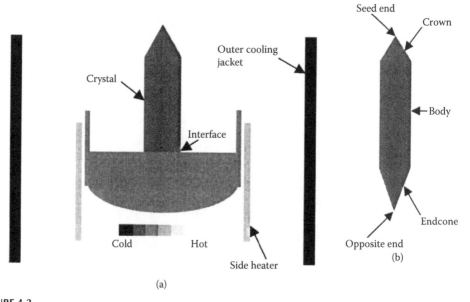

(a)

(b)

FIGURE 4.3
A schematic of (a) a CZ hot zone with a growing crystal; (b) a grown crystal with the crown, the body, and the end cone. (From M. S. Kulkarni, V. V. Voronkov, and R. Falster, *Journal of the Electrochemical Society*, **151**, G663–G678, 2004, and the references therein [4].)

the growth process. Heat is transferred by convection and conduction in the melt. But in the solid, heat is transferred only by conduction. Ambient heat loss from the surfaces of the melt and crystal is greatly affected by radiation. View factors among the ambient atmosphere, growing crystal, the free surface of the melt, and the exposed portion of the crucible must be calculated first to correctly simulate the radiation heat transfer conditions in the entire process.

Numerical simulations were carried out for the CZ method by Mihelčić et al. [6] with the time-dependent Navier-Stokes equations with Boussinesq approximation and the convective heat conduction equations using liquid copper. After several simulations with a low kinematic viscosity and characteristic rotational angular velocities of the crystal and crucible, the influence of free buoyancy-driven convection on the familiar forced convective flow patterns was studied, with special emphasis on the occurrence of flow and temperature oscillations. They also showed the combined free and forced convective flow patterns in the isorotational and the counterrotational CZ accelerated crucible rotation technique (CACRT) for a fluid of the viscosity of liquid silicon.

Heat transfer properties in CZ crystal growth were analyzed by Williams and Reusser [7]. They reported that the heat transport phenomenon plays an important role during the crystal growth process. Adverse thermal gradients may result in unsuitable crystal properties and less-than-optimum productivity performance. A heat transfer model has been developed for the silicon CZ process that predicts temperature distribution and provides information on heat transfer processes in other parts of the system. The model is useful to estimate the solid–liquid interface configuration.

Stefani et al. [8] used a computer-controlled eddy current system for the *in situ* monitoring of CZ silicon crystal growth. The principle involves the behavior of electromagnetic radiation penetrating the silicon crystal and its subsequent reflection. This behavior is outlined by Maxwell's equations. The solution of the electromagnetic field boundary value problem relates voltage measurements from the eddy current probe to the changing electrical conductivity of the crystal. Due to a strong temperature dependence of the electrical conductivity in solid silicon, it is possible to determine the thermal profiles within the growing crystal.

Accurate modeling of CZ crystal growth requires calculating the temperature distribution in the melt, crucible, surrounding components of the system, convection and dopant transport in the melt, shapes of the melt-crystal and melt-ambient interfaces, and the crystal shape. Brown et al. [9] reviewed a hierarchy of models designed to analyze transport processes in CZ and liquid-encapsulated CZ crystal growth. The models are based on a common thermal capillary description of the coupling between heat transfer in the melt and the crystal and the shapes of the unknown surfaces, as well as on a common finite-element/Newton analysis to simultanesously calculate the temperature field and interface shapes. This coupling between heat transfer and the surface shapes has been demonstrated by analyses based solely on conductive heat transfer in the silicon melt. The extensions of the thermal capillary model and the finite-element/Newton method to include melt convection, diffuse-gray radiation, and the heat transfer in the entire system are discussed. Atherton et al. [10] analyzed the effects of diffuse-gray radiation on the parametric sensitivity and stability of CZ growth process for growing single-crystal silicon in a thermal capillary model that governs heat transfer in the system, the shape of the melt-crystal and melt-gas interfaces, and the shape of the growing crystal. Their model demonstrates the sensitivity of the crystal radius, the shape of the crystal-melt interface, and pull rate for systems with high and low axial temperature gradients. The crystal radius exhibits decaying oscillations that dampen more slowly when the temperature gradient is low, indicating the incipient instability expected for an isothermal system. The oscillations are induced by the interactions of radiation with the shape of the crystal and are not predicted when an idealized model is used, which ignores this effect.

4.3.1 Crucible Choice for Molten Silicon

The crucible has an important role in CZ crystal growth because it is the container that keeps the molten silicon at temperatures above 1415°C. The crucible material should have a high melting point, thermal stability, and the necessary hardness. On the other hand, the crucible should be a refractory-type material and chemically unreactive, with minimum reaction or no reaction at all with molten silicon at operating temperatures. It should not leach any unwanted impurities into the molten silicon and must inherently have the best purity possible in all respects. Crucible material choices are limited to quartz (a high-purity SiO_2) [11, 12], silicon carbide (SiC with low carbon content), silicon nitride (Si_3N_4), and aluminum oxide (Al_2O_3). Many researchers have different preferences, such as silicon and silicon carbide coated graphite [13], silica crucibles with molybdenum retainers [14], and even crucibleless [15] growth.

In the majority of cases, high-purity fused quartz (SiO_2) is popular as a container for liquid or molten silicon. This material is structurally strong up to temperatures of 1450°C, but beyond that, it needs external support, which is normally done with the help of a mechanically strong supporter to maintain the required mechanical stability. Usage of graphite is a better option for this application and there are other refractory material choices were also experimented. High-purity and nuclear-grade graphite is the best choice at present. This combination of fused quartz supported with graphite provides better thermal distributions to the crucible. At high temperatures, quartz reacts with molten silicon and releases oxygen species into the silicon melt as per the following reaction:

$$Si + SiO_2 \rightarrow 2\,SiO \text{ or } [O]$$

In this reaction, the SiO is a volatile compound and will be carried away by the purge gas in the reactor chamber. Carbon monoxide is another gaseous species that is released at the graphite–SiO_2 interfaces. It can be removed through the proper flow of argon gas in the reactor environment and also from the molten silicon surface. The nascent oxygen [O] released in this reaction will continue to stay in the molten silicon and contaminate the growing silicon crystal. With time, the concentration of this oxygen becomes sufficiently large and leaches automatically into the growing crystal as an unwanted impurity. If the concentration levels become sufficiently large, such silicon crystals are unacceptable for VLSI and ULSI circuit fabrication. The oxygen concentration in molten silicon is a key parameter for the production of silicon crystals using the CZ technique; in excess concentration levels, it leads to the formation of SiO_2 precipitates and other defects during the cooling process. These defects act as gettering centers for residual impurities during device or circuit processing. The temperature dependence of oxidation-induced stacking faults (OISF) and oxygen incorporation into dislocation-free CZ silicon was studied by Bawa et al. [16]. According to them, the most dominant factor responsible for OISF is the temperature of the molten silicon. The other main limitation of the quartz crucible is its devitrification. It crystallizes into a silica formation, and cracks develop within the material. After crystal growth, remnants of molten silicon solidify inside the quartz crucible. Since molten silicon expands on solidification, the crucible develops cracks and becomes mechanically weak. Thus, a quartz silica crucible can be used only once, and a new crucible is necessary each time.

Oxygen is inevitably incorporated into the single crystals grown using the CZ method, since oxygen dissolves into the melt from the crucible. The dissolved oxygen is transferred to either the crystal-melt interface or the free surface of the melt by convection and diffusion. Most of the oxygen in melt evaporates at the free surface, while the rest incorporates into the growing crystal. Therefore, the oxygen concentration in grown crystals is determined by the following key factors: (1) dissolution at a melt-crucible interface, (2) diffusive and convective transport in the melt, and (3) evaporation from the free surface of the molten silicon and segregation at the crystal-melt interface. Yi et al. [17] have studied the details on oxygen transport mechanism on oxygen concentration in silicon crystals and the system operating parameters such as crucible and/or CR rates on oxygen incorporation in the growing silicon crystal. The team quantitatively estimated the contribution of flow modes, such as axisymmetric to nonaxisymmetric, to oxygen transfer in the melt. The role of oxygen in CZ-grown silicon crystals and how it can be used to improve the quality of silicon for integrated circuit processing are reviewed by Benson and Lin [18]. Their work also covers the fundamental aspects of oxygen incorporation during crystal growth and segregation issues.

The effects of molten silicon flow on the crucible and on the distribution of impurity atoms, especially oxygen, has been studied by many researchers from both a scientific point of view and for application in semiconductor device technology. The main source of oxygen is the reaction of the silicon melt with the silica crucible. In a real CZ crystal-growth process, the bulk melt is never saturated with oxygen because it is continuously extracted from the melt by evaporation of SiO and the removal of this SiO from the melt surface by forced gas convection maintained during the crystal growth process. It is estimated that 99% of the dissolved oxygen evaporates from the melt and only 1% is incorporated into the crystal. The transport of oxygen in the melt is fast due to the strong mixing by free and forced convection. The solubility concentration of oxygen in the silicon melt is only achieved at the crucible wall. Seidl and Müller [19] strongly recommend that within the bulk melt, the oxygen concentration must be below the solubility limit to avoid the formation of solids that would disturb the complete growth process. They also measured the oxygen solubility in molten silicon *in situ* using an electrochemical solid ionic sensor over a temperature range from the melting point to 1836 K, about 150°C above the melting point of silicon. The sensor used here is based on stabilized zirconia as the oxygen-conducting material. The evaluated oxygen concentration versus reciprocal temperature is plotted in Figure 4.4. The melt oxygen fraction a_o can be fitted by

$$\log a_o = \frac{-14036}{T} + (4.0 \pm 0.3)$$

from which the oxygen concentration is to be calculated by using the parameters $c_o = a_o$. 5.4×10^{22} cm^{-3}. The large error in the temperature-independent term is caused by the uncertainty in the free energy of the oxygen solution in liquid silicon ΔG^o. Apart from this remaining uncertainty, the results show that the temperature dependence of the oxygen solubility in silicon follows the Arrhenius equation.

Fraundorf and Shive [20] have studied a combination of conventional transmission electron microscope (TEM) imaging, high-resolution TEM imaging, energy dispersive x-ray analysis, electron energy loss spectroscopy, and electron diffraction to characterize the quartz crucibles used in CZ crystal growth techniques. The walls of several quartz crucibles obtained before and after use were examined in cross-section up to a depth. Submicron-sized crystalline silicon oxide inclusions were observed in the outer 20 μm of the crucible

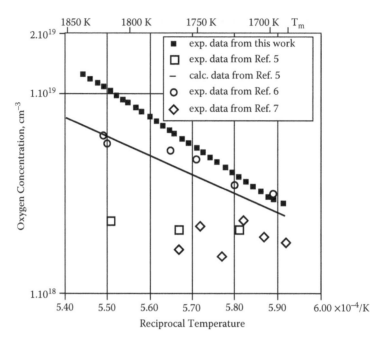

FIGURE 4.4

Arrhenius plot of the oxygen concentration in molten silicon. (From A. Seidl and G. Müller, *Journal of the Electrochemical Society*, **144**, 3243–3245, 1997, and the references therein [19].)

wall after the silicon growth process, with densities ranging from 1.0×10^{10} to 1.0×10^{12} per cm^3. This strongly supports the formation of silicon (sub)oxide during the process.

Silicon carbide (SiC) is another choice for crucible material. There are many crystalline forms, but α-SiC and β-SiC are commonly preferred. The segregation coefficient of carbon is low, and this leads to the formation of SiC at the silicon liquid–solid interface where the crystal is growing. Series and Barraclough [21] reported on the rate of carbon contamination of a silicon melt in a low-pressure CZ puller technique by repeated dips. They also studied the effect of placing a gas baffle between the heater and the melt and found that the contamination is mainly from the polysilicon charge. The origin of SiC impurities in silicon crystals grown from the melt in vacuum were analyzed by Schmid et al. [22]. The main source of high carbon levels in silicon crystals grown from the melt under reduced pressures and contained in quartz silica crucibles supported by a graphite retainer/susceptor was identified through thermodynamic analysis. Evidence of high carbon levels in silicon grown using the heat exchange method has led to a proposed mechanism for silicon carbide formation. It is associated with the use of graphite retainers in contrast with silica crucibles under reduced pressures. High carbon levels in silicon have been reported to cause a breakdown in crystallinity in the CZ as well as in silicon ribbon growth. The origin of high carbon levels in silicon processed in a vacuum is a result of reactions between silica crucibles and the graphite retainers. Further, it is suggested that if the graphite retainers are replaced by molybdenum, the carbon levels in the silicon can be considerably reduced. The calculations were verified by experimental results, and the carbon level was reduced by 50% with the use of molybdenum retainers. The purity of molybdenum and the emission properties may play a key role if graphite is replaced.

Silicon nitride–based crucibles are attractive because of the absence of oxygen. With slight nitrogen incorporation does not alter the electrical properties of silicon. Not much has been reported about silicon nitride crucibles in the open literature and the acceptance for their application in the industrial production of microcrystalline silicon. So far, there are patents that claim the idea of reusable silicon nitride crucibles and their application for directional solidification. Different production technologies and the impact on crucible material are described in Ref [23]. This material has problems similar to silicon carbide. Deike and Schwerdtfeger [24] reported that silicon nitride performed well as a crucible material during short times of up to 20 minutes. At these times, the silicon melt did not wet the silicon nitride crucible. Later, the melt penetrated into the ceramic matrix. This phenomenon finally led to the disintegration of the crucible. Silicon nitride gradually decomposes at 1500°C in streaming argon, forming liquid silicon in the pores. The initial nonwetting behavior was attributed to the protective silicon dioxide films on the surface of the nitride, and the wetting was initiated by the internal formation of silicon due to the decomposition reaction taking place at that temperature. Deike and Schwerdtfeger have measured the weight loss in the decomposition of silicon nitride according to the following reaction:

$$Si_3N_4 = 3\ Si + 2\ N_2$$

It is felt that in the streaming purified argon gas, nitrogen is readily removed from the silicon nitride crucible material.

Aluminum oxide is another choice for crucible material. Dissolution of this material releases aluminum into the silicon melt, and this enriches the molten silicon. Since aluminum is a p-type dopant material for silicon, this may lead to unexpected concentration variations in the grown silicon.

Silicon nitride (Si_3N_4) and fused silica (SiO_2) are preferred as crucible materials at present. However, the pieces of single-crystal silicon grown in these crucibles contain a substantial amount of interstitial oxygen in their lattices due to its inherent presence in the crucible material.

The melt motion in the CZ process consists of three components: (1) the thermocapillary convection driven by the decrease in surface tension along the free surface from the cold crystal edge to the hot crucible wall, (2) the buoyant convection driven by the decrease in density from the cold crystal-melt interface to the hottest point on the vertical crucible wall, and (3) the centrifugal pumping flows due to the rotations of the crystal and crucible about their common vertical axis [25]. The amounts and distribution of oxygen, dopants, and other impurities in the crystal depend on the aforementioned motions in the melt. Strong thermocapillary convection allows the excess oxygen to evaporate from the free surface, while a strong centrifugal pumping near the crystal-melt interface ensures radially uniform distributions of oxygen and dopants in the crystal.

Watanabe et al. [26] observed the molten silicon flow using a double-beam x-ray visualization technique with solid tracer particles. The melt height was monitored using an x-ray radiograph image. Under these conditions, the temperature gradient over the crystal-melt interface was not exactly equal to that under crystal growth conditions. However, this was the only way to observe the flow as the melt height changed. If the flow observation was simultaneously carried out during crystal growth, tracer particles would be incorporated into the growing crystal. This is one of the best options for visualizing the liquid silicon.

Kakimoto et al. [27] directly observed convection of molten silicon during CZ single-crystal growth by using x-ray radiography. The melt flow pattern was monitored using a tracer method whose density and wettability were adjusted to that of the molten silicon. The observed convection of the molten silicon in the crucible was not only steady, but also transient, and not axisymmetric, but asymmetric. This is attributed to the asymmetric temperature distribution within the crucible. The flow velocity of the molten silicon in the 75 mm diameter crucible was measured as 10 to 20 mm/s. Further studies were carried out by the team [28] on the natural and/or forced convection of molten silicon with solid traces for various crystals, CR speeds, and temperature distribution in a crucible holder. Downflow attributed to natural convection in the center of a crucible was simulated by numerical calculation, but was scarcely observed with and without CR. Numerical simulation of the molten silicon was carried out by a software package that measured nonaxisymmetric temperature distribution in a crucible holder. Unidirectional flow, with and without CRs, can be qualitatively explained by the numerical simulation with nonaxisymmetric temperature distribution in the crucible holder. The particle path attributed to natural convection near the solid–liquid interface was suppressed downward with an increase in crystal rotation speed. The phenomena were explained by the generation of forced convections beneath the rotating crystal. It is also shown that [29] flow mode was dependent on an aspect ratio of the melt. For a deep, low-aspect-ratio melt with a growing crystal that is identical to the shouldering process of the growth, the flow was unsteady and nonaxisymmetric. For a shallow melt without crystal and CRs, the flow was relatively steady and axisymmetric. However, flow became unsteady and nonaxisymmetric for a shallow melt with crystal rotation. The flow instability area, which was also thermally unstable, was found to be larger in the crystal-crucible isorotation condition.

Numerical simulation involving fluid flow, heat conduction, and heat exchange by radiation has been performed by Kakimoto et al. [30] using the geometry of a real CZ furnace for silicon single-crystal growth. The flow velocity fields of molten silicon were obtained from the extrapolation of the stream function, which was developed using the velocity boundary layer theory. They reported that the calculated flow velocity and particle path were semiquantitatively identical to the results obtained from the x-ray radiography experiment. The characteristic velocity was reported to be about 1.0×10^{-2} m/s. It has also become clear from a comparison of flow velocities between experimental and calculated results that the order of the volume expansion coefficient of the molten silicon is about 1.0×10^{-4} per K. The flow was almost axisymmetric and steady for a specific case with low crystal and CR rates and with a shallow melt. It was also reported that a flow with larger azimuthal velocity component exists just beneath a crystal, while that with opposite flow direction exists near the crucible wall. The simulation gave a clear picture of the molten flow and was compared with the experiment for validity.

Second-order turbulence simulation of the CZ growth melt was also studied for the flow. The statistics of the fluctuating velocity and temperature fields in a large-scale CZ silicon crystal grower were simulated using a Reynolds stress turbulence closure. The methodology and some of the ideas underlying such a simulation method are discussed by Ristorcelli and Lumley [31]. Simulations of the buoyancy-driven case are presented for Grashof numbers $Gr = 1.0 \times 10^9$ and 1.0×10^{10}, and Marangoni numbers $Ma = 2.0 \times 10^3$, 15.0×10^3, and 60.0×10^3. The solution procedures, which average over small-scale fluctuations, indicate the unsteady mean flow measure with time scales on the order of the integral time scale. It is felt that the source of the flow's unsteadiness is the occurrence, growth, and dissipation of eddies between the crucible wall and the free surface. This time scale corresponds to time scales seen in the dopant striations. The simulation results indicated two very active areas in the melt flow (1) at the outer edge of the crystal-melt interface and (2) at the bottom center of the crucible. Underneath the free surface, the temperature fluctuations are comparatively small. The material and structural inhomogeneities in the grown crystal are related to the variance and skewness of the temperature fluctuations and the vertical turbulent heat flux at the crystal-melt interface. Buoyantly driven flow of the molten silicon is the highlight of this simulation study.

Järvinen et al. [32] have developed a detailed mathematical model and numerical simulation tools for time-dependent silicon crystal growth and the crystal growth environment. All three heat transfer mechanisms—conduction, convection, and radiation—were applied to study the case, and the information regarding the temperature and velocity fields were evaluated. Modern supercomputers were used to solve the results to effectively assist the complete process. Steady-state temperature distribution is achieved by simulating pure heat transfer without melt convection. In Figure 4.5, the temperature distributions and

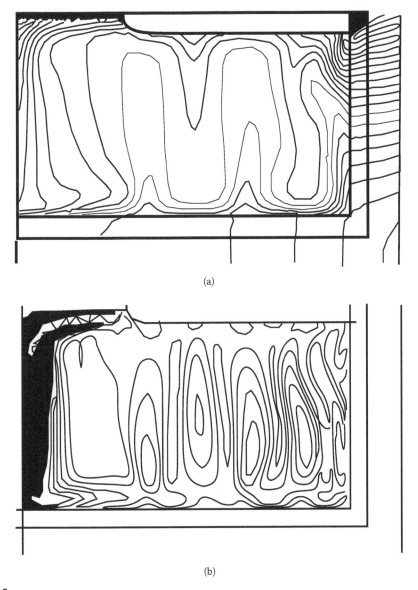

(a)

(b)

FIGURE 4.5
Temperature distribution and velocity field in the melt at $t = 3816$ s, Re* = Re/100. (From J. Järvinen, R. Nieminen, and T. Tiihonen, *Journal of Crystal Growth*, **180**, 468–476, 1997, and the references therein [32].)

70

velocity fields in the melt are presented at different time steps with the reduced Reynolds number Re* = Re/100. The case where the Reynolds number is two orders of magnitude lower than the real one leads to the loss of the steady-state solution, and time-dependent fluctuations of the velocity field develop. In the finite-element analysis, the authors have applied the penalty method together with the bilinear velocity-constant pressure approach in order to solve the Navier-Stokes equations. In fact, the system seems to find a periodic solution in a limit cycle of about 8 s.

Tanaka et al. [33] have used the melt surface temperature in CZ silicon growth studied by CCD camera observation. The thermal radiation energy from the melt surface was converted into temperature using the blackbody calibration method and was recorded with a VCR as two-dimensional color images. The experimental results without a crystal revealed that the temperature distribution at the melt surface can change in four patterns depending on the CR rate: there is an axisymmetric spoke pattern at low rotation rates, *n*-folded and island patterns at medium rotation rates, and cellular patterns at high rotation rates. To predict the fluid motion from the experimental observations, three-dimensional time-dependent numerical simulations of the silicon melt flow were executed. As a result, a qualitative transition model for the temperature distribution and CZ silicon melt flow was derived. The average temperature at the center of the melt surface and the temperature difference between the position at a radius of 74 mm and the center are shown in Figure 4.6. The increase of the average temperature at the CR rate of 0.7 rpm implies that the average heat flux to the center of the melt surface increases. The inversion of the temperature difference at CR rate ranges higher than 0.7 rpm indicates that the vertical heat flux to the center of the melt surface increases in comparison with the radial heat flux and makes the center hotter than the remaining area at the melt surface.

The calculations resulted in four temperature distribution patterns, corresponding to the different CR rate ranges, which qualitatively agreed well with the CCD observations. The evaluation of quantities such as the critical CR rate for the transition of temperature

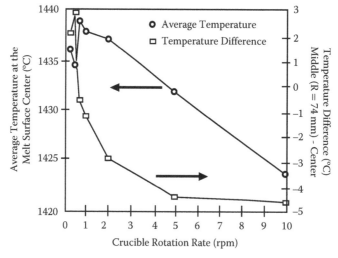

FIGURE 4.6
Crucible rotation rate dependence on the average temperature at the center of the melt surface, as well as temperature difference between the middle (*r* = 74 mm) and the center of the melt surface. (From M. Tanaka, M. Hasebe, and N. Saito, *Journal of Crystal Growth*, **180**, 487–496, 1997, and the references therein [33].)

distribution, the amplitude of temperature fluctuations, and the average temperature at the melt surface needs closer observations. Accurate boundary conditions are necessary for a quantitative agreement with the observations. The cellular pattern of the temperature distribution at the melt surface is obtained at the CR rate of 10 rpm. Most radial flows like natural convection, vortices are suppressed, and vertical convections remain. Many upward and downward flows can be seen in the vertical cross-section. In each hot cell, an upward flow exists. The experimental and numerical results show 3D time-dependent structures of the silicon melt flow, which depend on the CR rate. Figure 4.7 shows a qualitative model for the pattern formation and the associated transition of the CZ Si melt flow with respect to the change in CR rate. The axisymmetric spoke pattern region at low CR rates is produced by natural convection with alternate weak and strong flows from the crucible side-wall toward the center of the melt surface. These flows are caused by alternating weak and strong vertical convection streams near the crucible side-wall along the azimuthal direction. In this pattern region, heat is carried to the center of the melt surface mainly by natural convection. The *n*-folded pattern is formed as follows. At first, natural convection is suppressed by the Coriolis force caused by CR. After the cold area expands to a certain extent, radial melt streams of high temperature flow from the outer (hot) to the inner (cold) area and shrink the expanded cold area. In this pattern region, a hot vertical upward flow occasionally develops in the center. This is seen as high-temperature island formations in the cold area of the melt surface. Once the expanded cold area is heated by the radial or the vertical flow, the instability settles down and the center region cools again to form a

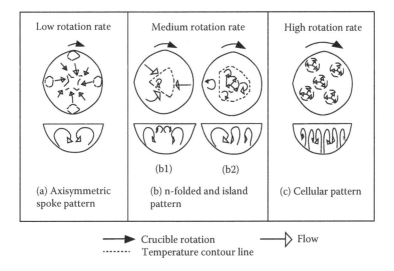

Low rotation rate   Medium rotation rate   High rotation rate

(b1)   (b2)

(a) Axisymmetric spoke pattern

(b) n-folded and island pattern

(c) Cellular pattern

→ Crucible rotation   ⇒ Flow
········ Temperature contour line

FIGURE 4.7
A model of melt flow patterns with respect to the CR rate change. Upper figures: horizontal view of the melt flow at the surface; lower figures: vertical view of the melt flow in the crucible. (a) Axisymmetric spoke pattern region at low CR rate. The melt flow is axisymmetric, and natural convection is dominant. Heat is carried mainly by the natural convection. (b) *n*-folded and island pattern region at medium CR rate. Natural convection is suppressed by the Coriolis force caused by crucible rotation. Vortices appear at the melt surface. Occasionally, vertical convection develops at the melt center. Heat is carried by the horizontal convection, the vertical convection, and the vortices. (c) Cellular pattern region at high CR rate. Most of the radial flows are suppressed by Coriolis force. Only vertical convection remains. Heat is carried mainly by the vertical convection. (From M. Tanaka, M. Hasebe, and N. Saito, *Journal of Crystal Growth*, **180**, 487–496, 1997, and the references therein [33].)

large cold area. Thus, in this CR rate range, heat is carried to the center of the melt surface by the radial and vertical convections. Spoke patterns are still present in the outer area of this region. In the cellular pattern region, most of the radial flows are suppressed by the Coriolis force and only vertical convection remains. In each cell, a hot vertical upward flow exists. Heat is carried to the center of the melt surface mainly by vertical convections. The model of melt patterns with CCD images gives a clear picture for studying flow dynamics in the molten silicon, and many patterns were visualized in this important study.

There are many interesting publications on the dissolution rate of crucible materials and their reactions with molten silicon [34–36], erosion of crucibles [37–42], behavior of oxygen and carbon [43–46], and geostrophic turbulence [47–49]. This is a major topic of research, and many research teams are working on these topics.

4.3.2   Chamber Temperature Profile

Crystal growth processes are somewhat slow, and the inside temperature of the entire chamber has an important role to play. It is necessary to maintain the temperature profile precisely over the entire range of the growth system with a high temperature (above the melting point of silicon) at the location of the crucible and at a slightly lower level for the grown crystal. The grown crystal experiences different zones of temperatures, and the crown area will be roughly around 400°C. The crucible area should be maintained very tightly, as the temperature fluctuations resulting from thermal convection during crystal growth are known to cause growth-rate fluctuations and impurity microsegregation. The thermohydrodynamics is an important factor in the growth of crystals, as this directly relates to the hydrodynamic behavior of the melt [50, 51] and to crystal growth and segregation characteristics. Thermal waves of a nonaxisymmetric flow are observed in CZ-type silicon melt. The wave number and pattern transition of the thermal waves depend on the crucible rotation rates [52]. To control the point defects in the growing crystals, one has to maintain an optimal surface temperature and radial uniformity [53] of the liquid. Surface tension of the melt also changes with the temperature and affects the transport properties of the liquid [54] toward the growing crystal.

Radio frequency (RF) heating or resistance heating is preferred to maintain the temperature in the crucible for holding silicon melt. Induction heating is useful for small melt sizes. For large crystals and for a large quantity of polysilicon charge, only resistance heating is preferred. In most of the cases, a direct current (DC) power supply is used. However, it is reported that dislocation-free silicon crystals of a moderate diameter were grown from an RF-heated cold crucible at pulling speeds of 1.0 to 2.8 mm/min in both [111] and [100] orientations by Ciszek [55] without using any auxiliary heating source. It is also claimed that the solar cells fabricated using these crystals have shown 4% to 8% higher efficiencies than the cells fabricated using conventionally CZ-grown wafers. In the case of inductively heated graphite crucibles, the temperature distribution in the crucible is affected by not only the radiative heat transfer, but also the conductive and natural convective heat transfer in the furnace. Numerical results of Du and Munakata [56] have revealed that the temperature distributions in the crucible are affected by the relative position between the crucible and induction coil due to the modification of the electromagnetic field in the CZ furnace.

When processing single-crystal silicon using the CZ method, hot zones with graphite parts placed above the silicon melt level produce crystals with higher edge-iron concentrations than those hot zones without graphite parts. Iron present in graphite, measured in ppm, diffuses into the crystal and increases the edge-iron concentrations of the grown crystals. To reduce this iron concentration in crystals, Sreedharamurthy [57] came out with a new technique to modify the crucibles. The graphite components were coated with two protective layers, including an initial protective layer of silicon carbide and a second protective layer of silicon. The two layers provide a barrier to iron emission by sealing the graphite surface. This makes it difficult for iron impurities to pass through the coatings by grain-boundary and bulk-diffusion mechanisms. The coating of silicon over and above the silicon carbide–coated graphite acts as a gettering sink for those unwanted contaminants. In addition, silicon has a high affinity for iron and thus readily reacts to form iron silicides in the coated layers. The iron concentrations were directly dependent on the iron concentrations of the graphite samples used, and thus iron present in the graphite samples is measured indirectly. Figures 4.8 and 4.9 compare the relative iron concentrations of the wafers exposed to the silicon carbide–coated graphite and silicon carbide–coated graphite

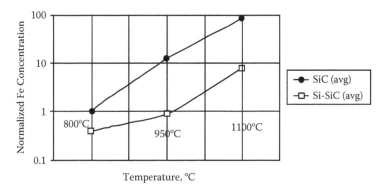

FIGURE 4.8
Iron concentrations of the wafer exposed to SiC-coated graphite and SiC-coated graphite samples with Si coating, as functions of temperature. (From H. Sreedharamurthy, "Reducing iron in single crystal silicon grown using CZ process," *210 ECE Meeting Cancun*, Mexico, 29 Oct–3 Nov, 2006 [57].)

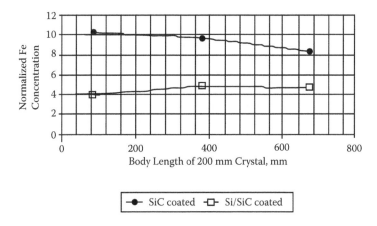

FIGURE 4.9
Iron concentration of wafers from crystals grown with SiC- and Si-SiC–coated graphite parts. (From H. Sreedharamurthy, "Reducing iron in single crystal silicon grown using CZ process," *210 ECE Meeting Cancun*, Mexico, 29 Oct–3 Nov, 2006 [57].)

with silicon coating on it. At higher temperatures, the vapor transport of iron from the test samples is higher. Wafers exposed to graphite and silicon carbide–coated graphite have higher iron concentrations than that of samples that were coated with silicon and silicon carbide. Application of uniform coatings on all graphite components is an issue in this case, particularly for varying silicon melt levels.

Experimental evaluations were carried out by Gilmore et al. [58] to study the impact of graphite furnace parts on radial impurity distributions for the CZ-grown crystals. A method was developed to examine the influence of the furnace parts as a contamination source during high-temperature processing. Graphite materials with and without silicon carbide coatings were evaluated. When silicon carbide–coated graphite was used, significant improvements in carrier lifetimes were observed. The silicon carbide film was found to act as an effective seal against impurities outgassing from the graphite parts. By analyzing the coatings, it was concluded that the attainment of high lifetime values is a function of the sealing capability of the film. This sealing capability is, in turn, dependent on film grain size and morphology, and thus is determined by the coating film chemical vapor deposition process parameters.

The application of numerical simulations of the crystal growth process became an important tool for the prediction of the thermal history and, hence, for the improvement of silicon crystal quality. Thermal simulations of the CZ growth process were analyzed by Dornberger et al. [59] in comparison with the experimental results. Inside temperatures were measured within an industrial CZ silicon puller and compared with the simulation results. The temperatures were measured by using thermocouples in the crystal along the axis, as well as inside the lateral and bottom insulations. For CZ crystal growth systems, thermocouples are known to be the most efficient method to experimentally validate numerically calculated temperature distributions within the relevant constituents of a silicon CZ puller of industrial size. However, for the growing crystal, a compromise between the dynamics of the real process and the experimental restrictions (fixed thermocouples) has to be found. The temperature distribution of the furnace was computed using three different software codes. They have demonstrated that complex heat transfer simulations were carried out, with the exception of melt convection issues. One possibility, which comes close to the process dynamics, is the ingrowth of thermocouples, but this is restricted to short crystals only. Instead of a growing crystal, prepulled 4-inch crystals of different lengths were prepared for different set of studies with a series of up to five PtRh30/PtRh6 thermocouples with proper protection in radial holes along the axis of the crystal, as shown in Figure 4.10.

The crystals were mounted in an industrial CZ silicon puller (Leybold EKZ 1300) with a 300 mm diameter crucible. After heating to process temperature, the prepared crystals were brought into contact with a silicon melt. The crucible rotation rate was 5 rpm, but the crystal was not rotated due to the thermocouple installation. After remelting the crystal-melt contact region, a slightly, concave crystal-melt phase boundary was established. As no crystal growth took place, no solidification heat was released, and the heat flux across the crystal-melt phase boundary was determined only by the thermal conditions in the surroundings. The temperature in the crystal was measured after a steady state was established. Then the crystal was quickly lifted from the melt to preserve the shaped-phase boundary. Eleven additional thermocouples, protected against the reducing graphite by small alumina tubes, were positioned at certain monitoring points along the lateral and bottom insulations of the furnace. Some thermocouples were installed between the graphite and graphite felt, and within the graphite felt. The total pressure in the chamber was 35 mbar with an argon flux of 700 l/min. In another experiment, a radiation shield was added as sketched in Figure 4.11. This radiation

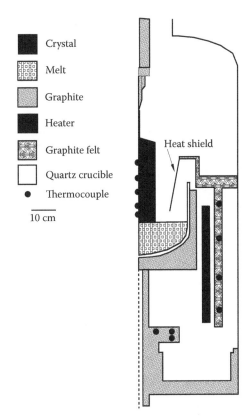

Crystal

Melt

Graphite

Heater

Graphite felt

Quartz crucible

Thermocouple

10 cm

Heat shield

FIGURE 4.10
Schematical drawing of a CZ silicon furnace with heat shield. Thermocouples are installed in a prepulled 100 mm diameter crystal and in the lateral and bottom insulations. (From E. Dornberger, et al., *Journal of Crystal Growth*, **180**, 461–467, 1997, and the references therein [59].)

shield was expected to have a large impact on the temperature distribution within the crystal because the direct radiation from the melt and from the upper heater is partly screened. This figure shows the global temperature distribution of the furnace with heat shield. This result is representative for all three models used for the present study of Dornberger [59]; some deviations occur in the melt region, which was handled differently by the codes. The team feels that the problem of melt convection has not yet been satisfactorily solved. This is especially true for modeling the mass transfer in the melt, which requires the exact handling of the convective transport even more than for modeling the heat transfer. Since for large melts the flow is three-dimensional and turbulent, more efforts have to be made to accurately predict the oxygen and dopant transport in the melt.

The axial thermal profiles of silicon crystals during the CZ crystal growth process were measured experimentally by Choe et al. [60] using an eddy current technology. The intrinsic conductivity changes in the crystal resulting from cooling were measured in terms of the eddy current amplitude and phase responses, and the axial thermal profiles of the growing crystal were subsequently derived from these results. Crystals up to 70 mm diameter and 406 mm in length were grown for this experimental observation. The eddy current sensor was initially positioned 100 mm above the melt, and then the measurements

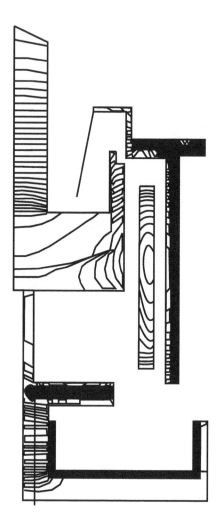

FIGURE 4.11
Temperature distribution of a 4-inch crystal growth furnace computed with simulation model M2. The isotherms are spaced in steps of 25 K. The dotted line corresponds to the melting point. (From E. Dornberger, et al., *Journal of Crystal Growth*, **180**, 461–467, 1997, and the references therein [59].)

were made as it was lowered to 25 mm above the melt at a constant speed of 25 mm/min. The experimental results indicate that the axial thermal profile in the region is nearly linear and quite transient during the initial phase of body growth. As the crystal gets longer, the increasing heat loss by radiation from the crystal surface causes the overall profile to shift downward. When the crystal reaches 200–250 mm in length, a steady-state condition is achieved, and the overall axial thermal profile stays nearly invariant for the remainder of the growth. It is estimated that the thermal gradient will be on the order of 4°C to 6°C per mm.

Numerical simulation of a complete crystal growth system includes the present-day approach to study the entire thermal profile in the crystal-growing units. Using finite-element methods and a 3D approach provides complete insight into the system on various temperature distributions, giving ample scope for one to estimate the profile [61–67]. Many teams are working on these simulations to produce perfect growth conditions to grow defect-free crystals. The internal environment changes with time due to continuous flushing with argon, and this affects the temperature profiles at each stage [68]; monitoring the temperature changes online yields good results.

4.3.3   Seed Selection for Crystal Pulling

Seed crystals are prepared to precise orientation tolerance and are inserted perpendicular to the melt surface. In the CZ method, crystals are grown by controlled withdrawal of a seed crystal that is initially touched to the free surface of the melt. The screw dislocation provides an easy growth mechanism for crystals since atoms/molecules find a ready place where they can position themselves and grow. No nucleation is necessary in such situations. If there are other dislocations in the seed crystal, then the crystal grows on these as well, but the dislocations will continue to grow at those points and may also multiply. To reduce this dislocation content, the growth rate is deliberately increased to reduce the diameter of the seed so that a neck forms in the growing crystal. As the neck forms, some of the dislocations will move out of the crystal at this location. The process can be repeated to obtain dislocation-free crystal. This technique is also known as the "Dash technique" after its discoverer W. C. Dash [69] who developed this method in 1959. Figure 4.12 shows the sequence of events to grow the single-crystal silicon ingots using a seed crystal. Once the ingot body part is reached, the growth rate has to be maintained to obtain a uniform diameter until all the molten silicon is consumed and the endcone terminates the process.

The most common single-crystal silicon growth orientations are <111>, <100>, and <110>. For silicon, the [111] crystalline planes exhibit the smallest separation (of 3.135 Å). Therefore, growth of the single crystal along this <111> direction is the slowest, since it results in the setting down of one atomic layer upon another in a close-packed form. Based on packing considerations, <111> oriented silicon is, therefore, the easiest to grow. It is also the least expensive. These <111> oriented crystals are widely used for many bipolar transistors, including high-power thyristors. Most of the VLSI and ULSI circuits are fabricated in <100> oriented silicon wafers. <110> oriented wafers find applications in microelectromechanical systems (MEMS) and specific crystal orientation–based microsensors [70].

Growing single-crystal silicon in the <110> orientation is rather difficult. Dyer [71] came out with a pragmatic solution to grow dislocation-free <110> silicon crystals by introducing one or more bulges into the stem between the seed and the growing crystal. Dislocations grow in a parallel direction away from the axis under the influence of an increasing stem diameter. The results are consistent and support the immobilization of dislocations at low stem diameters and their release at larger diameters. However, it was reported recently that dislocation-free, boron-doped silicon crystals were grown without the Dash method and thin necks [72–74]. Silicon multicrystals, with several large grains, have been grown by Hoshihawa et al. [75] by using two/three single crystal seeds in a single CZ growth method.

Murthy and Aubert [76] came out with a proposal to grow dislocation-free silicon crystals in the <110> orientation in a reproducible manner. Several flared-up bulges introduced into the stem are mainly responsible for effectively getting rid of the dislocations. The plausible mechanisms responsible for the diversion of axial dislocations and for eliminating them are thoroughly described. Studies of plastic deformation of these crystals by using *in situ* γ-ray diffractometry revealed that the mosaic distribution is even, which results in higher neutron reflectivity, and these are good starting materials for the preparation of efficient neutron monochromators.

(a)   (b)

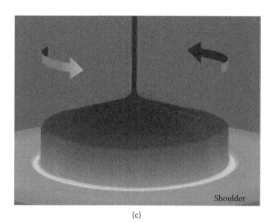

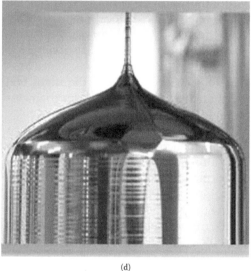

(c)   (d)

FIGURE 4.12
Sequence of CZ crystal growth method. (a) Seed crystal touching the molten silicon, (b) fast pulling to reduce the crystal diameter to implement the Dash process, (c) slow and steady pulling of crystal at a predetermined rate to maintain the diameter, and (d) top portion of the grown single-crystal silicon (Courtesy of Wacker Chemie AG, Munich, Germany.). (From S. K. Ghandhi, *VLSI Fabrication Principles: Silicon and Gallium Arsenide*, John Wiley & Sons, New York, 1994; C. W. Pearce, "Crystal growth and wafer preparation" in *VLSI Technology*, edited by S. M. Sze, McGraw-Hill, New York, 1988; O. Anttila, "Challenges of Silicon Materials Research: Manufacture of High Resistivity, Low Oxygen Czochralski Silicon," Okmetic Fellow, Okmetic Oyj, epp.final.gov/DocDB/0000/000007/001/CERN_RD-50_06_2005_Okmetic_OA.ppt, June 2005, and the references therein [70].)

As discussed earlier, in the CZ process, practical limitations imposed by the system dynamics require the growth of a conical section (known as the crown) in the beginning of the process, which is followed by the growth of the cylindrical portion (known as the crystal body or, simply, the body). The end of the process is marked by the growth of another long conical section termed the endcone. As discussed earlier, the substrate wafers for the microelectronic devices are manufactured from the crystal body, while the crown and the endcone are either recycled or discarded. Thus, defect dynamics that only occur during the body growth play a key role in the device yield [77]. It is this part of the crystal that is used for device fabrication.

Transport phenomena within the melt are important, as they affect the quality of the crystal. Momentum, energy, and mass transfer have been considered in the vicinity of the crystal—in other words, in the meniscus below the crystal. Driving forces due to buoyancy, crystal rotation, and thermocapillary effects have been included. Estimation of the nondimensional parameters involved reveals that the surface tension gradient and crystal rotations are important for the growth of silicon crystals. For pure thermocapillary flow with large Peclet number (Marangoni number) and small Prandtl number, an integral method has been used with the boundary-layer equations to determine the temperature, axial velocity, and concentration distribution in the meniscus region. The results of Balasubramaniam and Ostrach [78] show that in the ideal case when the energy transfer is one-dimensional, the thermocapillary flow causes the temperature and concentration distributions in the meniscus to be two-dimensional, which causes the interface to be curved and the solute to be segregated nonuniformly over the cross-section of the crystal. With the interface assumed to be flat initially, a correlated curved shape that protrudes into the melt has been obtained for the interface.

Motooka [79] applied molecular-dynamic simulations of crystal growth from melted silicon. It is reported that crystal growth in the [001] direction occurred by attaching the silicon atoms in melt at kink sites associated with {111} facets formed at the solid–liquid silicon interface, while in the [111] direction, double-layered two-dimensional nucleation was first created and then followed by double-step, layer-by-layer growth. This results in the remarkable difference of impurity atom effects between [001] and [111] crystal growth—that is, disturbance due to carbon atoms, present at site, tends to be localized in the former case, while it tends to be transmitted in the latter case affecting the crystal growth and defects formation.

To control the crystal diameter during CZ growth of silicon, a new technique based on auxiliary heating of the melt surface has been proposed by Ekhult et al. [80]. The experiments have been performed in a system used to grow 3-inch-diameter silicon crystals. By comparing two different crystals grown with and without this new technique, it was concluded that with the new technique, it is possible to control the diameter of the growing crystal. Further, it provided better microscope homogeneity, but macroscopic dopant profiles were affected by large changes in the heater power, indicating that convection in the system is influenced in this modification, hence the changes.

Van Run [81] came out with a critical pulling rate to suppress the remelt phenomena in CZ and FZ techniques. Remelt phenomena in crystals growing from the melt with a sinusoidally varying microscopic growth rate are suppressed if the pulling rate exceeds a critical value R_{crit}, which is proportional to the temperature gradient in the solid at the solid–liquid interface and is inversely proportional to the heat of fusion. For temperature gradients in a growing silicon crystal typical for CZ (at 120°C/cm) and for FZ (at 230°C/cm), R_{crit} is 2.7 and 5.3 mm/min, respectively, and is close to the experimentally observed values. If not properly maintained, the remelt process may lead to many unwanted issues.

In the CZ technique, the diameter of the crystal is always maintained at a predetermined value by changes in the pull rate. A new technique was proposed by Ekhult and Carlberg [82] to control the diameter of the growing crystal. The technique involves the use of fast, adjustable infrared (IR) heaters directed toward the melt surface and facilitates growth at a constant pull rate. Using this method, two crystals were grown and compared to a reference crystal grown with the normal CZ technique. The axial oxygen distribution in the reference crystal could be understood reasonably well by applying a steady-state model of the dynamic oxygen equilibrium in the melt. The other crystals showed a more even oxygen concentration than predicted, and it was found that power adjustment of the IR heaters was followed by variations that could not be accounted for by the steady-state model. Time dependent model for oxygen incorporation was applied and verified with the standard CZ grown crystals. Based on that the deviations from the steady-state model were explained, and so were the short-range oxygen variations. There were higher levels of axial carbon distributions in the crystal grown using IR heating. It was pointed out that the radial carbon and oxygen measurements did not reveal any significant differences between the reference crystal and the two crystals grown using the new technique.

Choe et al. [83] studied the effects of growth condition on thermal profiles during CZ growth. They adapted an eddy current testing method to continuously monitor the crystal growth process and investigated the effects of growth conditions on thermal profiles. The experimental concept involved monitoring the intrinsic electrical conductivities of the growing crystal and deducing temperature values from them. In terms of the experiments, the effects of changes in growth parameters, which include the crystal and CR rates, crucible position, pull rate, and hot-zone geometries, were investigated. The results show that the crystal thermal profile could shift significantly as a function of crystal length if the closed-loop control fails to maintain a constant thermal condition. As direct evidence to the effects of the melt flow on the heat transfer processes, a thermal gradient minimum was observed when the crystal-CR combination is maintained at 20/–10 rpm (clockwise and counterclockwise). The thermal gradients in the crystal near the growth interface were reduced most either by decreasing the pull rate or by reducing the radiant heat loss to the environment; a nearly constant axial thermal gradient was achieved when the pull rate was decreased by half, the height of the exposed crucible wall was doubled, or a radiation shield was placed around the crystal. Under these conditions, the average axial thermal gradient along the surface of the crystal was about 4°C to 5°C per mm. When compared to theoretical results, the axial profiles correlated well with the results of the models that included the radiant interactions.

A transient simulation of the CZ crystal growth of silicon is presented that is based on the integrated thermal-capillary model (ITCM), which includes conductive heat transport in the melt, crystal, and all other components of a prototypical CZ system; diffuse-grey radiation between these components; and the dynamics of the melt-crystal interface, the meniscus, and the growing crystal shape. Zhou et al. [84] carried out time-dependent simulations using the finite-element method and fully implicit time integration. The utility of the transient analysis is demonstrated by calculating the operating states and their temporal stability for a quasi-steady-state model. Multiple operating states are computed for a given crystal growth rate, and a limiting value of the power to the heater is found, below which no steady-state solutions exist. Transient simulations connect the existence of multiple states to long time-scale instabilities in the CZ system, which have implications for the control of the batch-wise process.

Molten silicon flow instability during crystal growth is a major topic to simulate the temperature contours using numerical simulations. Kakimoto [85] has studied the details of temperature distributions and velocity vector at different stages of the molten liquid. Simulation results were used to estimate the temperature distributions close to both the molten silicon and the bottom portion of the crucible, where both the crystal and the crucibles revolve in opposite directions. Thermal waves of nonaxisymmetric flow were reported by Nakamura et al. [86] by studying the wave number and pattern transition of the thermal waves at various crucible rotation rates. It is further reported that the thermal wave number increased as the CR rate increased. Many research groups have developed melt convection and turbulence models developed to understand crystal growth behavior under normal and undercooled situations. Details are discussed elsewhere [87–90]. Melt turbulence is a major issue for the CZ crystal growth process.

Enger et al. [91] studied numerical simulations of melt convection in CZ crystal growth. The predictions are based on three-dimensional time-dependent simulations of the flow and heat transfer in identical crucible geometry. The simulations were performed using a fine grid with 1,945,600 control volumes and six geometric blocks. To improve the convergence rate and reduce the simulation time, the multigrid method was applied. The analysis of the transformed data showed that by applying a high resolution in space and time, it is possible to predict the impact of growth parameters on the flow field. Rujano et al. [92] reported simulation results of transient oscillatory flow. By using the global simulation method, solid–liquid interface shapes were predicted by Shiraishi et al. [93]. Such simulations greatly help one to understand the situations taking place in a sealed crystal growth environment.

Kim and Smetana [94] estimated the maximum length of large-diameter CZ single-crystal silicon at the fracture stress limit. The growth of large-diameter CZ silicon crystals requires complete elimination of dislocations by means of the Dash technique, where the seed diameter is reduced to a small size—typically 3 mm—in conjunction with an increase in the pull rate. The maximum length of large CZ silicon is estimated at the fracture stress limit of the seed neck diameter (d). The maximum lengths for 200 and 300 mm CZ crystals amount to 197 and 87 cm, respectively, with $d = 0.3$ mm; the estimated maximum weight is 144 kg. This will help to design the total height of any crystal pulling system and the charge one can select.

Miyazaki et al. [95] reported on thermal stress analysis of a silicon-bulk single crystal with diameters of 6 and 8 inches performed in the cases of the [001] and [111] pulling directions. The team has used a three-dimensional finite-element program developed for calculating the thermal stress in bulk crystals grown during CZ process. Elastic anisotropy and temperature dependence of material properties are taken into account in this program. The temperature distribution and shape of a silicon-bulk single crystal, which are required for the thermal stress analysis, are obtained for a computer program for a transient heat conduction analysis and converted into parameters related to dislocation density. Miyazaki et al. also reported on the time variations of these parameters. The relation between these parameters and the shape of the crystal-melt interface issues are discussed.

The high-speed melt growth behavior of silicon crystals in (001), (011), (112), and (111) orientations was assessed by Cullis et al. [96] in detail. Rapid growth conditions were established by transient melting of silicon samples using nanosecond laser radiation pulses. Very fast cooling after irradiation leads to the formation of highly undercooled melts and yields large crystal growth rates. Under the most extreme conditions, the melt undercooling becomes so large that crystal growth breaks down, and an amorphous final solid phase is produced. Digges Jr. and Shima [97] demonstrated that the maximum growth rates of CZ crystals up to 80% of the theoretical limit can be attained while maintaining the single crystal property. It is difficult to attain the remaining 20% [98]. Copper was utilized as a test impurity to check the crystal growth parameters. At diameters greater than 7.5 cm, concentrations of greater than 1 ppm copper were attained in the solid (in a liquid that has a concentration of 45,000 ppm) without breakdown at maximum crystal growth speeds. For smaller diameters, the sensitivity of impurities is much more apparent.

4.3.4   Environmental and Ambient Control in the Crystal Chamber

The furnace structure must be airtight to prevent contamination from the atmosphere. Conditions have to be maintained such that no part of the unit experiences hot spots and releases vapor pressure to contaminate the growing crystal. During crystal growth, the ambient is maintained in an inert atmosphere of helium or argon gas. In some cases, hydrogen has also been used, whereas others have preferred vacuum conditions to avoid surface reactions on the grown crystal. Argon gas passing through the feeding area was exhausted to the outside of the furnace through the exhaust tube on both sides of the furnace. Thus, CO gas or amorphous particles did not exert any negative influence on the single crystal [99]. The total pressure in the CZ crystal growth chamber was recorded as 35 mbar with an argon flux of 700 l/min [100]. Heat transport by inert gas convection was neglected due to the low pressure involved [101]. The control system is used to maintain temperature, crystal diameter, pull rate, and rotation speeds. The control could be an open-loop or closed-loop configuration. An infrared temperature sensor output is linked to the pulling mechanism and controls the pull rate and the diameter of the growing crystal. The grown crystal leaves the furnace through a purge tube, where the ambient gas is directed along the surface of the crystal to cool it. From here, the crystal is removed for further processing. The convection currents inside the silicon melt change direction as the silicon crystal grows since heat flow direction through the supporting stem changes. This changes the curvature of the solid–liquid interface affecting the growth and impurity segregation. Figure 4.13 shows typical flow patterns inside the crystal growth units.

Most of the unwanted impurities in the crystal come from those present in the polycrystalline silicon starting material, quartz crucible, and the ambient gas. The polycrystalline silicon contains a considerable amount of carbon, with values ranging between 5.0×10^{15} and 3.0×10^{17} atoms/cm^3. Schmid et al. [102] reported that from experimental observations, it was found that for the levels of carbon generally encountered, contamination from the ambient gas is minor when compared to the contamination of the charge that occurs before growth starts. Suzuki et al. [103] investigated out-diffusion of boron impurity under various ambient conditions. Boron out-diffusion is significant in an H_2 ambient, while the out-diffusion is negligible in nitrogen or helium ambient. Comparing the analytical model to the experimental data, they identified the diffusion coefficient of boron in hydrogen ambient and found that it is smaller than that in nitrogen ambient. The significant out-diffusion in hydrogen ambient is attributed to the enhancement of the boron transport coefficient at the silicon surface. Negligible out-diffusion in the nitrogen or helium ambient is attributed to the negligible transport coefficient at the surface. According to Choe and Strudwick [104], the majority of crystals grown from state-of-the-art CZ pullers contain less than 1 ppma carbon. Consistent control of carbon contamination to less than 0.4 ppma may require stringent quality control of the polycrystalline silicon starting material and the development of growth processes specifically geared toward reducing the contamination via gas-phase transport from graphite parts within the furnace.

In a CZ-grown single-crystal silicon, antimony doping decreases the oxygen concentration due to enhancement of the oxygen evaporation. To minimize the evaporation of oxide species from the silicon melt, the atmospheric pressure of the background gas over the silicon melt has to be increased. While studying the control of oxygen concentration in heavily antimony doped crystals, Izunome et al. [105] studied the effect of atmospheric pressure in the growth chamber and clarified when heavily Sb-doped CZ silicon crystals were grown in different argon pressures of 30, 60, and 100 Torr. The oxygen and antimony concentrations of the melt surface were found to increase in response to argon pressure.

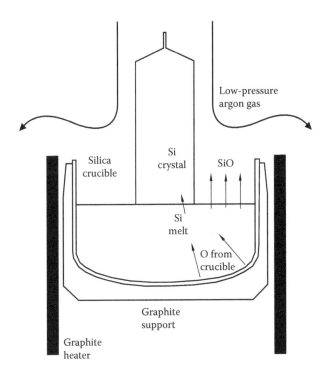

FIGURE 4.13

Typical environment inside the CZ crystal pulling system and removal of outgassed species using high-purity argon gas flushing. (From S. K. Ghandhi, *VLSI Fabrication Principles: Silicon and Gallium Arsenide*, John Wiley & Sons, New York, 1994; C. W. Pearce, "Crystal growth and wafer preparation" in *VLSI Technology*, edited by S.M. Sze, McGraw-Hill, New York, 1988; O. Anttila, "Challenges of Silicon Materials Research: Manufacture of High Resistivity, Low Oxygen Czochralski Silicon," Okmetic Fellow, Okmetic Oyj, epp.final.gov/DocDB/0000/000007/001/CERN_RD-50_06_2005_Okmetic_OA.ppt, June 2005, and the references therein [70].)

The oxygen concentration under an argon pressure of 100 Torr was 1.2 times that under 30 and 60 Torr when the solidified fractions were 0.5 or larger. The oxygen evaporation rate is controllable by gas-phase transport of Sb_2O at high argon pressures.

The effects of argon gas flow rate and the furnace pressure on the oxygen concentration values were reported by Machida et al. [106] through experimental crystal growth studies. The results were analyzed in terms of calculated velocity of the argon gas flow along the gas controller over the silicon melt surface. The flow velocity of argon gas was affected by both the argon gas flow rate and the furnace pressure. The flow velocity of argon gas was found to increase in response to flow rate under a fixed furnace pressure or with a decrease in furnace pressure under a fixed argon gas flow rate. The oxygen concentration in the CZ silicon crystals was found to be proportional to the flow velocity of argon gas. In the presence of a transverse magnetic (TM) field, CZ-grown crystals showed a completely different relationship from that of the CZ gas-controlled crystals [107]. With the increase in the argon gas flow velocity, the oxygen concentration of the TMCZ crystals is decreased. The surface temperature model could not explain the discrepancy between the TMCZ and the CZ crystals at the equivalent flow velocity of the argon gas. In contrast with the CZ gas-controlled crystals, the oxygen concentration was decreased with an increase in the flow velocity of argon gas in the TMCZ gas-controlled crystals. Incorporation of oxygen from the ambience is a major active topic of research, and many teams are working to interpret the partial pressure, evaporation of SiO, and its distribution [108–110]. It is difficult to play with this unintentional impurity when growing single-crystal silicon.

The presence of nitrogen in growing crystals has a lot of influence on the formation of vacancies, self-interstitials, and microdefects in CZ crystals. Defect formation mechanisms in the presence of nitrogen are a major research topic at present. Nitrogen monomers and dimmers, nitrogen and vacancies, and formation of clusters are studied by Kulkarni [111] in growing CZ crystals. Yu et al. [112] further investigated nitrogen behavior during growth. It was found that the nitrogen impurities in silicon mainly exist as a nitrogen pair and nitrogen-oxygen complex. The nitrogen concentration can be determined exactly by Fourier Transform Infrared (FTIR) spectroscopy after eliminating the thermal donors (TDs). Above a critical concentration of $4.0 \times 10^{15}/cm^3$, the nitrogen impurities easily form Si_3N_4 particles, causing the dislocations, grain boundary, and cellular structure in the crystal. Efforts are being made to reduce the concentration levels to below 1.0×10^{14} per cm^3 [113, 114] and are one of the main concerns for the research teams involved in crystal growth.

Kalaev et al. [115] presented an engineering model of the global heat transfer in CZ systems, including the self-consistent calculation of silicon melt turbulent convection, inert gas flow, and the melt-crystal interface geometry. For different growth conditions, the model is used to study the inert gas flow effect on turbulent melt convection and the global heat transfer in an industrial CZ system. Recommendations for engineering applications of global heat transfer models of CZ silicon growth have been derived from the numerical analysis with respect to the sensitivity of melt convection and the melt-crystal interface geometry to the inert gas flow. Using computational algorithms, the team has simulated the complete pulling system for temperature distribution and the argon flow velocity within the system.

4.3.5 Crystal Pull Rate and Seed/Crucible Rotation

The CZ process is a liquid-solid monocomponent growth system, and crystal growth involves the solidification of atoms from a liquid phase at the solid–liquid interface. The speed of growth is determined by the number of sites on the face of the crystal and the specifics of heat transfer at the interface. Pull rate is inversely proportional to the square root of the crystal diameter [70]:

$$L\frac{dm}{dt} + K_l \frac{dT}{dx_1} A_1 = K_s \frac{dT}{dx_2} A_2$$

where L is the latent heat of fusion, $\frac{dm}{dt}$ is the amount of silicon freezing (mass solidification rate), K_l is the thermal conductivity of liquid, $\frac{dT}{dx_1}$ is the temperature gradient (near the interface on the liquid side), A_1 is the cross-sectional area (on the liquid side), K_s is the thermal conductivity of the solid, $\frac{dT}{dx_2}$ is the temperature gradient (near the interface on the solid side), and A_2 is the cross-sectional area (on the solid side). The maximum pull rate of a crystal under the condition of zero thermal gradient in the melt is $\frac{dT}{dx_1} = 0$.

By converting the mass solidification rate to a growth rate using density and area, we obtain

$$V_{max} = \frac{k_s}{Ld}\frac{dT}{dx}$$

V_{max} is pull speed and d is the density of solid silicon.

The complete crystal-growing unit has to work without any mechanical vibrations, and other movements are to be maintained to great precision. The two most important parameters in CZ crystal pulling are (1) the pull rate and (2) crystal rotation. Many studies have been done on the dynamic behavior of the CZ melt, which is in a chaotic state. The experimental systems are typically based on a crucible with a circular shape and a specific depth, depending upon the polysilicon charge and the ingot diameter intended to be grown. The molten silicon is first prepared by melting lumps of poly-silicon crystals in the crucible, and then the lower parts of the heaters surrounding the crucible are turned off. This results in the formation of a solid silicon layer over the bottom of the crucible. Thus, the melt is sandwiched between the growing crystal on the melt and the solid layer over the bottom [116]. The crucible rotation rate is set to $R_c = 1$ and 5 rpm. In many systems, the crystal and crucible rotate in opposite directions. All the thermocouples will have fixed locations, and they do not rotate with the crucible.

Defect generation, diffusion, interaction, and agglomeration of voids and defects during crystal growth, along with the thermal history to which the ingots are exposed, are closely linked to the pulling parameter V/G (V = pulling speed, G = the thermal gradient in the crystal at the solidification interface of melt-crystal) [117]. Tight control of V/G is key to producing consistent homogeneous material properties. Today's state-of-the-art CZ processes for volume production are based on hot zone designs such that the ratio of the V/G value is optimized.

During the growth process of single-bulk crystals from melt, the defect density is strongly affected by the shape of the melt/crystal interface. The shape of the interface is governed by the construction of the growth equipment, including the heating system and the convection in the melt. The temperature field during the growth process is governed globally by the construction of the equipment and the heating system. Locally, in the melt and near the interface, the temperature field is strongly influenced by the thermal convection. To reduce the effect of the buoyancy convection, the crucible and/or the crystal is rotated. Miller et al. [118] studied the influence of melt convection on the interface. They developed a simple model to describe the phase-change problem in the weak form. Figure 4.14 shows clearly that the convection roll caused by buoyancy increases the deflection of the interface and leads to a multicurved interface. To reduce the negative effect of the buoyancy convection on the shape of the interface, one has to rotate the crucible and/or the crystal.

Kakimoto et al. [119] used x-ray radiography and *in situ* observation of the solid–liquid interface shape during growth of single-crystal silicon using the CZ method. The contrast attributed to the existence of the solid–liquid interface was obtained. Simulation of the transmitted x-ray image by absorption calculation was carried out using absorption coefficients for both molten and solid silicon, and supports the fact that the solid–liquid interface can be observed. They also observed the change of the interface shape from convex to concave during the shouldering process.

Understanding the dynamic interaction between crystal diameter, pulling speed, and heater power is a key requirement for the development and control of the growth technique. A more methodical approach was taken by Dornberger et al. [120] on this key issue. According to them, an important task in CZ growth is to control the crystal diameter by adjusting the power supply and pulling speed. The simulated results show how the heater power responds to the frequent modifications of the pull rate in order to maintain the imposed diameter. Short time-scale diameter variations are controlled by the pull rate, which immediately influences the heat balance at the solid–liquid interface. However, the pull rate has to be kept in a certain range in order to maintain crystal quality. Therefore, long-range deviations of the nominal pulling speed must be compensated by adjusting the power supply, which is much more difficult to perform than controlling the pull rate. During the growth process, the length of the crystal increases and the height of the melt decreases. The evaluation of radiative exchanges in the furnace enclosures requires the computation of the view factors at each time step due to the varying geometrical configuration inside the furnace. Convective heat transfer in the melt plays a dominant role in the entire process, and extra precautions are required to deal with this.

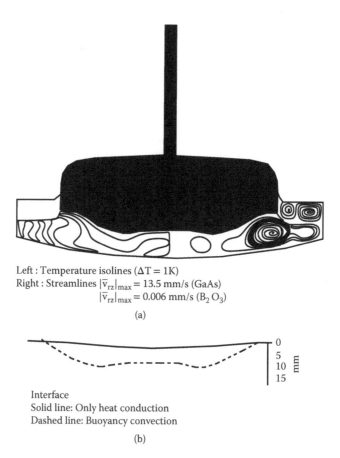

Left : Temperature isolines ($\Delta T = 1K$)
Right : Streamlines $|\bar{v}_{rz}|_{max} = 13.5$ mm/s (GaAs)
$|\bar{v}_{rz}|_{max} = 0.006$ mm/s ($B_2 O_3$)

(a)

Interface
Solid line: Only heat conduction
Dashed line: Buoyancy convection

(b)

FIGURE 4.14
Influence of the convection on the shape of the solid-liquid interface with a value of buoyancy convection (Grashof number (Gr) = 5.2×10^7). (From W. Miller, U. Rehse, and K. Böttcher, *Solid-State Electronics*, 44, 825–830, 2000 [118].)

Crucible rotation has some influence on the incorporation of impurity such as oxygen. A three-dimensional numerical simulation by Yi et al. [17], clarified with a modified crucible with graphite sheet, at the bottom, that a high crucible rotation rate, roughly about 20% of the oxygen in the grown crystals was transferred by convection in the melt from the bottom of the crucible. With a low CR rate, a melt with a small oxygen concentration was directly transferred from the gas-melt interface to the crystal-melt interface; therefore, the team concluded that the oxygen concentration in crystals grown at a low CR rate was lower than that for crystals grown at a high rotation rate. CR dependence of the oxygen concentration at the center of the crystals grown by the two different crucibles is shown in Figure 4.15. Here, the experimental results are indicated by broken lines, while the numerical results are plotted by solid lines. Open and closed circles represent the results obtained from the crucibles with and without a carbon sheet. The figure indicates an abrupt increase in the oxygen concentration between CR rates of 4 and 6 rpm with a normal crucible. Flow visualization using x-ray radiography clarified that the flow mode of the melt was axisymmetric below a 4 rpm CR rate, while the mode was nonaxisymmetric above 6 rpm, as shown in the figure. The oxygen concentrations in the crystals grown from the crucible with a carbon sheet are less than those in the crystals grown in the crucible without a carbon sheet. The discrepancy in the concentration between both cases is dependent on the CR rates. This is about 25% of the total concentration for a low CR rate, while the value for a high rotation rate is about 45% of the total concentration, as indicated in the figure.

Oxygen transport from the silica crucible was analyzed both experimentally and numerically by Togawa et al. [121]. They used a crucible with a limited oxygen source to determine the dominant flow of the species in silicon melt. Oxygen concentration is usually adjusted by the crucible rotation during growth, and it is well known that a high rotation rate yields a crystal high in oxygen. The reason for this phenomenon is that a high rotation rate creates a well-stirred melt and during crystal pulling, oxygen is supplied from the crucible bottom by the rising current. The oxygen concentration was found

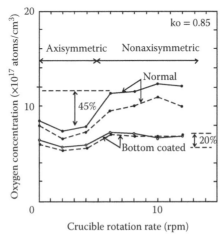

FIGURE 4.15
Oxygen concentration as a function of crucible rotation rates for two different cases indicated by (○) and (●), respectively. (—) and (---) indicate numerical and experimental results. (From K.-W. Yi, K. Kakimoto, M. Eguchi, and H. Noguchi, *Journal of Crystal Growth*, **165**, 358–361, 1996, and the references therein [17].)

to be controlled mainly by the upward flow below the growth interface, with the crucible bottom as the dominant source. The region under the growth interface is assumed to remain rich in oxygen. It is further indicated that the concentration of this region strongly depends on the free surface region, whose concentration is, to a large extent, determined by dissolution from the crucible corner. As per the results shown in Figure 4.16, up to 0.15 of g (solidified fraction), the oxygen concentration in the crystal grown using the conventional crucible is higher than that using the side-masked crucible, decreasing linearly with g. As indicated, the oxygen concentration in the crystal grown from the side-masked crucible is low in the region $g < 0.15$. This suggests that oxygen transport depends not only on the crucible bottom, but also on the crucible side-wall in the early stage of crystal growth.

Figure 4.17 shows the comparison of the radial distribution of oxygen in both experimental results to that in numerical simulation data. Experimental results show that oxygen concentration at the center of the crystal is almost the same for both the conventional and side-masked crucibles. At the periphery of the crystal, however, the oxygen concentration is considerably lower in the side-masked crucible, meaning that the oxygen from the crucible side-wall affects the peripheral region of the crystal even at a point where the dominant flow of the oxygen transport is vertical ($g > 0.15$). The TM field only suppresses vertical movement, producing low and radially uniform oxygen crystals. When the bottom-masked crucible is used, the oxygen concentration of the whole melt becomes low and becomes uniform, as with the TM field. The radial oxygen profile in the grown

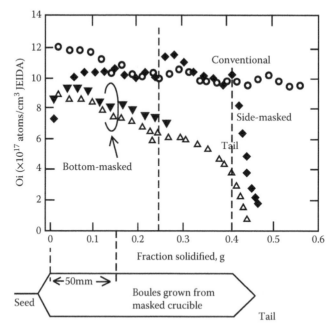

FIGURE 4.16

Longitudinal oxygen distribution in silicon crystals for each of the growth conditions. (▼) First and (Δ) second runs of the bottom-masked experiments. The dotted line at $g = 0.25$ shows the position of the numerical simulation. At the point $g = 0.41$ (indicated with dotted line), the tail process is started, except for the conventional case. (From S. Togawa, K. Izunome, S. Kawanishi, S.-I. Chung, K. Terashima, and S. Kimura, *Journal of Crystal Growth*, **165**, 362–371, 1996, and the references therein [121].)

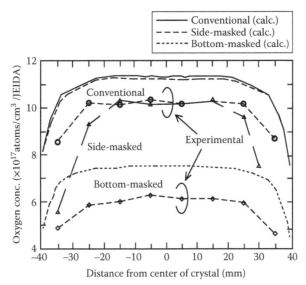

FIGURE 4.17
Comparison of the calculated radial oxygen distribution and the experimental measurement. (From S. Togawa, K. Izunome, S. Kawanishi, S.-I. Chung, K. Terashima, and S. Kimura, *Journal of Crystal Growth*, **165**, 362–371, 1996, and the references therein [121].)

crystal then becomes low and uniform. The part exerting the greatest effect on oxygen concentration of the whole bulk melt is the crucible corner, which is the hottest point during the crystal growth process. Dissolution from the crucible corner remains high in the oxygen concentration in the whole bulk melt, and oxygen concentration below the growth interface region reaches an equilibrium with the free surface area, resulting in crystals with a high oxygen content. Togawa et al. [121] confirmed that the upward flow from the crucible bottom transports oxygen to the growth interface, and dissolution from the crucible corner determines the overall oxygen concentration of the silicon melt in the crucible.

The influence of crystal and crucible rotation rate on the oxygen-concentration distribution in 8-inch crystals grown from 18-inch crucibles was investigated by Kanda et al. [122], both by growth experiments and by water-model melt-flow simulation. After the shoulder transition region, increasing seed rotation (SR) decreases oxygen concentration, and this effect is enhanced by lowering the CR. Under SR >9 rpm, radial variation of the oxygen concentration increases as CR increases (15 > 10 > 5 rpm), but under SR <9 rpm, this order changes to CR = 10 > 15 > 5 rpm. The axial distribution of the oxygen concentration is shown in Figure 4.18. Oxygen concentration is high under high CR because of the high crucible wall temperature, and it decreases with the body length because the contact areas of the silicon melt as the crucible position decreases. For the crystal with CR = 5 rpm, the concentration actually increased. The axial distribution of the radial variation of oxygen concentration is shown in Figure 4.19. The radial variation of oxygen concentration is defined as

$$\text{Radial variation (\%)} = \frac{(A-B)}{B} \times 100$$

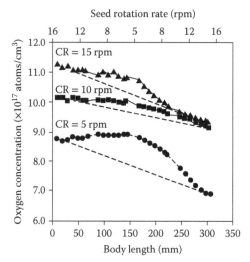

FIGURE 4.18
Axial distribution of the oxygen concentration at the center of the crystals. In the first half of the body, the decrease in oxygen concentration is suppressed as seed rotation rate decreases. (From I. Kanda, T. Suzuki, and K. Kojima, *Journal of Crystal Growth*, **166**, 669–674, 1996, and the references therein [122].)

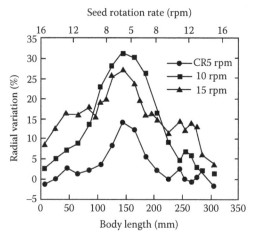

FIGURE 4.19
Radial variation of oxygen concentration as a function of SR and CR. (From I. Kanda, T. Suzuki, and K. Kojima, *Journal of Crystal Growth*, **166**, 669–674, 1996, and the references therein [122].)

where A and B are the oxygen concentration at the center and 15 mm from the edge, respectively. Under higher SR, the radial variation decreases. Under SR >9 rpm, the radial variation increases as CR increases (15 > 10 > 5 rpm), but under SR <9 rpm, this order changes to 10 > 15 > 5 rpm. Increasing CR suppresses the effect of crystal rotation. With careful consideration for the balance between natural and forced convection, the team simulated the internal motion of the silicon melt using a water-model method. In the axial section, stable, radially outward flow was observed below the free surface. The sweep length of this flow on the free surface increases as the ratio of the Reynolds number of crystal rotation to that of CR increases. By combining the near-surface flow patterns and evaporation of SiO, Kanda et al. [122] proposed a model that describes well the oxygen concentration behavior in the grown silicon crystals.

Sugawara and Tochikubo [123] came out with specific parameter values to improve the radial impurity distribution in CZ silicon crystals. According to them, the standard growth condition was a pull rate of 1.4 mm/min, and the rotation speed of the seed and that of the crucible should be 15 rpm and –5 rpm (in reverse rotation with respect to the seed), respectively. To study the radial distribution, the grown crystals were cut normal to the grown axis into 5 mm thick sections, and the lapped samples were tested for resistivity measurements using the four-point probe technique in an interval of 1 mm spacing. It was proved that the impurity distribution along the radius was generally convex and the concentration at the edge of the crystals was smaller by a factor of 4% to 20% than the corresponding concentration at the middle of the crystal.

The conventional CZ process has several inherent limitations associated with it [124]. Since it is a batch process, the silicon melt height changes from the beginning to the end and the aspect ratio reduces from a value of about unity to below 0.1. This brings continuous changes in the convective flow area, the melt flow and temperature fields, and the melt-crystal interface shape. Unsteady, oscillatory flows of molten silicon produce defects in the grown crystal because of the interface oscillations and the fluctuations in the transport of impurities present in the silicon. Because of the reduction in the melt height, the wetted surface area of the crucible changes, thereby changing the oxygen impurity addition into the melt. As a result, no two grown crystals have identical properties. Moreover, once the melt level is very low, the process has to be stopped for a fresh charge. For a given run, the length of the grown crystal is limited by the initial charge selected for the batch.

Many of the inhomogeneities and defects in the CZ crystal grown from a pool of melt are because of the inherently unsteady growth kinetics and flow instabilities. A scaled-up version of the CZ process induces oscillatory and turbulent conditions in the melt, resulting in the production of nonuniform crystals. A numerical study by Jafri et al. [124] revealed that a crucible partition shorter than the melt height can significantly improve the melt conditions. The obstruction at the bottom of the crucible is helpful, but the variations in heat flux and flow patterns remain random. When the obstruction is introduced at the top of the melt, the flow conditions become much more desirable and oscillations are suppressed to a greater extent. An optimal size of the blockage and its location to produce the most desirable process conditions will depend on the growth parameters, including the melt height, and on the crucible diameter. The numerical study lays the foundation for the possibility of adding a short crucible partition in the CZ growth of silicon crystals. The numerical results indicate that for the melt aspect ratio of 0.5 and conditions closer to reality (Pr = 0.015, Gr = 10^8, Re$_c$ = 10^4, Re$_s$ = -10^4, Ma = 10^3 and Rr = 0.5), a short obstruction, $h_p \approx 0.25$, at the top of the crucible can greatly improve the melt conditions. A short partition at the bottom of the crucible, on the other hand, does not provide the same benefits. A partition at the top weakens the surface tension–induced flows, which are generally strong in the upper region of the melt. This is a desirable condition for the melt. On the contrary, a partition promotes the Benard convection conditions in the inner region and produces stronger flows and instabilities in the crystal region. These effects are much stronger when the partition is tall. The team also suggested that the partition should not be placed close to the crucible side-wall; otherwise, the benefits of adding a blockage are either completely diminished or at least reduced. The details are shown in Figure 4.20. The team opined that the results presented here show only a qualitative trend, and a suitable partition size and location will have to be determined for realistic process conditions.

The critical V/G ratio is of fundamental importance in the growth of dislocation-free silicon crystals with controlled microdefect properties. The reported numbers for the critical ratio scatter considerably, in the range from 0.12 to 0.2 mm^2/min K. One of the reasons for such a scatter is an uncertainty in the measured or calculated value of G in the vicinity of the crystal-melt interface. By analyzing the intrinsic point defects and the impurities in silicon on the V/G ratio, Voronkov and Falster [125] came out with the following fundamental reasons for a scattered critical V/G ratio: (1) the effect of impurities on the incorporation of intrinsic point defects into the growing crystal, and (2) the effect of the interface shape on the diffusion field of point defects in the vicinity of the interface. The incorporation of intrinsic point defects into a growing crystal is affected by the presence of impurities that can react with vacancies and interstitials. The critical value of the ratio of the growth rate V to the axial temperature gradient G (V/G ratio) that separates the interstitial growth mode from the vacancy growth mode is shifted by impurities, and this effect can be described by simple analytical expressions. Some impurities, such as oxygen, nitrogen, and hydrogen, trap vacancies, and others cause a downward shift in the critical V/G ratio. Impurities like carbon trap self-interstitials cause an upward shift in the critical V/G ratio. The impurities affect both the incorporation and agglomeration stages of microdefect production. We shall discuss the details of these impurities in another chapter.

A well-known configuration for crystal growth using the CZ technique involves a cylindrical crucible filled with a melt that is placed inside a cylindrical heating furnace providing an axial-symmetric heat field. The mechanisms of heat and mass transfer within the melt are strongly correlated with the growth of high-structural-quality crystals. Hence, it is very important to study the possible ways in which the heat mass transfer processes can be controlled [126]. The circular movement of the heat wave by the perimeter of a growing crystal greatly differs from a simple periodical change of temperature–heat field pulsation. The former, unlike the latter, provides the conditions for the renewal of a crystallization medium near the interface due to forced convective flow. Another advantage of this method is the opportunity of melt-solution homogenization without mechanical stirring. A modified (electromagnetic stirring) CZ technique has been developed by Brückner and Schwerdtfeger [127] to pull single crystals without mechanical rotation of the crystal and the crucible by bringing forth the controlled convection in the melt solely by a rotating electromagnetic field. Silicon crystals were successfully grown with this new technique.

Crystal rotation and CR at different angular velocities is one of the major topics that, over many decades, has drawn a large number of teams to study the behavior of defects, incorporation of unintentional impurities, and quality of crystals grown. Several good research publications are available in the open literature [128–140]. Similarly, the melt-solid interface is another key issue for growing good quality crystals [141–144]. Numerical simulation of the entire configuration is being explored to study the hydrodynamic behavior of the melt in crucible [145, 146] to understand and predict the crystals grown by this technique.

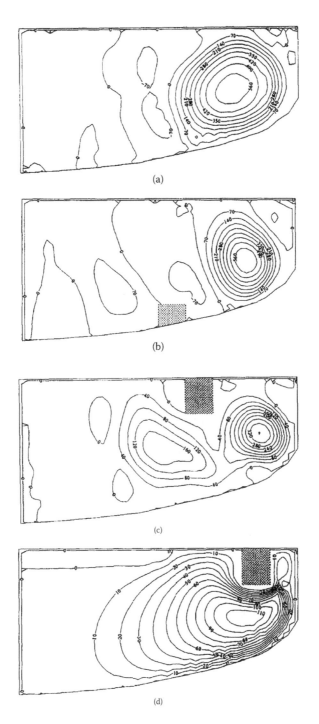

FIGURE 4.20
(a) Oscillatory melt flow fields for Gr = 10^8, Re_c = 10^4, Re_s = -10^4, Ma = 10^3, Rr = 0.5, and Ar = 0.5 without a blockage. (b) Oscillatory melt flow fields in a crucible with a short partition in the bottom, h_p = 0.25, d_p = 0.55, and t = 0.1. (c) Oscillatory melt flow fields in a crucible with a short partition in the upper region, h_p = 0.25, d_p = 0.65, and t = 0.1. (d) Representative flow fields for a partition in the upper region, h_p = 0.25, d_p = 0.85. (From I. H. Jafri, V. Prasad, A. P. Anselmo, and K. P. Gupta, *Journal of Crystal Growth*, **154**, 280–292, 1995, and the references therein [124].)

4.3.6   Dopant Addition for Growing Doped Crystals

Dopant species are generally added in the initial stages of the melt process along with the polysilicon blocks. In some cases, dopant is added directly to the charge, and in other cases, heavily doped polysilicon blocks are added. The latter option is better suited for crystals where lightly doped silicon crystals are desired. Direct addition of dopant has a disadvantage when the silicon is being melted. If the dopant vapor pressure is high, the dopant could escape from the charge before the silicon is totally melted. Once the dopant is inside the melt, proper mixing is important to get a uniformly doped silicon crystal. Incorporation depends on the segregation coefficient of dopant species between the solid–liquid silicon. This segregation behavior of impurities is an important factor in the semiconductor single-crystal growth process. Different dopant impurities exhibit the behavior differently. Each impurity segregation coefficient changes with temperature and will be affected by other parameters as well, particularly when they are in the molten silicon. Diffusion coefficients of impurities in silicon melt are reported by Kodera [147] for boron, aluminum, gallium, indium, phosphorus, arsenic, and antimony. It is further reported that the diffusion coefficient of Group V impurities are found to increase as the tetrahedral covalent radius of the impurity atom decreases. Adding impurity at a later stage influences the silicon melt, particularly with antimony [148–151]. In addition to the added dopant species, the silicon melt receives unintentional dopants from other sources, such as quartz crucible dissolution, and from the impurities associated with the dopant itself. Here, the purity of the dopant has a key role.

According to Liaw [152], the radial dopant distribution in silicon has been correlated with the interface shape and its radius of curvature. No correlation was found between the interface shape and radial distribution in the case of oxygen or carbon species. These studies on the interface shape and radial distribution of impurities found that the solid–liquid interface shape at the shoulder of ingot growth is affected by the taper angle. The interface changes from concave to convex with an increase of taper angle. The interface shape at the main body of ingot growth depends on the total crystal length. The interface becomes more concave as the length of the grown crystal increases. The radial dopant distribution has been correlated with the interface shape and its radius of curvature. No correlation is found between the interface shape and radial distribution of the other unintentional impurities. Their distribution is mainly decided by other growth parameter values.

New experimental data by de Kock and van de Wigert [153] strongly suggests that the formation of the different types of swirl defects is due to parallel condensation processes involving silicon interstitials and vacancies. It is proposed that during CZ growth, three types of swirl defects—type A, type B, and type C—can form. Doping with donor concentration species (phosphorus, arsenic, and antimony) more than 1.0×10^{17} cm^{-3} suppresses the formation of type A swirl defects. Doping with acceptors (boron and gallium) with concentration values more than 1.0×10^{17} cm^{-3} eliminates the formation of type B and type C swirl defects. The observed doping effects are explained in terms of complex formation as a result of coulomb attraction between dopants and charged thermal point defects. More details on these defects are discussed in other chapters.

4.3.6.1 Boron

Boron is the only acceptor impurity with adequate solubility for use as an efficient emitter in transistors or for other applications requiring extremely high impurity concentrations.

Boron diffuses at about the same rate as phosphorus. The surface concentration of boron during the diffusions ranged from 10^{20} to 10^{21} atoms/cm^3. The solid solubility of boron is as large as 4.0×10^{20} atoms/cm^3 and it has a tetrahedral radius of 0.88 Å, giving it a misfit factor of 0.254. As a result, the presence of large amounts of boron in the silicon lattice is accompanied by strain-induced defects, which lead to considerable crystal damage. This sets an upper limit of nearly 5.0×10^{19} atoms/cm^3 on the impurity concentration that can actually be achieved in practical device structures, with the rest being electronically inactive and residing in silicon.

The incorporation of boron atoms into the silicon lattice is expected to change the thermodynamic properties of self-interstitials and vacancies, due to the differences in size and nature of chemical bonding. Each boron atom affects the silicon lattice in the same way, as C_{crit} linearly depends on the boron concentration. Although the effects of boron atoms on equilibrium concentrations and diffusivities of interstitials and vacancies have been discussed in many studies, no quantitative data is available concerning the temperature range near the melting point of silicon [154]. Almost all boron atoms reside on substitutional sites, but only a small fraction will get with self-interstitials or crystal lattice vacancies. As the temperature decreases, boron atoms on defect sites will be converted to boron atoms on substitutional sites, which may lead to increased silicon self-interstitial or decreased vacancy concentration. Due to the different sizes and electron configurations of silicon and boron atoms, the chemical bonds are somewhat strained at those locations. Relaxation can occur by changing the intrinsic point defect concentrations. A reduction of the diffusivities of self-interstitials and vacancies is plausible as well, as interstitial boron and vacancy boron complexes may be formed and trap the intrinsic point defects for short durations. According to Dornberger et al. [154]. the smaller size of boron, as compared to silicon, reduces the lattice constant—for example, a boron concentration of 2.0×10^{19} cm^{-3} accounts for a decrease of the lattice constant by a factor of 1.2×10^{-4}, which results in a volume decrease by a factor of 3.6×10^{-4}. The theoretical relative volume decrease of a silicon crystal is 2.3×10^{-4}, if 2.0×10^{19} cm^{-3} silicon atoms are substituted by boron atoms and the volume is calculated according to the hard sphere model with covalent radii of 1.17 Å and 0.88 Å for a silicon and a boron atom, respectively. Hence, the lattice constant and the volume density of the crystal change slightly when boron atoms are incorporated, which may influence the equilibrium concentration of interstitials and vacancies. It is opined that no quantitative analysis is available for the physical mechanisms as discussed earlier.

4.3.6.2 Phosphorus

The misfit factor is considerably lower with the phosphorus impurity, at 0.068. Active carrier concentration of 3.0×10^{20} cm^{-3} can be achieved in practical structures. The solid solubility limit in silicon is close to 10^{21} cm^{-3}. The surface concentrations of phosphorus during diffusions ranged from 6.0×10^{20} to 9.0×10^{21} atoms/cm^3. Anomalous diffusions are observed with phosphorus dopant at high concentration levels.

Izumi et al. [155] created a segregation model based on the existing theories for single-crystal silicon growth from the melt in order to explain the typical phenomena. The phosphorus model is based on the combination of the temperature dependence of impurity diffusivity in liquid and the kinetic undercooling. The model can quantitatively explain not only the deviations, but also the difference due to the orientations. In addition, the following two ideas, both being important to understand the crystal growth of silicon, can be introduced through the analysis. One possibility is that the existence of large kinetic undercooling, even in the <100> continuous growth. The energy barrier is estimated at 1.24 eV. The other is a mixture of two "aggregates," both with different diffusivities, and the large change of the ratio in temperature.

4.3.6.3 Arsenic

The tetrahedral radius of arsenic is identical to that of silicon, so it can be introduced in large concentrations without causing lattice strain. Its maximum solid solubility in silicon is 2.0×10^{21} cm^{-3}, and active electron concentrations as high as 5.0×10^{20} cm^{-3} can be achieved with this dopant. Arsenic has a diffusivity that is about one-tenth that of boron or phosphorus. Consequently, it is used in situations where it is important that the dopant be relatively immobile with subsequent processing. It is also desirable for shallow diffusions, where its abrupt doping profile and low diffusivity make control of the junction depth more precise [70]. Arsenic movement is primarily substitutional in nature, with a small interstitialcy component. It is relatively free from anomalous diffusion tails of the type observed with phosphorus at high doping levels. Consequently, it is the preferred dopant for *n*-type regions in VLSI circuits.

4.3.6.4 Gallium

Gallium diffuses somewhat more rapidly in silicon. Gallium-doped CZ silicon wafers are gaining importance as a better substrate for solar cell applications in place of boron-doped crystals due to the better performance and also to control the minority carrier lifetime. Hoshikawa et al. [156] have reported that when the dopant gallium was placed with polysilicon nuggets before making the silicon melt, the gallium concentration in the crystals decreased considerably below the designed values; however, when it was added directly to the molten silicon in a rapid operation, the concentration was in good agreement with the designed values. The team evaluated the gallium concentration in the grown crystals and proposed that the doping procedure in which the gallium dopant is added directly to the molten silicon to control the concentration more precisely. The segregation coefficient of this impurity and its effect with boron codoping are being pursued at present by many research teams [157–159].

4.3.6.5 Nitrogen

In recent years, nitrogen doping of CZ-grown silicon substrates has attracted much attention. It is becoming a key technology for growing defect-free silicon crystals [160]. Appropriate nitrogen doping significantly improves the mechanical strength against slip and increases bulk microdefect generation, which are essential for high-temperature processing and intrinsic gettering, respectively [161]. Now the industry has started using these wafers for VLSI and ULSI circuits. Furthermore, nitrogen is reported to suppress the vacancy agglomeration, leading to smaller crystal-oriented particles (COPs). It strongly enhances oxygen precipitation, which improves the gettering potential of CZ silicon. It is believed that nitrogen-oxygen defects form stable nuclei at high temperatures, where oxygen-only nuclei are unstable. There have been suggestions these are nitrogen-vacancy-oxygen defects [162]. Huber et al. [163] have demonstrated that in MOSFET device gate-oxide integrity testing, fabricated using these crystals, manifest in a high defect density and also shift of COP related breakdown to lower the Q_{bd} parameters especially for thicker gate oxides. The silicon wafers used for this study were boron doped and (001) oriented with 3.0-18.0 Ω-cm resistivity. Nitrogen doping was in the range of 1.0×10^{15} atoms/cm^3. They concluded that nitrogen doping strongly modifies the defect species' morphology and density, leading to numerous extrinsic breakdowns for thick and thin gate oxides. It has become indispensable to establish a method for measuring nitrogen concentration in CZ silicon crystals. A conventional notion is that the major form of nitrogen is dimeric interstitial N_2; under certain conditions, a small amount of substitutional nitrogen N_s also can be present. The data on the evolution of implanted nitrogen profiles by a rapid thermal annealing shows that the diffusivity of the major nitrogen form is very low; the profile evolves by dissociation of the major species into minor but highly mobile species, within the conventional notion, into monomers: $N_2 \rightarrow 2\,N_I$ [164]. However, this simple treatment of nitrogen species turns out to be oversimplified, as the nitrogen concentration is low and nitrogen has various configurations.

Yu et al. [165] has investigated the effect of nitrogen doping on grown-in COPs in CZ silicon and COP annihilation during annealing in hydrogen at high temperature. It was found that nitrogen doping leads to denser, smaller COPs during CZ silicon crystal growth. The grown-in COPs in the nitrogen-doped CZ (NCZ) silicon can be annihilated at a much lower temperature compared to those in the CZ silicon; furthermore, the denude zone (DZ) with lower COP density in NCZ silicon was much wider than that in CZ silicon. The team concluded that nitrogen doping reduces the supersaturation of vacancies prior to the void formation and thus decreases the onset temperature of void formation. Thus, the size of the void is smaller and the oxide films on the inner walls of the void are assumed to be thinner in NCZ silicon, compared with those in CZ silicon. Therefore, the voids in NCZ silicon can be annihilated more easily by hydrogen annealing. The contemporary advanced device processes that adopt the CZ silicon wafers with diameters of 200 mm and larger are moving toward low initial oxygen concentration ($[O_i]$) and thermal budgets. Therefore, in such a case, there is less opportunity to achieve a reliable internal gettering (IG) effect. For the 300 mm CZ silicon wafers, a double-sided polishing technology is conventionally employed so that the external gettering processes cannot be employed again. Accordingly, how to enhance the IG capability of 300 mm CZ silicon wafers is a critical matter [166]. Obviously, nitrogen doping is an effective way to address this issue and may become a standard approach to obtain wafers for VLSI and ULSI circuits.

4.3.6.6 Antimony

Antimony doping is used as an alternative to arsenic because of its comparable diffusion coefficient. Its misfit factor in silicon is 0.153, and its electronically active surface concentration is limited to about 5.0×10^{19} atoms/cm^3. Antimony diffusion is almost purely substitutional in nature, with no interstitialcy component. Thus, it is completely free from anomalous effects and is superior to arsenic impurity in this regard. It is occasionally used in VLSI applications because of this reason.

With the development of new types of electronic power devices, the demand has increased for large-diameter, heavily antimony-doped silicon wafers. The atomic diameter of antimony is 15% larger than that of silicon, resulting in a very large lattice mismatch. The segregation coefficient of antimony in silicon is low, making it difficult to obtain crystals with the required resistivity. Crystal growth of 200 mm heavily doped antimony silicon with low resistivity and high yield remains one of the toughest challenges for the ULSI industry. Zhou et al. [167] designed a new quartz tool to steadily add vapor-phase antimony into silicon melt. Antimony vapor created under high temperature flows directly onto the silicon melt surface through a specially designed quartz mouthpiece. Antimony was absorbed by the silicon melt surface and diffused into the bulk melt. By controlling the inside temperature of the quartz tool and furnace pressure, all antimony is smoothly doped into the silicon melt without any splash. They demonstrated that the vapor-doping technique is feasible and effective for heavily antimony-doped silicon crystal growth with good yield. With better hot zone design, including heat shields, along with pressure control and other optimized growth parameters, heavily antimony-doped 200 mm silicon with resistivity less than 0.02 Ω-cm can be produced with high efficiency.

Many researchers have investigated the oxygen segregation coefficient in a silicon crystal during the pulling process, as it is an important parameter in the study of oxygen incorporation [168, 169]. The reported results included several calculations from thermodynamic data and the analysis of the oxygen distribution in CZ silicon crystal. However, the results were inconsistent and difficult to analyze, varying widely from 0.21 to 1.4. The problem probably resulted from the fact that it is difficult to obtain experimental data not influenced by other effects and from the fact that some uncertain physical property values were used for analyzing the data. The preliminary results of Huang et al. [169] showed that the oxygen segregation coefficient decreased as the antimony concentration increased. The reduction of the oxygen segregation coefficient due to Sb-doping is considered to result from the existence of Sb_2O ordering in the Sb-doped silicon melt. The crystals were grown in a closed quartz ampoule using a normal freezing method. The oxygen segregation coefficient was obtained by comparing the oxygen concentration in the head of the silicon crystal with the oxygen concentration in the initial silicon melt. Although there are many reports about the oxygen segregation coefficient in silicon crystal growth, there is not a lot of data on the effect of impurity doping on the oxygen segregation coefficient. It was found that the oxygen segregation coefficient was markedly affected by the presence of antimony atoms in the silicon melt. The results showed that for an antimony concentration less than 1 at%, the oxygen segregation coefficient decreased slowly, and that for an antimony concentration above 1 at%, the oxygen segregation coefficient decreased rapidly.

In CZ silicon crystals, antimony doping decreases the oxygen concentration by enhancing the oxygen species evaporation from the melt surface. The oxygen segregation coefficient in the growth of Sb-doped silicon crystal has been investigated by Izunome et al. [170] using the normal freezing method, and it was found to be nearly unity when the antimony concentration was less than 0.5 at% and reduced slightly with higher doping levels. The team has reported that the oxygen evaporation has also been measured when antimony dopant was added, and it was found that the oxygen concentration in heavily Sb-doped CZ silicon crystals was reduced due to increased oxygen evaporation. Analysis of the evaporated species showed that evaporation of Sb_2O and SiO significantly increased the oxygen loss from the silicon melt, clearly being the main reason for the oxygen reduction in Sb-doped crystals. It has been suggested that the atmospheric pressure of the background gas over the silicon melt may be adjusted to control the evaporation of oxide species and regulate the doping concentrations in the grown crystals. It is further reported that the antimony concentration increases as the solidified fraction at each argon pressure increases due to the fact that this impurity has a very small segregation coefficient (roughly 0.023). The antimony concentration under argon pressure of 100 Torr is higher than that under pressures of 30 Torr and 60 Torr in the same solidified fractions, clearly suggesting that the antimony evaporation is suppressed by increasing the argon pressure in the chamber, just as with the oxygen evaporation. This regulation of argon pressure has shown good dependency on the incorporation of oxygen in growing CZ silicon crystals.

4.3.6.7 Germanium

With the reduction in size of the features of ULSI circuits, microdefects such as voids in CZ silicon crystals play roles that are more important in the reliability and yield of these devices. As one of the tetravalent atoms, the behavior of germanium (Ge) doped in CZ silicon crystals has attracted considerable attention in recent years. Yang et al. [171] have reported that the grown-in characteristics of voids in a germanium-doped CZ wafer, including flow pattern defects (FPDs) and COPs, suggested that germanium can suppress large voids, resulting in denser and smaller voids. Meanwhile, it has been found that the density of voids can be decreased by germanium doping and they can be eliminated easily in CZ silicon crystals through high-temperature annealing. It is therefore speculated that the gate oxide integrity (GOI) of semiconductor devices can be improved by doping the crystals with germanium impurity.

In recent years, germanium-doped CZ (GCZ) silicon has drawn more attention because germanium atoms can result in easier elimination of void defects and a significant increase of internal gettering capability. In addition, the mechanical strength of the wafers can be improved by the Ge doping. The as-grown oxygen precipitation in GCZ Si can be enhanced in comparison with the conventional CZ silicon at high temperatures, even above the formation temperature of the void. Supersaturated interstitial oxygen atoms in CZ silicon usually accumulate to generate oxygen precipitation during thermal cycles, which have a significant impact on the device quality. The formation of TDs can be suppressed, while the formation of new donors (NDs) can be enhanced in GCZ Si. Moreover, it has been speculated that Ge can combine with vacancies and form a Ge-V complex in Si crystals due to the electron radiation at low temperature. Chen et al. [172] has reported the details of germanium's effect on as-grown oxygen precipitation in CZ-grown silicon crystals. Germanium atoms situate at the substitutional sites in CZ silicon crystals, and they can induce distortion and local stresses in the silicon lattice due to their larger radius than the silicon atoms. So the lattice sites where germanium atoms located are provided with potential activities and are inclined to interact with other structural defects or impurity atoms mainly through the overlapping of their strain fields. To relieve the lattice stress, the excess vacancies are inclined to accumulate around the substitutional germanium atom, and Ge-related complexes will form. These Ge-related complexes will form at high temperatures above 1200°C, which are higher than the formation temperature for the void (1100°C–1020°C). It is speculated that these complexes act as the heterogeneous nuclei for oxygen precipitation.

Anomalous resistivities were reported by Voltmer and Digges Jr. [173] in the upper portions of CZ-grown silicon crystals of over a length of about 15–25 inches due to the presence of TDs. These donors are formed in regions that are far enough from the actual growth interface in each crystal to spend appreciable periods at temperatures in the 300°C and 500°C range before the growth process is completed and the system is allowed to cool to room temperature. These TDs are unstable and disappear when the wafers made from that part of the grown crystal are heated to high temperatures.

Salnick et al. [174] developed a new technology approach to produce *n*-type, high-resistivity (up to 150 Ω-cm), large-diameter (up to 125 mm), low-cost silicon crystals for high-power devices such as diodes, thyristors, and transistors. The desirable parameters of this material were achieved through both silicon crystal growth and power device manufacturing processes. At first, the silicon melt was doped lightly with phosphorus to grow relatively high-resistivity crystals. After that, the silicon wafers cut from these crystals were also doped with TDs at the final stages of power device fabrication at ~450°C. It is reported that axial and radial resistivity uniformities of the crystals can be achieved in this approach.

Impurity engineering is a major issue for CZ-grown single-crystal silicon, particularly for the present VLSI and ULSI circuits, and is another major research area for getting better and better crystals and more reliable circuits. Potential mechanisms of impurity doping and better uniformity over the entire silicon wafer are challenging. Many research teams are active in this key area of crystal growth [175–185].

4.3.7 Methods for Continuous Czochralski Crystal Growth

Several schemes have been proposed to make the CZ growth process continuous and keep the silicon melt height constant by continuously recharging the melt externally [124]. In such processes, the buoyancy effects can be reduced significantly by maintaining a shallow layer of the melt region. By keeping the melt height fixed, many kinds of inhomogeneities, unsteady kinetics, and instabilities can be suppressed. The oxygen impurity will also be decreased, since the wetted surface area will remain at a constant location. By continuously charging the melt, the crystal quality can be tightly controlled; the dopants can be added continuously to the desired concentration levels. The main parameters and issues governed at this stage are melt aspect ratio, crystal-to-crucible radius ratio, Grashof number (Gr), Prandtl number (Pr), Marangoni number (Ma), crucible Reynolds number (Re$_c$), and crystal Reynolds number (Re$_s$).

Anselmo et al. [186] proposed and investigated a process whereby polysilicon pellets ($\cong$1 mm diameter) feed a continuous CZ growth process for single-crystal silicon. Experiments in an industrial puller with a 14- to 18-inch-diameter crucible successfully demonstrate the feasibility of this process. The advantages of the proposed scheme are a steady-state growth process, a low aspect ratio melt, uniformity of heat addition, and a growth apparatus with a single crucible and no baffle(s). The addition of dopant with the solid charge will allow better control of oxygen concentration, leading to crystals with uniform properties and better quality. They also presented theoretical results on the melting of fully and partially immersed silicon spheres and numerical solutions on temperature and flow fields in low-aspect ratio melts, with and without the addition of solid pellets. This approach is the best way to achieve a continuous CZ crystal production process.

In the continuous CZ growth system with a shallow and replenished melt proposed by Wang et al. [187], large-diameter crystals may be grown at a high pull rate and reduced melt convection. This system consists of two heaters. An interface control algorithm achieves the desired interface shape by adjusting the power level of the bottom heater. The control algorithm is incorporated into an existing process model, and the efficiency of the control algorithm was also tested. In the present system, the shape of the solidification interface is critical for high-quality silicon growth. For the newly designed continuous CZ system, the silicon diameter was controlled by the pull rate, and the solidification interface shape was controlled by the power levels of two heaters. The schematic of the continuous CZ system is presented in Figure 4.21. Here, the efficient way to suppress turbulent flow in a large-diameter silicon growth system is through either an applied magnetic field or melt height reduction. In the new design, the crucible consists of a shallow growth compartment in the center and a deep feeding compartment around the periphery. Two compartments are separated by a velocity partition and connected with a narrow annular channel. Two heaters heat up the silicon melt in two compartments. The main heater is set at the outside of the crucible, similar to a conventional CZ crystal growth system. The bottom heater provides supplemental heat to the central growth compartment. The figure also shows the temperature distribution in the system furnace when the main and bottom heater powers are at 60 kW and 20 kW, respectively. Temperature in the melt is critical for the interface shape and impurity incorporation in this

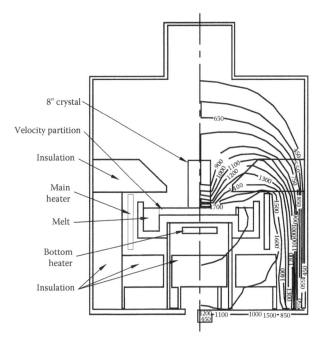

FIGURE 4.21
Geometric configuration of an 8-inch-diameter continuous CZ system for silicon growth and temperature distribution with a main heater power of 60 kW and a bottom heater power of 20 kW. (From C. Wang, H. Zhang, T. Wang, and L. Zheng, *Journal of Crystal Growth*, **287**, 252–257, 2006 [187].)

case. During crystal growth, the melt temperature should be higher than the melting point to avoid any cold spots in the melt. Temperature differences in the melt should be minimized to reduce temperature fluctuation. The fluid flow and temperature distribution in the silicon melt should be carefully managed to avoid any additional hot points. In the present case, melt convection is weak due to the shallow melt height. Three cells are observed in the melt. One counterrotated cell in the deep feeding compartment makes the temperature uniform, which is beneficial for pellet melting. The flow field under the growing crystal is weak, which is beneficial for high-quality crystal growth. A control algorithm controls the solidification interface shape by adjusting the bottom heater power. A predefined solidification interface can be achieved using the proposed control algorithm with different control gains.

Two long-term, solid-pellet-feed, continuous CZ growth experiments were performed in an industrial CZ crystal puller by Anselmo et al. [188]. The goals of these experiments were to see if the polysilicon pellets would melt, to discover the thermal effects the pellets would have on the overall melt, and to discover if pellet addition could be an effective melt replenishment technique. These experiments demonstrate that the quality of the melt for the continuous CZ growth is based heavily on the surface temperature of the melt. Several critical issues need to be addressed to develop a successful continuous CZ process. In this process, silicon pellets of about 1 mm diameter are added continuously to the melt, from which more than one crystal can be grown in one furnace cycle. Since the charge is added continuously, a shallow, constant height melt may be sufficient to grow the crystal, reducing buoyancy effects significantly and suppressing many kinds of unsteady kinetics, instabilities, and inhomogeneities. Numerical simulations show that oscillations can be reduced significantly for low-aspect-ratio melts in a crucible with a curved bottom. The heat input in this process can be constant, and the crucible will not need to be moved up or down for most of the growth process. Above all, by continuously charging the polysilicon and dopant, the crystal quality can be tightly controlled. To minimize problems during the pellet addition process, a 101 mm (4-inch) diameter crystal was grown using a 406 mm (16-inch) diameter crucible. The major variables of interest here are the pellet feed rate, the heater power, the pull rate, and the melt temperature. The location of the feed tube is shown schematically in Figure 4.22. The system had an initial charge of 50 kg. Crystal and crucible rotations were kept counter to each other, at 8 rpm.

Another approach to achieve continuous CZ growth has been developed by Shiraishi et al. [189]. This new method for growing silicon crystals is not based on the use of granular raw materials, but by the liquid feeding method. In this new technology, drops of molten silicon are supplied as raw materials. The silicon drops are made by heating polycrystalline silicon rods. Because the rod materials can be melted without contacting other substances, high-purity silicon can be fed directly into the crucible. Polycrystalline silicon rods were used for charging and were melted by carbon heaters over a crucible without any contact between the raw material and the other substances. The structure of the feeding heater is spiral in shape, as shown in Figure 4.23. To reduce the thermal influence on the crystal, carbon insulators shielded the radiation from the heaters. When molten silicon was dropped into the crucible, it caused a vibration on the melt surface. Two silica tubes were dipped into the molten silicon to prevent the propagation of the surface vibration to the growing single crystal. Silicon drops were placed inside the tubes. Using this method, silicon crystals with diameters as large as 6 or 8 inches and good uniformity along the growth direction were successfully grown. The team reported that fluctuation of the measured resistivity and oxygen concentration values along the growth direction were found to be less than 5.4% and 5.2%, respectively, in these grown crystals. Many other methods and evaluations are reported in the open literature [190–197] on this interesting method of continuous CZ crystal growth techniques.

4.3.8 Impurity Segregation Between Liquid and Grown Silicon Crystals

The quality of a CZ crystal, which is determined by its oxygen content and defect distribution, is affected by the flow, heat, and mass transfer in the silicon melt from which the

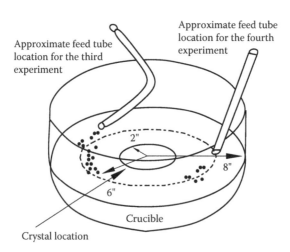

FIGURE 4.22
Location of pellet-feed nozzle for the first and second long-term continuous CZ experiments (not to scale). The feed tube locations are shown in opposing locations for clarity. (From A. Anselmo, J. Koziol, and V. Prasad, *Journal of Crystal Growth*, **163**, 359–368, 1996, and the references therein [188].)

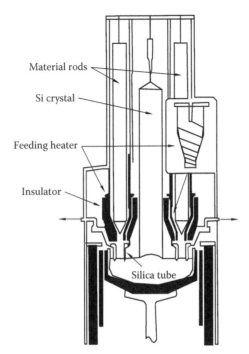

FIGURE 4.23
Schematic diagram of liquid feeding CZ crystal growth method. (From Y. Shiraishi, S. Kurosaka, and M. Imai, *Journal of Crystal Growth*, **166**, 685–688, 1996, and the references therein [189].)

crystal is grown. Buoyancy force is present in the CZ melt because of temperature differences in the crucible. Marangoni forces act at the gas-melt interface due to variations in surface tension with temperature. Generated flow under these conditions is turbulent or transition-to-turbulent in nature. External forces such as Coriolis and centrifugal force, due to crystal and crucible rotation, and Lorentz forces, due to the applied magnetic field, are used for stabilizing the flow. Careful balance between Lorentz force and Coriolis force is essential to achieve better control over melt flow. Knowledge of interactions among these forces is important to study the effect of flow on heat and mass transfer rates related to the crucible [198]. Many multiscale efforts, both experimental and theoretical, are underway to understand the complexities involved in the crystal growth process. Numerical simulation is gaining a lot of attention due to the improved computational speed provided by technological evolutions. In view of the continuous efforts to predict flow characteristics of the melt in the presence of the magnetic field, it is essential to study the predictive capabilities of such numerical models. Prediction of interface shape is still a crucial step in the CZ crystal growth process and depends upon the accuracy of quantification of the effects of magnetic fields and turbulence in the system. Although 2D axisymmetric models do not mimic the actual flow, these models are useful for preliminary analysis and parametric sensitivity studies. Time-dependent 3D simulations are more computationally demanding, but essential for accurate prediction of the flow field of CZ silicon melts. Figure 4.24 shows the interface shape prediction for 300 mm diameter silicon crystals at different rpm levels. Wada et al. [199] investigated the nucleation mechanism of large oxide precipitates in as-grown CZ silicon. The precipitates are square-shaped platelets with a half-diagonal length of about 500 Å to 2600 Å. The nucleation temperature is determined to be around

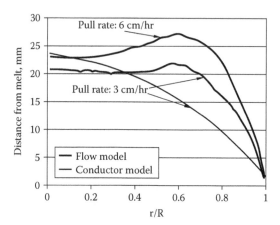

FIGURE 4.24
Interface shape prediction for 300 mm crystal (crystal speed = 15 rpm, crucible speed = 4 rpm, and 3.0 cm/hr pull rate). (From P. R. Gunjal, M. S. Kulkarni, and P. A. Ramachandran, "Melt dynamics in Czochralski crystal growth of silicon," *210 ECE Meeting Cancun*, Mexico, 29 Oct–3 Nov 2006, and the references therein [198].)

1250°C to 1000°C. The supercooling degree for the nucleation is about 50°C to 300°C and is smaller than that for homogeneous nucleation.

Kuroda [200] reported the temperature oscillations in the melt and in the solid, which were measured with a fine thermocouple. The oscillation amplitude is large in the bulk melt, small near the growth interface, and nearly zero in the crystal. The amplitude near the growth interface varies from 0.8°C to 4°C when the heater position is changed. The gradient and oscillation amplitude of temperature were accurately measured by Kuroda and Kozuka [201] in both the melt and solid during silicon crystal growth using the CZ technique. According to them, the temperature oscillations near the growth interface are closely related to the growth of high-quality crystals. The oscillation amplitude decreases as the heater position is raised or the pulling rate is increased. The amplitude also decreases with a reduction of the crystal rotation rate, but is independent of the CR rate. The amplitude depends on the melt temperature gradient near the growth interface and on the shape of the growth interface. The effect of temperature oscillations at the growth interface was further studied by Kuroda et al. [202], and they found a correlation between the microdefect density in crystals and the oscillations near the interface. The microdefect density decreases with higher heater position, increased pulling rate, decreased crystal rotation rate, and oscillation amplitude. The defect density in the grown crystal varies if the annealing temperature is more than 1000°C immediately after the completion of the growth process. Minimum oscillations provide better-quality crystals.

Analyzing the residual stresses in crystal growth, Kim and Kwon [203] proposed a scheme consisting of 1.8 mm/min ingot pulling speed and 9.8°C/min cooling rate, particularly in the crystal cooling schedule range of 1000°C to 1200°C, to suppress the growth of COPs by fast cooling rate after supersaturating vacancies by increasing pulling speed. On the other hand, the conventional scheme, consisting of 0.6 mm/min ingot pulling speed and 0.9°C/min cooling rate, cannot supersaturate a vacancy that is not suppressing further growth. Suppressing the further growth of COPs by a fast cooling rate after supersaturating vacancies by increasing the pulling speed leads to a COP-rich region with a high density and smaller size extending to the wafer edge. Upon subsequent annealing in oxygen ambient, initially higher COPs decrease through the vacancy-interstitial annihilation mechanism. The pulling speed and cooling rates specified earlier improve the high-density dynamic access memory (DRAM) retention time by nearly 40%.

Flow visualization and crystal growth experiments by Watanabe et al. [26] confirm that the crystal-melt interface shape of the CZ-grown silicon crystal is changed by the flow-mode transition of molten silicon caused by the baroclinic instability. They also confirmed that the oxygen distribution in the grown crystal is modified by the flow-mode transition. The flow visualization and crystal growth experiments were carried out by the team under the same conditions. The shape of the crystal-melt interface was identified from the growth striation shape in the grown crystal by x-ray topography. The oxygen distribution in the grown crystal was studied using micro-FTIR absorption spectroscopy. The crystal-melt interface was changed from a convex to a gull-winged shape by the transition from a nonaxisymmetric flow with vortices to an axisymmetric flow. The oxygen distribution along the growth direction in the crystal grown under the axisymmetric flow was more homogeneous than that grown under the nonaxisymmetric flow. Experimental results further suggest that an inhomogeneous oxygen distribution could result from thermal asymmetry generated by rotating vortices in the nonaxisymmetric flow conditions. Figure 4.25 shows the observed molten silicon flows for melt heights of 40 mm and 25 mm. The particle paths were obtained from a rotating viewpoint. The figure indicates that the flow-mode transition caused by the baroclinic instability takes place, not only because of the rotation, but also because of changes in the melt height.

Watanabe et al. [26] developed a double-beam x-ray radiography system to achieve a three-dimensional view of silicon flow during CZ crystal growth. Using this system, they have successfully observed the transition of silicon from axisymmetric to nonaxisymmetric flow in a rotating crucible due to baroclinic instability. The instability results from a

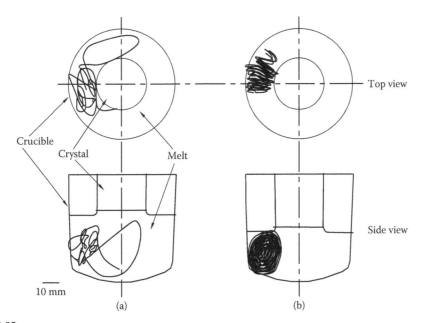

FIGURE 4.25
Particle paths of top and side views for (a) nonaxisymmetric flow with vortex, and (b) axisymmetric flow, in a rotating coordinate frame. (From M. Watanabe, M. Eguchi, K. Kakimoto, H. Ono, S. Kimura, and T. Hibiya, *Journal of Crystal Growth*, **151**, 285-290, 1995 [26].)

breakdown of the balance between buoyancy force, caused by the difference in density of the melt, and the Coriolis force generated by the rotation of the crucible. This breakdown is introduced by the following two operations: (1) decrease of the buoyancy force caused by modulation of temperature distribution in the melt, and/or (2) increase in the Coriolis force caused by an increase of the rotation rate of the crucible. Since the change of melt height and/or crucible position against the heater modulates the temperature field, it is likely that the baroclinic instability is generated during the CZ crystal growth process. If the mode change between axisymmetric and nonaxisymmetric flow takes place during crystal growth, it is likely that two different regions of oxygen atom distribution will exist in single grown crystal. If a change in the crystal-melt interface shape by the flow-mode transition during crystal growth is detected, modification of oxygen distribution in the grown crystal will be clear. The growth rate fluctuation under nonaxisymmetric flow originates from thermal asymmetry in the azimuthal direction, which was introduced by the rotating vortices in the silicon flow, as shown in Figure 4.26. Thermal asymmetry in the azimuthal direction could modify the temperature at the crystal-melt interface. Therefore, the growth rate could be modulated, and consequently, the concentration of oxygen incorporated from the melt into the crystal could fluctuate. On the other hand, the temperature field in the axisymmetric flow of molten silicon was almost symmetric to the point that the amplitude of the fluctuation at the crystal-melt interface is less than that for the nonaxisymmetric flow. Consequently, the oxygen in the crystal grown under axisymmetric flow conditions is more homogeneously distributed than that under nonaxisymmetric flow conditions.

Numerical study of CZ growth of silicon in an axisymmetric magnetic field was carried out by Sabhapathy and Salcudean [204]. The amounts and distribution of oxygen, dopants, and other contaminants in CZ-grown silicon depend critically on motions in the melt during crystal growth. The motions in the melt, which are induced mainly by temperature gradients and by crystal and crucible rotations, can be suppressed and the crystal quality can be improved by applying a magnetic field. The flow fields for a 75 mm diameter crystal growth in a steady, axisymmetric (axial or cusp) magnetic field were numerically obtained for crystal rotational rates from 0 to 30 rpm and for crucible counterrotational

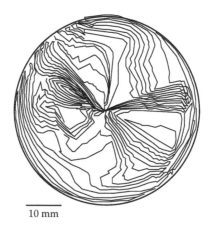

10 mm

FIGURE 4.26

Calculated temperature distribution of the horizontal plane of the melt. (From M. Watanabe, M. Eguchi, K. Kakimoto, H. Ono, S. Kimura, and T. Hibiya, *Journal of Crystal Growth*, **151**, 285–290, 1995, and the references therein [26].)

rates from 0 to 15 rpm. For small axial magnetic fields (= 0.1 T), the meridional flow field depended significantly on the CR rate, and for large axial magnetic fields, it depended significantly on the crystal rotational rate. For cusp magnetic fields, the meridional flow field was dominated by thermocapillarity. The flow fields indicated that the oxygen incorporation into the crystal would be less for the growth in a cusp magnetic field than in an axial magnetic field. The study provided good insight into the distribution of contaminants.

Yen and Tiller [205] developed a model for calculating the dynamic oxygen concentration change in silicon bulk melts during CZ crystal growth via consideration of the balance among competing oxygen fluxes in the system. In the study, a single oxygen species in the melt was assumed. In this model, the important parameters were found to be (1) the equilibrium oxygen concentration at the crucible wall, (2) the surface area ratio between the free melt surface and the crucible-melt contacting surface, (3) the SiO partial pressure in the main gas stream, and (4) a lumped melt convection parameter Q. The effect of these parameters on the oxygen content in silicon melts was examined, and the possible use of this model for online control of oxygen content in silicon crystals was discussed in detail. The crystal periphery generally becomes a low-oxygen region, compared to the high-oxygen region near the crystal center [206]. This difference in the radial oxygen content degrades the wafer quality. Therefore, it is necessary to clarify the radial oxygen distribution immediately beneath the growth interface, as this relates to the influence of both forced convection due to crystal rotation and fluid motion from the melt surface.

Series and Barraclough [207] carried out a detailed measurement of axial distribution of carbon content in silicon crystals grown in a low-pressure CZ system. Contamination rates during growth were measured as 0.1 µg/s in the absence of other sources of contamination. This would limit the carbon content of the crystals to a value of about 2.0×10^{15} atoms/cm^3 at the seed and 2.0×10^{16} atoms/cm^3 at a fraction solid of 0.7. The measured carbon contents of the crystals were generally higher than this value, indicating that contamination of the crucible, starting charge, and melt prior to growth were the dominant sources of carbon. No evidence was found to suggest a loss of carbon from the melt through the formation of carbon monoxide.

Dopant impurity incorporation into the growing silicon is an important topic, as it is at this stage they get into the silicon lattice and liquid–solid interface, and the dopant segregation coefficients play a crucial role at this point. Impurity distribution decides many issues relating to the device/circuit fabrication. As indicated earlier, the interface also admits unwanted impurities into the growing silicon. Both intentional and unintentional dopant species getting into the growing silicon has been evaluated by numerous studies [208–213]. By applying molecular dynamics in both two and three dimensions, studies have been carried out on melt convection and heat transfer statistics to evaluate and examine the grown CZ silicon crystals [214–222]. Many scientific groups are seriously involved in this area of crystal growth.

4.3.9 Crystal Growth Striations

Impurity striations reflect the shape of the solid–liquid interface present during CZ silicon crystal growth. They appear in both CZ- and FZ-grown crystals. In general, they exhibit impurity structures that are related to fluctuations in the growth rate. Jindal et al. [223] reported both the experimental and theoretical studies of interface shapes and striations on aluminum-doped single-crystal silicon. Based on theoretical calculations of variations in aluminum dopant concentration in the crystal as a function of variations in the crystal growth rate, they have deduced that the likely mechanism of interface atomic attachment is by layer motion. Effects of thermal fluctuations and fluid flow are also considered to explain the striations. There is no evidence of a single central facet, and the striations appear as straight lines of varying lengths and are not necessarily perfectly parallel to each other, as shown in Figure 4.27a. By using the exact solution of hydrodynamic equations for an infinite disk concept, they have formulated the liquid rotational motion. It was assumed that the solute concentration is only a function of distance from the disk surface and not a function of radius r or the angle around the disk. However, the distribution of streamlines at the disk surface has been shown to exhibit a spiral pattern, as shown in Figure 4.27b. These streamlines move across the surface, and for a flat untransforming disk, give a constant time-average momentum and solute distribution that is independent of r and φ. For a transforming interface, the rotating spiral-shaped oscillations of the momentum and solute boundary layers occur and influence the local growth conditions. Such growth striations were also reported by De Kock et al. [224] for CZ and FZ crystals. The influence of remelt phenomena and solid-state diffusion on striation formation has been analyzed, and directives are obtained to eliminate striations during the crystal growth process.

The dynamic behavior of striations is described by Shintani et al. [225]. According to them, the striations can be described as a nonchaotic motion with some nonstationarity in high-dimensional phase space, which is ascribed basically to temporal fluctuations in the melt temperature condition, including the temperature gradient induced by thermal convection. The mechanical rotations in the CZ system have no directly matched correspondence to the striations' periodicities. In addition, the pulling rate fluctuations show no direct correspondence to striation formation. The effect of a large TM field on growth striations and impurity concentrations in CZ silicon crystals was studied by Choe [226]. Silicon crystals of 125 mm in diameter were pulled in the [100] direction from 30 kg of melt. A magnetic induction of B = 2000 G was used to suppress melt convection. The crystals were not found to be striation-free: periodic striations spaced about 42 μm due to 25/–0.3 rpm cw crucible/crystal rotation in asymmetric radial thermal field were observed. Preferential chemical etching of heavily doped samples was used to delineate the growth striations. The crystal growth rate estimated from the periodic striations was about 1.1 mm/min, somewhat lower than the nominal pull rate of 1.5 mm/min. The growth striations in the (110) cross-sections of the heavily boron-doped horizontal magnetic-field-assisted CZ crystal are shown in Figure 4.28. From the results, it was concluded that convective flows in the melt and associated thermal fluctuations at the crystal-melt interface were suppressed, but the radial thermal asymmetry in the melt became large enough to cause the crystal/crucible rotation to form growth striations. Quantitative measurement and surface-tension-driven flow are some of the other issues related to growth striations that are being studied [227–230].

4.3.10   Use of a Magnetic Field in the Czochralski Growth Technique

Applying a magnetic field to the melt is one way of reducing thermal convection currents and the oxygen. The magnetic fields at which the suppression occurs were found to be 4.0 kG for an axially applied field and 2.0 kG for transverse fields. The feasibility of using magnetohydrodynamic effects to improve the quality of silicon crystals obtained by CZ growth from the melt has been demonstrated by several experiments [231]. The effect of a uniform magnetic field on the uptake of oxygen impurity by the growing crystal from CZ melt growth was investigated by Organ and Riley [232]. A reduction is forecast

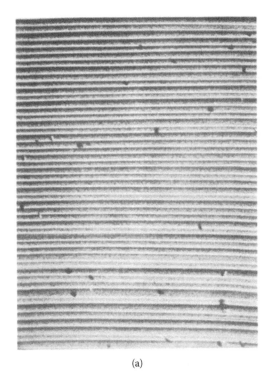

(a)

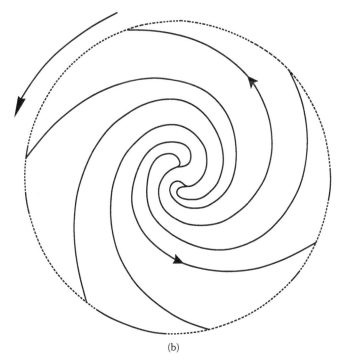

(b)

FIGURE 4.27
(a) Striation patterns in the outer regions of the grown crystal. (b) Spiral streamlines at the surface of a rotating disk. (From B. K. Jindal, V. V. Karelin, and W. A. Tiller, *Journal of the Electrochemical Society*, **120**, 101–105, 1973 [223].)

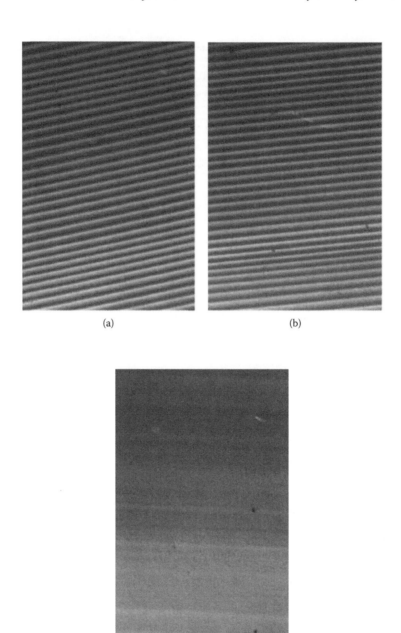

(a)   (b)

(c)

FIGURE 4.28
The (110) cross-sections of an HMCZ crystal grown under $B = 2000$ G and crystal/crucible rotation of 25/−0.3 rpm cw, after 10 min Yang etch: (a) near the edge, (b) at the half-radius point, and (c) in the center of the crystal. The average peak-to-peak distance of the growth striations shown is about 42 μm. (From K. S. Choe, *Journal of Crystal Growth*, **262**, 35–39, 2004 [226].)

in the amount of oxygen assimilated by the crystal when a magnetic field is applied. Hjellming and Walker [233] studied the mass transport of oxygen in a CZ silicon crystal puller with an extremely strong, uniform, and axial magnetic field. According to them, the problem is intrinsically unsteady and the oxygen concentration of a crystallizing particle depends on its trajectory since the beginning of crystal growth. The results provide the base solution for the mass transport problem with much weaker magnetic fields. Hirata et al. [234] investigated the effects of an axially symmetric vertical magnetic field of up to 2500 Oe on the thermal symmetry and crystal homogeneity in the CZ silicon growth process. The temperature difference around the pulling axis first increased at a magnetic field of about 500 Oe and then decreased to a level below that with no magnetic field, then with a magnetic field of 1000 Oe, or more. Further, it was noticed that an improvement in dopant concentration homogeneity is achieved due to an improvement in thermal symmetry caused by the vertical magnetic field of 1000 Oe. A vertical field of 1500 Oe or more, on the other hand, degrades the crystal homogeneity.

Series [235] undertook a systematic study of an axial magnetic field on the incorporation of oxygen, carbon, and phosphorus into the CZ silicon crystals. A quantitative comparison between the observed changes in the effective distribution coefficient for carbon and phosphorus and an analytical theory were part of the study. A simple analytical model to describe the conditions under good radial uniformity of dopant species was proposed based on the observations. The effect of varying the magnetic field at constant crystal and crucible rotations and the effect of varying crystal and crucible rotations at fields of 0 and 2000 G were investigated. A series of experiments on the effect of a shaped magnetic field on the CZ growth of silicon was also reported [236]. The radial uniformity of both dopant and oxygen is comparable to that of crystals grown under the corresponding zero-field conditions. The effect of the shaped magnetic field was also studied by means of two opposing superconducting solenoids and was configured to produce a field that is predominantly normal to the crucible wall. This configuration was designed to lead to a high degree of damping of the flows up the crucible wall, thus causing a significant reduction in the oxygen content of the melt. By choosing appropriate conditions, crystals can be grown with an oxygen content that varies over a wide range. The magnetic field configuration is amenable to scaling to large diameters.

The amounts and distribution of oxygen, impurity dopants, and other contaminants in CZ-grown silicon depend critically on motions in the melt during crystal growth. The motions in the melt, induced mainly by temperature gradients and by crystal and crucible rotations, can be suppressed and the crystal quality can be improved by applying a magnetic field. The flow fields for a large-diameter crystal grown in a steady, axisymmetric (axial or cusp) magnetic field were numerically obtained for crystal rotational rates from 0 to 30 rpm and for crucible counterrotational rates from 0 to 15 rpm [204]. For small axial magnetic fields (= 0.1 T), the meridional flow field depended significantly on the crucible rotational rate, and for large axial magnetic fields, it depended significantly on the crystal rotational rate. Numerical analysis of oxygen transport in the magnetic CZ growth of silicon was studied by Kobayashi [237] to interpret the different kinds of behaviors between the transverse and axial magnetic CZ on the oxygen contents in grown silicon crystals. They were low in the transverse-field CZ, but very high in axial-field-grown crystals. Explanation for such observations was explained. Hicks et al. [238] studied the effect of a configured magnetic field on the uptake of oxygen by a CZ-grown silicon crystal. According to them, the amount of oxygen assimilated by the crystal is reduced as the magnetic field strength increases, and a more uniform distribution within the crystal than can be achieved with a uniform magnetic field was realized.

Oxygen transfer in silicon melts during crystal growth under vertical magnetic fields was investigated numerically and experimentally by Kakimoto et al. [239]. A 3D numerical simulation, including melt convection and oxygen transport, was carried out to understand how oxygen transfers in the melt under magnetic fields. The experimental results obtained were compared to results from a numerical simulation. By using a 3D numerical simulation and a silicon crystal growth experiment, evaluations were carried out for these changes. An axisymmetric flow pattern was obtained with the type A heating system within the magnetic field from 0 to 0.3 T. However, a nonaxisymmetric flow pattern was observed in the simulation with the type B heating system, as shown in Figure 4.29a and b, which indicates profiles of the velocity and temperature distribution at the top of the melt under a magnetic field of 0.1 T. The velocity profile in the $z - \theta$ plane is shown in Figure 4.29c. An anomalous increase is reported in the oxygen concentration of the grown crystals under a magnetic field of about 0.03 T. The cause of this anomaly is identified as the Benard instability, since the temperature at the bottom of the crucible is higher than that at the interface. When the temperature at the bottom is decreased, the Benard cell can be removed, and a monotonic decrease in the oxygen concentration in the single silicon crystals can be observed. When using a vertical magnetic field, one can control the incorporation of oxygen, as suggested by the team earlier. A significant amount of progress has been made in understanding heat and mass transfer under vertical and horizontal magnetic fields. Cusp-shaped magnetic fields have excellent potential to modify fluid flow because they offer desired homogeneous concentrations of oxygen in grown crystals [240]. Since magnetic fields are able to control the flow more effectively for large-diameter melts, the use of magnetic fields has become more important to obtaining crystals with low and homogeneous oxygen concentrations by controlling the fluid flow.

CZ silicon crystal growth in the presence of an axially symmetric cusp magnetic field was reported for the first time by Hirata and Hoshikawa [241]. The free surface of the melt is centered between two superconducting coils. In this way, the oxygen concentration was successfully controlled from 1.0×10^{18} to 2.0×10^{17} atoms/cm^3 by increasing the cusp magnetic field strength up to 3500 Oe at the center of the bottom melt-silica crucible interface, while keeping the crystal rotation constant at 30 rpm and CR at –10 rpm. Both the oxygen and dopant concentrations were homogenized by the presence of the magnetic field. The controllability of oxygen concentration is due to the advantageous characteristics of the cusp magnetic field to realize localized control of thermal convection at the melt-crucible interface, independent of that at the melt free surface. Good crystal homogeneity results because there is no need to change the crystal and crucible rotations adequately to control the oxygen concentration, in contrast to the previous use of a transverse or vertical magnetic field in which extreme changes in the crucible or crystal rotation rate are required.

Hirata and Hoshikawa [242] reported the effects of a cusp magnetic field on the flows, oxygen transport, and heat transfer in an actual CZ silicon melt. These were quantitatively analyzed using boundary conditions based on experimental measurements and a numerical method involving the direct solution of the Navier-Stokes and Maxwell equations for three-dimensional unsteady flows. The unstable and instantaneously asymmetric flow components and forced convections emanating from the rotating crucible bottom in the absence of a magnetic field are effectively suppressed when a cusp magnetic field is applied. On the other hand, the rotational flows caused by CR become uniform and the circumferential velocity increases upon application of a cusp magnetic field. The combination of these rotating melt flows and crystal rotation in the opposite direction produces strong, stable forced convections that move toward the center directly below the growth interface. Due to these changes in the melt flows, the oxygen concentration at the growth interface is decreased and the radial uniformity is improved, temperature fluctuations

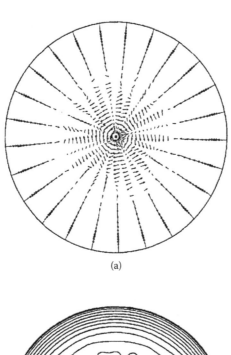

(a)

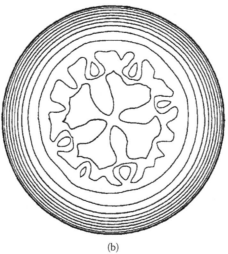

(b)

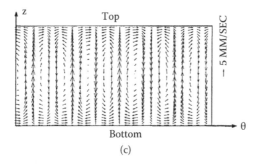

(c)

FIGURE 4.29
(a) Asymmetric fields of velocity. (b) Temperature distribution at the top of the melt. (c) Velocity profile in the $\theta - z$ plane at one-third of the crucible radius. (From K. Kakimoto, K.-W. Yi, and M. Eguchi, *Journal of Crystal Growth*, **163**, 238–242, 1996, and the references therein [239].)

within the melt are significantly reduced, and the radial temperature gradient is increased (whereas the temperature gradient decreases with respect to depth). These results regarding temperature and oxygen concentration are all in good agreement with experimental observations.

Hjellming and Walker [233] studied the mass transport of dopants and impurities in the silicon melt of a CZ crystal puller with a uniform axial magnetic field. For magnetic field strengths $1 = B_o = 5T$, the dominant flow fields are those driven by buoyancy and the growth of the crystal (the pulling of the crystal and the raising of the crucible bottom). According to them, the mass transport is intrinsically unsteady, and the concentration of a crystallizing particle depends on the entire history of it. The buoyancy-driven flow provides a mechanism for the recirculation of fluid particles. However, the character of the flow in the diffusion boundary layers and the convection-dominated cores depends on the competition between the crystal growth and buoyancy-driven velocities, which in turn, depend on the strength and depth of the magnetic field.

In CZ growth, fluid motion within the melt affects the transport of heat and solutes to the growth interface, and has therefore been the subject of much study. CZ silicon melts are excellent conductors of electricity, so the flow can be modified hydromagnetically. Flow under the influence of an axial magnetic field has received the most attention because the configuration remains rotationally symmetric. In the case of a uniform field, the theory has been extensively developed, using both asymptotic and numerical methods, by Langlois et al. [243]. The advantage of quiescent flow is offset by an unfavorable radial distribution of solutes in the finished crystal. A transverse field destroys the axial symmetry, but can sometimes be investigated by a combination of asymptotic and numerical methods. A nonuniform axial field is another possibility that may offer advantages to growing uniform crystals.

While earlier research focused on uniform axial (vertical) or transverse (horizontal) magnetic fields, recent experiments and numerical studies have indicated that a nonuniform, axisymmetric magnetic field may lead to better control of crystal properties. Most recent research has focused on a "cusp" field produced by two identical solenoids, which are placed symmetrically above and below the common horizontal plane of the crystal-melt interface and free surface, and which are carrying equal but opposite electric currents. The axial magnetic fields from the two solenoids cancel at the free surface and crystal-melt interface, so the local magnetic field is purely radial. For the CZ process, the melt motion has three components: the buoyant convection, the thermocapillary or Marangoni convection, and the rotationally driven flow associated with the rotations of the crystal and crucible about their common vertical axis. The three components of the melt motion are decoupled for simulation purposes and can be treated independently. According to Khine and Walker [244], the three solutions must be superimposed for the mass transport problems, since the convective transport of dopants and impurities is never negligible.

Centrifugal pumping flows are produced in the melt by crystal and crucible rotations during the CZ growth. By using a family of magnetic fields, Khine and Walker [25] derived numerial solutions for the growth conditions. Since molten silicon has large electrical conductivity, magnetic fields can be used to control the melt motion. Early research focused on uniform axial or transverse fields. However, the desirable centrifugal pumping flow is strongly suppressed by the uniform axial fields. It was also demonstrated that the centrifugal pumping flows are not strongly suppressed with a uniform transverse field, but the associated deviations from axisymmetry in the temperature and oxygen or dopant concentrations produce undesirable striations in the crystal. This is reviewed in the literature on CZ silicon growth with externally applied, steady magnetic fields. Due to the undesired effects of both uniform axial and uniform transverse fields, most recent research has concentrated on a particular nonuniform, axisymmetric cusp field. Since the solenoids carry equal but opposite currents, the axial fields at the crystal-melt interface and free surface cancel, resulting in a purely radial field along that plane. The transport of dopants is different from that of oxygen. There is no dopant transfer at either the free surface or the crucible surfaces. The segregation coefficients for most dopants in silicon are small, so that rejected dopants elevate concentrations in a mass-diffusion boundary layer adjacent to the crystal-melt interface. The melt motion transports dopants from the mass-diffusion layer so that concentrations in the bulk of the melt increase during crystal growth.

Variation of crystal growth parameters, such as the crystal rotation rate, the CR rate, and the configuration of cusp magnetic fields, results in the increase or decrease of critical growth rate V^*, the transition from interstitial to vacancy, of CZ-grown silicon crystals. It can also make axially asymmetric distributions of grown-in microdefect regions. V^* is remarkably increased in some range of the crystal rotation rate and the CR rate, which are likely to depend on crucible shape, size, and melt volume. The crystal rotation rate, as well as the CR rate, affect the incorporation of point defects into the growing crystal by modifying the melt convection. In addition, the novel unbalanced magnetic technique proposed by Cho et al. [245] controls heat transfer and oxygen transfer separately. A melt-focused technique, such as the electromagnetic CZ (EMCZ) method, was proposed for the purpose of preparing single-crystal silicon with high V^* by controlling oxygen concentration. Despite EMCZ's advantages, such as the possibility of a high pulling rate and easy control of oxygen transfer, there are still disadvantages, such as an inconvenient apparatus and its limitation on process variability.

Due to the asymmetry of the melt temperature distribution, vacancy and interstitial point defects are incorporated into the growing crystal with considerably asymmetric concentration distribution at the moment of crystallization. As a result, a substantially axially asymmetric V/I boundary shape is formed, despite the radial Fickian-typed diffusion of such point defects and their pair annihilation. The melt flow-control experiments showed that growth parameters such as single crystal rotation rate (SR), crucible rotation rate (CR), and unbalanced magnetic ratio are effective means for growing oxygen-controlled and defect-free crystals at a high rate. In particular, the unbalanced magnetic technique is useful for obtaining CZ silicon crystals of low oxygen concentration. If the previously mentioned growth parameters are optimal, this promotes heat transfer to the crystal-melt interface. In view of melt convection, the outer melt convection cell should be balanced with the core melt convection cell to obtain defect-free crystals at a high growth rate. In addition, crystals with low oxygen concentration and high defect-free growth rate can be grown at moderately high unbalanced magnetic ratios, which suppress mass transfer in the lower part of the melt contained in the quartz glass crucible. Figure 4.30 shows 2D simulation results of the melt convection under two different growth conditions: one is a conventional growth condition of SR of 20 rpm, CR = −4.0 rpm, and the other is the new growth condition of SR of 17 rpm, CR = −0.5 rpm. Under the conventional growth condition, the outer melt convection cell seems strong compared with the core melt convection cell.

Watanabe et al. [246] came out with a modification of the crystal-melt interface shape by using the EMCZ method to grow large-diameter, high-quality silicon wafers by suitably modifying the magnetic field to reduce the thermal convection in the silicon melt and by modifying the crystal-melt interface. In this case, the required Lorentz force was created by combining the static magnetic field with an electric current passing through the melt from a growing crystal. The controlled melt flow in this method makes it easy to

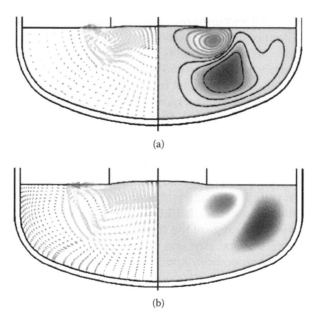

(a)

(b)

FIGURE 4.30

The two-dimensional simulation results of the melt convection: (left side) the velocity vector of melt flow, (right side) the stream function of melt flow, (a) crystal rotation rate 20 rpm and crucible rotation rate = −4 rpm, (b) crystal rotation rate = 17 rpm and crucible rotation rate = −0.5 rpm (assuming symmetric distribution of the melt temperature). (From H.-J. Cho, B.-Y. Lee, and J. Y. Lee, *Journal of Crystal Growth*, **292**, 260–265, 2006, and the references therein [245].)

control temperature distribution around the crystal-melt interface. In this way they were successful in effectively modifying the interface shape to eliminate grown-in defects in 200 mm diameter silicon crystals. Three-dimensional, time-dependent numerical simulations of temperature distribution in both the crystals and melt, including melt flow, were performed with the parallelized finite-volume code STAMAT3D, with the same boundary conditions as the experiments. The melt flow was governed by 3D equations describing mass, momentum, and heat transport using the Boussinesq approximation for an incompressible fluid. The 3D, time-dependent numerical simulation results of the velocity field in a horizontal section close to the free surface and temperature distribution in the melt are shown in Figure 4.31. The velocity profile shown in the figure indicates that the melt spontaneously rotates with a small rotation rate. This rotational flow distribution in the melt is completely different from that induced by crucible rotation in the conventional magnetic CZ method: in the latter case, melt rotates like the solid body rotation. The temperature distribution in the melt is parallel to the growth direction and the interface shape is flat or slightly concave in nature. On the other hand, with EMCZ, the interface shape of a convex curve toward the crystal was observed in the simulation results, as shown in Figure 4.32. This is similar to the shape observed in the experiments. The team concluded that there is a good qualitative agreement between the simulations and experimental results. This method is expected to be advantageous for the next generation of crystal growth technology to make high-quality silicon crystals—in particular, to grow large-diameter wafers for VLSI and ULSI circuits.

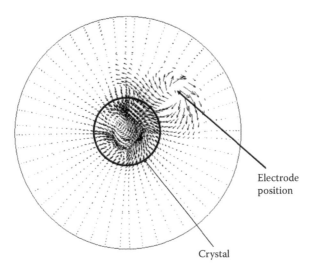

FIGURE 4.31
Velocity field in a horizontal section very close to the free surface of the melt in the electromagnetic CZ growth conditions. The electrode position and growing crystal are also indicated in the figure. (From M. Watanabe, D. Vizman, J. Friedrich, and G. Müller, *Journal of Crystal Growth*, **292**, 252–256, 2006 [246].)

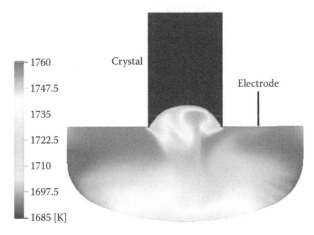

FIGURE 4.32
Temperature field in a vertical section in the electromagnetic CZ crystal growth conditions. Temperature of the isothermal color is shown in the figure. The difference of temperature in the crystal is smaller than the difference of temperature in the melt, so that the crystal part was the same color. (From M. Watanabe, D. Vizman, J. Friedrich, and G. Müller, *Journal of Crystal Growth*, **292**, 252–256, 2006 [246].)

A 300 mm, *p*-type,<100>-oriented nitrogen-doped CZ (NCZ) silicon crystal was grown with a cusp magnetic field as reported by Tian et al. [166]. The oxygen precipitation within the P-band was stronger than that in the region outside of the P-band, which is generally not the case for the conventional CZ wafer with a P-band. Based on this result, it is deduced that the grown-in oxygen precipitates existing in the P-band are relatively smaller, but with higher density compared to those in the CZ crystal.

Present-day CZ crystal growth depends heavily on the use of a magnetic field to produce large-diameter single-crystal silicon. This method provides better dopant impurity distributions and controls the oxygen incorporation into the growing silicon crystals. DC, electromagnetic, and traveling magnet fields [247–249] are applied in this method. Axial [250–254], transverse [255–263], and cusp magnetic fields [264–271] are used for this application, and there are a number of good publications covering a wide range of topics on intentional dopants and unintentional dopant species. Computer simulations were carried out on the transport of dopant species and molten silicon movement under the influence of the magnetic fields. These are widely applied to grow CZ silicon crystals, and several efforts are being explored [272–277] to obtain a perfect silicon crystal for the fabrication of VLSI and ULSI circuits.

4.3.11   Large-Area Silicon Crystals for VLSI And ULSI Applications

Technological advancements in the field of VLSI and ULSI that require a large chip surface area on the wafer, and the desire to reduce growth costs for single crystals of silicon have pushed industrial efforts toward designing systems that can grow crystals 200–300 mm in diameter using 500 to 600 mm diameter crucibles with initial charges upwards of 150 kg. However, simply scaling up the system does not solve the growth problems and introduces many more process complexities [124]. Indeed, a scaled-up version of the system results in much stronger oscillations, chaotic behavior, and/or turbulent flows, leading to higher levels of inhomogeneities in the grown crystals. This creates a lot of other issues at the cost of increasing the crystal diameter.

Time-dependent temperature distribution on a single crystal of silicon has been step-wise simulated by Virzi [278] during the CZ growth of a 6-inch rod from a 45 kg polysilicon charge, taking into account the physical and geometrical constraints to the heat transfer given by the overall pulling furnace architecture. For each selected growth step, a conduction-dominated Laplace's model with detailed boundary conditions and careful material properties has been solved by means of the finite-element technique.

At present, 200 mm wafers have an oxygen level of 19–31 ppma (as per American Society for Testing and Materials, ASTM 1979). Oxygen in CZ silicon crystal comes from dissolution of the quartz crucible; it is transported to the crystal-melt growth interface and segregated into the growing crystal. Transfer of oxygen in the melt is affected substantially by the melt convection. As crucible diameters increase, the melt convection fluid flow becomes increasingly more turbulent. Melt convection is suppressed most effectively by applying a magnetic field. For 200 mm crystal growth, magnetic CZ with a cusp-type magnetic field has been developed successfully; it controls oxygen concentration to 19–31 ±2 ppma (or less) with a radial gradient of 5% or less [279]. For 300 mm silicon crystals, and subsequently 450 mm, cusp magnetic CZ will be effectively used to control oxygen. Melt temperature fluctuates severely due to turbulent melt convection in a large-size melt for 300 mm crystal growth, and may not provide stable conditions in normal CZ methods. A further benefit of magnetic CZ is that it minimizes the emission of SiO_2 particles from the quartz crucible wall in contact with the molten silicon melt. The currently required oxygen level of 18–31 ±2.0 (or less) ppma, with a radial gradient less than 5%, can be achieved effectively for 300 mm wafers by cusp magnetic CZ by applying the proper magnetic field strength in conjunction with adjusting crystal growth conditions to get wafers for ULSI processing.

A silicon ingot grown using the CZ crystal puller technique (300 mm diameter) is shown in Figure 4.33 [280]. The crystal is grown by applying the Dash necking method and using varying growth rates. The thin suspended portion of the narrow-diameter silicon crystal takes the complete weight of the grown crystal. Although there are several methods for growing single crystals of silicon, the CZ process has virtually dominated the production of commercial silicon wafers. Demand for large silicon wafers has driven the growth of silicon crystals as large as 400 mm in diameter. Also, larger and larger diameter single-crystal silicon wafers are being used in photovoltaic applications. However, the technological progress of CZ silicon crystal growth has stalled, and simply scaling up the process may not produce crystals of good quality. The undesirable effects of a much stronger melt convection and a slow pull rate are usually unavoidable when growing large-diameter crystals.

In terms of industrial silicon production, the growth of large crystals with 300 and 400 mm diameter is under investigation. This means an increase in the melt volumes and in the crucible diameters. Since the temperature of the growing crystal is fixed at the interface, the temperature of the crucible wall tends to increase with larger melt volumes. At the same time, the incorporation of oxygen into the growing crystal from the crucible has to be controlled precisely at a rate of 5.0×10^{17} cm^{-3}, which is far below the solubility in the silicon melt [19]. Changes in the oxygen concentration within the molten silicon will tend to form silicon oxide precipitates and lead to other unwanted issues with crystal growth.

FIGURE 4.33
Fully grown, 300 mm diameter, single-crystal silicon ingot by CZ method. (From: Photograph courtesy Kayex Corporation. 300 mm silicon ingot. http://depts.washington.edu/chemcrs/bulkdisk/chem364A_spr05/handout_Silicon.pdf, and the references therein [280].)

Large-diameter, high-quality silicon wafers are required to further advance ULSI device processing. Therefore, a new crystal growth technique is needed to obtain these crystals with homogeneously distributed oxygen and a reduction of grown-in defect density in the concentration required for ULSI device processing. To address this requirement, Watanabe et al. [246] developed a new crystal growth technique using the electromagnetic force (EMF) CZ method. Using this method, the team was able to grow defect-free Si crystals of 200 mm diameter with a higher pulling rate. High-speed pulling of defect-free crystals using the EMCZ method is due to large modifications of the crystal-melt interface. In this method, the interface shape is largely modified to the upward convexly towards the grown single crystal. The large upward convex shape of the crystal-melt interface during the growth results from the temperature distribution at the interface by the controlled melt flow generated by the EMF. Experiments and numerical simulations confirmed this large modification of the interface shape.

A finite-element, quasi-steady-state modeling of heat transfer during the growth of a single crystal of silicon using the CZ method was presented by Virzi [243]. Computations were performed for a comprehensive geometrical model of an actual puller, accounting for detailed radiation heat exchange and introducing a reviewed set of temperature-dependent thermophysical properties. The implemented problem-solving procedure goes through the recursive scaling of the power yield in order to guarantee that the melt-crystal-gas tri-junction temperature matches the melting one. An effective, time-saving numerical scheme was employed for the coupled convergence of both the melt-crystal interface location and the temperature distribution. Many different growth conditions were simulated to analyze the sensitivity of the heat flow pattern versus changes of pull rate, gas convection, and material properties. The results stress the strong influence of growth rate on the phase-change interface deflection, while the input of a correct melt emissivity value is shown to be imperative for working out reliable predictions of the crystal thermal field. The computed temperatures were compared with the corresponding thermographic data collected by means of an infrared imaging system, allowing a satisfactory validation of the adopted mathematical model.

Predictions of the integrated hydrodynamic thermal-capillary model (IHTCM) were compared to experimental measurements of the thermal field from growth of 83 mm diameter crystals and the melt-crystal interface shape of a 100 mm diameter crystal obtained in a conventional CZ system. The temperature measurements, obtained by Kinney et al. [281], showed good quantitative agreement with predictions, irrespective of the model for melt convection. Calculated values of the thermal stress in the crystal exceed the critical resolved shear stress near the melt-solid interface for all flow conditions. The maximum stress at the melt-solid interface depends on the interface shape and varies between 2.8 and 6.2 times the critical resolved shear stress, depending on the flow.

Accuracy in the prediction of the thermal field in a CZ crystal growth system is crucial for quantitative application of models. Predictions of the integrated hydrodynamic thermal-capillary model (IHTCM) are compared to experimental measurements of the thermal field from growth of 83 mm diameter crystals and the melt-crystal interface shape of a 100 mm diameter crystal obtained in a conventional CZ system. The temperature measurements, measured by Kinney et al. [281], showed good quantitative agreement with predictions irrespective of the model for melt convection. However, the predicted melt-crystal interface shape is much more sensitive to the impurity type and state of convection in the silicon melt. The IHTCM includes steady-state laminar flows driven by crystal and crucible rotation, natural convection, and thermocapillary. Although flows at the correct intensity can be computed for each mechanism separately, solutions for the combined driving forces only are found when the viscosity of the melt is set artificially to a high value. Calculated values of the thermal stress in the crystal exceed the critical resolved shear stress near the melt-solid interface for all flow conditions. The maximum stress at the melt-solid interface depends on the interface shape and varies widely to the critical times the critical resolved shear stress, depending on the flow conditions.

Over the past three decades, numerical simulations of CZ crystal growth have been performed, with varying degrees of complexity. Both finite-difference and finite-element methods have been used to simulate first the two-dimensional and then the three-dimensional melt flows with many different kinds of boundary conditions. A regular cylindrical geometry has usually been considered in most of the papers available in the open literature. Many of the inhomogeneities and defects in the crystal grown from a pool of melt are a direct result of the nonsteady nature of the growth kinetics and complex heat transfer and flow mechanisms. The transport phenomena become more complicated in a high-pressure system because of the thermal interaction between melt and gas flows and require special numerical treatment. A high-resolution computer model based on multizone adaptive grid generation and curvilinear finite volume discretization (MASTRAPP2d) has been developed by Zhang and Prasad [282] to simulate the crystal growth processes at low and high pressures. The scheme allows consideration of a multiphase system for more than one material in one single domain, which may consist of irregular and moving boundaries, interfaces, and free surfaces. CZ growth of silicon crystals is modeled to demonstrate the robustness and effectiveness of this innovative scheme.

Though significant progress has been made in the computer simulation of CZ silicon crystal growth, further developments are needed before simulation can be effective in developing crystal growth processes for 300 to 400 mm silicon wafers. Kim [283] solved a two-dimensional, steady-state turbulence simulation with finite element method (FEM) using an efficient Newton method with quadratic convergence. This method is more general and versatile. Among other advantages, it allows complicated realistic geometries, such as the crucible shape and the crystal-to-melt interface, to be entered and solved. Computer simulation of the CZ silicon crystal growth system guides optimization of the growth parameters, thus allowing oxygen and defect control in large-diameter silicon crystals. Recently, comprehensive models have been developed for heat transfer, melt convection, and oxygen/dopant transfer in large CZ silicon growth systems. Software simulates fluid flow in the CZ silicon melt, temperature distribution (both in the melt and crystal), and the shape of the crystal-to-melt interface. Turbulence is the most complicated of fluid motions. It differs from laminar flow in that streamlines fluctuate randomly over small distances with high temporal frequencies. It is thus inherently three-dimensional, and is characterized by a self-reinforcing cascade of energy from larger to smaller flow structures through a continuous spectrum. A standard turbulence model has been applied to the fluid flow in CZ silicon crystal growth, and it has proved useful. Further, close agreement was found between the simulated and experimental measurements in the thermal fields and for the crystal-melt interface shape. There was qualitative agreement for the radial distribution of oxygen in the crystal.

Remarkable progress in CZ silicon crystal size, oxygen uniformity, and purity has led to high minority-carrier lifetimes and low OISF densities. In addition, we now have an improved understanding and better control of point defects and secondary defects, and their impacts on device performance. The 200 mm wafers meet all the starting material requirements for subsequent device applications. As we approach the ULSI era, and as wafer sizes increase to 300 mm or larger, the linkage between the crystal growth conditions, defects in silicon material, and their impact on device performance needs to be continually identified [283]. Then we shall be able to develop and optimize crystal growth processes, wafer preparation, and device fabrication processes to achieve superior performance, reliability, and yields for semiconductor chips. The growth this field is experiencing draws more researchers to respond to the ever-increasing demands of VLSI and ULSI applications.

The demand for larger CZ-grown single-crystal silicon has brought lot of competition among experts in this area to grow perfect crystals that satisfy the requirement of VLSI and ULSI process engineers. The future requirement to grow wafers with a 450 mm diameter are inviting more troubles too. Maybe it is too early to conclude anything on the size and quality of these wafers. Both debates and achievements have been reported [284–292]. Hopefully, the developments in this field will satisfy the demands of VLSI and ULSI process engineers.

4.3.12   Post-Growth Thermal Gradient and Crystal Cooling after Pull-Out

The temperature inside a growing crystal sharply drops as the distance from the melt-crystal interface increases, which decreases the equilibrium concentrations of the point defects. Under such conditions, the Frenkel reaction decreases the concentrations of both the point defects inside. The point defect concentration gradients drive the diffusion of the point defects into the crystal. The physical displacement or convection of the crystal also contributes to the flux of the point defects relative to a fixed coordinate system. In the absence of external sources and sinks, such as the crystal surface or the thermomechanically induced dislocations, the established point defect concentration difference remains almost the same. After the initial incorporation beyond the recombination length, the established concentration difference determines the type of the prevailing point defect species at a distance far away from the interface. As the pull rate of the crystal increases, the convection of the point defects dominates their diffusion from the interface, leading to vacancy-rich conditions because the equilibrium concentration of vacancies at the interface is higher than the equilibrium concentration of interstitials. As the magnitude of the axial temperature gradient near the interface increases, the sharp temperature drop near the interface dramatically increases the concentration gradients of the point defects and allows the diffusion to dominate the convection. Self-interstitials diffuse faster than vacancies and lead to the interstitial-rich conditions [293]. The surviving point defects precipitate at a lower temperature to form microdefects.

A theoretical study has been carried out by Capper and Wilkes [294] on the cooling rates of silicon crystals, with diameters between 50 and 100 mm, from 650°C to room temperature, under natural and forced convection conditions. The significance of the results is discussed in terms of actual changes in donor concentration of a CZ-grown crystal after annealing at this temperature. The annealing step is required for ingots, particularly in the case of materials required to meet close tolerances of resistivity. They have described an empirical procedure of silicon manufacturers, typically 3 h at 650°C in nitrogen ambient, followed by quenching in air or with the assistance of forced cooling by fans. However, the core of large ingots is a problem, and reheating of slices may be necessary.

Nakanishi et al. [295] studied the influence of annealing in CZ crystals by quenching the crystals (from 1000°C) and compared them with those gradually cooled to room temperatures. The quenched samples were supersaturated with oxygen and differed greatly from the standard samples. Further, in the quenched crystal, the density of defects (silicon oxide precipitates) were detected even after heat treatment at 1000°C for 12 h. The concentration came down by three orders of magnitude when compared with standard crystals.

As the ULSI process has entered a period of nanotechnology, the design rule of the semiconductor device has scaled down to tens of nanometers. Therefore, there are rising demands for silicon wafers free of grown-in defects, as well as with super flatness. The defect-free wafer now produced by some suppliers is characterized by no-vacancy-rich and no-self-interstitial-rich defects. Cho et al. [296] made a critical analysis on the microdefects and self-interstitials depending on the growth rate issues. According to them, the Voronkov finding is well depicted in Figure 4.34: a higher V/G ratio than the critical value results in vacancy-rich silicon, whereas a lower V/G ratio results in interstitial-rich silicon. The analyses of this correlation are as follows: at the moment of crystallization, vacancies and self-interstitials are generated with thermal equilibrium concentrations (the equilibrium concentration of vacancy is a little higher than that of self-interstitial), and then they are rapidly reduced by the recombination reaction during crystal cooling. A large concentration of gradients is accordingly formed near the crystal-melt interface and are the cause of the diffusion of point defects from the crystal-melt interface to the crystal. A higher G produces a larger concentration of gradients, and a lower V induces sufficient diffusion influx of point defects from the interface into the crystal. Therefore, a low V/G ratio results in self-interstitial–dominant silicon, since the diffusion coefficient of self-interstitials is much greater than that of vacancies. On the contrary, a high V/G ratio results in vacancy-dominant silicon because the contribution of diffusion of point defects is small. In other words, variation of the V/G ratio changes the effect of the diffusion influx of self-interstitials against the convection influx of vacancies, and consequently determines the dominant type of point defects. However, these analyses are not always available. As an example, Voronkov's theory does not account for the various critical V/G ratios reported in the open literature. Therefore, another aspect of the defect formation mechanism should be considered to explain such melt convection–involved defect formation behaviors. Some reports have stated that the shape of the crystal–melt interface governs the axial temperature gradients and the critical V/G ratio.

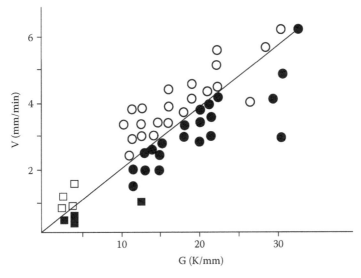

FIGURE 4.34
Observed microdefect types at various growth rates (V) and axial temperature gradients (G). Circles: FZ crystals, squares: CZ crystals; open: vacancy-rich defects, filled: self-interstitial-rich defects. (From H.-J. Cho, B.-C. Sim, and J. Y. Lee, *Journal of Crystal Growth*, **289**, 458–463, 2006, and the references therein [296].)

The crystals grown by the CZ method are slowly allowed to cool to room temperature to avoid the creation of point defects. Fast cooling of crystal arrests many atomic movements, and this may create randomness within the grown crystal, leading to many unexpected defects. Several studies have been carried out on the influence of cooling conditions and isotherm shapes within the grown crystals. Optimization of cooling conditions were reported from time to time. Crystal cooling at this stage is one of the crucial issues to terminating the crystal growth. References [297–302] provide insight into this topic on the thermal history of the ingot when the growth process is complete.

4.4 Float-Zone Crystal Growth Technique

In the FZ process, the polysilicon rod is held vertically and is allowed to melt by using an induction coil. This molten part of the silicon is carefully held by the surface tension without any spill-out. A seed crystal is put at the bottom of the polysilicon rod, and the liquid is slowly moved up to grow the crystal. As seen in the CZ process, the seed will guide the condensation for solidification, and the crystal grows as per the atomic arrangement of the seed crystal. This method has an inherent advantage because of the absence of a crucible and crucible-related contaminations. In this method, there is no way for the crystal to come into physical contact with the quartz crucible and graphite heaters, and hence it is free from oxygen and carbon contamination. The only disadvantage is the sharp temperature gradient present between the molten silicon and the solid condensed crystal. This temperature gradient produces many point defects, and cooling the crystal to low temperatures is an issue. Because heating occurs within a small and limited area, the largest diameter crystal one can grow is limited to several inches only. This process produces a small fraction of the silicon crystals for the modern microelectronic industry. Electronic-grade and metallurgical-grade silicon will be initially cast into graphite molds. It is advised to use crack-free rods for this FZ technique. One of the main drawbacks of this method is axial and radial uniformities of resistivity of the grown crystals.

Dislocation-free silicon crystals made by the FZ technique contain vacancy clusters that form during the cooling of the crystal after growth. As a result, the crystal becomes supersaturated with vacancies. This may lead to the formation of vacancy clusters because no vacancy sinks are present in the bulk. Impurities may play a part in the cluster formation. The distribution and concentration of these defects have been determined and measured using standard techniques. de Kock [303] has presented a model describing the formation of these clusters. The influence of vacancy clusters on the leakage current of planar diodes is explained based on these results.

Through different experimental observations, Collins [304] reported on growth parameters of large-diameter silicon crystals and outlined different factors affecting the ability to grow dislocation-free FZ silicon crystals up to 80 mm in diameter. The highest yield is obtained for 80 mm diameter crystals by starting with 68 mm to 74 mm diameter polycrystal rod stock. Lower transport speeds for crystal growth of (111) orientations were 3 to 4 mm/min, and for (100) 2 to 3 mm/min. Rotation rates of both upper and lower shafts were found to have an effect on growth at the solid–liquid interface. Rates established for the lower shaft were 6 to 8 rpm for the (111) crystals and 3 to 4 rpm for (100), counterclockwise. Upper rotation rates were 2 rpm on (111) crystals and 3 to 5 rpm on (100), clockwise. Seed orientation, which is critical, was held to within $\pm(\frac{1}{2})°$ of perfect orientation.

The minimum seed growth length was 50 to 70 mm. To assist in reducing the side lobes on (111) dislocation-free crystals, a cooling ring with a flow of argon was used. For best (100) growth, the lower side of a one-turn copper RF work coil was made conical.

FZ crystal growth with a pancake induction coil was introduced in the mid-1900s to produce single silicon crystals with large diameters (>100 mm). For the molten zone, the induced electric current, the temperature gradient, and the feed and crystal rotation speeds determine the free surface shape and cause the fluid motion. The presence of the relatively small hole in the center of the inductor causes the open melting front on the feed rod and has a strong influence on the heat transfer and interface shapes of the molten zone. The interface shapes are strongly coupled with the distribution of the electromagnetic field, and as a consequence, with the shape of the inductor [305]. The quality of the growing crystal depends on the shape of the growth interface, the temperature gradients, and especially on the fluid flow near this interface. It is difficult to measure and control the shape of the molten zone and the melt convection by simple experimental methods.

Measurement of temperature fluctuations, molten liquid flow and the interface, transient behavior of molten liquid and crystalline properties, and the angle and melt meniscus are key research topics [306–311] in FZ crystal growth. This continues to be a topic of research.

4.4.1   Seed Selection

Seed selection is important, as this seed determines the crystal to be grown. The standard procedure is to go for a seed that is free from all possible defects. Select the proper orientation as per the requirement and the crystal to be grown. As mentioned, in FZ crystal growth, a seed crystal is introduced to the end of the polycrystalline silicon rod. The seed is brought up from below to make contact with the drop of melt formed at the tip of the polysilicon rod. This seed sets the crystal structure, similar to the CZ crystal seed. The polycrystal is first molten, but during cooling, assumes the structure of the seed crystal. The heated region is slowly guided along the rod, and the polycrystalline silicon rod slowly transforms into a single crystal due to the phase change from a liquid to solid state as guided by the seed crystal.

4.4.2   Environment and Chamber Ambient Control

The ambient in FZ is somewhat similar to that of the CZ crystal environment. Oxygen- and moisture-free argon is preferred for the environment inside the chamber. Bhihe et al. [293] investigated out-diffusion of boron under various ambient conditions. Boron out-diffusion is significant in an H_2 ambient, while the out-diffusion is negligible in an N_2 or He ambient. Comparing the analytical model to the experimental data, they identified the diffusion coefficient of boron in an H_2 ambient and found that it is smaller than in an N_2 ambient. The significant out-diffusion in H_2 ambient is attributed to the enhancement of the boron transport coefficient at the Si surface. Negligible out-diffusion in an N_2 or He ambient is attributed to the negligible transport coefficient at the surface.

4.4.3   Heating Mechanisms and RF Coil Shape

The FZ process involves a molten silicon zone between a melting polycrystalline feed rod above and a growing single crystal below. Frequently, the feed rod is melted, and the float zone is kept molten by the Joulean heating from alternating current (AC) electric currents induced by an RF induction coil around the float zone. The shape of the melt's free surface is determined by a balance of the hydrostatic pressure in the melt, the surface tension force, and the inward electromagnetic body force due to the AC electric currents near the surface of the melt. For a given geometry of the induction coil, the distribution of AC electric currents depends on the free surface shape. Therefore, the determination of the free surface shape and the determination of the distribution of the AC electric currents due to the induction coil are intrinsically coupled. Lie et al. [312] presented an iterative method to determine both the free surface shape and the AC electric current distribution. The method is illustrated with four cases combining two different induction coil shapes and two different AC voltages applied to the induction coil, assuming that the zone's radial and axial dimensions are known.

The shape of the high-frequency inductor plays a decisive role in the development of any FZ crystal growth process. The temperature field in the hot zone is, to a large extent, determined by the distribution of the induced Joulean heat or EM power. Since the Joulean heat distribution can be controlled by the shape of the inductor, its modification allows the temperature field to be influenced on a comparatively large scale. Changes in the inductor geometry as small as 1 mm can have a significant impact on the Joulean heat distribution and the temperature field [313]. The inductor must ensure proper, homogeneous melting of the feed rod and maintain stability of the molten zone at the same time. A pancake-shaped inductor, with or without lateral slits, is commonly used in the FZ growth processes for crystals with large diameters. The inductor is water cooled, and sharp corners are avoided, as this may lead to current crowding at the edges and cause local melting of the inductor. Copper is the most widely used material, but silver, with its higher reflectivity and electrical conductivity, is also an option. The inductor surfaces are polished to achieve the best possible reflection of radiation from the molten zone and increase efficiency. A typical FZ heating arrangement is shown in Figure 4.35 [314]. Here, the needle-eye inductor is shown, which creates a short melt region of silicon. It also shows the sharp liquid–solid interface on top of the crystal. Numerical modeling on the shape of the RF coil to study the RF field, heat transfer, and thermal stress for various FZ crystals is being explored [315–317] for this important silicon crystal growth. As discussed earlier, crystal size is one limitation in this case.

Lan [318] used a robust finite-volume/Newton method to simulate the heat transfer, fluid flow, and interface shapes of FZ silicon growth under an axisymmetric magnetic field generated by coils around the growth axis. The global method allows all the field variables—both the heat flow and magnetic fields—and the interface variables—both the interfaces and the free surface—to be solved simultaneously in an efficient manner. Calculated results show that the suppression of convection by the applied magnetic fields is effective for both natural convection and forced convection due to rotation, but less effective for thermocapillary convection. As a result, the melt in the core region is quiescent and the effect of rotation becomes trivial, while in the periphery, the flow is still vigorous. The core size and flow patterns depend on the applied magnetic strength and distribution. The calculated core size and dopant segregation for growth in a monoellipsoid furnace are in good agreement with the measured ones. The effects of rotation and finite electrical conductivity on current distribution are explained in the report.

4.4.4 Crystal Growth Rate and Seed Rotation

For typical FZ geometries, the stability criterion is based on the finding that the steady-state growth of crystals with a constant cross-section in a meniscus-controlled process requires that the angle φ between the meniscus and the growth axis equal φ_o (a constant; for silicon $\varphi_o = 11°$) [319]. An analytical solution for the temperature distribution in freely radiating crystals was derived by Kuiken and Roksnoer [320]. Within the growing crystal, the heat transfer due to radiation is small compared with the heat transfer due to conduction. The

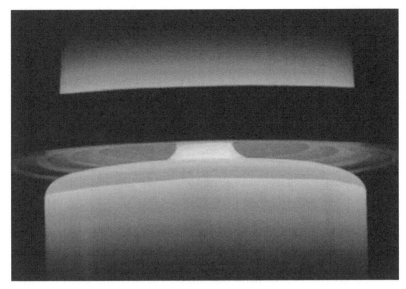

FIGURE 4.35
FZ growth of silicon; visible are the feed rod (above) and the single crystal (below); the dark ring is the RF needle-eye inductor. Note the very sharp liquid–solid interface on top of the solidified crystal. Courtesy of Leibniz Institute for Crystal Growth Berlin (From http://www.ikz-berlin.de [314].)

influence of the crystal growth rate on the temperature profile must be included when it is higher than 1–2 mm/min. The longitudinal temperature gradient at a concave solid–liquid interface was found to vary in a characteristic manner with the distance from the axis. A similar variation was observed for the height of the defect area formed at the solid–liquid interface in quenched silicon crystals. Eyer et al. [321] used a closed double-ellipsoid mirror heating facility to grow dislocation-free single crystals of relatively smaller diameter, touching upon the zone stability and the formation of various micro-inhomogeneities. The focused radiation of two halogen lamps served as a heat energy source for the growing crystals. These experiments served as preparation for growing silicon crystals under micro-gravity aboard Spacelab. However, it was reported that the crystals have been grown at 4 mm/min at rotation rates in the range 0–13 rpm [282].

Mühlbauer et al. [322] investigated the significance of the rotation scheme of the feed rod and crystal on the dopant distribution. The calculated dopant concentration directly at the growth interface is used to determine the normalized lateral resistivity distribution in the grown single crystal. The experimental setup considered has a crystal diameter of 103 mm, a growth rate of 3.3 mm/min, and an inductor current $I_o = 945$ A. For the calculations, the basic growth stage was chosen with a crystal length of 195 mm and a feed rod length of 390 mm. The rotation rate of the single crystal is 5 rpm, and the feed rod rotates at 20 rpm in the opposite direction. The example of the calculated shape of the molten zone with the corresponding electromagnetic, temperature, and flow fields is shown in Figure 4.36 [322]. The calculated resistivity distributions are compared with lateral spreading resistivity measurements in the single crystal. According to them, the scheme of rotation of the feed rod and the single crystal has a strong influence on the velocity field in the molten zone. This melt convection determines the dopant concentration field and, as a consequence, the resistivity distribution in the grown crystal. By using optimum rotation parameters, it is possible to get homogeneous resistivity distribution in the grown crystals.

4.4.5   Dopant Distribution in Growing Crystals

The dopant distributions are almost similar as those seen with CZ crystal growth. The same segregation issues are applicable in this case as well. The difference has to do with the interface shape of the solid–liquid silicon melt and the temperature gradient existing close to this phase.

Modern FZ silicon crystals are often doped with nitrogen to suppress grown-in micro-defects. The doping level can be up to 5.0×10^{15} cm^{-3}, and nitrogen can affect to some extent the electrical properties of the material. The dominant nitrogen form, dimeric interstitial N_2, is inactive, but the crystal can contain an appreciable amount of substitutional nitrogen N_s—a center with deep levels. In the as-grown state, no electrical activity of nitrogen is normally found, but deep centers are produced after annealing at 900°C or 1000°C, resulting in a huge increase in resistivity. This effect is most likely due to the N_s centers that were initially present in a hidden (or deactivated) state but were transformed back to N_s by annealing. The deactivation reaction is most likely between the two major species—by attaching N_2 to N_s. The activation by annealing occurs, accordingly, by the backward dissociation. Voronkova et al. [323] found that a drastic change in resistivity and in the apparent carrier concentration also occurs after annealing in a range of moderate temperatures (450°C to 700°C). Studies on three different samples clearly showed this effect. The most pronounced change was found in the boron-doped samples with initial higher resistivity: ≥1500 Ω-cm. In particular, annealing at 680°C for 90 min resulted in a huge resistivity increment by a factor of up to 50. The initial *p*-type conductivity was changed to an *n*-type that corresponds to the generation of deep donors in a concentration well above 1.0×10^{13} cm^{-3}. The temperature dependence of the carrier concentration below room temperature

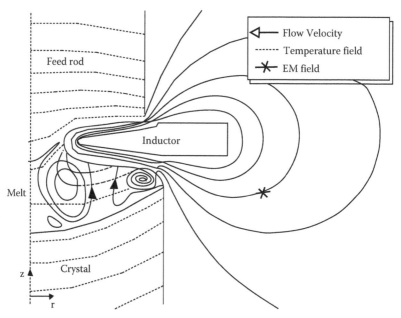

FIGURE 4.36
Calculated interface shape and electromagnetic, thermal, and hydrodynamic fields for FZ crystal growth with a pancake induction coil. (From A. Mühlbauer, A. Muiznieks, and J. Virbulis, *Journal of Crystal Growth*, **180**, 372–380, 1997, and the references therein [322].)

was deduced from the two-probe resistivity ρ and the Hall coefficient R measured down to the liquid helium temperature. In uniform samples, both techniques gave almost identical curves for the carrier concentration. A fast drop of the electron concentration below room temperature was found in the annealed samples, as shown in Figure 4.37. This is a clear manifestation of deep donor production by annealing. The samples turned out to be strongly nonuniform, as evidenced by an extremely low Hall mobility of the electrons: μ_n at room temperature is about 100 cm^2/V.s, as compared to the normal value of about 1600 cm^2/V.s found in uniform n-type samples of pure silicon. Also, the temperature dependence of μ_n is anomalous in comparison to a simple power law found in uniform, pure n-type silicon.

Ultra-high-purity silicon with high concentrations of interstitial oxygen has been proposed by Crouse et al. [324] using the FZ technique. In this material, interstitial oxygen concentration [O] approximately equal to 9.0×10^{17} cm^{-3} was achieved while maintaining good crystalline structure and a ratio of oxygen to shallow impurities greater than 10^5. This method provides efficient oxygen "doping" by exposing only the melted zone during boule growth. Results of characterization using IR absorption, photoluminescence spectroscopy, and transport and crystal quality measurements were also presented by the team. This method is of fundamental importance in the investigation of thermal donors in silicon, where contamination from Group III and Group V impurities, as well as carbon, complicates the role of oxygen in the formation of thermal donors. Good numerical modelings [325, 326] are carried out on the microscopic inhomogeneities and the resistivity distribution in the FZ-grown crystals.

During the growth of FZ crystals, two types of swirl defects (type A and type B clusters) are formed as a result of point defect condensation. Section topographic analysis indicated that the type A clusters rapidly increase in size with a decreasing crystal cooling rate [327]. Evidence was presented that the predominant cause for the loss of dislocation-free growth, particularly of large-diameter crystals, is due to emission of dislocation arrays from the type A clusters. We shall discuss this more in later chapters.

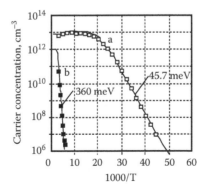

FIGURE 4.37
Temperature dependence of the carrier concentration deduced from the two-probe resistivity; curve "a": reference boron-doped sample of resistivity 1500 Ω-cm; curve "b": after annealing at 680°C for 90 min. (From G. I. Voronkova, A. V. Batunina, V. V. Voronkov, R. Falster, L. Moiraghi, and M. G. Milvidski, "Effects of annealing on the electrical properties of nitrogen-doped float-zoned silicon," *210 ECE Meeting Cancun*, Mexico, 29 Oct–3 Nov, 2006 [323].)

4.4.6 Impurity Segregation between Liquid and Grown Silicon Crystals

The influence of important growth parameters on the shape of the solid–liquid interfaces of the FZ and melt shape was explained by different groups. The interdependence of all the parameters is so great that it is difficult to single out any one. The macroscopic radial distribution of phosphorus dopant was found to vary by a factor of 40% in the case of (100) and 70% in the case of (111). Growth striations were also observed in the FZ due to inhomogeneous incorporation of doped impurities; this indicates that the formation is mainly due to remelt phenomena of grown crystal and solid-state diffusion at the solid–liquid interface [328]. The FZ silicon crystals were quenched from the melt to prevent impurity clustering and precipitation, and to minimize their effects on the swirl defect nucleation. In this case, the silicon self-interstitials are the dominant point defects during this process. Dislocation-free silicon crystals grown by FZ techniques generally contain two types of vacancy clusters located in a strained pattern. It was established that the formation of clusters predominantly depends on the crystal cooling rate [329]. This is mainly due to crystal rotation and melt turbulence, which promote the trapping of nonequilibrium vacancy concentrations at the solid–liquid interface.

4.4.7 Use of Magnetic Fields for Float-Zone Growth

Robertson and O'Connor [282] have studied the magnetic field effects on FZ silicon crystals. Transverse magnetic fields up to 5500 G have been applied during the growth of Ga-doped silicon crystals. Crystals have been grown at 4 mm/min at rotation rates in the range of 0–13 rpm. It was found that the strong fields have pronounced effects on the growth interface shape, on the Ga impurity distribution, and on the crystal morphology. For nonrotating crystals, the magnetic field produced a crystal with an elliptical cross-section, with the major axis aligned with the field direction. In these crystals, the interface assumes a cylindrical shape, with the generatrix aligned along the field direction. Oscillations in dopant concentration with a frequency of approximately two per minute are seen in the nonrotating crystals. It was also found that the strong magnetic fields have pronounced effects on the growth interface shape, on the Ga impurity distribution, and on the crystal morphology. They also carried out axial magnetic field studies on FZ crystals ranging in strength up to 5000 G for Ga-doped <100> crystals [330]. The crystals were grown at a pull rate of 4 mm/min, with rotation rates ranging from 0–12 rpm. Evaluation by spreading resistance showed a marked reduction in the impurity dopant concentration fluctuations for certain growth conditions in axial fields greater than 3000 G. However, in the strong fields, the growth axis wandered randomly and made such crystals difficult to grow, especially when the rotation rates were reduced. A crystal growth without rotation in a 5000 G field gave the most uniform doping, but only in a central core region of the crystal extending over about one-third of the grown diameter.

De Leon et al. [331] investigated the effects of applying a magnetic field during the crystal growth of zero-etch-pit-density (0-EPD) high-resistivity silicon crystals using FZ-inverted pedestal techniques. They demonstrated that magnetic fields as low as 80 G have a significantly effect on radial resistivity distribution and the resistivity ratios approaching unity in 42 mm diameter *n*-type crystals. In contrast, the nucleation process governing the striation content is not affected by fields below 180 G.

Morthland and Walker [332] presented numerical solutions for the thermocapillary convection during the FZ growth of silicon crystals with a steady, externally applied magnetic field. Three FZ shapes were treated: the "barrel shape" corresponding to growth in microgravity, the "bottle shape" corresponding to terrestrial growth without an induction coil, and a "necked shape" corresponding to growth with an RF induction coil. The magnetic field is either uniform or axial, or it is nonuniform and axisymmetric. Two nonuniform fields were considered. The results for the bottle shape with a uniform magnetic field help explain recent experimental results presented earlier in the published data. The nonuniform fields show how the radial dopant nonuniformities in these experimental results can be avoided. The results for the barrel shape and for both uniform and nonuniform magnetic fields point to the benefits of combining micro-gravity and a magnetic field. Since molten silicon is a good electrical conductor, an externally applied, a steady magnetic field can eliminate the unsteadiness in the melt motion and can reduce the magnitude of the residual steady motion. According to their findings, the best results are obtained with the field lines fringing outward and the field strength decreasing, as one can move from the crystal face to the feed rod face for all three FZ shapes that were considered here.

The feasibility of modulating dopant segregation using rotation for FZ silicon growth in axisymmetric magnetic fields was investigated by Lan and Liang [333] through computer simulation. In the model, heat and mass transfer, fluid flow, magnetic fields, melt-solid interfaces, and the free surface were solved globally by a robust finite-volume/Newton method. Different rotation modes, single and counterrotations, were applied to the growth under both axial and cusp magnetic fields. The schematic of the system is shown in Figure 4.38. The magnetic fields were generated by superconducting wire coils carrying different currents. Because the convection in the molten zone is greatly suppressed by magnetic fields, the occurrence of oscillatory and three-dimensional flows is less likely. Therefore, the FZ crystal growth is described by an axisymmetric convective heat and mass transfer model. The flow, temperature, dopant concentration, and free surface ($R_m(z)$), as well as the shapes of the feed front ($h_f(r)$) and the growth front ($h_c(r)$), are represented in a cylindrical coordinate system (r,z). The dopant is assumed to be uniformly distributed in the feed rod, and its concentration (C_0) is small; highly doped situations can also be considered if necessary. Under the magnetic fields, dopant mixing is poor in the quiescent core region of the molten zone, and the weak convection there is responsible for the segregation. Under an axial magnetic field, moderate counterrotation or crystal rotation improves dopant uniformity. However, excess counterrotation or feed rotation alone results in more complicated flow structures and thus induces larger radial segregation. For the cusp fields, rotation can enhance more easily the dopant mixing in the core melt and thus improve dopant uniformity. Application of different magnetic fields in the FZ has drawn a lot of attention as a means to control the oxygen segregation and dopant uniformity issues. Many numerical simulations [334–339] are being carried out to study the entire process to obtain good-quality, perfect single silicon crystals.

4.4.8 Large Area Silicon Crystals and Limitations of Shape and Size

The interface shape, heat transfer, and fluid flow in the FZ growth of large (>100 mm) silicon crystals using the needle-eye technique along with feed and crystal rotation were reported by Mühlbauer et al. [305]. Natural convection, thermocapillary convection, EM forces, and rotation in the melt parameters were considered for this study. The unknown shape of the molten zone was calculated as a coupled thermal-electromagnetic-hydrodynamic problem and compared with that observed during the experiments. The effects of the growth rate and the process stage on the shape of the interface were demonstrated with practical results. It was observed that natural convection and rotation dominate over thermocapillary and EM convection, at least for conditions corresponding to industrial FZ silicon production with the needle-eye technique. It was further shown that under these

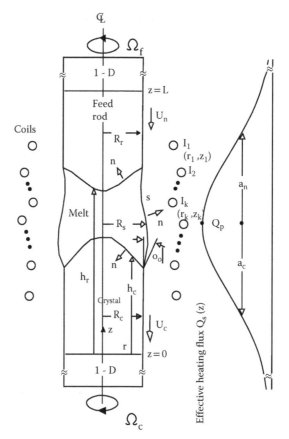

FIGURE 4.38
Schematic sketch of FZ silicon crystal growth in magnetic fields. (From C. W. Lan and M. C. Liang, *Journal of Crystal Growth*, **180**, 381–387, 1997, and the references therein [333].)

conditions, the rotation destabilizes the flow and only unsteady flows exist in the molten zone. The calculated distributions of the oscillation amplitude of the tangential velocity at the growing interface correspond to the radial resistivity distributions measured in the single crystal by the photo-scanning technique.

A single picture was taken from the video record to obtain the experimental shapes of the melting front and the free surface, as shown in Figure 4.39a. To obtain the resistivity distribution and the shape of the growth interface, the grown crystal was cut longitudinally and photo-scanning measurements in this cross-section of the grown crystal were taken. The surface of the cut was scanned in the vertical direction with a resolution of 0.1 mm with a radial step of 1.0 mm. The measured signals were proportional to the vertically differentiated local resistivity in the crystal. An example of the measured signal field in the vertical cross-section of the grown crystal is shown in Figure 4.39b [305]. The intensity of the measured signal is shown by the brightness of the measured point. It can be seen that the resistivity striations describe the shape of the growing interface. The design of the inductor has a distinct influence on the shape of all the interfaces of the molten zone with the needle-eye technique.

The photograph taken during crystal growth and the photo-scanning measurements are used to obtain the experimental shape of the melting front, the free surface, and the growth

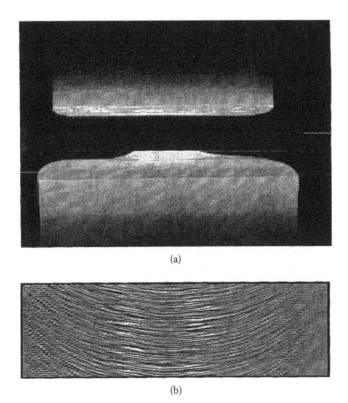

(a)

(b)

FIGURE 4.39
Experiment: (a) photograph of the FZ process with a pancake induction coil; (b) photo-scanning resistivity measurements in a vertical cross-section. (From A. Mühlbauer, A. Muiznieks, J. Virbulis, A. Lüdge, and H. Riemann, *Journal of Crystal Growth*, **151**, 66–79, 1995, and the references therein [305].)

front. The corresponding interfaces are shown in Figure 4.40 as dots. The figure reveals that the calculated shapes of the interfaces are in good agreement with the observed ones. It also shows the influence of the melt convection on the shape of the growth interface. In the system considered, the rotation destabilizes the flow and only oscillatory flows exist in the molten zone. As a consequence, the heat transfer through the melt is also unsteady. The "a" shape calculated by taking into account the unsteady heat transfer from the melt (averaged over time) is compared with the calculated "b" shape of the interface in the case without convection in the melt. The interface "c" calculated for the melt flow without rotation (i.e., steady-state flow with $Re_{F,C} = 0$) is also shown in the figure. The growth interface in the case of unstable solution "a" is similar to the interface calculated by neglecting the melt convection "b." For more details, see the original research paper by the author.

The influence of the crystal growth rate on the shape of the interfaces is shown in Figure 4.41 [305]. The growth rate varies from $v = 2.5$ to 3.6 mm/min, and the current in the inductor changes so that the upper edge of the crystal (melt/crystal/gas triple point) is a constant distance from the inductor. At a higher growth rate, the growing interface is more concave and the melting interface is located closer to the inductor. Obviously, when the growth rate is higher, it becomes more complicated to melt the center of the feed rod. In the worst case, direct contact between the feed rod and the crystal may occur. This behavior is well known with the FZ growth technique. Natural convection and rotation dominate in the molten zone for the conditions examined in the present study (i.e., corresponding

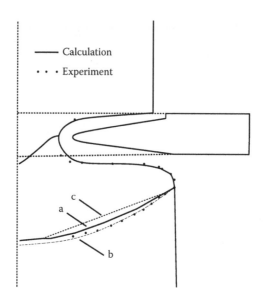

FIGURE 4.40
Comparison between experimental (dots) and calculated (lines) interfaces of the molten zone in the case of the basic growth stage. Calculated growth interface: (a) in the real case with the unsteady melt convection; (b) by neglecting the melt convection; (c) in the case with stable convection by neglecting the crystal rotation. (From A. Mühlbauer, A. Muiznieks, J. Virbulis, A. Lüdge, and H. Riemann, *Journal of Crystal Growth*, **151**, 66–79, 1995, and the references therein [305].)

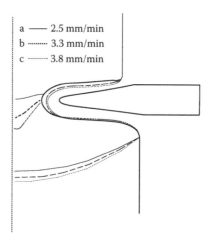

FIGURE 4.41
Effect of the growth rate on the shape of interfaces: (a) $v_c = 2.5$; (b) $v_c = 3.3$; (c) $v_c = 3.6$ mm/min. (From A. Mühlbauer, A. Muiznieks, J. Virbulis, A. Lüdge, and H. Riemann, *Journal of Crystal Growth*, **151**, 66–79, 1995, and the references therein [305].)

to commercial FZ production of large silicon crystals with the needle-eye technique). The flow is unstable for these conditions, and the velocity and temperature fields have oscillations in the melt. It is possible to obtain a stable flow for typical parameters if the growing crystal is rotated faster. The convection in the molten zone influences the heat transfer through the growth front and the shape of this interface. The averaged-in-time unsteady heat flux through the growth front differs from that in the case of steady-state flow without rotation, and is similar to the heat flux without any convection at all in the melt. The results are quite accurate and provided authentic information for growing large-size FZ crystals.

A computer simulation was carried out to study the dopant concentration fields in the molten zone and in the growing crystal for the FZ growth of large (>100 mm) Si crystals with the needle-eye technique and with feed/crystal rotation [322]. The mathematical model developed in the previous work was used to calculate the shape of the molten zone and the velocity field in the melt. The influence of melt convection on the dopant concentration field was considered. The significance of the rotation scheme of the feed rod and crystal on the dopant distribution was investigated. The calculated dopant concentration directly at the growth interface was used to determine the normalized lateral resistivity distribution in the single crystal. The calculated resistivity distributions were compared with lateral spreading resistivity measurements in the single crystal.

4.4.9 Thermal Gradient and Post-Growth Crystal Cooling

Joshi [340] has reported the effects of fast cooling on the silicon crystals. According to the practical observations, the misfit dislocations in phosphorus-diffused silicon suffer rearrangement during fast cooling. These dislocations extend in specific directions and act as centers of phosphorus precipitates. There are also other areas of precipitation where the particles are of irregular shape and size. The motion of a dislocation network in the depths of the diffusion profile is mainly by glide, although a less dominant climbing motion is observed during cooling. Rohatgi and Rai-Choudhury [341] reported process-induced effects on carrier lifetime and defects in FZ silicon. There are several other issues in terms of defect formation and the nature of defects formed during the thermal gradient of these crystals. The defects and defect-related issues are discussed in Chapter 9 of this book.

4.5 Zone Refining of Single-Crystal Silicon

Zone refining is quite similar to the FZ method in which a narrow region of a crystal is molten, and this molten zone is moved along the crystal. The molten region melts impure solid at its forward edge and leaves a wake of purer material solidified behind it as it moves through the ingot. The impurities present in the crystal get into the molten silicon, and they concentrate in the melt and are moved to one end of the ingot because of the segregation properties between solid and liquid phases. This technique was developed to purify semiconductors and is one of the more common methods now used in many scientific fields.

Metallic impurities in metallurgical-grade silicon were removed substantially by zone refining because of their low segregation coefficients in silicon. Chu et al. [342] reported that solar cells prepared from the metallurgical silicon purified by two zone-pass refining, though somewhat inferior to those from semiconductor-grade silicon, have reasonable conversion efficiencies. Zone melting recrystallizations are adapted to grow single-crystal silicon layers on silicon dioxide layers to obtain silicon-on-insulator (SOI) substrate wafers [343]. This method is also useful to avoid precipitation of carbon and nitrogen present in multicrystalline silicon [344]. Prakash et al. [345] reported on a new process of preparing shaped multicrystalline silicon ingots in graphite molds based on the directional solidification technique. The mold has a coating of silicon oxynitride instead of the more commonly used silicon nitride on the inner side of its walls. The coating acts as a mold-releasing layer and makes the mold reusable. It also enables the use of demountable molds made by assembling two or more parts together. It is expected that the repeatable use of molds made of graphite sheets will promote a considerable savings in the production cost of multicrystalline silicon ingots in industrial practice. Three-dimensional simulation approaches [346–348] are being implemented to better understand this technique and use it to better refine silicon crystals useful for VLSI and ULSI devices.

4.6 Other Silicon Crystalline Structures and Growth Techniques

Gereth [349] grew silicon bicrystals in an inert atmosphere, avoiding quartz crucibles. The oxygen content of the bicrystals was reported to be below the limit of detection with the 9.0-μm absorption line. The Dash pedestal method was modified, and the bicrystals were pulled from the melt by means of a "bi-seed," which contains a small angle grain boundary. An FZ scanner was converted into a crystal grower. A drill press etching apparatus facilitates the etching of the seed crystals to a needlelike shape, with the grain boundary in the center of the 0.1 mm diameter of the tip. The physical dimensions of these crystals were small, and no devices have been reported based on these crystals.

4.6.1 Silicon Ribbons

The ribbons are uniform and have flat faces that approximate parallel planes. It has been shown that at least two twin planes parallel to and approximately midway between the flat faces are necessary for continuous ribbon growth to take place. Ribbon growth involves two propagation mechanisms: one for lengthwise growth and one for lateral growth. Dermatis et al. [350] first reported the growth of silicon ribbons and their surface morphology. They were grown from a super-cooled melt and examined by various techniques. From the morphology of the ribbons, dislocation patterns, twin structures, and microsegregation traces, it was shown that the basic growth sequence is similar to that of germanium. Some differences were found, however. Micro-segregation traces were used to elucidate the growth sequence of degenerate ribbons. With flexible electronics around, there is intense work surrounding these silicon ribbons at present [351–355], and this has become an important field of research.

Sarma and Rice Jr. [356] have reported a new high-pressure plasma (HPP) deposition process, operating at a pressure of about 1 atm, characterized for its potential for producing polycrystalline silicon ribbons. They have examined silicon tetrachloride, trichlorosilane, dichlorosilane, and silane as silicon source gases. Use of silane was found to be impractical due to excessive gas-phase nucleation and it led to powder growth. Improvements in deposition efficiencies were observed with HPP—they were found to be highest at lower reactant gas concentration values. Modeling and computational simulations are reported [357, 358] on these silicon ribbons, which are expected to take a lead over flat polycrystalline silicon panels for photovoltaic applications. The silicon ribbon technology offers a less expensive manufacturing technique, but is not capable of achieving the same electrical performance as the standard wafer technology.

4.6.2 Silicon Sheets

Sanjurjo [359] has reported that the cost of photovoltaic devices can be lowered substantially if efficient photovoltaic solar cells can be fabricated from polycrystalline silicon, rather than from single-crystal silicon. This was achieved by solidifying molten silicon as unsupported sheets. A silicon sheet was formed in a graphite die without the undesirable formation of SiC. Silicon was melted in the presence of a molten salt that wets the silicon, preventing it from reacting with the graphite walls. The silicon coalesced into a pool that was flattened to a disk by a graphite piston. Silicon was then solidified in this condition to form flat sheets. Ciszek et al. [360] have reported that by using the edge-supported pulling method for silicon sheet growth, parallel filaments immersed in the liquid localize the meniscus edges and provide exceptionally stable sheet growth conditions. The filaments are withdrawn from the liquid integral with the solidifying silicon. The silicon sheets grown by this method were made into solar cells with AM1 conversion efficiencies greater than 13%. These sheets continued to be studied, particularly in terms of solar cell applications [361–365] and for low-cost and low-efficiency flexible panels.

4.6.3 Silicon Whiskers and Fibers

Silicon fibers were grown by Osada et al. [366] by thermal decomposition of silane at temperatures of 580°C to 700°C on a silicon substrate that had been contaminated by dust in air or aluminum in HF + NH_4F solution. They were about 50 to 60 microns in length and a few thousand angstroms in diameter for a growth time of 3 min. According to electron-diffraction analyses, the fiber was composed of an almost single-crystalline pith grown in the direction of <110> by a vapor-liquid-solid (VLS) mechanism, with an amorphous shell deposited around it. Similarly, silicon whiskers were reported and were grown by the molecular beam epitaxy method [367–369]. Not much information is available on the use of these thin, stand-alone structures.

4.6.4 Silicon in Circular and Spherical Shapes

Zhang et al. [370] have reported the growth of single-crystalline silicon cylinders (SCSC) on a silicon (100) surface by using a novel surface melting and recrystallizing method. The structure was standing sturdily in a molten silicon spot and could grow to several millimeters in height and several hundred micrometers in diameter. Analysis showed that the crystal orientation of the SCSC was the same as that of the silicon (100) substrate. Spherical single crystals for solar cell substrates have been grown successfully by Huang et al. [371], with a yield of almost 100%. Spherical silicon multicrystals with diameters of approximately 400 μm in a teardrop shape were initially fabricated by a dropping method. The as-dropped spherical silicon multicrystals were melted into droplets on a silica plate in an oxygen atmosphere, and the droplets were then recrystallized to form single crystals by supercooling within a specific temperature range. Similar results are also reported by Liu et al. [372] to grow spherical single crystals of silicon. These structures are designed for solar cell applications, but it is too early to conclude if they will dominate this industry in the near future.

4.6.5   Silicon Hollow Tubes

Preparation of silicon tubes and boats was suggested as a semiconductor device technology to replace the quartz material presently being used. Silicon tubes and boats can be made from monocrystalline or polycrystalline bulk silicon by coring, sawing, and grinding methods. This is, however, an expensive and difficult process due to the brittle nature of the material. Dietze et al. [373] have reported that silicon tubes with dimensions up to 1.8 m long by 100 mm inside diameter can be prepared by the chemical vapor deposition (CVD) route on the correspondingly shaped surface of heated graphite, which can be separated from the silicon afterward. Similarly, silicon hollow bodies with a noncircular cross-section can be produced, making possible the preparation of a variety of different silicon boats for supporting silicon slices during high-temperature diffusion and oxidation processes. CVD silicon is considered superior to conventional quartz in these applications for reasons of purity, permeability to trace impurities, and mechanical stability at high temperatures.

4.6.6   Casting of Polycrystalline Silicon for Photovoltaic Applications

Polycrystalline silicon cast by the ubiquitous crystallization process is a suitable candidate sheet material for low-cost solar cell applications. Hyland et al. [374] reported fabrication of solar cells from silicon ingots using this process. Impurities and defects, including grain boundaries, particles, and dislocations, control the quality of solar cell performance. In the bottom of the ingot, impurities either are incorporated into the grain boundaries or are rejected into the melt during the growth. Vizman et al. [375] have reported on the influence of important process parameters, such as the size of the mold, crystallization rate, and slight tilting of the mold (1°–3°), on the shape of the solid–liquid interface. This was studied for an industrial silicon ingot casting process by 3D, time-dependent simulations using the STHAMAS3D software package. It turns out that melt convection has a strong influence on the results, and these performances are to be considered in the simulations prior to casting. Wang et al. [376] came out with a novel recycling approach to utilize the cutting kerf loss of silicon for a directional solidification process, thus minimizing the losses of high-purity silicon material. Presently, there are efforts to utilize the polysilicon casting for solar cell fabrication activity [377, 378] by using metallurgical silicon to achieve the best possible efficiency.

4.7   Summary

In this chapter, we have discussed different techniques used to grow the single crystals of silicon. The Bridgman, Czochralski, float-zone, and zone-refining methods were discussed. The Czochralski (CZ) pulling method is the main technique used at present to grow single crystals of silicon, and very large crystals are possible only with this method. Polysilicon feed is used as a raw material, and the entire charge is kept inside the quartz crucible supported by graphite. By using the graphite heater, the charge is melted to the desired temperature and dopant species are added at this stage to the silicon melt to get the desired resistivity values. When using a seed crystal, it is first allowed to touch the molten silicon and then is slowly pulled up, allowing the molten silicon to solidify according to the seed crystal. In this process, the liquid silicon container (quartz) and graphite heaters become the source of oxygen and carbon contamination for the growing silicon crystal.

With the float-zone (FZ) process, crucible contamination is eliminated and the crystal is grown vertically with the help of a seed and an RF induction heating element to melt the silicon locally. The crystal is allowed to grow by holding the molten silicon, and a polysilicon rod feeds the growing silicon in this process. This method eliminates the crucible contamination and provides better crystals than does the CZ method. Large crystals are difficult to grow with this method, and there are limitations with the induction coil shape.

Zone refining is used to get better crystals similar to the FZ method. Several repetitions provide better silicon crystals. In addition to these techniques, we covered the other crystalline structures of silicon, including ribbons, sheets, whiskers, fibers, and circular and spherical shapes, and the casting of polycrystalline silicon for photovoltaic applications.

References

1. J. D. Plummer, M. D. Deal and P. B. Griffin, *Silicon Crystal Structure and Growth*, Chapter 3. www.usna.edu/EE/ee452/LectureNotes/05-Processing_Technology/18_Silicon.ppt.
2. P. S. Ravishankar, L. P. Hunt, and R. W. Francis, "Effective segregation coefficient of boron in silicon ingots grown by the Czochralski and Bridgman techniques," *Journal of the Electrochemical Society*, **131**, 872–874, 1984.
3. G. Fisher, M. R. Seacrist, and R. W. Standley, "Silicon crystal growth and wafer technologies," *Proceedings of the IEEE*, **100**, 1454–1474, 2012, and the references therein.
4. M. S. Kulkarni, V. V. Voronkov, and R. Falster, "Quantification of defect dynamics in unsteady-state and steady-state Czochralski growth of monocrystalline silicon," *Journal of the Electrochemical Society*, **151**, G663–G678, 2004, and the references therein.
5. H.-T. Chung, S.-C. Lee, and J.-K. Yoon, "Numerical prediction of operational parameters in Czochralski growth of large-scale Si," *Journal of Crystal Growth*, **163**, 249–258, 1996, and the references therein.
6. M. Mihelčić, C. Schröck-Pauli, K. Wingerath, H. Wenzl, W. Uelhoff, and A. van der Hart, "Numerical simulation of free and forced convection in the classical Czochralski method and in CACRT," *Journal of Crystal Growth*, **57**, 300–317, 1982.
7. G. Williams and R. E. Reusser, "Heat transfer in silicon Czochralski crystal growth," *Journal of Crystal Growth*, **64**, 448–460, 1983.
8. J. A. Stefani, J. K. Tien, K. S. Choe, and J. P. Wallace, "Eddy current monitoring system and data reduction protocol for Czochralski silicon crystal growth," *Journal of Crystal Growth*, **88**, 30–38, 1988.
9. R. A. Brown, T. A. Kinney, P. A. Sackinger, and D. E. Bornside, "Towards an integrated analysis of Czochralski growth," *Journal of Crystal Growth*, **97**, 99–115, 1989.
10. L. J. Atherton, J. J. Derby, and R. A. Brown, "Radiative heat exchange in Czochralski crystal growth," *Journal of Crystal Growth*, **84**, 57–78, 1987.
11. D. R. Hamilton, D. L. Barrett, H. Wehrli, and A. I. Bennett, "Surface tensions, moving melts and the harmful effects on crystal growth," *Journal of Crystal Growth*, **7**, 296–300, 1970.
12. G. K. Fraundorf and L. Shive, "Transmission electron microscope study of quartz crucibles used in growth of Czochralski silicon," *Journal of Crystal Growth*, **102**, 157–166, 1990.
13. H. Sreedharamurthy, "Reducing iron in single crystal silicon grown using CZ process," *210 ECE Meeting Cancun*, Mexico, 29 Oct–3 Nov, 2006, 43085_0_art_file_0_1149187433.
14. F. Schmid, C. P. Khattak, T. G. Digges, Jr., and L. Kaufman, "Origin of SiC impurities in silicon crystals grown from the melt in vacuum," *Journal of the Electrochemical Society*, **126**, 935–938, 1979.
15. T. F. Ciszek, "Growth of 40-mm diameter silicon crystals by a pedestal technique using electron beam heating," *Journal of Crystal Growth*, **12**, 281–287, 1972.

16. M. S. Bawa, W. J. Bell, H. M. Grimes, and T. J. Shaffner, "Temperature dependence of oxidation induced stacking faults and oxygen incorporation in dislocation-free Czochralski silicon," *Journal of Crystal Growth*, **94**, 803–806, 1989.

17. K.-W. Yi, K. Kakimoto, M. Eguchi, and H. Noguchi, "Oxygen transport mechanism in Si melt during single crystal growth in the Czochralski system," *Journal of Crystal Growth*, **165**, 358–361, 1996, and the references therein.

18. K. E. Benson and W. Lin, "The role of oxygen in silicon for VLSI," *Journal of Crystal Growth*, **70**, 602–608, 1984.

19. A. Seidl and G. Müller, "Oxygen solubility in silicon melt measured in situ by an electrochemical solid ionic sensor," *Journal of the Electrochemical Society*, **144**, 3243–3245, 1997, and the references therein.

20. G. K. Fraundorf and L. Shive, "Transmission electron microscope study of quartz crucibles used in growth of Czochralski silicon," *Journal of Crystal Growth*, **102**, 157–166, 1990.

21. R. W. Series and K. G. Barraclough, "Control of carbon in Czochralski silicon crystals," *Journal of Crystal Growth*, **63**, 219–221, 1983.

22. F. Schmid, C. P. Khattak, T. G. Digges, Jr., and L. Kaufman, "Origin of SiC impurities in silicon crystals grown from the melt in vacuum," *Journal of the Electrochemical Society*, **126**, 935–938, 1979.

23. M. P. Bellmann, E. A. Meese, M. Syvertsen, A. Solheim, H. Sørheim, and L. Arnberg, "Silica versus silicon nitride crucible: Influence of thermophysical properties on the solidification of multi-crystalline silicon by Bridgman technique," *Journal of Crystal Growth*, **318**, 265–268, 2011, and the references therein.

24. R. Deike and K. Schwerdtfeger, "Reactions between liquid silicon and different refractory materials," *Journal of the Electrochemical Society*, **142**, 609–614, 1995.

25. Y. Y. Khine and J. S. Walker, "Centrifugal pumping during Czochralski silicon growth with a strong, non-uniform, axisymmetric magnetic field," *Journal of Crystal Growth*, **165**, 372–380, 1996, and the references therein.

26. M. Watanabe, M. Eguchi, K. Kakimoto, H. Ono, S. Kimura, and T. Hibiya, "Flow mode transition and its effects on crystal-melt interface shape and oxygen distribution for Czochralski-grown Si single crystals," *Journal of Crystal Growth*, **151**, 285–290, 1995, and the references therein.

27. K. Kakimoto, M. Eguchi, H. Watanabe, and T. Hibiya, "Direct observation by x-ray radiography of convection of molten silicon in the Czochralski growth method," *Journal of Crystal Growth*, **88**, 365–370, 1988.

28. K. Kakimoto, M. Eguchi, H. Watanabe, and T. Hibiya, "Natural and forced convection of molten silicon during Czochralski single crystal growth," *Journal of Crystal Growth*, **94**, 412–420, 1989.

29. K. Kakimoto, M. Eguchi, H. Watanabe, and T. Hibiya, "Flow instability of molten silicon in the Czochralski configuration," *Journal of Crystal Growth*, **102**, 16–20, 1990.

30. K. Kakimoto, P. Nicodème, M. Lecomte, F. Dupret, and M. J. Crochet, "Numerical simulation of molten silicon flow: Comparison with experiment," *Journal of Crystal Growth*, **114**, 715–725, 1991.

31. J. R. Ristorcelli and J. L. Lumley, "A second-order turbulence simulation of the Czochralski crystal growth melt: The buoyantly driven flow," *Journal of Crystal Growth*, **129**, 249–265, 1993.

32. J. Järvinen, R. Nieminen, and T. Tiihonen, "Time-dependent simulation of Czochralski silicon crystal growth," *Journal of Crystal Growth*, **180**, 468–476, 1997, and the references therein.

33. M. Tanaka, M. Hasebe, and N. Saito, "Pattern transition of temperature distribution at Czochralski silicon melt surface," *Journal of Crystal Growth*, **180**, 487–496, 1997, and the references therein.

34. A. Mühe and G. Müller, "Optical in-situ measurement of the dissolution rate of a silica-Czochralski-crucible with silicon melt and comparison to ex-situ measurements," *Microelectronic Engineering*, **56**, 147–152, 2001.

35. S.M. Schnurre and R. Schmid-Fetzer, "Reactions at the liquid silicon/silica glass interface," *Journal of Crystal Growth*, **250**, 370–381, 2003.

36. T. Minami, S. Maeda, M. Higasa, and K. Kashima, "In-situ observation of bubble formation at silicon melt-silica glass interface," *Journal of Crystal Growth*, **318**, 196–199, 2011.
37. R.E. Chaney and C.J. Varker, "The erosion of materials in molten silicon," *Journal of the Electrochemical Society*, **123**, 846–852, 1976.
38. R.E. Chaney, "Comparison of the erosion of vitreous carbon and high density graphite in molten silicon," *Journal of the Electrochemical Society*, **124**, 1460–1461, 1977.
39. B. Bathey, H.E. Bates, and M. Cretella, "Effect of carbon on the dissolution of fused silica in liquid silicon," *Journal of the Electrochemical Society*, **127**, 771–772, 1980.
40. M. Ishimaru, S. Munetoh, T. Motooka, K. Moriguchi, and A. Shintani, "Behavior of impurity atoms during crystal growth from melted silicon: carbon atoms," *Journal of Crystal Growth*, **194**, 178–188, 1998.
41. X. Huang, S. Koh, K. Wu, M. Chen, T. Hoshikawa, K. Hoshikawa, and S. Uda, "Reaction at the interface between Si melt and a Ba-doped silica crucible," *Journal of Crystal Growth*, **277**, 154–161, 2005.
42. L. Raabe, O. Pätzold, I. Kupka, J. Ehrig, S. Würzner, and M. Stelter, "The effect of graphite components and crucible coating on the behavior of carbon and oxygen in multicrystalline silicon," *Journal of Crystal Growth*, **318**, 234–238, 2011.
43. U. Ekhult, "A time dependent model for oxygen behavior in a Czochralski silicon melt," *Journal of the Electrochemical Society*, **136**, 3494–3501, 1989.
44. U. Ekhult and T. Carlberg, "Oxygen solubility in liquid silicon in equilibrium with SiO and SiO$_2$," *Journal of the Electrochemical Society*, **136**, 551–556, 1989.
45. A. Seidl, R. Marten, and G. Müller, "Development of an electrochemical oxygen sensor for Czochralski silicon melts," *Journal of the Electrochemical Society*, **141**, 2564–2566, 1994.
46. S. Maeda, K. Takeuchi, M. Kato, K. Abe, H. Nakanishi, K. Hoshikawa, and K. Terashima, "Morphology variations on inner surface of silica crucibles depending on oxygen concentration in silicon melts," *Journal of Crystal Growth*, **194**, 70–75, 1998.
47. Y. Kishida and K. Okazawa, "Geostrophic turbulence in CZ silicon crucible," *Journal of Crystal Growth*, **198–199**, 135–140, 1999.
48. H. Nakanishi, M. Watanabe, and K. Terashima, "Dependence of Si melt flow in a crucible on surface tension variation in the Czochralski process," *Journal of Crystal Growth*, **236**, 523–528, 2002.
49. G. N. Kozhemyakin, "Imaging of convection in a Czochralski crucible under ultrasound waves," *Journal of Crystal Growth*, **257**, 237–244, 2003.
50. J. R. Carruthers, "Temperature oscillations in Czochralski crystal growth," *Journal of the Electrochemical Society*, **114**, 1077, 1967.
51. K. M. Kim, A. F. Witt, and H. C. Gatos, "Crystal growth from the melt under destabilizing thermal gradients," *Journal of the Electrochemical Society*, **119**, 1218–1226, 1972.
52. S. Nakamura, M. Eguchi, T. Azami, and T. Hibiya, "Thermal waves of a nonaxisymmetric flow in a Czochralski-type silicon melt," *Journal of Crystal Growth*, **207**, 55–61, 1999.
53. H. S. Woo, J. H. Jeong, and I. S. Kang, "Optimization of surface temperature distribution for control of point defects in the silicon single crystal," *Journal of Crystal Growth*, **247**, 320–332, 2003.
54. Y.-R. Li, N. Imaishi, Y. Akiyama, L. Peng, S.-Y. Wu, and T. Tsukada, "Effects of temperature coefficient of surface tension on oxygen transport in a small CZ furnace," *Journal of Crystal Growth*, **266**, 48–53, 2004.
55. T. F. Ciszek, "Some applications of cold crucible technology for silicon photovoltaic material preparation," *Journal of the Electrochemical Society*, **132**, 963–968, 1985.
56. D.-X. Du and T. Munakata, "Temperature distribution in an inductively heated CZ crucible," *Journal of Crystal Growth*, **283**, 563–575, 2005.
57. H. Sreedharamurthy, "Reducing iron in single crystal silicon grown using CZ process," *210 ECE Meeting Cancun*, Mexico, 29 Oct–3 Nov, 2006.
58. D. Gilmore, T. Arahori, M. Ito, H. Murakami, and S.-i. Miki, "The impact of graphite furnace parts on radial impurity distribution in Czochralski-grown single-crystal silicon," *Journal of the Electrochemical Society*, **145**, 621–628, 1998.

59. E. Dornberger, E. Tomzig, A. Seidl, S. Schmitt, H.-J. Leister, Ch. Schmitt, and G. Müller, "Thermal simulation of the Czochralski silicon growth process by three different models and comparison with experimental results," *Journal of Crystal Growth*, **180**, 461–467, 1997, and the references therein.

60. K. S. Choe, J. A. Stefani, J. K. Tien, and J. P. Wallace, "Eddy current measurement of crystal axial thermal profiles during Czochralski silicon crystal growth," *Journal of Crystal Growth*, **88**, 39–52, 1988.

61. A. Lipchin and R. A. Brown, "Hybrid finite-volume/finite-element simulation of heat transfer and melt turbulence in Czochralski crystal growth of silicon," *Journal of Crystal Growth*, **216**, 192–203, 2000.

62. F. Barvinschi, T. Duffar, and J. L. Santailler, "Numerical simulation of heat transfer in transparent and semitransparent crystal growth processes," *Journal of Optoelectronics and Advanced Materials*, **2**, 327–331, 2000.

63. K. Takano, Y. Shiraishi, J. Matsubara, T. Iida, N. Takase, N. Machida, M. Kuramoto, and H. Yamagishi, "Global simulation of the CZ silicon crystal growth up to 400 mm in diameter," *Journal of Crystal Growth*, **229**, 26–30, 2001.

64. D. Vizman, J. Friedrich, and G. Müller, "Comparison of the predictions from 3D numerical simulation with temperature distributions measured in Si Czochralski melts under the influence of different magnetic fields," *Journal of Crystal Growth*, **230**, 73–80, 2001.

65. M. Li, Y. Li, N. Imaishi, and T. Tsukada, "Global simulation of a silicon Czochralski furnace," *Journal of Crystal Growth*, **234**, 32–46, 2002.

66. M. H. Tavakkoli and H. Wilke, "Numerical study of induction heating and heat transfer in a real Czochralski system," *Journal of Crystal Growth*, **275**, e85–e89, 2005.

67. O. V. Smirnova, N. V. Durnev, K. E. Shandrakova, E. L. Mizitov, and V. D. Soklakov, "Optimization of furnace design and growth parameters for Si CZ growth, using numerical simulation," *Journal of Crystal Growth*, **310**, 2185–2191, 2008.

68. V. V. Kalaev, I. Yu. Evstratov, and Yu. N. Makarov, "Gas flow effect on global heat transport and melt convection in Czochralski silicon growth," *Journal of Crystal Growth*, **249**, 87–99, 2003.

69. W. C. Dash, "Growth of silicon crystals free from dislocations," *Journal of Applied Physics*, **30**, 459, 1959.

70. S. K. Ghandhi, *VLSI Fabrication Principles: Silicon and Gallium Arsenide*, John Wiley & Sons, New York, 1994; C. W. Pearce, "Crystal growth and wafer preparation" in VLSI Technology, edited by S.M. Sze, McGraw-Hill, New York, 1988; O. Anttila, "Challenges of Silicon Materials Research: Manufacture of High Resistivity, Low Oxygen Czochralski Silicon," Okmetic Fellow, Okmetic Oyj, epp.final.gov/DocDB/0000/000007/001/CERN_RD-50_06_2005_Okmetic_OA.ppt, June 2005, and the references therein.

71. L. D. Dyer, "Dislocation-free Czochralski growth of <110> silicon crystals," *Journal of Crystal Growth*, **47**, 533–540, 1979.

72. X. Huang, T. Taishi, I. Yonenaga, and K. Hoshikawa, "Dislocation-free B-doped Si crystal growth without Dash necking in Czochralski method: Influence of B concentration," *Journal of Crystal Growth*, **213**, 283–287, 2000.

73. T. Taishi, X. Huang, I. Yonenaga, and K. Hoshikawa, "Dislocation-free Czochralski Si crystal growth without a thin neck: Dislocation behavior due to incomplete seeding," *Journal of Crystal Growth*, **258**, 58–64, 2003.

74. K. Hoshikawa, X. Huang, and T. Taishi, "Heavily doped silicon crystals: Neckless growth and robust wafers," *Journal of Crystal Growth*, **275**, 276–282, 2005.

75. T. Hoshikawa, T. Taishi, X. Huang, S. Uda, M. Yamatani, K. Shirasawa, and K. Hoshikawa, "Si multicrystals grown by the Czochralski method with multi-seeds," *Journal of Crystal Growth*, **307**, 466–471, 2007.

76. M. R. L. N. Murthy and J. J. Aubert, "Growth of dislocation-free silicon crystals in the <110> direction for use as neutron monochromators," *Journal of Crystal Growth*, **52**, 391–395, 1981.

77. M. S. Kulkarni, V. Voronkov, and R. Falster, "Dynamics of point defects and formation of microdefects in Czochralski crystal growth: Modeling, simulation and experiments," *Future Fab International*, Issue 14, Section 7, and the references therein.

78. R. Balasubramaniam and S. Ostrach, "Transport phenomena near the interface of a Czochralski-grown crystal," *Journal of Crystal Growth*, **88**, 263–281, 1988.
79. T. Motooka, "Molecular dynamics simulations of crystal growth from melted silicon," 25–29_end.pdf. *Proceedings of the First Symposium on Atomic-scale Surface and Interface Dynamics*, March 13–14, 1997, Tokyo, and the references therein.
80. U. Ekhult, T. Carlberg, and M. Tilli, "Infra-red assisted Czochralski growth of silicon crystals," *Journal of Crystal Growth*, **98**, 793–800, 1989.
81. A. M. J. G. Van Run, "A critical pulling rate for remelt suppression in silicon crystal growth," *Journal of Crystal Growth*, **53**, 441–442, 1981.
82. U. Ekhult and T. Carlberg, "Oxygen and carbon incorporation during infra-red assisted Czochralski growth of silicon crystals," *Journal of Crystal Growth*, **98**, 801–809, 1989.
83. K. S. Choe, J. A. Stefani, T. B. Dettling, J. K. Tien, and J. P. Wallace, "Effects of growth conditions on thermal profiles during Czochralski silicon crystal growth," *Journal of Crystal Growth*, **108**, 262–276, 1991.
84. W. Zhou, D. E. Bornside, and R. A. Brown, "Dynamic simulation of Czochralski crystal growth using an integrated thermal-capillary model," *Journal of Crystal Growth*, **137**, 26–31, 1994.
85. K. Kakimoto, "Flow instability during crystal growth from the melt," *Progress in Crystal Growth and Characterization of Materials*, **30**, 191–215, 1995.
86. S. Nakamura, M. Eguchi, T. Azami, and T. Hibiya, "Thermal waves of a nonaxisymmetric flow in a Czochralski-type silicon melt," *Journal of Crystal Growth*, **207**, 55–61, 1999.
87. A. Lipchin and R. A. Brown, "Comparison of three turbulence models for simulation of melt convection in Czochralski crystal growth of silicon," *Journal of Crystal Growth*, **205**, 71–91, 1999.
88. K. Fujiwara, K. Nakajima, T. Ujihara, N. Usami, G. Sazaki, H. Hasegawa, S. Mizoguchi, and K. Nakajima, "In situ observations of crystal growth behavior of silicon melt," *Journal of Crystal Growth*, **243**, 275–282, 2002.
89. Z. Jian, K. Nagashio, and K. Kuribayashi, "Direct observation of the crystal-growth transition in undercooled silicon," *Metallurgical and Materials Transactions A*, **33A**, 2947–2953, 2002.
90. Y.-R. Li, N. Imaishi, T. Azami, and T. Hibiya, "Three-dimensional oscillatory flow in a thin annular pool of silicon melt," *Journal of Crystal Growth*, **260**, 28–42, 2004.
91. S. Enger, O. Gräbner, G. Müller, M. Breuer, and F. Durst, "Comparison of measurements and numerical simulations of melt convection in Czochralski crystal growth of silicon," *Journal of Crystal Growth*, **230**, 135–142, 2001.
92. J. R. Rujano, R. A. Crane, M. M. Rahman, and W. Moreno, "Numerical analysis of stabilization techniques for oscillatory convective flow in Czochralski crystal growth," *Journal of Crystal Growth*, **245**, 149–162, 2002.
93. Y. Shiraishi, S. Maeda, and K. Nakamura, "Prediction of solid-liquid interface shape during CZ Si crystal growth using experimental and global simulation," *Journal of Crystal Growth*, **266**, 28–33, 2004.
94. K. M. Kim and P. Smetana, "Maximum length of large diameter Czochralski silicon single crystals at fracture stress limit of seed," *Journal of Crystal Growth*, **100**, 527–528, 1990.
95. N. Miyazaki, H. Uchida, T. Munakata, K. Fujioka, and Y. Sugino, "Thermal stress analysis of silicon bulk single crystal during Czochralski growth," *Journal of Crystal Growth*, **125**, 102–111, 1992.
96. A. G. Cullis, N. G. Chew, H. C. Webber, and D. J. Smith, "Orientation dependence of high speed silicon crystal growth from the melt," *Journal of Crystal Growth*, **68**, 624–638, 1984.
97. T. G. Digges, Jr. and R. Shima, "The effect of growth rate, diameter and impurity concentration on structure in Czochralski silicon crystal growth," *Journal of Crystal Growth*, **50**, 865–869, 1980.
98. S. Yasuami, M. Ogino, and S. Takasu, "The swirl formation of defects in Czochralski-grown silicon crystals," *Journal of Crystal Growth*, **39**, 227–230, 1977.
99. Y. Shiraishi, S. Kurosaka, and M. Imai, "Silicon crystal growth using a liquid-feeding Czochralski method," *Journal of Crystal Growth*, **166**, 685–688, 1996, and the references therein.

100. E. Dornberger, E. Tomzig, A. Seidl, S. Schmitt, H.-J. Leister, Ch. Schmitt, and G. Müller, "Thermal simulation of the Czochralski silicon growth process by three different models and comparison with experimental results," *Journal of Crystal Growth*, **180**, 461–467, 1997, and the references therein.

101. E. Dornberger, W. von Ammon, N. Van den Bogaert, and F. Dupret, "Transient computer simulation of a CZ crystal growth process," *Journal of Crystal Growth*, **166**, 452–457, 1996, and the references therein.

102. R. W. Series and K. G. Barraclough, "Control of carbon in Czochralski silicon crystals," *Journal of Crystal Growth*, **63**, 219–221, 1983.

103. K. Suzuki, H. Yamawaki, and Y. Tada, "Boron out diffusion from Si substrates in various ambients," *Solid-State Electronics*, **41**, 1095–1097, 1997.

104. K. S. Choe and T. H. Strudwick, "Pregrowth ambient gas analysis of Czochralski Si crystal puller," *Journal of the Electrochemical Society*, **135**, 706–710, 1988.

105. K. Izunome, X. Huang, S. Togawa, K. Terashima, and S. Kimura, "Control of oxygen concentration in heavily antimony-doped Czochralski Si crystals by ambient argon pressure," *Journal of Crystal Growth*, **151**, 291–294, 1995, and the references therein.

106. N. Machida, Y. Suzuki, K. Abe, N. Ono, M. Kida, and Y. Shimizu, "The effects of argon gas flow rate and furnace pressure on oxygen concentration in Czochralski-grown silicon crystals," *Journal of Crystal Growth*, **186**, 362–368, 1998.

107. N. Machida, K. Hoshikawa, and Y. Shimizu, "The effects of argon gas flow rate and furnace pressure on oxygen concentration in Czochralski silicon single crystals grown in a transverse magnetic field," *Journal of Crystal Growth*, **210**, 532–540, 2000.

108. T. Azami and T. Hibiya, "Interpreting the oxygen partial pressure around a molten silicon drop in terms of its surface tension," *Journal of Crystal Growth*, **233**, 417–424, 2001.

109. S. Maeda and K. Terashima, "Macroscopic uniformity of oxygen concentration in Czochralski silicon crystals closely related to evaporation of SiO from the free surface of melts," *Journal of the Electrochemical Society*, **150**, G319–G326, 2003.

110. Y.-Y. Teng, J.-C. Chen, C.-W. Lu, H.-I Chen, C. Hsu, and C.-Y. Chen, "Effects of the furnace pressure on oxygen and silicon oxide distributions during the growth of multicrystalline silicon ingots by the directional solidification process," *Journal of Crystal Growth*, **318**, 224–229, 2011.

111. M. S. Kulkarni, "Defect dynamics in the presence of nitrogen in growing Czochralski silicon crystals," *Journal of Crystal Growth*, **310**, 324–335, 2008.

112. X. Yu, D. Yang, and K. Hoshikawa, "Investigation of nitrogen behaviors during Czochralski silicon crystal growth," *Journal of Crystal Growth*, **318**, 178–182, 2011.

113. T. Fukuda, M. Koizuka, and A. Ohsawa, "A Czochralski silicon growth technique which reduces carbon to the order of 10^{14} per cubic centimeter," *Journal of the Electrochemical Society*, **141**, 2216–2220, 1994.

114. D. E. Bornside, R. A. Brown, T. Fujiwara, H. Fujiwara, and T. Kubo, "The effects of gas-phase convection on carbon contamination of Czochralski-grown silicon," *Journal of the Electrochemical Society*, **142**, 2790–2804, 1995.

115. V. V. Kalaev, I. Yu. Evstratov, and Yu. N. Makarov, "Gas flow effect on global heat transport and melt convection in Czochralski silicon growth," *Journal of Crystal Growth*, **249**, 87–99, 2003.

116. T. Miyano, S.-i. Inami, A. Shintani, T. Kanda, and M. Hourai, "Dynamical behaviour of Czochralski melts and its influence on crystal growth," *Journal of Crystal Growth*, **166**, 469–475, 1996, and the references therein.

117. W. Sittenthaler, "Wafer suppliers face dilemma in doing much more with less," *Wafer News*, July 2005.

118. W. Miller, U. Rehse, and K. Böttcher, "Influence of melt convection on the interface during Czochralski crystal growth," *Solid-State Electronics*, **44**, 825–830, 2000.

119. K. Kakimoto, M. Eguchi, H. Watanabe, and T. Hibiya, "In-situ observation of solid-liquid interface shape by x-ray radiography during silicon single crystal growth," *Journal of Crystal Growth*, **91**, 509–514, 1988.

120. E. Dornberger, W. von Ammon, N. Van den Bogaert, and F. Dupret, "Transient computer simulation of a CZ crystal growth process," *Journal of Crystal Growth*, **166**, 452–457, 1996, and the references therein.

121. S. Togawa, K. Izunome, S. Kawanishi, S.-I. Chung, K. Terashima, and S. Kimura, "Oxygen transport from a silica crucible in Czochralski silicon growth," *Journal of Crystal Growth*, **165**, 362–371, 1996, and the references therein.

122. I. Kanda, T. Suzuki, and K. Kojima, "Influence of crucible and crystal rotation on oxygen-concentration distribution in large-diameter silicon single crystals," *Journal of Crystal Growth*, **166**, 669–674, 1996, and the references therein.

123. K. Sugawara and H. Tochikubo, "A method to improve the radial impurity distribution in Czochralski silicon crystals," *Journal of the Electrochemical Society*, **124**, 951–952, 1977.

124. I. H. Jafri, V. Prasad, A. P. Anselmo, and K. P. Gupta, "Role of crucible partition in improving Czochralski melt conditions," *Journal of Crystal Growth*, **154**, 280–292, 1995, and the references therein.

125. V. V. Voronkov and R. Falster, "Intrinsic point defects and impurities in silicon crystal growth," *Journal of the Electrochemical Society*, **149**, G167–G174, 2002, and the references therein.

126. A. E. Kokh and N. G. Kononova, "Crystal growth under heat field rotation conditions," *Solid-State Electronics*, **44**, 819–824, 2000, and the references therein.

127. F.-U. Brückner and K. Schwerdtfeger, "Single crystal growth with the Czochralski method involving rotational electromagnetic stirring of the melt," *Journal of Crystal Growth*, **139**, 351–356, 1994.

128. K.-M. Kim, "Materials: Silicon-pulling technology for 2000+," *Solid State Technology*, January 2000, and the references therein.

129. O. A. Louchev, "The influence of radiative heat transfer on the limit pull rate in Czochralski crystal growth of silicon," *Journal of Crystal Growth*, **129**, 179–190, 1993.

130. J. R. Carruthers and K. E. Benson, "Solute striations in Czochralski-grown silicon crystals: Effect of crystal rotation and growth rates," *Applied Physics Letters*, **3**, 100–102, 1963, and the references therein.

131. C. T. Yen and W. A. Tiller, "Oxygen partitioning analysis during Czochralski silicon crystal growth via a dopant marker and a simple transfer function modeling technique. I. Rotation rate transients," *Journal of Crystal Growth*, **109**, 142–148, 1991.

132. H. H. Lee, "Ultimate limitation on Czochralski pull rate due to constitutional supercooling," *Journal of the Electrochemical Society*, **134**, 971–975, 1987.

133. A. Kokh, "Crystal growth through forced stirring of melt or solution in Czochralski configuration," *Journal of Crystal Growth*, **191**, 774–778, 1998.

134. J. Furukawa, H. Tanaka, Y. Nakada, N. Ono, and H. Shiraki, "Investigation on grown-in defects in CZ-Si crystal under slow pulling rate," *Journal of Crystal Growth*, **210**, 26–30, 2000.

135. J.-S. Kim and T.-y. Lee, "Numerical study on the effect of operating parameters on point defects in a silicon crystal during Czochralski growth. I. Rotation effect," *Journal of Crystal Growth*, **219**, 205–217, 2000.

136. A. Natsume, N. Inoue, K. Tanahashi, and A. Mori, "Dependence of temperature gradient on growth rate in CZ silicon," *Journal of Crystal Growth*, **225**, 221–224, 2001,

137. S.-S. Son and K.-W. Yi, "Experimental study on the effect of crystal and crucible rotations on the thermal and velocity field in a low Prandtl number melt in a large crucible," *Journal of Crystal Growth*, **275**, e249–e257, 2005.

138. S.-S. Son, P.-O. Nam, and K.-W. Yi, "The effect of crystal rotation direction on the thermal and velocity fields of a Czochralski system with a low Prandtl number melt," *Journal of Crystal Growth*, **292**, 272–282, 2006.

139. S. Nakano, L. J. Liu, X. J. Chen, H. Matsuo, and K. Kakimoto, "Effect of crucible rotation on oxygen concentration in the polycrystalline silicon grown by the unidirectional solidification method," *Journal of Crystal Growth*, **311**, 1051–1055, 2009.

140. P. K. Kulshreshtha, Y. Yoon, K. M. Youssef, E. A. Good, and G. Rozgonyi, "Oxygen precipitation related stress-modified crack propagation in high growth rate Czochralski silicon wafers," *Journal of the Electrochemical Society*, **159**, H125–H129, 2012.

141. R. K. Srivastava, P. A. Ramachandran, and M. P. Duduković, "Czochralski growth of crystals: Simple models for growth rate and interface shape," *Journal of the Electrochemical Society*, **133**, 1009–1015, 1986.
142. B. M. Park, G. H. Seo, and G. Kim, "Effects of pulling rate fluctuation on the interstitial-vacancy boundary formation in CZ-Si single crystal," *Journal of Crystal Growth*, **203**, 67–74, 1999.
143. L. Liu and K. Kakimoto, "Effects of crystal rotation rate on the melt-crystal interface of a CZ-Si crystal growth in a transverse magnetic field," *Journal of Crystal Growth*, **310**, 306–312, 2008.
144. O. A. Noghabi, M. M'Hamdi, and M. Jomâa, "Effect of crystal and crucible rotations on the interface shape of Czochralski grown silicon single crystals," *Journal of Crystal Growth*, **318**, 173–177, 2011.
145. K. M. Beatty and K. A. Jackson, "Monte Carlo modeling of silicon crystal growth," *Journal of Crystal Growth*, **211**, 13–17, 2000.
146. A. Yu. Gelfgat, A. Rubinov, P. Z. Bar-Yoseph, and A. Solan, "Numerical study of three-dimensional instabilities in a hydrodynamic model of Czochralski growth," *Journal of Crystal Growth*, **275**, e7–e13, 2005.
147. H. Kodera, "Diffusion coefficients of impurities in silicon melt," *Japanese Journal of Applied Physics*, **2**, 212–219, 1963.
148. Z. Liu and T. Carlberg, "On the mechanism of oxygen content reduction by antimony doping of Czochralski silicon melts," *Journal of the Electrochemical Society*, **138**, 1488–1492, 1991.
149. S. Maeda, M. Kato, K. Abe, H. Nakanishi, K. Terashima, and K. Hoshikawa, "Temperature variations of the surface of a silicon melt due to evaporation of chemical species: I. Antimony addition," *Journal of the Electrochemical Society*, **144**, 3185–3188, 1997.
150. K. Abe, K. Terashima, T. Matsumoto, S. Maeda, and H. Nakanishi, "Fused quartz dissolution rate in silicon melts: Influence of boron addition," *Journal of Crystal Growth*, **186**, 557–564, 1998.
151. S. Maeda, M. Kato, H. Nakanishi, K. Hoshikawa, and K. Terashima, "Evaporation of oxygen-containing species from boron-doped silicon melts," *Journal of the Electrochemical Society*, **145**, 2548–2552, 1998.
152. H. M. Liaw, "Interface shape and radial distribution of impurities in <111> silicon crystals," *Journal of Crystal Growth*, **67**, 261–270, 1984.
153. A. J. R. de Kock and W. M. van de Wijgert, "The effect of doping on the formation of swirl defects in dislocation-free Czochralski-grown silicon crystals," *Journal of Crystal Growth*, **49**, 718–734, 1980.
154. E. Dornberger, D. Gräf, M. Suhren, U. Lambert, P. Wagner, F. Dupret, and W. von Ammon, "Influence of boron concentration on the oxidation-induced stacking fault ring in Czochralski silicon crystals," *Journal of Crystal Growth*, **180**, 343–352, 1997, and the references therein.
155. T. Izumi, "Model analysis of segregation phenomena for silicon single crystal growth from the melt," *Journal of Crystal Growth*, **181**, 210–217, 1997, and the references therein.
156. T. Hoshikawa, T. Taishi, S. Oishi, and K. Hoshikawa, "Investigation of methods for doping CZ silicon with gallium," *Journal of Crystal Growth*, **275**, e2141–e2145, 2005.
157. T. Hoshikawa, X. Huang, S. Uda, and T. Taishi, "Segregation of Ga during growth of Si single crystal," *Journal of Crystal Growth*, **290**, 338–340, 2006.
158. X. Huang, M. Arivanandhan, R. Gotoh, T. Hoshikawa, and S. Uda, "Ga segregation in Czochralski-Si crystal growth with B codoping," *Journal of Crystal Growth*, **310**, 3335–3341, 2008.
159. M. Arivanandhan, R. Gotoh, K. Fujiwara, T. Ozawa, Y. Hayakawa, and S. Uda, "The impact of Ge codoping on grown-in O precipitates in Ga-doped Czochralski-silicon," *Journal of Crystal Growth*, **321**, 24–28, 2011.
160. N. Inoue, K. Shingu, and K. Masumoto, "Measurement of nitrogen concentration in CZ silicon," presented at *N1-Ninth International Symposium on Silicon Materials Science and Technology*, 201st Meeting, #621, Philadelphia, PA, May 12–17, 2002, and the references therein.
161. Q. Ma, D. Yang, X. Ma, and D. Que, "Oxygen precipitation of nitrogen-doped Czochralski silicon subjected to multi-step thermal process," *210 ECE Meeting Cancun*, Mexico, 29 Oct–3 Nov, 2006.

162. N. Fujita, R. Jones, S. Öberg, and P. R. Briddon, "Identification of nitrogen-oxygen defects in silicon," *210 ECE Meeting Cancun*, Mexico, 29 Oct–3 Nov, 2006.
163. A. Huber, M. Kapser, J. Grabmeier, U. Lambert, W. v. Ammon, and R. Pech, "Impact of nitrogen doping in silicon onto gate oxide integrity," presented at *N1-Ninth International Symposium on Silicon Materials Science and Technology*, 201st Meeting, #573, Philadelphia, PA, May 12–17, 2002.
164. V.V. Voronkov and R. Falster, "Nitrogen diffusion in Silicon: A multi-species process," *210 ECE Meeting Cancun*, Mexico, 29 Oct–3 Nov, 2006, and the references therein.
165. X. Yu, D. Yang, X. Ma, L. Li, and D. Que, "Hydrogen annealing of grown-in voids in nitrogen-doped Czochralski grown silicon," *Semiconductor Science and Technology*, **18**, 399–403, 2003.
166. D. Tian, D. Yang, X. Ma, L. Li, and D. Que, "Crystal growth and oxygen precipitation behavior of 300 mm nitrogen-doped Czochralski silicon," *Journal of Crystal Growth*, **292**, 257–259, 2006, and the references therein.
167. Q. Zhou, F. Qin, J. Zhou, F. Fang, J. Wang, and H. Tu, "Growth technology for 200 mm antimony heavily doped silicon single crystals," #0576, presented at *N1-Ninth International Symposium on Silicon Materials Science and Technology*, 201st Meeting, Philadelphia, PA, May 12–17, 2002.
168. K. M. Kim, "Interface morphological instability in Czochralski silicon crystal growth from heavily Sb-doped melt," *Journal of the Electrochemical Society*, **126**, 875–878, 1979.
169. X. Huang, K. Terashima, K. Izunome, and S. Kimura, "Effect of antimony-doping on the oxygen segregation coefficient in silicon crystal growth," *Journal of Crystal Growth*, **149**, 59–63, 1995, and the references therein.
170. K. Izunome, X. Huang, S. Togawa, K. Terashima, and S. Kimura, "Control of oxygen concentration in heavily antimony-doped Czochralski Si crystals by ambient argon pressure," *Journal of Crystal Growth*, **151**, 291–294, 1995, and the references therein.
171. D. Yang, J. Chen, H. Li, X. Ma, D. Tian, L. Li, and D. Que, "Micro-defects in Ge doped Czochralski grown Si crystals," *Journal of Crystal Growth*, **292**, 266–271, 2006, and the references therein.
172. J. Chen, D. Yang, H. Li, X. Ma, and D. Que, "Germanium effect on as-grown oxygen precipitation in Czochralski silicon," *Journal of Crystal Growth*, **291**, 66–71, 2006, and the references therein.
173. F. W. Voltmer and T. G. Digges, Jr., "Anomalous resistivity profiles in long silicon crystals grown by the Czochralski method," *Journal of Crystal Growth*, **19**, 215–217, 1973.
174. Z.A. Salnick, Y.A. Miklyaev, O.S. Salnick, A.V. Naumov, and A.V. Phomichev, "Untraditionally doped CZ-grown silicon for power devices," *Journal of Crystal Growth*, **172**, 120–123, 1997, and the references therein.
175. Z. A. Salnick, Y. A. Miklyaev, O. S. Salnick, A. V. Naumov, and A. V. Phomichev, "Untraditionally doped CZ-grown silicon for power devices," *Journal of Crystal Growth*, **172**, 120–123, 1997, and the references therein.
176. S. J. Baek, J. S. Chang, E. S. Choi, and H. H. Lee, "Axial dopant distribution and its control in bulk crystal growth," *Journal of Crystal Growth*, **131**, 481–485, 1993.
177. A. F. Witt and H. C. Gatos, "Homogeneous impurity incorporation during crystal growth from the melt," *Journal of the Electrochemical Society*, **116**, 511–513, 1969.
178. G. F. Wakefield, "Inhomogeneities in silicon crystals grown from the melt," *Journal of the Electrochemical Society*, **127**, 1139–1143, 1980.
179. P. S. Ravishankar, L. P. Hunt, and R. W. Francis, "Effective segregation coefficient of boron in silicon ingots grown by the Czochralski and Bridgman techniques," *Journal of the Electrochemical Society*, **131**, 872–874, 1984.
180. M. Ishimaru, S. Munetoh, T. Motooka, K. Moriguchi, and A. Shintani, "Behavior of impurity atoms during crystal growth from melted silicon: carbon atoms," *Journal of Crystal Growth*, **194**, 178–188, 1998.
181. H.-D. Chiou, "Phosphorus concentration limitation in Czochralski silicon crystals," *Journal of the Electrochemical Society*, **147**, 345–349, 2000.

182. D. Yang, C. Li, M. Luo, J. Xu, and D. Que, "Reduction of oxygen during the crystal growth in heavily antimony-doped Czochralski silicon," *Journal of Crystal Growth*, **256**, 261–265, 2003.

183. A. Murgai, A. F. Witt, and H. C. Gatos, "Elimination of random compositional inhomogeneities in Czochralski grown silicon," *Journal of the Electrochemical Society*, **122**, 1276–1277, 1975.

184. J. H. Wang, "Resistivity distribution of silicon single crystals using codoping," *Journal of Crystal Growth*, **280**, 408–412, 2005.

185. D. Yang, J. Chen, X. Ma, and D. Que, "Impurity engineering of Czochralski silicon used for ultra large-scaled-integrated circuits," *Journal of Crystal Growth*, **311**, 837–841, 2009.

186. A. Anselmo, V. Prasad, J. Koziol, and K. P. Gupta, "Numerical and experimental study of a solid pellet feed continuous Czochralski growth process for silicon single crystals," *Journal of Crystal Growth*, **131**, 247–264, 1993.

187. C. Wang, H. Zhang, T. Wang, and L. Zheng, "Solidification interface shape control in a continuous Czochralski silicon growth system," *Journal of Crystal Growth*, **287**, 252–257, 2006.

188. A. Anselmo, J. Koziol, and V. Prasad, "Full-scale experiments on solid-pellets feed continuous Czochralski growth of silicon crystals," *Journal of Crystal Growth*, **163**, 359–368, 1996, and the references therein.

189. Y. Shiraishi, S. Kurosaka, and M. Imai, "Silicon crystal growth using a liquid-feeding Czochralski method," *Journal of Crystal Growth*, **166**, 685–688, 1996, and the references therein.

190. N. Ono, M. Kida, Y. Arai, and K. Sahira, "Thermal analysis of the double-crucible method in continuous silicon Czochralski processing: I. Experimental analysis," *Journal of the Electrochemical Society*, **140**, 2101–2105, 1993.

191. N. Ono, M. Kida, Y. Arai, and K. Sahira, "Thermal analysis of the double-crucible method in continuous silicon Czochralski processing: II. Numerical analysis," *Journal of the Electrochemical Society*, **140**, 2106–2111, 1993.

192. J. H. Wang, D. H. Kim, and H.-D. Yoo, "Two-dimensional analysis of axial segregation in batchwise and continuous Czochralski process," *Journal of Crystal Growth*, **198–199**, 120–124, 1999.

193. B. Fickett and G. Mihalik, "Multiple batch recharging for industrial CZ silicon growth," *Journal of Crystal Growth*, **225**, 580–585, 2001.

194. A. Voigt, C. Weichmann, and K.-H. Hoffmann, "Multiscale simulations of industrial crystal growth, *Proceedings of Algoritmy*, Conference on Scientific Computing, 1–13, 2002.

195. K. Fujiwara, K. Nakajima, T. Ujihara, N. Usami, G. Sazaki, H. Hasegawa, S. Mizoguchi, and K. Nakajima, "In situ observations of crystal growth behavior of silicon melt," *Journal of Crystal Growth*, **243**, 275–282, 2002.

196. N. V. Abrosimov, V. N. Kurlov, and S. N. Rossolenko, "Automated control of Czochralski and shaped crystal growth processing using weighing techniques," *Progress in Crystal Growth and Characterization of Materials*, **46**, 1–57, 2003.

197. C. Wang, H. Zhang, T. H. Wang, and T. F. Ciszek, "A continuous Czochralski silicon crystal growth system," *Journal of Crystal Growth*, **250**, 209–214, 2003.

198. P. R. Gunjal, M. S. Kulkarni, and P. A. Ramachandran, "Melt dynamics in Czochralski crystal growth of silicon," *210 ECE Meeting Cancun*, Mexico, 29 Oct–3 Nov, 2006, and the references therein.

199. K. Wada, H. Nakanishi, H. Takaoka, and N. Inoue, "Nucleation temperature of large oxide precipitates in as-grown Czochralski silicon crystal," *Journal of Crystal Growth*, **57**, 535–540, 1982.

200. E. Kuroda, "Temperature oscillation at the growth interface in silicon crystals," *Journal of Crystal Growth*, **61**, 173–176, 1983.

201. E. Kuroda and H. Kozuka, "Influence of growth conditions on melt interface temperature oscillations in silicon Czochralski growth," *Journal of Crystal Growth*, **63**, 276–284, 1983.

202. E. Kuroda, H. Kozuka, and Y. Takano, "The effect of temperature oscillations at the growth interface on crystal perfection," *Journal of Crystal Growth*, **68**, 613–623, 1984.

203. I.-G. Kim and J.-S. Kwon, "Reduction of grown-in defects by vacancy-assisted oxygen precipitation in high density dynamic access memory," *Applied Physics Letters*, **83**, 4863–4865, 2003.

204. P. Sabhapathy and M. E. Salcudean, "Numerical study of Czochralski growth of silicon in an axisymmetric magnetic field," *Journal of Crystal Growth*, **113**, 164–180, 1991.

205. C. T. Yen and W. A. Tiller, "Dynamic oxygen concentration in silicon melts during Czochralski crystal growth," *Journal of Crystal Growth*, **113**, 549–556, 1991.
206. S. Kawanishi, S. Togawa, K. Izunome, K. Terashima, and S. Kimura, "Melt quenching technique for direct observation of oxygen transport in the Czochralski-grown Si process," *Journal of Crystal Growth*, **152**, 266–273, 1995, and the references therein.
207. R. W. Series and K. G. Barraclough, "Carbon contamination during growth of Czochralski silicon," *Journal of Crystal Growth*, **60**, 212–218, 1982.
208. A. F. Witt and H. C. Gatos, "Homogeneous impurity incorporation during crystal growth from the melt," *Journal of the Electrochemical Society*, **116**, 511–513, 1969.
209. A. Murgai, H. C. Gatos, and A. F. Witt, "Quantitative analysis of microsegregation in silicon grown by the Czochralski method," *Journal of the Electrochemical Society*, **123**, 224–229, 1976.
210. S. N. Rea, J. D. Lawrence, and J. M. Anthony, "Effective segregation coefficient of germanium in Czochralski silicon," *Journal of the Electrochemical Society*, **134**, 752–753, 1987.
211. P. J. Ribeyron and F. Durand, "Oxygen and carbon transfer during solidification of semiconductor grade silicon in different processes," *Journal of Crystal Growth*, **210**, 541–553, 2000.
212. B.-C. Sim, K.-H. Kim, and H.-W. Lee, "Boron segregation control in silicon crystal ingots grown in Czochralski process," *Journal of Crystal Growth*, **290**, 665–669, 2006.
213. Y. Rosenstein and P. Z. Bar-Yoseph, "Three-dimensional instabilities in Czochralski process of crystal growth from silicon melt," *Journal of Crystal Growth*, **305**, 185–191, 2007.
214. W. Wijaranakula, "A real time simulation of point defect reactions near the solid and melt interface of a 200 mm diameter Czochralski silicon crystal," *Journal of the Electrochemical Society*, **140**, 3306–3316, 1993.
215. K. Kakimoto, S. Kikuchi, and H. Ozoe, "Molecular dynamics simulation of oxygen in silicon melt," *Journal of Crystal Growth*, **198–199**, 114–119, 1999.
216. J. H. Wang, D. H. Kim, and H.-D. Yoo, "Two-dimensional analysis of axial segregation in batchwise and continuous Czochralski process," *Journal of Crystal Growth*, **198–199**, 120–124, 1999.
217. K. Nishihira, S. Munetoh, and T. Motooka, "Uniaxial strain observed in solid/liquid interface during crystal growth from melted Si: A molecular dynamics study," *Journal of Crystal Growth*, **210**, 60–64, 2000.
218. I. Yu. Evstratov, V. V. Kalaev, A. I. Zhmakin, Yu. N. Makarov, A. G. Abramov, N. G. Ivanov, A. B. Korsakov, E. M. Smirnov, E. Dornberger, J. Virbulis, E. Tomzig, and W. v. Ammon, "Numerical study of 3D unsteady melt convection during industrial-scale CZ Si-crystal growth," *Journal of Crystal Growth*, **237–239**, 1757–1761, 2002.
219. D. P. Lukanin, V. V. Kalaev, Yu. N. Makarov, T. Wetzel, J. Virbulis, and W. von Ammon, "Advances in the simulation of heat transfer and prediction of the melt-crystal interface shape in silicon CZ growth," *Journal of Crystal Growth*, **266**, 20–27, 2004.
220. Y. Shiraishi, S. Maeda, and K. Nakamura, "Prediction of solid-liquid interface shape during CZ Si crystal growth using experimental and global simulation," *Journal of Crystal Growth*, **266**, 28–33, 2004.
221. Th. Wetzel, J. Virbulis, A. Muiznieks, W. von Ammom, E. Tomzig, G. Raming, and M. Weber, "Prediction of the growth interface shape in industrial 300 mm CZ Si crystal growth," *Journal of Crystal Growth*, **266**, 34–39, 2004.
222. L. Liu, K. Kakimoto, T. Taishi, and K. Hoshikawa, "Computational study of formation mechanism of impurity distribution in a silicon crystal during solidification," *Journal of Crystal Growth*, **265**, 399–409, 2004.
223. B. K. Jindal, V. V. Karelin, and W. A. Tiller, "Impurity striations in Czochralski grown Al-doped Si single crystals," *Journal of the Electrochemical Society*, **120**, 101–105, 1973.
224. A. J. R. De Kock, P. J. Roksnoer, and P. G. T. Boonen, "Formation and elimination of growth striations in dislocation-free silicon crystals," *Journal of Crystal Growth*, **28**, 125–137, 1975.
225. A. Shintani, T. Miyano, and M. Hourai, "A novel approach to the characterization of growth striations in Czochralski silicon crystals," *Journal of the Electrochemical Society*, **142**, 2463–2469, 1995.
226. K. S. Choe, "Growth striations and impurity concentrations in HMCZ silicon crystals," *Journal of Crystal Growth*, **262**, 35–39, 2004.

227. M. Imai, Y. Shiraishi, M. Shibata, H. Noda, and Y. Yatsurugi, "Quantitative measuring method of growth striations in Czochralski-grown silicon crystal," *Journal of the Electrochemical Society*, **135**, 1779–1783, 1988.

228. T. Azami, S. Nakamura, M. Eguchi, and T. Hibiya, "The role of surface-tension-driven flow in the formation of a surface pattern on a Czochralski silicon melt," *Journal of Crystal Growth*, **233**, 99–107, 2001.

229. J. Zhang, C. Liu, Q. Zhou, J. Wang, Q. Hao, H. Zhang, and Y. Li, "Investigation of flow pattern defects in as-grown and rapid thermal annealed CZ Si wafers," *Journal of Crystal Growth*, **262**, 1–6, 2004.

230. T. Taishi, X. Huang, I. Yonenaga, and K. Hoshikawa, "Behavior of the edge dislocation propagating along the growth direction in Czochralski Si crystal growth," *Journal of Crystal Growth*, **275**, e2147–e2153, 2005.

231. W. E. Langlois and K.-J. Lee, "Czochralski crystal growth in an axial magnetic field: Effects of joule heating," *Journal of Crystal Growth*, **62**, 481–486, 1983.

232. A. E. Organ and N. Riley, "Oxygen transport in magnetic Czochralski growth of silicon," *Journal of Crystal Growth*, **82**, 465–476, 1987.

233. L. N. Hjellming and J. S. Walker, "Mass transport in a Czochralski puller with a strong magnetic field," *Journal of Crystal Growth*, **85**, 25–31, 1987.

234. H. Hirata, K. Hoshikawa, and N. Inoue, "Improvement of thermal symmetry in CZ silicon melts by the application of a vertical magnetic field," *Journal of Crystal Growth*, **70**, 330–334, 1984.

235. R. W. Series, "Czochralski growth of silicon under an axial magnetic field," *Journal of Crystal Growth*, **97**, 85–91, 1989.

236. R. W. Series, "Effect of a shaped magnetic field on Czochralski silicon growth," *Journal of Crystal Growth*, **97**, 92–98, 1989.

237. S. Kobayashi, "Numerical analysis of oxygen transport in magnetic Czochralski growth of silicon," *Journal of Crystal Growth*, **85**, 69–74, 1987.

238. T. W. Hicks, A. E. Organ, and N. Riley, "Oxygen transport in magnetic Czochralski growth of silicon with a non-uniform magnetic field," *Journal of Crystal Growth*, **94**, 213–228, 1989.

239. K. Kakimoto, K.-W. Yi, and M. Eguchi, "Oxygen transfer during single silicon growth in Czochralski system with vertical magnetic fields," *Journal of Crystal Growth*, **163**, 238–242, 1996, and the references therein.

240. K. Kakimoto, M. Eguchi, and H. Ozoe, "Use of an inhomogeneous magnetic field for silicon crystal growth," *Journal of Crystal Growth*, **180**, 442–449, 1997, and the references therein.

241. H. Hirata and K. Hoshikawa, "Silicon crystal growth in a cusp magnetic field," *Journal of Crystal Growth*, **96**, 747–755, 1989.

242. H. Hirata and K. Hoshikawa, "Three-dimensional numerical analyses of the effects of a cusp magnetic field on the flows, oxygen transport and heat transfer in a Czochralski silicon melt," *Journal of Crystal Growth*, **125**, 181–207, 1992.

243. W. E. Langlois, K. M. Kim, and J. S. Walker, "Hydromagnetic flows and effects on Czochralski silicon crystals," *Journal of Crystal Growth*, **126**, 352–372, 1993.

244. Y. Y. Khine and J. S. Walker, "Buoyant convection during Czochralski silicon growth with a strong, non-uniform, axisymmetric magnetic field," *Journal of Crystal Growth*, **147**, 313–319, 1995, and the references therein.

245. H.-J. Cho, B.-Y. Lee, and J. Y. Lee, "The effects of several growth parameters on the formation behavior of point defects in Czochralski-grown silicon crystals," *Journal of Crystal Growth*, **292**, 260–265, 2006, and the references therein.

246. M. Watanabe, D. Vizman, J. Friedrich, and G. Müller, "Large modification of crystal-melt interface shape during Si crystal growth by using electromagnetic Czochralski method (EMCZ)," *Journal of Crystal Growth*, **292**, 252–256, 2006.

247. M. Watanabe, M. Eguchi, W. Wang, T. Hibiya, and S. Kuragaki, "Controlling oxygen concentration and distribution in 200 mm diameter Si crystals using the electromagnetic Czochralski (EMCZ) method," *Journal of Crystal Growth*, **237–239**, 1657–1662, 2002.

248. V. V. Kalaev, "Combined effect of DC magnetic fields and free surface stresses on the melt flow and crystallization front formation during 400 mm diameter Si CZ crystal growth," *Journal of Crystal Growth*, **303**, 203–210, 2007.

249. A. Krauze, N. Jēkabsons, A. Muižnieks, A. Sabanskis, and U. Lācis, "Applicability of LES turbulence modeling for CZ silicon crystal growth systems with traveling magnetic field," *Journal of Crystal Growth*, **312**, 3225–3234, 2010.

250. L. N. Hjellming, "A thermal model for Czochralski silicon crystal growth with an axial magnetic field," *Journal of Crystal Growth*, **104**, 327–344, 1990.

251. N. Kobayashi, "Oxygen transport under an axial magnetic field in Czochralski silicon growth," *Journal of Crystal Growth*, **108**, 240–246, 1991.

252. Z.A. Salnick, "Oxygen in Czochralski silicon crystals grown under an axial magnetic field," *Journal of Crystal Growth*, **121**, 775–780, 1992.

253. K. M. Kim and P. Smetana, "Oxygen segregation in CZ silicon crystal growth on applying a high axial magnetic field," *Journal of the Electrochemical Society*, **133**, 1682–1686, 1986.

254. L. N. Hjellming and J. S. Walker, "Mass transport in a Czochralski crystal puller with an axial magnetic field: Melt mouton due to crystal growth and buoyancy," *Journal of Crystal Growth*, **92**, 371–389, 1988.

255. J. S. Walker and M. G. Williams, "Effects of the crystal's non-zero electrical conductivity on the rotationally driven melt motion during Czochralski silicon growth with a uniform, transverse magnetic field," *Journal of Crystal Growth*, **132**, 31–42, 1993.

256. J. S. Walker and M. G. Williams, "Centrifugal pumping during Czochralski silicon growth with a strong transverse magnetic field," *Journal of Crystal Growth*, **137**, 32–36, 1994.

257. K. Hoshi, N. Isawa, T. Suzuki, and Y. Ohkubo, "Czochralski silicon crystals grown in a transverse magnetic field," *Journal of the Electrochemical Society*, **132**, 693–700, 1985.

258. J. M Hirtz and N. Ma, "Dopant transport during semiconductor crystal growth: Axial versus transverse magnetic fields," *Journal of Crystal Growth*, **210**, 554–572, 2000.

259. K. Kakimoto and H. Ozoe, "Oxygen distribution at a solid-liquid interface of silicon under transverse magnetic fields," *Journal of Crystal Growth*, **212**, 429–437, 2000.

260. N. Machida, K. Hoshikawa, and Y. Shimizu, "The effects of argon gas flow rate and furnace pressure on oxygen concentration in Czochralski silicon single crystals grown in a transverse magnetic field," *Journal of Crystal Growth*, **210**, 532–540, 2000.

261. L. Liu, S. Nakano, and K. Kakimoto, "An analysis of temperature distribution near the melt-crystal interface in silicon Czochralski growth with a transverse magnetic field," *Journal of Crystal Growth*, **282**, 49–59, 2005.

262. K. Kakimoto and L. Liu, "Partly three-dimensional calculation of silicon Czochralski growth with a transverse magnetic field," *Journal of Crystal Growth*, **303**, 135–140, 2007.

263. L. Liu and K. Kakimoto, "Effects of crystal rotation rate on the melt-crystal interface of a CZ-Si crystal growth in a transverse magnetic field," *Journal of Crystal Growth*, **310**, 306–312, 2008.

264. H. Hirata and K. Hoshikawa, "Homogeneous increase in oxygen concentration in Czochralski silicon crystals by a cusp magnetic field," *Journal of Crystal Growth*, **98**, 777–781, 1989.

265. Y. Y. Khine and J. S. Walker, "Melt-motion during the Czochralski growth of silicon crystals with a cusp magnetic field," *Journal of the Electrochemical Society*, **144**, 1861–1866, 1997.

266. M. Watanabe, M. Eguchi, and T. Hibiya, "Flow and temperature field in molten silicon during Czochralski crystal growth in a cusp magnetic field," *Journal of Crystal Growth*, **193**, 402–412, 1998.

267. M. Ma, T. Irisawa, T. Tsuru, T. Ogawa, M. Watanabe, and M. Eguchi, "Study on defects in CZ-Si crystals grown by normal, cusp magnetic field and electromagnetic field techniques using multi-chroic infrared light scattering tomography," *Journal of Crystal Growth*, **218**, 232–238, 2000.

268. K. Kakimoto, A. Tashiro, H. Ishii, and T. Shinozaki, "Mechanism of heat and oxygen transfer under electromagnetic CZ crystal growth with cusp-shaped magnetic fields," *Journal of the Electrochemical Society*, **150**, G648–G652, 2003.

269. B.-C. Sim, I.-K. Lee, K.-H. Kim, and H.-W. Lee, "Oxygen concentration in the Czochralski-grown crystals with cusp-magnetic field," *Journal of Crystal Growth*, **275**, 455–459, 2005.

270. Y. Kishida, T. Tamaki, K. Okazawa, and W. Ohashi, "Geostrophic turbulence in CZ silicon melt under CUSP magnetic field," *Journal of Crystal Growth*, **273**, 329–339, 2005.

271. Y.-H. Hong, B.-C. Sim, and K.-B. Shim, "Distribution coefficient of boron in Si crystal ingots grown in cusp-magnetic Czochralski process," *Journal of Crystal Growth*, **310**, 83–90, 2008.

272. K. M. Kim and W. E. Langlois, "Computer simulation of boron transport in magnetic Czochralski growth of silicon," *Journal of the Electrochemical Society*, **133**, 2586–2590, 1986.

273. M. Watanabe, K. W. Yi, T. Hibiya, and K. Kakimoto, "Direct observation and numerical simulation of molten silicon flow during crystal growth under magnetic fields by X-ray radiography and large-scale computation," *Progress in Crystal Growth and Characterization of Materials*, **38**, 215–238, 1999.

274. J. Virbulis, Th. Wetzel, A. Muiznieks, B. Hanna, E. Dornberger, E. Tomzig, A. Mühlbauer, and W. v. Ammon, "Numerical investigation of silicon melt flow in large diameter CZ-crystal growth under the influence of steady and dynamic magnetic fields," *Journal of Crystal Growth*, **230**, 92–99, 2001.

275. D. Vizman, J. Friedrich, and G. Müller, "Comparison of the predictions from 3D numerical simulation with temperature distributions measured in Si Czochralski melts under the influence of different magnetic fields," *Journal of Crystal Growth*, **230**, 73–80, 2001.

276. D. Vizman, O. Gräbner, and G. Müller, "3D numerical simulation and experimental investigations of melt flow in an Si Czochralski melt under the influence of a cusp-magnetic field," *Journal of Crystal Growth*, **236**, 545–550, 2002.

277. A. Krauze, A. Muiznieks, A. Mühlbauer, Th. Wetzel, L. Gorbunov, A. Pedchenko, and J. Virbulis, "Numerical 2D modelling of turbulent melt flow in CZ system with dynamic magnetic fields," *Journal of Crystal Growth*, **266**, 40–47, 2004.

278. A. Virzi, "Finite element analysis of the thermal history for Czochralski growth of large diameter silicon single crystals," *Journal of Crystal Growth*, **97**, 152–161, 1989.

279. K.-M. Kim, "Materials: Silicon-pulling technology for 2000+," *Solid State Technology*, January 2000, and the references therein.

280. Photograph courtesy Kayex Corporation. 300 mm silicon ingot. http://depts.washington.edu/chemcrs/bulkdisk/chem364A_spr05/handout_Silicon.pdf, and the references therein.

281. T. A. Kinney, D. E. Bornside, R. A. Brown, and K. M. Kim, "Quantitative assessment of an integrated hydrodynamic thermal-capillary model for large-diameter Czochralski growth of silicon: Comparison of predicted temperature field with experiment," *Journal of Crystal Growth*, **126**, 413–434, 1993.

282. G. D. Robertson, Jr. and D. J. O'Connor, "Magnetic field effects on float-zone Si crystal growth: II. Strong transverse fields," *Journal of Crystal Growth*, **76**, 100–110, 1986.

283. K.-M. Kim, "Growing improved silicon crystals for VLSI/ULSI," *Solid State Technology*, November 1996, and the references therein.

284. K. Takada, "400-mm development on track," *Solid State Technology*, **41**, September 1998.

285. S. Kawado, "Large-area x-ray topography to observe 300-mm-diameter silicon crystals," Industial Applications, p89.pdf, X-ray Research Laboratory, Rigaku Corporation, Japan.

286. R. Takeda, P. Xin, J. Yoshikawa, Y. Kirino, Y. Matsushita, Y. Hosoki, N. Tsuchiya, and O. Fujii, "300-mm diameter hydrogen annealed silicon wafers," *Journal of the Electrochemical Society*, **144**, L280–L282, 1997.

287. M. Akatsuka, K. Sueoka, H. Katahama, and N. Adachi, "Calculation of slip length in 300-mm silicon wafers during thermal processes," *Journal of the Electrochemical Society*, **146**, 2683–2688, 1999.

288. E. Dornberger, J. Virbulis, B. Hanna, R. Hoelzl, E. Daub, and W. von Ammon, "Silicon crystals for future requirements of 300 mm wafers," *Journal of Crystal Growth*, **229**, 11–16, 2001.

289. Y. Shiraishi, K. Takano, J. Matsubara, T. Iida, N. Takase, N. Machida, M. Kuramoto, and H. Yamagishi, "Growth of silicon crystal with a diameter of 400 mm and weight of 400 kg," *Journal of Crystal Growth*, **229**, 17–21, 2001.

290. K. Takano, Y. Shiraishi, J. Matsubara, T. Iida, N. Takase, N. Machida, M. Kuramoto, and H. Yamagishi, "Global simulation of the CZ silicon crystal growth up to 400 mm in diameter," *Journal of Crystal Growth*, **229**, 26–30, 2001.

291. Th. Wetzel, J. Virbulis, A. Muiznieks, W. v. Ammon, E. Tomzig, G. Raming, and M. Weber, "Prediction of the growth interface shape in industrial 300 mm CZ Si crystal growth," *Journal of Crystal Growth*, **266**, 34–39, 2004.

292. Z. Lu and S. Kimbel, "Growth of 450 mm diameter semiconductor grade silicon crystals," *Journal of Crystal Growth*, **318**, 193–195, 2011.

293. C. K. Bhihe, P. A. Mataga, J. W. Hutchinson, S. Rajendran, and J. P. Kalejs, "Residual stresses in crystal growth," *Journal of Crystal Growth*, **137**, 86–90, 1994.

294. P. Capper and J. G. Wilkes, "On the cooling rates of large-diameter silicon crystals," *Applied Physics Letters*, **32**, 187–189, 1978.

295. H. Nakanishi, H. Kohda, and K. Hoshikawa, "Influence of annealing during growth on defect formation in Czochralski silicon," *Journal of Crystal Growth*, **61**, 80–84, 1983.

296. H.-J. Cho, B.-c. Sim, and J. Y. Lee, "Asymmetric distributions of grown-in microdefects in Czochralski silicon," *Journal of Crystal Growth*, **289**, 458–463, 2006, and the references therein.

297. S. S. Kim and W. Wijaranakula, "The effect of the thermal history of Czochralski silicon crystals on the defect generation and refresh time degradation in high density memory devices," *Journal of the Electrochemical Society*, **142**, 553–559, 1995.

298. T. Iwasaki, Y. Tsumori, K. Nakai, H. Haga, K. Kojima, and T. Nakashizu, "Influence of cooling condition during crystal growth of CZ-Si on oxide breakdown property," *Journal of the Electrochemical Society*, **143**, 3383–3388, 1996.

299. T. Ebe, "Effects of isotherm shapes on the point-defect behavior in growing silicon crystals," *Journal of Crystal Growth*, **244**, 142–156, 2002.

300. Z. Jian, K. Nagashio, and K. Kuribayashi, "Direct observation of the crystal-growth transition in undercooled silicon," *Metallurgical and Materials Transactions A*, **33A**, 2947–2953, 2002.

301. A. I. Prostomolotov, N. A. Verezub, M. V. Mezhennii, and V. Ya. Reznik, "Thermal optimization of CZ bulk growth and wafer annealing for crystalline dislocation-free silicon," *Journal of Crystal Growth*, **318**, 187–192, 2011.

302. A. Sarikov, V. Litovchenko, I. Lisovskyy, M. Voitovich, S. Zlobin, V. Kladko, N. Slobodyan, V. Machulin, and C. Claeys, "Mechanisms of oxygen precipitation in CZ-Si wafers subjected to rapid thermal anneals," *Journal of the Electrochemical Society*, **158**, H772–H777, 2011.

303. A. J. R. de Kock, "Vacancy clusters in dislocation-free silicon," *Applied Physics Letters*, **16**, 100–102, 1970.

304. R. L. Collins, "Growth parameters for large diameter float zone silicon crystals," *Journal of Crystal Growth*, **42**, 490–492, 1977.

305. A. Mühlbauer, A. Muiznieks, J. Virbulis, A. Lüdge, and H. Riemann, "Interface shape, heat transfer and fluid flow in the floating zone growth of large silicon crystals with the needle-eye technique," *Journal of Crystal Growth*, **151**, 66–79, 1995, and the references therein.

306. M. Schweizer, A. Cröll, P. Dold, Th. Kaiser, M. Lichtensteiger, and K. W. Benz, "Measurement of temperature fluctuations and microscopic growth rates in a silicon floating zone under microgravity," *Journal of Crystal Growth*, **203**, 500–510, 1999.

307. C. W. Lan and J. H. Chian, "Three-dimensional simulation of Marangoni flow and interfaces in floating-zone silicon crystal growth," *Journal of Crystal Growth*, **230**, 172–180, 2001.

308. G. Raming, A. Muižnieks, and A. Mühlbauer, "Numerical investigation of the influence of EM-fields on fluid motion and resistivity distribution during floating-zone growth of large silicon single crystals," *Journal of Crystal Growth*, **230**, 108–117, 2001.

309. A. Rudevičs, A. Muižnieks, G. Ratnieks, A. Mühlbauer, and Th. Wetzel, "Numerical study of transient behavior of molten zone during industrial FZ process for large silicon crystal growth," *Journal of Crystal Growth*, **266**, 54–59, 2004.

310. M. Kitamura, N. Usami, T. Sugawara, K. Kutsukake, K. Fujiwara, Y. Nose, T. Shishido, and K. Nakajima, "Growth of multicrystalline Si with controlled grain boundary configuration by the floating zone technique," *Journal of Crystal Growth*, **280**, 419–424, 2005.

311. M. Wünscher, A. Lüdge, and H. Riemann, "Growth angle and melt meniscus of the RF-heated floating zone in silicon crystal growth," *Journal of Crystal Growth*, **314**, 43–47, 2011.

312. K. H. Lie, J. S. Walker, and D. N. Riahi, "Free surface shape and AC electric current distribution for float zone silicon growth with a radio frequency induction coil," *Journal of Crystal Growth*, **100**, 450–458, 1990.

313. R. Menzel, "Growth conditions for large-diameter FZ Si single crystals," http://opus.kobv.de; genehmigte Dissertation, Technischen Universität Berlin, Berlin 2013, and the references therein.

314. FZ photo website. http://www.ikz-berlin.de.

315. H. Riemann, A. Lüdge, K. Böttcher, H.-J. Rost, B. Hallmann, W. Schröder, W. Hensel, and B. Schleusener, "Silicon floating zone process: Numerical modeling of RF field, heat transfer, thermal stress, and experimental proof for 4-inch crystals," *Journal of the Electrochemical Society*, **142**, 1007–1014, 1995.

316. C. W. Lan, "Heat transfer, fluid flow, and interface shapes in zone melting processing with induction heating," *Journal of the Electrochemical Society*, **145**, 3926–3935, 1998.

317. N. Ma, J. S. Walker, A. Lüdge, and H. Riemann, "Combining a rotating magnetic field and crystal rotation in the floating-zone process with a needle-eye induction coil," *Journal of Crystal Growth*, **230**, 118–124, 2001.

318. C. W. Lan, "Effect of axisymmetric magnetic fields on heat flow and interfaces in floating-zone silicon crystal growth," *Modelling and Simulation in Materials Science and Engineering*, **6**, 423–445, 1998.

319. T. Surek and S. R. Coriell, "Shape stability in float zoning of silicon crystals," *Journal of Crystal Growth*, **37**, 253–271, 1977.

320. H. K. Kuiken and P. J. Roksnoer, "Analysis of the temperature distribution in FZ silicon crystals," *Journal of Crystal Growth*, **47**, 29–42, 1979.

321. A. Eyer, B. O. Kolbesen, and R. Nitsche, "Floating zone growth of silicon single crystals in a double-ellipsoid mirror furnace," *Journal of Crystal Growth*, **57**, 145–154, 1982.

322. A. Mühlbauer, A. Muiznieks, and J. Virbulis, "Analysis of the dopant segregation effects at the floating zone growth of large silicon crystals," *Journal of Crystal Growth*, **180**, 372–380, 1997, and the references therein.

323. G. I. Voronkova, A. V. Batunina, V. V. Voronkov, R. Falster, L. Moiraghi, and M. G. Milvidski, "Effects of annealing on the electrical properties of nitrogen-doped float-zoned silicon," *210 ECE Meeting Cancun*, Mexico, 29 Oct–3 Nov, 2006.

324. A. G. Crouse, J. E. Huffman, C. S. Tindall, and M. L. W. Thewalt, "Float zone growth of high purity and high oxygen concentration silicon," *Journal of Crystal Growth*, **109**, 162–166, 1991.

325. A. Mühlbauer, A. Muiznieks, G. Raming, H. Riemann, and A. Lüdge, "Numerical modelling of the microscopic inhomogeneities during FZ silicon growth," *Journal of Crystal Growth*, **198–199**, 107–113, 1999.

326. G. Raming, A. Muižnieks, and A. Mühlbauer, "Numerical investigation of the influence of EM-fields on fluid motion and resistivity distribution during floating-zone growth of large silicon single crystals," *Journal of Crystal Growth*, **230**, 108–117, 2001.

327. A. J. R. de Kock, P. J. Roksnoer, and P. G. T. Boonen, "The introduction of dislocations during the growth of floating-zone silicon crystals as a result of point defect condensation," *Journal of Crystal Growth*, **30**, 279–294, 1975.

328. W. Zulehner, "Czochralski growth of silicon," *Journal of Crystal Growth*, **65**, 189–213, 1983.

329. A. J. R. de Kock, P. J. Roksnoer, and P. G. T. Boonen, "Effect of growth parameters on formation and elimination of vacancy clusters in dislocation-free silicon crystals," *Journal of Crystal Growth*, **22**, 311–320, 1974.

330. G. D. Robertson, Jr. and D. J. O'Connor, "Magnetic field effects on float-zone Si crystal growth: III. Strong axial fields," *Journal of Crystal Growth*, **76**, 111–122, 1986.

331. N. De Leon, J. Guldberg, and J. Salling, "Growth of homogeneous high resistivity FZ silicon crystals under magnetic field bias," *Journal of Crystal Growth*, **55**, 406–408, 1981.

332. T. E. Morthland and J. S. Walker, "Thermocapillary convection during floating-zone silicon growth with a uniform or non-uniform magnetic field," *Journal of Crystal Growth*, **158**, 471–479, 1996, and the references therein.

333. C. W. Lan and M. C. Liang, "Modulating dopant segregation in floating-zone silicon growth in magnetic fields using rotation," *Journal of Crystal Growth*, **180**, 381–387, 1997, and the references therein.

334. A. Cröll, P. Dold, Th. Kaiser, F. R. Szofran, and K. W. Benz, "The influence of static and rotating magnetic fields on heat and mass transfer in silicon floating zones," *Journal of the Electrochemical Society*, **146**, 2270–2275, 1999.

335. K. Kakimoto and H. Ozoe, "Oxygen distribution at a solid-liquid interface of silicon under transverse magnetic fields," *Journal of Crystal Growth*, **212**, 429–437, 2000.

336. N. Ma, J. S. Walker, A. Lüdge, and H. Riemann, "Combining a rotating magnetic field and crystal rotation in the floating-zone process with a needle-eye induction coil," *Journal of Crystal Growth*, **230**, 118–124, 2001.

337. K. Li and W. R. Hu, "Magnetic field design for floating zone crystal growth," *Journal of Crystal Growth*, **230**, 125–134, 2001.

338. K. Kakimoto, "Effects of rotating magnetic fields on temperature and oxygen distributions in silicon melt," *Journal of Crystal Growth*, **237–239**, 1785–1790, 2002.

339. C. W. Lan and B. C. Yeh, "Three-dimensional simulation of heat flow, segregation, and zone shape in floating-zone silicon growth under axial and transversal magnetic fields," *Journal of Crystal Growth*, **262**, 59–71, 2004.

340. M. L. Joshi, "Effect of fast cooling on diffusion-induced imperfections in silicon," *Journal of the Electrochemical Society*, **112**, 912–916, 1965.

341. A. Rohatgi and P. Rai-Choudhury, "Process-induced effects on carrier lifetime and defects in float zone silicon," *Journal of the Electrochemical Society*, **127**, 1136–1139, 1980.

342. T. L. Chu, S. S. Chu, R. W. Kelm, Jr., and G. W. Wakefield, "Solar cells from zone-refined metallurgical silicon," *Journal of the Electrochemical Society*, **125**, 595–597, 1978.

343. P. M. Zavracky, D.-P. Vu, and M. Batty, "Silicon-on-insulator wafers by zone melting recrystallization," *Solid State Technology*, **34**, 55–57, April 1991.

344. M. Trempa, C. Reimann, J. Friedrich, and G. Müller, "The influence of growth rate on the formation and avoidance of C and N related precipitates during directional solidification of multi crystalline silicon," *Journal of Crystal Growth*, **312**, 1517–1524, 2010.

345. P. Prakash, P. K. Singh, S. N. Singh, R. Kishore, and B. K. Das, "Use of silicon oxynitride as a graphite mold releasing coating for the growth of shaped multicrystalline silicon crystals," *Journal of Crystal Growth*, **144**, 41–47, 1994.

346. C. W. Lan and B. C. Yeh, "Three-dimensional simulation of heat flow, segregation, and zone shape in floating-zone silicon growth under axial and transversal magnetic fields," *Journal of Crystal Growth*, **262**, 59–71, 2004.

347. X. J. Chen, S. Nakano, L. J. Liu, and K. Kakimoto, "Study on thermal stress in a silicon ingot during a unidirectional solidification process," *Journal of Crystal Growth*, **310**, 4330–4335, 2008.

348. Y.-Y. Teng, J.-C. Chen, C.-W. Lu, H.-I Chen, C. Hsu, and C.-Y. Chen, "Effects of the furnace pressure on oxygen and silicon oxide distributions during the growth of multicrystalline silicon ingots by the directional solidification process," *Journal of Crystal Growth*, **318**, 224–229, 2011.

349. R. Gereth, "Growth of silicon bicrystals by the Dash pedestal-method," *Journal of the Electrochemical Society*, **109**, 1068–1070, 1962.

350. S. N. Dermatis, J. W. Foust, Jr., and H. F. John, "Growth and morphology of silicon ribbons," *Journal of the Electrochemical Society*, **112**, 792–796, 1965.

351. T. F. Ciszek, "Techniques for the crystal growth of silicon ingots and ribbons," *Journal of Crystal Growth*, **66**, 655–672, 1984.

352. G. Willeke, B. Bitnar, M. Wendl, C. Kloc, E. Bucher, and A. Vallêra, "Electromagnetic ribbon: Proposal of a novel method for silicon sheet generation," *Semiconductor Science and Technology*, **13**, 440–443, 1998.

353. T. Surek, C. B. H. Rao, J. C. Swartz, and L. C. Garone, "Surface morphology and shape stability in silicon ribbons grown by the edge-defined, film-fed growth process," *Journal of the Electrochemical Society*, **124**, 112–123, 1977.

354. A. Schönecker, L. Laas, A. Gutjahr, M. Goris, P. Wyers, G. Hahn, and D. Sontag, "Ribbon-growth-on-substrate: Status, challenges and promises of high-speed silicon wafer manufacturing," *12th Workshop on Crystalline Silicon Solar Cells, Materials and Processes*, rx02038.pdf, and the references therein.

355. J. P. Kalejs and L.-Y. Chin, "Modeling of ambient-meniscus melt interactions associated with carbon and oxygen transport in EFG of silicon ribbon," *Journal of the Electrochemical Society*, **129**, 1356–1361, 1982.

356. K. R. Sarma and M. J. Rice, Jr., "High-pressure plasma (HPP) deposition of polycrystalline silicon ribbons," *Journal of the Electrochemical Society*, **128**, 2647–2655, 1981.

357. J. P. Kalejs, "Modeling contributions in commercialization of silicon ribbon growth from the melt," *Journal of Crystal Growth*, **230**, 10–21, 2001.

358. H.-M. Jeong, H.-S. Chung, and T.-W. Lee, "Computational simulations of ribbon-growth on substrate for photovoltaic silicon wafer," *Journal of Crystal Growth*, **312**, 555–562, 2010.

359. A. Sanjurjo, "Silicon sheet for solar cells," *Journal of the Electrochemical Society*, **128**, 2244–2247, 1981.

360. T. F. Ciszek, J. L. Hurd, and M. Schietzelt, "Filament materials for edge-supported pulling of silicon sheet crystals," *Journal of the Electrochemical Society*, **129**, 2838–2843, 1982.

361. E. Yablonovitch and T. Gmitter, "Wetting angles and surface tension in the crystallization of thin liquid films," *Journal of the Electrochemical Society*, **131**, 2625–2630, 1984.

362. M. Tanielian, S. Blackstone, and R. Lajos, "A new technique of forming thin free standing single-crystal films," *Journal of the Electrochemical Society*, **132**, 507–509, 1985.

363. K. C. Lee, "The fabrication of thin, freestanding, single-crystal, semiconductor membranes," *Journal of the Electrochemical Society*, **137**, 2556–2574, 1990.

364. G. Willeke, B. Bitnar, M. Wendl, C. Kloc, E. Bucher, and A. Vallêra, "Electromagnetic ribbon: Proposal of a novel method for silicon sheet generation," *Semiconductor Science and Technology*, **13**, 440–443, 1998.

365. J. Lu and G. Rozgonyi, "Oxygen and carbon precipitation in crystalline sheet silicon: Depth profiling by infrared spectroscopy, and preferential defect etching," *Journal of the Electrochemical Society*, **153**, G986–G991, 2006.

366. Y. Osada, H. Nakayama, M. Shindo, T. Odaka, and Y. Ogata, "Growth and structure of silicon fibers," *Journal of the Electrochemical Society*, **126**, 31–36, 1979.

367. R. S. Wagner and C. J. Doherty, "Mechanism of branching and kinking during VLS crystal growth," *Journal of the Electrochemical Society*, **115**, 93–99, 1968.

368. P. Rai-Choudhury and W. J. Takei, "Formation of silicon whiskers by aluminum-quartz interaction," *Journal of the Electrochemical Society*, **121**, 1228–1229, 1974.

369. N. Zakharov, P. Werner, L. Sokolov, and U. Gösele, "Growth of Si whiskers by MBE: Mechanism and peculiarities," *Physica E*, **37**, 148–152, 2007.

370. Y. Zhang, G. Li, C. S. Lee, and S.-T. Lee, "Silicon cylinder grown on the surface of a silicon wafer," *Journal of Crystal Growth*, **182**, 337–340, 1997.

371. X. Huang, S. Uda, H. Tanabe, N. Kitahara, H. Arimune, and K. Hoshikawa, "In-situ observations of crystal growth of spherical Si single crystals," *Journal of Crystal Growth*, **307**, 341–347, 2007.

372. Z. Liu, A. Masuda, and M. Kondo, "Investigation on the crystal growth process of spherical Si single crystals by melting," *Journal of Crystal Growth*, **311**, 4116–4122, 2009.

373. W. Dietze, L. P. Hunt, and D. H. Sawyer, "The preparation and properties of CVD-silicon tubes and boats for semiconductor device technology," *Journal of the Electrochemical Society*, **121**, 1112–1115, 1974.

374. S. Hyland, D. Leung, A. Morrison, K. Stika, and H. Yoo, "Correlation of solar cell electrical properties with material characteristics of silicon cast by the ubiquitous crystallization process," *Journal of the Electrochemical Society*, **130**, 1373–1376, 1983.

375. D. Vizman, J. Friedrich, and G. Mueller, "3-D time-dependent numerical study of the influence of the melt flow on the interface shape in a silicon ingot casting process," *Journal of Crystal Growth*, **303**, 231–235, 2007.

376. T. Y. Wang, Y. C. Lin, C. Y. Tai, R. Sivakumar, D. K. Rai, and C. W. Lan, "A novel approach for recycling of kerf loss silicon from cutting slurry waste for solar cell applications," *Journal of Crystal Growth*, **310**, 3403–3406, 2008.

377. A. Karoui, G. A. Rozgonyi, R. Zhang, and T. Ciszek, "Silicon crystal growth and wafer processing for high efficiency solar cells and high mechanical yield," *Conference at NCPV Prog. Review Meeting*, Lakewood, Colorado, 14–17, October, 2001.

378. M. Lipiński, P. Panek, and R. Ciach, "The industrial technology of crystalline silicon solar cells," *Journal of Optoelectronics and Advanced Materials*, **5**, 1365–1371, 2003.

5

From Silicon Ingots to Silicon Wafers

5.1 Introduction

Once the silicon crystal is grown, the silicon ingot is sliced into thin wafers—most of them no greater than 1 mm thickness. Since silicon is a hard, brittle material, special materials are needed to shape it to the desired levels. Industrial-grade diamonds, SiC, and Al_2O_3 are often used for this purpose. The conversion of silicon ingots into final polished wafers requires at least six machining operations, two chemical operations, and one or two polishing operations, depending on whether the wafers are to be polished only on one side or both sides. Figure 5.1 shows fully grown single-crystal silicon ingots [1] ready to be converted to standard wafers for very large scale integration (VLSI) and ultra large scale integration (ULSI) circuit processing.

First, the ingot seed and tang ends are removed, as this part of the grown crystal either will have many more defects or may be rich in many metallic impurities, which could get into the growing crystal toward the end of growth operation. Some portions of the crystal may not produce the required resistivity values, and it is better to discard such portions of the grown crystal. Surface grinding of the ingot is carried out to smooth the surface and get a uniform shape. This step is also carried out to reduce the diameter of the ingot. Thus, it is always advised to grow a slightly oversized ingot so it can be shaped to the desired crystal diameter. The ingots are ground in a lathe-like machine tool to the required diameter. A rotating cutting tool makes multiple passes down a rotating ingot until the chosen diameter is achieved. Figure 5.2 shows the steps involved in this operation [2]. Figure 5.3 shows the grinding operation to reduce the diameter of the crystal ingot.

5.2 Radial Resistivity Measurements

The resistivity of the cylindrical ingots can be determined using a two-point probe resistivity measurement by applying suitable correction factors [4]. A standard linear four-point probe method on the top surface of the ingot can also be used [5]. It is widely reported that the surface resistivity analysis nearly matches wafer resistivity values.

The procedure for determining the resistivity along the length of a cylinder involves placing current probes at each end of the cylinder and measuring ΔV between two

FIGURE 5.1
Fully grown silicon ingots using the Czochralski (CZ) technique. (From: http://www.google.co.in and www.sz-wholesaler.com/userimage/1229/1230sw2/silicon-ingot-p.176.jpg [1]).

traveling probes spaced a distance (*D-E*) from each other [4]. The resistivity at any given position is then calculated from

$$\rho = \frac{\pi^2 C}{4F} \frac{\Delta V}{I}$$

where *C* is a constant and the appropriate *F* is determined by *X*, the average distance of the probes from the end of the cylinder.

5.3 Boule Formation, Identification of Crystal Orientation, and Flats

The crystal orientation and the direction of the crystal planes needs to be confirmed based on the symmetry of the ingot. The best option is to use the Laue back-reflection x-ray technique to identify the crystal orientation. Presently, high-intensity reflectograms are used to identify the crystal planes. All silicon ingots, irrespective of their doping concentrations and orientations, look alike. Thus, selecting the correct wafer size, dopant type, and crystal orientation is crucial; any simple mistake in selection may not yield the correct device and circuit functionality planned. The complete processing effort is lost if you start with the incorrect wafer.

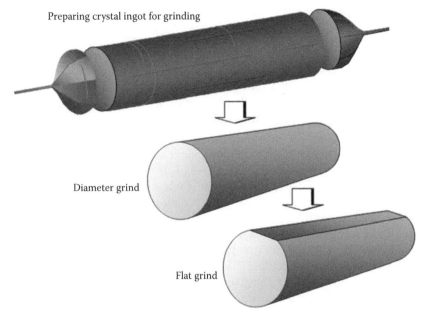

FIGURE 5.2
Preparation of ingot at different stages for wafer slicing, such as removal of seed and tang end, diameter grinding, and primary cut. (From Lecture notes on silicon crystal structures and growth (Plummer: Chap. 3), 18_Silicon.ppt, EE-452, www.powershow.com/view/abbbc-ODQ3O/Silicon_Structure_and_Growth_ Powerpoint_ppt_presentation, 13-26 [2].)

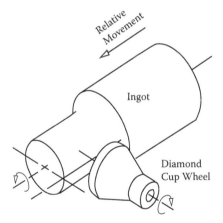

FIGURE 5.3
Grinding a grown silicon ingot to the desired diameter. (From C. W. Pearce, "Crystal growth and wafer preparation" in *VLSI Technology*, edited by S. M. Sze, McGraw-Hill, New York, 1988, and the references therein [3].)

Following the diameter grinding and resistivity confirmation, one or more flats are ground along the length of the ingot. The largest flat is identified as the "major," or "primary," flat, and the second one is called the "secondary flat." After confirming the crystal resistivity, the ingot will move further to subsequent operations. Previously—up to the 150 mm wafer diameter era—wafers had flats, and the flats had two things to indicate (1) the doping type of the wafer (*n*- or *p*-type doping) and (2) the orientation of the wafer:

[100] or [111]. Figure 5.4 shows the different crystal primary and secondary flats for {111} *n*-type, {111} *p*-type, {100} *n*-type, and {100} *p*-type wafers. In practice, fewer *n*-type silicon wafers are produced when compared to *p*-type wafers. In larger wafers, the flat serves to align the wafer crystallographically to guide the wafer dicing, which falls in the easy cut directions. With a single flat in the <110> direction, it is easy to locate and align and cut the wafer to chips in parallel and perpendicular directions.

For wafers with diameters equal to or larger than 100 mm (4 inches) and 150 mm (6 inches), just one flat for the {100} *p*-type and {111} *p*-type are applied. Wafers with diameters equal to or larger than 200 mm (8 inches) will have no flat, but just a small notch is provided, because we lose too much expensive wafer area by cutting a flat. The details are shown in Figure 5.5. There are other simple ways of checking the wafer details. The wafer doping type is generally checked with the help of a hot probe. Depending on the polarity induced, the wafer is identified as *n*- or *p*-type. Orientations are decided by normal cleavage paths. More details are discussed elsewhere [6] on these issues. Breakage of wafers is a direct loss of yield. In most cases, such wafers are considered "dead" and are not useful either for evaluation purposes or for any parameter estimation. Such wafers are no longer usable for making integrating circuits [7] and are discontinued from the lot.

Schwuttke [8] has reported on the use of high-intensity reflectograms to determine crystal orientations. A practical instrument was developed for routine crystal orientation work. With the help of this instrument, it is possible to orient a crystal in any direction and then cut along this predetermined direction without removing the crystal from its original mount. The reflectograms produced are very bright, as shown in Figure 5.6. The accuracy

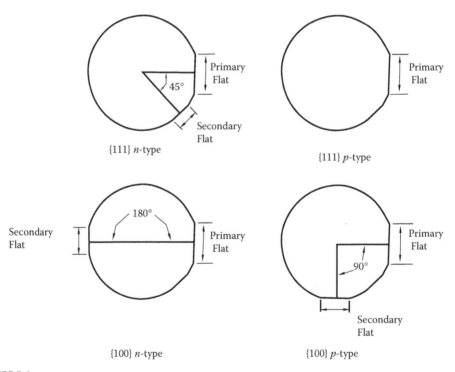

FIGURE 5.4
Different crystal orientations and identification of primary and secondary flats for {111} and {100} wafers. (From C. W. Pearce, "Crystal growth and wafer preparation" in *VLSI Technology*, edited by S. M. Sze, McGraw-Hill, New York, 1988, and the references therein [3].)

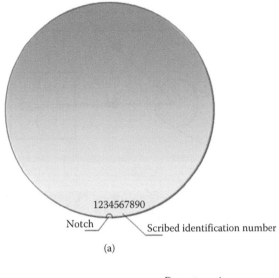

(a)

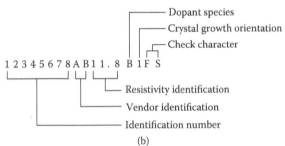

(b)

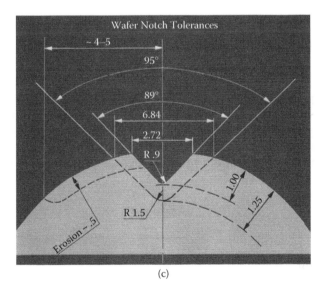

(c)

FIGURE 5.5
(a) Notch on silicon wafer and scribed identification number. (From Lecture notes on silicon crystal structures and growth (Plummer: Chap. 3), 18_Silicon.ppt, EE-452, 13–1 to 13-60 Internet and https://www.google.co.in and www.powershow.com/view/abbbc-ODQ3O/Silicon_Structure_and_Growth_Powerpoint_ppt_presentation [2].) (b) General wafer-marking patterns followed by different crystal growing industries. (c) Wafer notch locators (dimensions in mm; not to scale). (From S. Peredy, *Semiconductor International*, 128–131, June 1989 [6].)

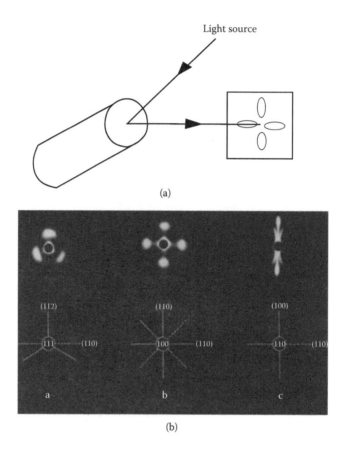

(a)

(b)

FIGURE 5.6
(a) Reflection of high-intensity light source on selectively treated ingot surface and the reflection beam. (b) Directional relationship in (111) plane, (100) plane, and (110) plane. (From G. H. Schwuttke, *Journal of the Electrochemical Society*, **106**, 315–317, 1959 [8].)

of orientation is reported to be 12 min of arc for standard silicon crystal reflections. Based on these principles, commercial systems are available—one such unit is shown in Figure 5.7 [9]. A sample is prepared by using a water-based slurry of 500-mesh silica that was subsequently etched with a freshly prepared 50% NaOH or KOH solution for 6–12 min. After cleaning the surface, high-intensity light is used to locate the correct crystal orientations.

5.4 Ingot Slicing

Silicon is very hard, and special tools are needed to slice the ingot into smaller wafers. Slicing is important, as it determines four important parameters: surface orientation, wafer thickness, taper, and bow. The mechanical property of the silicon wafer differs with the orientation, but also slightly with the dopant present in it. The presence of nitrogen and germanium doping [10, 11] in crystals increases the fracture strength.

(a)

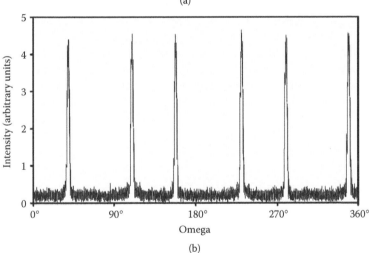

(b)

FIGURE 5.7

(a) Measurement of a 6-inch silicon wafer for ingot and raw crystal orientation determination used in wafer measurement mode. (b) Example of a measured diagram. (From EFG International, Germany, Technical Pamphlets. For more details on this unit, visit www.efg-berlin.de [9].)

Wafers of <100> and <110> orientations are usually cut on the orientation without any deviation. Wafers of <111> orientation are cut with a slight off-orientation. This value depends on the specific application, and this off-orientation may vary from 1° to 3°. This off-orientation is used to create nucleation sites necessary for silicon homoepitaxial growth. The wafer thickness is typically determined by the mechanical strength of the material under study. The wafer must be sufficiently thick enough to support its own weight without cracking during handling, and its self-weight should not multiply the existing defect density. Chen [12] estimated the minimum silicon wafer thickness values for internal diameter (ID) wafering. According to this, the relationship equation describing the wafer thickness versus the diameter can be expressed in the following form:

$$t^2 = \frac{3PD}{b} \frac{\sqrt{\pi a_c}}{K_{IC}} (1 + \cos \theta)$$

Here, t is wafer thickness, D is diameter, b is width of the rigid support, P is applied force, a_c is critical flaw size, and K_{IC} is the material constant, which includes the tensile and elastic stress parameters. The value of the angle is defined as $\theta = \sin^{-1}\left(\frac{b}{D}\right)$. The minimum thickness is dependent upon the depth of surface damage (flaw size a_c) to the wafer to the ¼ power. Wafer thickness is essentially fixed by taking into account losses that occur during the processing operation. Here, monitoring the blade vibration and position of the blade is also important. Excessive curvature of the blade may result in more losses.

Two types of slicing methods have been used in the silicon wafer industry: the ID saw and the wire saw [13]. ID slicing is the common mode of slicing silicon ingots. The saw blade is a thin sheet of stainless steel approximately 325 μm thick with a diamond-bonded inner rim. The rotation speed is generally maintained at 2000 rpm. The blade is moved relative to the stationary ingot, and the operation continues until the step is complete. ID saws cut only one wafer at a time, and the saw is water cooled to avoid excess temperature generation. In this process, the kerf loss due to blade width and vibration is a direct loss of one-third of the crystal grown as saw dust. In this case, this loss is unavoidable. A wire saw, on the other hand, takes several hours to cut through a crystal ingot, but makes several hundred parallel saw cuts simultaneously, wafering the entire ingot in a single operation. The details of the wire saw are shown in Figure 5.8. For crystal diameters larger than 150 mm, wire saws have proven to be more economical than ID saws, with higher throughput and the potential for lower kerf losses by using thinner wires.

5.5 Mechanical Lapping of Wafer Slices

Both wafer-slicing methods produce a lot of unevenness on the surface of the wafer and mechanical polishing is necessary to remove these surface irregularities. The wafers are placed on a lapping pad, and fine alumina powder (Al_2O_3) mixed in glycerin is applied under constant pressure. This is a two-side lapping method, and approximately 20 μm thickness is removed on both sides. This step removes all uneven shapes present on the surface depending on the powder fineness. Most of the time, this step produces a uniform flatness within a 2 μm range. A typical lapping and grinding machine is shown in Figure 5.9. A typical surface finish of a lapped silicon wafer measured using a surface profiler is shown in Figure 5.10. Here, the wafer was verified in a scan area of 2 mm and the average height is recorded as 5490 Å. The entire wafer may be verified at different locations to get the average height. This one is too rough for integrated circuit fabrication work and needs further polishing to improve the surface finish.

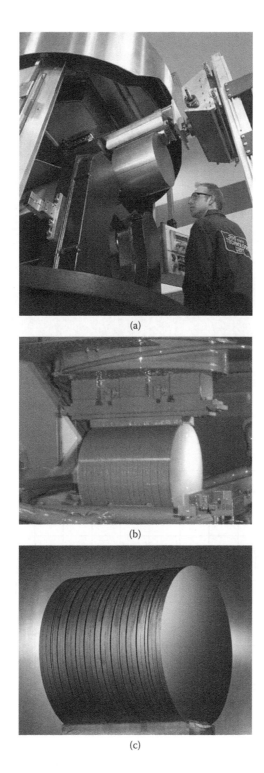

(a)

(b)

(c)

FIGURE 5.8
(a) Uncut ingot being loaded into the saw. (b) Ingot being raised after passing through the wire slicing web. (c) The finished sliced ingot still mounted on the platform. (From G. Fisher, M. R. Seacrist, and R. W. Standley, *Proceedings of the IEEE*, **100**, 1454–1474, 2012, and the references therein [13].)

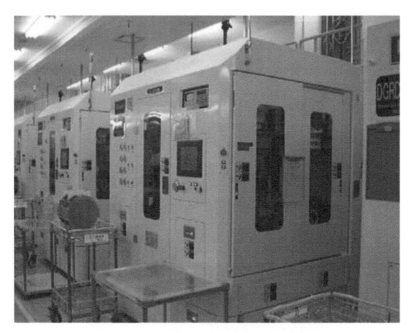

FIGURE 5.9
Typical lapping and grinding machine used for silicon wafers. (From G. Fisher, M. R. Seacrist, and R. W. Standley, *Proceedings of the IEEE*, **100**, 1454–1474, 2012, and the references therein [13].)

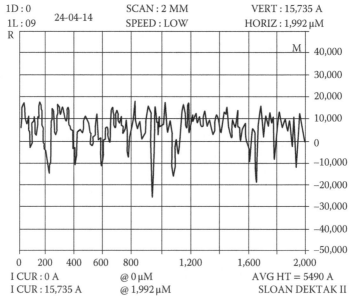

FIGURE 5.10
A typical silicon wafer surface finish after the wafer lapping stage. Scanned area length: 2 mm.

5.6 Edge Profiling of Slices

After mechanical lapping, the wafers are separated and are taken for edge contouring, where the edge of the silicon wafer is changed from a sharp to a round shape. This is usually done in cassette-fed, high-speed equipment, and each wafer is handled separately, as shown in Figure 5.11. The advantages are that the sharp-edged silicon wafer does not damage the polishing pads and aids in controlling the buildup of photoresist at the wafer edge while applying the photoresist coating for the photolithography step. Figure 5.12 shows one type of commercial equipment used for this purpose. The edge-contouring step is necessary to ensure better exposure conditions for the wafers, as this aids in applying a uniform photoresist coating.

5.7 Chemical Etching and Mechanical Damage Removal

The previously described shaping operations leave the wafer surface and edges damaged and contaminated with chemicals and fine particles. The mechanical damages are roughly 10 μm deep from the surface and are removed by using wet chemical etching. However, as a safety precaution, a 20 μm thickness is removed. Before etching, a laser is commonly used to engrave an alphanumeric identification mark on each wafer.

The chemical etching involves a combination of hydrofluoric acid (HF), nitric acid (HNO_3), and acetic acid (CH_3COOH):

$$3\,Si + 4\,HNO_3 + 18\,HF \leftrightarrow 3\,H_2SiF_6 + 4\,NO\,(\uparrow) + 8\,H_2O$$

HF dissolves the product. Acetic acid does not participate in the reaction, but dilutes the system so that etching can be better controlled. It also improves the silicon surface and leaves it shining.

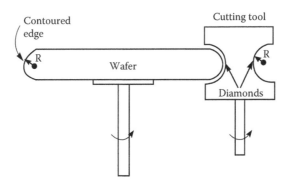

FIGURE 5.11
Edge contouring of silicon wafer. (From C. W. Pearce, "Crystal growth and wafer preparation" in *VLSI Technology*, edited by S. M. Sze, McGraw-Hill, New York, 1988, and the references therein [3].)

FIGURE 5.12
Setup used for edge contour profiling of silicon wafers. (From G. Fisher, M. R. Seacrist, and R. W. Standley, *Proceedings of the IEEE*, **100**, 1454–1474, 2012, and the references therein [13], https://www.sitrigroup.com/content/files/20131206142232_7705.pdf.)

5.8   Chemimechanical Polishing for Planar Wafers

Chemical polishing is the final step. Depending on the application, polishing is carried out on either one or both sides. For VLSI and ULSI applications, polishing one side is sufficient. For microelectomechanical system (MEMS) applications, polishing both sides is preferred. The purpose of polishing is to provide a smooth, perfect surface on which device features can be photoengraved. A main concern for any process engineer is to produce a surface with a high degree of surface flatness and minimum local slope to meet the requirements of optical projection lithography and in some critical etching steps. A schematic diagram for chemical polishing is shown in Figure 5.13. Here, the polishing pad is in touch with the wafer to be polished and is allowed to rotate continuously. Chemical slurry is fed to the polishing pad, and this leads to reactions on the surface of the wafer that remove the unevenness present on the surface. For this purpose, only lapped silicon wafers are preferred. Figure 5.14 shows the typical polished pad and the equipment used for this purpose.

When chemically polishing large wafers, a single-wafer process is preferred. This will help in applying external pressure precisely on the wafer and control the chemical polishing/surface reactions and removal rate. This way, wafer breakage can be avoided. The polishing pads are made of an artificial fabric, such as polyester felt or polyurethane laminate, and are chemically inert; in addition, they will not contaminate the polishing process. Because of the external pressure and the surface chemical reactions that are taking place, this step yields a better silicon surface. The slurry used here is a colloidal suspension of fine SiO_2 particles, typically 100 Å in diameter, in an aqueous solution of sodium hydroxide. The typical polishing process removes a 25 µm thickness of the silicon wafer.

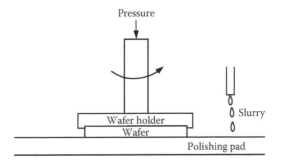

FIGURE 5.13
Schematic diagram of single-wafer chemical polishing on one side using chemical slurry. (From C. W. Pearce, "Crystal growth and wafer preparation" in *VLSI Technology*, edited by S. M. Sze, McGraw-Hill, New York, 1988, and the references therein [3].)

FIGURE 5.14
Chemical polishing of silicon wafers using chemical slurry and polishing pad. (From www.fullman.com [14].)

5.9 Surface Roughness and Overall Wafer Topography

As mentioned, the chemical polishing removes all uneven shapes present on the lapped surface, depending on the polishing duration and the pressure applied to the wafers. In these cases, the process produces a uniform flatness within 100 Å range. The typical surface finish of a chemically polished silicon wafer measured using a surface profiler is shown in Figure 5.15. The wafer also was verified in a scan area of 2 mm and the average height recorded as 71 Å, which is widely accepted for integrated circuit fabrication. The entire wafer may be verified at different locations for confirmation. No further polishing is required once the value has reached this stage of polishing.

5.10 Megasonic Cleaning

After the chemical polishing, the wafers are chemically cleaned with acid, base, and/or solvent mixtures to remove the slurry residue from the surface. If any glue layers are used for single-side polishing, the residue is removed with a proper solvent that dissolves the material. Slight ultrasonic agitation is also applied here to free any colloidal particles that are adhered to the surface. Thorough inspections are carried out on these wafers before they move to the next processing step. This is carried out in batches, and care is taken that the wafers do not touch each other. Figure 5.16 shows a typical setup for this megasonic cleaning.

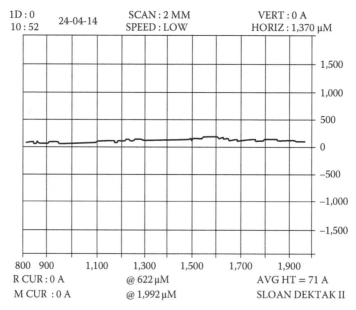

FIGURE 5.15
Typical silicon wafer surface finish after the wafer chemical polishing stage. Scanned area length: 2 mm.

FIGURE 5.16
Chemical cleaning of silicon wafers to remove slurry residues used for final polishing. (From Internet and open literature available at https://www.google.co.in and https://indico.cern.ch/event/a053148/session/0/contribution/s0t1/material/1/2.pdf [15].)

5.11   Final Cleaning and Inspection

The final cleaning of the silicon wafers is carried out in different stages, such as standard piranha cleaning, based on a mixture of sulfuric acid and hydrogen peroxide and/or RCA cleaning steps. Figure 5.17 shows one such chemical bench where these cleaning steps are carried out. This is again a batch process, and several wafers are handled at a time. The details of the cleaning steps are discussed elsewhere in other chapters.

5.12   Summary

In this chapter, we have discussed how the grown single-crystal ingot is sliced into smaller discs and how they take shape to become a complete silicon wafer for integrated circuit fabrication. First, the resistivity values of the ingot are verified, and only that part of the ingot is selected for slicing into small thicknesses. Boule formation, identification of crystal orientation, and flats are defined on the ingot. The ingot is cut into wafer slices using a mechanical saw or wire. Mechanical lapping and chemical polishing steps then are carried out on these wafers to get a finely polished surface. Finally, chemical etching and cleaning are performed on these wafers to free the silicon wafer surface from contamination.

FIGURE 5.17
Final cleaning of silicon wafers for packaging. (From G. Fisher, M. R. Seacrist, and R. W. Standley, *Proceedings of the IEEE*, **100**, 1454–1474, 2012, and the references therein [13].)

References

1. http://www.google.co.in and www.sz-wholesaler.com/userimage/1229/1230sw2/silicon-ingot-p.176.jpg.
2. J. D. Plummer, M. D. Deal and P. B. Griffin, Chapter-3: Lecture notes on silicon crystal structures and growth, 18_Silicon.ppt, EE-452, 13-1 to 13-60. (Source Internet and https://www.google.co.in and www.powershow.com/view/abbbc-ODQ3O/Silicon_Structure_and_Growth_Powerpoint_ppt_presentation.)
3. C. W. Pearce, "Crystal growth and wafer preparation" in *VLSI Technology*, edited by S. M. Sze, McGraw-Hill, New York, 1988, and the references therein.
4. H. H. Gegenwarth, "Correction factors for a two-point probe resistivity measurement of cylindrical crystals," *Journal of the Electrochemical Society*, **116**, 1166–1167, 1969.
5. F. Padovani and G. Valant, "Resistivity characterization of semiconductor crystal ingots," *Journal of the Electrochemical Society*, **120**, 585–587, 1973.
6. S. Peredy, "Wafer orientation: In search of an intelligent prealigner," *Semiconductor International*, 128–131, June 1989.
7. www.tf.uni-kiel.de/matwis/amat/elmat_en/kap_5/illustr/i5_2_4.html.
8. G. H. Schwuttke, "Determination of crystal orientation by high-intensity reflectograms," *Journal of the Electrochemical Society,*, **106**, 315–317, 1959.
9. EFG International, Germany, Technical Pamphlets. For more details on this unit, visit www.efg-berlin.de.
10. L. Jastrzebski, G. W. Cullen, R. Soydan, G. Harbeke, J. Lagowski, S. Vecrumba, and W. N. Henry, "The effect of nitrogen on the mechanical properties of float zone silicon and on CCD device performance," *Journal of the Electrochemical Society*, **134**, 466–470, 1987.
11. P. Wang, X. Yu, Z. Li, and D. Yang, "Improved fracture strength of multicrystalline silicon by germanium doping," *Journal of Crystal Growth*, **318**, 230–233, 2011.

12. C. P. Chen, "Minimum silicon wafer thickness for ID wafering," *Journal of the Electrochemical Society*, **129**, 2835–2837, 1982.
13. G. Fisher, M. R. Seacrist, and R. W. Standley, "Silicon crystal growth and wafer technologies," *Proceedings of the IEEE*, **100**, 1454–1474, 2012, and the references therein.
14. Photo courtesy www.fullman.com.
15. Taken from Internet and open literature available at https://www.google.co.in and https://indico.cern.ch/event/a053148/session/0/contribution/s0t1/material/1/2.pdf.

6

Evaluation of Silicon Wafers

6.1 Introduction

The performance of silicon-based semiconductors has been greatly improved due to the development of ultra large scale integration (ULSI) technology up to the submicron design rule. Therefore, the quality of silicon wafers has become even more important, since near-surface defects, which are generated during a mechanical polishing process, greatly affect and deteriorate device performance. The silicon wafer industry requires methods to evaluate and characterize single-crystal semiconductors that are nondestructive and relatively simple to analyze a structurally damaged thin layer near the surface [1] so that the devices and circuits fabricated in them provide the intended results. Sometimes, the evaluation report may provide feedback to the crystal growth team to alter certain parameters and produce better crystal ingots and, of course, the perfect wafers for processing.

In general, the evaluations are widespread. They include (1) a visual inspection for cracks, twinning, and shape irregularity; (2) a test for electrical resistivity values and dopant type; (3) evaluation of crystal perfection; (4) mechanical properties, such as size, thickness, and mass; and (5) the presence of oxygen or carbon, as well as the metal impurity type and the concentration. Some tests are destructive in nature and are generally used to test wafers or samples prepared for the purpose of evaluation. Many of the tests covered in this chapter are both destructive and nondestructive. Table 6.1 lists some of the tests carried out on silicon wafers to verify the quality and check their suitability for fabricating very large scale integration (VLSI) and ULSI circuits. Although this chapter has a wide scope, limited results are provided. More references are cited for those who wish to look for specific information.

6.2 Acoustic Laser Probing Technique

Ruiz et al. [39] analyzed silicon slices using a helium-neon (He-Ne) laser. Design criteria and performance characteristics were reported by using the laser to identify wafer imperfections—particularly the fractures, scratches, etc.—on the polished surfaces. This method provides fast and accurate defect counting and distribution on polished, etched semiconductor surfaces. The basic technique involves a scanning He-Ne laser beam that is focused on the surface of the slice and analyzes the reflected and scattered light via polarizers and photomultipliers. Using a similar approach, it is possible to evaluate the particles resting on the silicon surfaces. Locke and Donovan [40] evaluated the intensity of light scattering

TABLE 6.1

Various Evaluation Techniques Used to Study Silicon Wafers

Evaluation/Defect Type	Technique Used	Ref.
Cobalt impurity doping	Deep-level transient spectroscopy	[2]
Concentration of vacancies	Fast-diffusing elements (Pt, Au)	[3]
Concentration of vacancies	Excess carrier lifetime	[3]
Convection in molten silicon	X-ray radiography	[4]
Crystallization and defect formation processes	High-resolution electron microscope	[5]
Defect formation and propagation	High-resolution electron microscope	[6]
Dislocation concentration	Etch-pit-count technique	[7]
Dislocation density	Secco etch	[8]
Dislocation movement	X-ray sensing cine camera with television	[9]
Dislocation striations	Spreading resistance technique	[10]
Dislocation striations	Preferential etching	[10]
Inclination fringes	Transmission electron microscopy	[11]
Interstitial oxygen striations	Micro-FTIR	[12]
Iron impurity concentration	Surface photo voltage measurement	[13]
Melt flow pattern	Tracer method	[4]
Molten silicon flow	Double-beam x-ray visualization	[14]
Nitrogen distribution	Secondary ion mass spectroscopy	[15]
Nitrogen doping	Infrared absorption technique	[16]
Nitrogen doping	Secondary ion mass spectroscopy	[16]
Nitrogen doping	Charged particle activation analysis	[16]
Nitrogen-oxygen-vacancy defects	FTIR	[17]
Oxidation-induced stacking faults	Wright etch	[18]
Oxygen concentration/doping	Infrared absorption technique	[19]
Oxygen concentration/doping	Photoluminescence spectroscopy	[19]
Oxygen concentration distribution in melt	Micro-FTIR	[20]
Oxygen distribution	Micro-FTIR	[14]
Oxygen distribution	X-ray topography	[14]
Oxygen micro-segregation	Interface demarcation	[21]
Oxygen micro-segregation	Spreading resistance technique	[21]
Oxygen micro-segregation	Infrared absorption	[21]
Oxygen micro-segregation	High-resolution etching	[21]
Oxygen micro-segregation	X-ray topography	[21]
Oxygen precipitation	Transmission electron microscopy	[22]
Oxygen precipitation	Selective etching	[22]
Oxygen precipitation	Fourier infrared spectroscopy	[22]
Oxygen precipitation	Laser scanning tomography	[22]
Oxygen precipitation/distribution	Transmission electron microscopy	[23]
Oxygen precipitation/distribution	Differential infrared absorption	[23]
Oxygen solubility in melt	Electrochemical solid ionic sensor	[24]
Pendellösunge fringes	Transmission electron microscopy	[11]
Plastic deformation	Gamma-ray diffractometry	[25]
Quartz crucible	TEM imaging	[26]
Quartz crucible	High-resolution TEM	[26]
Quartz crucible	Energy dispersive x-ray analysis	[26]

TABLE 6.1 *(Continued)*

Quartz crucible	Electron energy loss spectroscopy	[26]
Quartz crucible	Electron diffraction	[26]
Silicon ingot material	Secco etchant	[8]
Solid–liquid interface	X-ray radiography	[27]
Structure of dislocations	Synchrotron white x-ray topography combined with topotomography	[28]
Swirl defect formation	Interface demarcation	[21]
Swirl defect formation	Spreading resistance technique	[21]
Swirl defect formation	Infrared absorption	[21]
Swirl defect formation	High-resolution etching	[21]
Swirl defect formation	X-ray topography	[21]
Swirl defects	Chemical etching	[29]
Swirl defects	Chemical etching	[30]
Swirl defects	Metal (Cu and Li) decoration	[31]
Swirl defects	Preferential etching	[31]
Swirl defects	SEM in EBIC-mode	[31]
Swirl defects	Transmission electron microscopy	[32]
Swirl defects	X-ray topography	[33]
Swirl defects	X-ray topography	[30]
Swirl defects	X-ray topography	[29]
Swirl defects	X-ray transmission topography	[31]
Thermal profiles of crystals	Eddy current sensor	[34, 35]
Thermal profiles of crystals	Infrared imaging system	[36]
Thermally induced microdefects	Infrared absorption technique	[37, 38]
Thermally induced microdefects	Etching and optical microscopy	[37]
Thermally induced microdefects	Transmission electron microscopy	[37, 38]
Thermally induced microdefects	Preferential etching	[38]
Thickness fringes	Transmission electron microscopy	[11]
Vacancy concentration	Fast-diffusing metals (Pt, Au)	[3]
Vacancy concentration	Excess carrier lifetime study	[3]

from the particles resting on the surface of the polished surfaces. Metallic contaminations were evaluated using laser microwave photoconductance studies on ultraviolet-irradiated silicon wafers. Lee and Khong [41] reported that the metallic contamination enhances the minority carrier recombination lifetime, and surface contaminations on the wafer can be extracted using this technique. With the help of a bright-field infrared laser interferometer, Nakai et al. [42] have characterized the stacking faults and perfect dislocations in the grown-in Czochralski (CZ) silicon crystals. They were successful in quantifying the density of these stacking faults.

Rutherford backscattering spectroscopy and high-resolution x-ray diffraction have been used as a structural characterization method to evaluate the finished silicon wafers. However, these techniques are limited in terms of complexity and sensitivity; moreover, the evaluation takes a long time and requires some effort on the part of the analyzer. Along with conventional optical spectroscopic methods, photoacoustic methods [1] have been looked at as a means to realize a practical (i.e., an easy-to-operate, nondestructive approach), fast, and highly sensitive evaluation.

When a material is excited with an intensity-modulated energy source, part or all of the excitation energy is converted to thermal energy, and the thermal properties can be altered by absorping the incident photon energy. The heating of the illuminated region leads to a temperature rise in the interaction volume, which in turn causes a thermal expansion of the sample surface called photoacoustic displacement. Since the degradation of thermal conductivity due to subsurface damage leads to significant thermal expansion, measuring displacement of the sample surface caused by thermal expansion provides useful data on possible defects near the silicon wafer surface, especially in terms of buckling and displacement. This means a unit of photoacoustic displacement amplitude is generally very small. Choi et al. [1] realized that this method is sensitive to the presence of lattice defects in the surface layer of a silicon wafer; in addition, the entire wafer can be scanned in a short time.

6.3   Atomic-Force Microscope Studies on Surfaces

Atomic-force microscopes are useful in exhibiting the quantitative roughness, root mean square, roughness average, and peak to valley on surfaces of typical polished silicon wafers. They can be operated either in contact mode or tapping mode. This tool can be used to scan a small selective area to assess the pit density, distribution, and many other defects present on the wafer surface. It is also useful in determining the degree of polishing on the finished silicon wafers. Figure 6.1 shows the selected surface of a polished silicon wafer scanned for an area of 10 µm × 10 µm [1]. The quality of the wafer and number of defects present on the surface can be viewed to decide whether the wafer is fit for high-density VLSI and ULSI circuits. Araki et al. [43] have used this tool to analyze the atomically flat wafer-surface roughness changes on a hydrogen-terminated silicon (100) surface during high-temperature argon annealing. By using the contact mode and tapping mode, Usuda and Yamada [44] evaluated the silicon surface morphologies, minutely analyzing the changes that took place on the wafer surfaces during treatment with deionized water. This evaluation method is time consuming, especially when scanning a large wafer, but is a powerful tool for analyzing wafer surfaces.

6.4   Auger Electron Spectroscopic Studies

Auger electron spectroscopy (AES) is a common analytical technique used specifically in the study of surfaces and material properties. The method is based on the analysis of energetic electrons emitted from an excited atom after a series of internal relaxation events. The method provides a powerful, practical, and straightforward technique to probe chemical and compositional surface environments. Information can be obtained on the condition of silicon wafers that undergo different chemical treatments and cleaning processes [45]. On silicon surfaces, oxygen and carbon are typically present because of the nascent oxide. By using the AES method, Fontana et al. [46] investigated the carbon contamination on processed silicon wafers and showed that the concentration of carbon is higher on the *n*-type silicon surfaces than on the *p*-type surfaces. However, no difference was found in oxygen content.

Yang et al. [47] used AES to evaluate the chemical state of silicon (111) surfaces cleaned by various liquid reagents, ion sputtering, and plasma treatment. The two most common impurities observed on the silicon surfaces were carbon and oxygen. Experimental

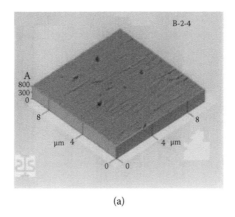

(a)

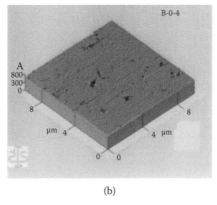

(b)

FIGURE 6.1
Analysis of silicon wafer using atomic force microscopy observations. (From C.-Y. Choi, J.-H. Lee, and S.-H. Cho, *Solid-State Electronics*, **43**, 2011–2020, 1999, and the references therein [1].)

evaluations indicate that liquid reagent cleaning produced surfaces that are heavily contaminated with carbon impurity. When oxidizing reagents were used, carbon contamination was reduced to a minimum value. Sputtering was found to be an unsatisfactory method of cleaning because slices became activated after the sputtering process and adsorbed impurities from the atmosphere when removed from the vacuum system. Plasma treatment was the most effective of the three types of cleaning techniques used, producing surfaces with the lowest carbon concentrations. One exception was found with CF_4 plasma, which was shown to deposit carbon on the surface, possibly due to the dissociation of CF_4 leaving active carbon in the plasma chamber. Typical AES spectra are shown in Figure 6.2 recorded after each surface cleaning procedure.

By using a scanning AES, it is also possible to identify different impurities present in silicon. Thomas III et al. [48] have adapted this approach to evaluate metallurgical-grade silicon for elemental analysis. Figure 6.3 shows AES of an ingot, and the spectra were focused for a specific defect analysis. The recorded composition was found to be position sensitive. Through this study, the team has demonstrated the utility of scanning AES as applied to silicon defect characterization in metallurgical-grade silicon. In addition, impurities detected by bulk analysis methods segregate into various silicide-like particles during solidification. In this way, it is possible to analyze crystal defects and the metallic precipitates that often agglomerate into precipitations.

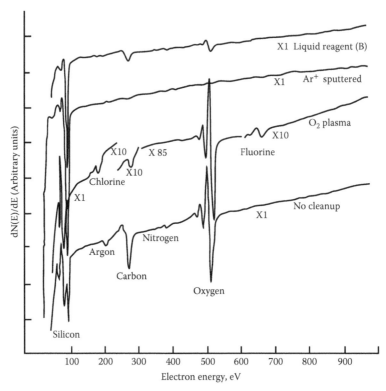

FIGURE 6.2
Typical Auger spectra after various surface cleaning procedures. (From M. G. Yang, K. M. Koliwad, and G. E. McGuire, *Journal of the Electrochemical Society*, **122**, 675–678, 1975, and the references therein [47].)

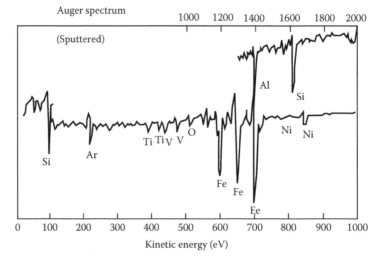

FIGURE 6.3
AES spectra of metallurgical-grade silicon and the elemental analysis. (From J. H. Thomas III, R. V. D'Aiello, and P. H. Robinson, *Journal of the Electrochemical Society*, **131**, 196–200, 1984 [48].)

6.5   Chemical Staining and Etching Techniques

Chemical staining permanently colors and/or alters the appearance of any specific shape or defect that exists in the crystals. Chemical etching is particularly useful, as certain chemical etchants are strongly dependent on crystal defects and defect structures existing in the crystal lattice. These chemical etchants are commonly used to highlight or delineate any defects that are present, thereby magnifying their appearance. Many liquid etchants are in use [49]. Some of them are often used in silicon processing technology to delineate crystal defects present both in CZ and float-zone (FZ)–grown crystals [50]. The development of a rapid, reliable method to evaluate defects in silicon crystals is of major importance to crystal growth technology, wafer preparation, and processing. Chemical etching is the simplest means of detecting the presence of defects in silicon crystals [18]. The most common preferential etchants presently in use are those of Sirtl, Dash, and Secco. Sirtl etchant has been used to delineate defects on most crystal orientations; however, it favors the (111) orientation. Although the Dash etch reveals defects on all planes, the 4–16 h etch time required is unsuitable for rapid evaluation. The Secco etch produces elliptical-shaped dislocation pits on both the (100) and (111) planes, making it more difficult to determine the orientation of etch pits. Wright etchant can be used for oxidation-induced stacking faults, dislocations, swirl, and striations, which are clearly defined by minimum surface roughness or extraneous pitting. A relatively slow etch rate (~1.0 µm/min) at room temperature provides etch control. The long shelf life of this etch allows the solution to be stored in large quantities. Table 6.2 lists some of the popular chemical etchant mixtures for defect delineation applications.

TABLE 6.2
Various Chemical Wet Etchants Used for Defect Delineation in Silicon

Etchant Name	Chemical Composition	Application	Shelf Life	Ref.
Dash	1 ml HF 3 ml HNO_3 10 ml CH_3COOH	(111) or (100) orientation silicon wafers; known to be best for *p*-type silicon wafers (15 h typical etch duration)	8 h	[50]
Sirtl	1 ml HF 1 ml Cr_2O_3	(111) orientation silicon wafers (5 min etch duration)	5 min	[50, 51]
Secco	2 ml HF 1 ml $K_2Cr_2O_7$ (5 M in H_2O)	(100) or (111) orientation silicon wafers	5 min	[50, 52]
Secco	2 ml HF 1 ml Cr_2O_3 (5 M in H_2O)	(100) or (111) orientation silicon wafers	5 min	[49, 50]
Wright	60 ml HF 30 ml HNO_3 60 ml CH_3COOH 60 ml H_2O 30 ml solution of 1g Cr_2O_3 in 2 ml H_2O 2 g $(CuNO_3)_2 \cdot 3H_2O$	(100) or (111) orientation silicon wafers (5–30 min etch duration)	Long life	[18, 50]
White	1 ml HF 3 ml HNO_3	*p-n* junction etch	–	[49]
CP-4	3 ml HF 5 ml HNO_3 3 ml CH_3COOH	*p-n* junction etch (10 s to 3 min)	–	[49, 53]

The concentration of dislocations in single crystals of silicon was investigated using the etch-pit-count technique. The large concentration of etch pits at the peripheries of the etched wafers is explained by spontaneous generation of dislocations during the growth of the crystals. Impurity doping striations can also be detected with preferential etching [10].

Wierzchowski et al. [22] used Wright etchant to evaluate large oxygen precipitates in silicon. The selective etching patterns obtained using this etchant solution revealed several forms of etching pictures, as shown in Figure 6.4. In the sample annealed for 5 h at 750°C (Figure 6.4a) one can distinguish the following structures: the "boat-shaped" etch pit, denoted "a"; the "rhombic," denoted "b"; and the "symmetric elongated," denoted "c." In the sample annealed with a combined cycle, including 5–6 h heating with the temperature increasing from 600° to 1050°C and further 6 h of annealing at 1050°C (as shown in Figure 6.4b), a generally higher concentration of dislocations is observed, as well as

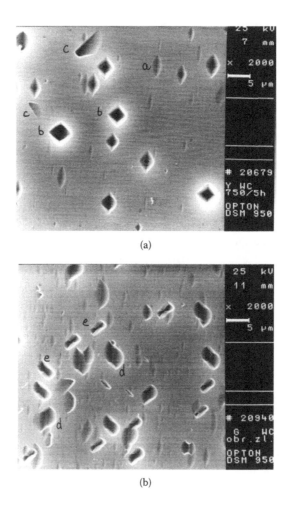

(a)

(b)

FIGURE 6.4
Representative SEM images of selective etching patterns of thermally annealed CZ silicon obtained using Wright etchant at <110> cleavage faces: (a) sample annealed for 5 h at 750°C revealing three characteristic types of etch pits (denoted "a," "b," and "c"); (b) sample annealed with combined cycle, including 5–6 h heating with increasing temperature from 600 to 1050°C, where two other types of complex etch pits, denoted "d" and "e," are visible. (From W. Wierzchowski, K. Wieteska, W. Graeff, M. Pawłowska, B. Surma, and S. Strzelecka, *Journal of Alloys and Compounds*, **362**, 301–306, 2004, and the references therein [22].)

two other kinds of more complex etching pits denoted as "d" and "e." The evaluations were supported by x-ray topographic investigations. The method of using etchants clearly shows the strength of this approach in evaluating the quality of silicon crystals. However, the approach is destructive in nature.

de Kock and van de Wijgert [31] investigated the effect of impurity doping on CZ-grown crystals with electrically active impurities (such as boron, gallium, antimony, phosphorus, and arsenic) and isoelectronic impurities (such as tin) on the formation of swirl defects using preferential etching, copper and lithium decoration, x-ray transmission topography, and electron beam-induced current (EBIC)-mode scanning electron microscopy (SEM). The configuration of the decorated defects corresponded to the swirl pattern revealed by the Sirtl etching reported by Yasuami et al. [33]. Micro-inhomogeneities such as swirl defects were evaluated by etching and x-ray topography [29] in FZ crystals. Oxygen micro-segregation and swirl defect formation in CZ-grown silicon were studied by employing interface demarcation, spreading resistance measurements, infrared absorption, high-resolution etching, and x-ray topography [21]. Thermally induced microdefects in CZ silicon crystals were investigated using an infrared absorption technique, etching/optical microscopy, and transmission electron microscopy [37].

Sirtl etchant has been widely used to reveal crystal defects in silicon, particularly for (111) surface orientations. Secco has formulated an etch composed of two parts hydrofluoric acid (HF) to one part 0.15 M $K_2Cr_2O_7$ to reveal dislocations in (100) silicon surfaces. Ultrasonic agitation is required to minimize etching artifacts due to gas bubble formation during the process and to minimize the etching time necessary to achieve easily visible and identifiable dislocation etch pits [8], especially dislocation densities in <100> silicon. Experiments have shown that under certain conditions, cavitation of the ultrasonics can generate anomalous defect etch pits. A new defect etch formulation was developed for evaluating <100> silicon ingot material that does not require ultrasonic agitation. It consists of two parts HF and one part 1M CrO_3. The etching time with this formulation compared to Secco etch with ultrasonics is reduced by one-half in developing etch pits of equivalent size. This composition is useful for revealing dislocations in 0.6–15.0 Ω-cm n-type and p-type silicon. In addition, dislocation etch pits are readily developed in more heavily doped <100> silicon crystals using a mixture of 2 parts HF, 1 part 1M CrO_3, and 1.5 parts H_2O. Heavily doped <100> silicon ingot material can also be evaluated using a simple water dilution of the etchant to 2 parts HF, 1 part 1M CrO_3, and 1.5 parts H_2O.

In the double-layered CZ process, the melt temperature under the growing crystal is expected to be slightly higher than the melting point and to have a small vertical temperature gradient (~1°C/cm). This facilitates the quenching of the melt under the growth interface. Regarding the solidification rate, the segregation of oxygen is an important factor to consider, mainly because the oxygen concentration distribution in the quenched melt depends strongly on the fluctuation in the solidification rate when the oxygen segregation coefficient K_0 does not equal to 1. Many reports have been published on the values of the equilibrium oxygen segregation coefficient K_0 (e.g., 0.21, 0.25, 1, and 1.25). Kawanishi et al. [20] reported that the oxygen segregation coefficient K_0 is close to unity. This fluctuation leads to striated patterns. Figure 6.5 shows that the striated pattern in the melt was raised by the difference in oxygen concentration. These findings suggest that the low oxygen content at the melt surface flowed into the growth region, forming oxygen striations together with the high oxygen content introduced by what is known as the Cochran flow.

Crystal evaluation is a major research area, drawing many chemical engineers to the field of crystal growth and evaluation methodologies. Many scientific groups actively participate in studying chemical staining [54–59], discovering new etchants for silicon crystal

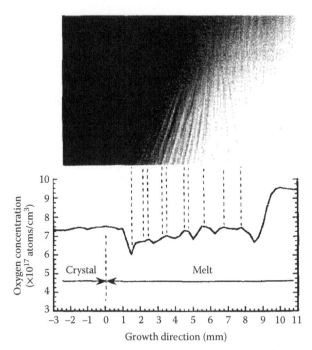

FIGURE 6.5
Striated patterns revealed by Wright etching are attributable to differences in oxygen concentration. (From S. Kawanishi, S. Togawa, K. Izunome, K. Terashima, and S. Kimura, *Journal of Crystal Growth*, **152**, 266–273, 1995, and the references therein [20].)

evaluation [60–65], and studying defects and defect-related engineering [66–75]. The only issue is that this approach is destructive and cannot be carried out on process-bound silicon wafers. These methodologies are nonreversible, and only test wafers should be analyzed.

6.6 Contactless Characterization

Contactless measurements are attractive because they do not require any sample preparation. Some parameters that are considered contactless are electrical resistivity, sheet resistance, wafer flatness (global and local), epitaxial layer thickness, junction depth, implantation dose, surface-adsorbed impurities, presence of surface particles, bulk and localized impurities, bulk lifetime, surface-oxide interface charge, insulating layer thickness, critical dimensions of patterns defined, crystal defects, built-in stress, dielectric/metal voids, metal electromigration, metal thickness, and dielectric layer porosity [76]. A contactless capacitance doping profiling measurement technique uses a contact held in close proximity to the semiconductor wafer. A novel carrier lifetime characterization approach uses infrared radiation from a blackbody transmitted through the semiconductor wafer. Photoluminescence is most commonly used at low temperatures to identify impurities and to determine low doping densities in silicon at different locations. Localized stress is typically measured with the help of Raman spectroscopy. A more recent method produces two-dimensional stress maps through polariscopy—a nondestructive, highly sensitive, and quantitative mapping technique for residual strains in semiconductor process–bound wafers [77]. Recently, a noncontact sheet resistance and leakage current measurement technique has been developed. The big advantage with this is that even process-bound wafers can be analyzed for any of the issues listed earlier. For a silicon wafer, only the electrical parameters are important, along with dopant and defect evaluations.

Stefani et al. [34] introduced a computer-controlled eddy current system for *in situ* monitoring of CZ silicon crystal growth. Eddy current testing involves observing the behavior of electromagnetic radiation penetrating the silicon crystal and its subsequent reflection. This behavior is outlined by Maxwell's equations, and it is the solution of the electromagnetic field boundary-value problem that relates voltage measurements from the eddy current probe to the changing electrical conductivity of the crystal. Due to strong temperature dependence of the electrical conductivity in solid silicon, it is possible to determine thermal profiles within the growing crystal. Similarly, the axial thermal profiles of silicon crystals were measured experimentally by Choe et al. [35] using eddy current technology. The intrinsic conductivity changes in the crystal resulting from the cooling were measured in terms of eddy current amplitude and phase responses, and the axial thermal profiles of the growing crystal were subsequently derived from the results.

Zeni et al. [78] have shown the feasibility of a new nondestructive, contactless, optical tomography approach to the one-dimensional doping profile reconstruction in semiconductor materials. The method does not require the realization of any specific test structure on the sample under investigation. The problem is formulated as an inverse one, and the choice of data and unknowns permits the investigator to avoid local minima. Furthermore, no *a priori* assumption on the doping profile shape is required. Numerical results indicate that the proposed approach is useful for reconstructing typical doping profiles with relatively high accuracy. It can be used to measure dopant concentration profiles and the lateral distribution of dopant species in fresh silicon wafers.

Nonlinear optical characterization of the surface of silicon wafers is reported by Reif et al. [79] for the detection of external stress. A simple technique was proposed to discriminate those sections of the wafer that are ready for use in further applications from those that are not useable for proper device fabrication, thus enabling the selection of appropriate material from as-grown crystals. Due to ever-increasing integration of semiconductor devices, requirements have become more stringent in terms of the quality of the substrate wafer, the gate oxide layers, and the metallic contacts. Thus, methods for preselecting appropriate substrates, as well as online growth monitoring and nondestructive characterization of subsequent layers, are highly desired. Optical techniques are a good fit for this field because they are essentially nondestructive and can be applied *in situ*. By using photoluminescence, metallic contamination can also be analyzed [80] as a part of contactless characterization methodology.

6.7 Deep-Level Transient Spectroscopy

Deep-level transient spectroscopy (DLTS) is an experimental tool for studying electrically active defects in semiconductors and for measuring defect parameters and their concentrations. DLTS investigates the defects present in the space charge region, and for this either Schottky or simple *p-n* junctions are commonly used. This technique has a higher sensitivity than any other tool and can detect impurities and defects at concentration levels of 1 in 10^{12} of the host silicon atoms.

To better understand the origin of defects that affect device performance, such as large-area charge coupled devices (CCDs), we need to characterize deep levels in silicon wafers at various stages of processing. Derivative surface photovoltage spectroscopy (DSPS) measurements can reveal the presence of deep centers (in the range of 10^{11}–10^{13} cm^{-3}) around the middle of the energy gap in the as-grown and the heat-treated FZ silicon wafers. Jastrzebski and Lagowski [81] revealed the presence of two deep levels—0.23 and 0.56 eV above the valence band in the as-grown and oxidized wafers. Their concentration was found to be about 10^{11}–10^{12} cm^{-3} in the virgin crystal and was increased about one order of magnitude with the oxidation step. The effect of these deep centers on the quantum efficiency and the fixed pattern noise in CCD imagers was evaluated for better performance. Simoen et al. [82] extended these studies further to evaluate the behavior of oxygen in oxygen-doped, high-resistivity n-type FZ silicon using a combination of analytical techniques. In the as-doped material, a large number of deep levels has been observed using this DLTS method.

Interstitial iron [Fe$_i$] is a slow diffuser compared to copper and nickel, and can be frozen in the interstitial form in silicon wafers by fast cooling after an iron homogenization annealing cycle at 1100°C for 15 min. Measurement of Fe$_i$ is an excellent tool for surface contamination detection, and is important to the VLSI and ULSI wafer industry. Using DLTS methods, Ryoo and Socha [83] have shown that the room-temperature transformation of iron, with time, from the interstitial form to iron-acceptor complexes was clearly observed by measuring the decrease in concentration of iron interstitials and the increase of iron-boron complexes. Figure 6.6 shows the typical DLTS spectrum for Fe$_i$ and FeB, which appears at 270 and 60 K, respectively. The minimum Fe$_i$ concentration that affects the minority carrier diffusion length significantly is 1.0×10^{11} atoms/cm^3. Nauka and Gomez [84] reported on the iron impurities present in the front-end operations of a 0.35 μm complementary metal-oxide semiconductor (CMOS) technology with DLTS, and its detection is mostly in the wafer bulk around 1 μm below the surface region.

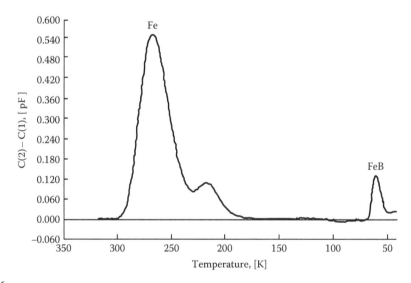

FIGURE 6.6
DLTS spectrum of *p*-type silicon wafers showing Fe and FeB. (From K. Ryoo and W. E. Socha, *Journal of the Electrochemical Society*, **138**, 1424–1426, 1991 [83].)

One of the most challenging problems is the fabrication of large-area crystals free from metallic concentrations to fabricate perfect devices. The interstitial and substitutional defects of transition metals generate deep energy levels in the forbidden gap of silicon. In cobalt-doped crystals, it is generally assumed that the energy levels observed with DLTS are primarily caused by the substitutional defects with a donor level at $E_v + 0.4$ eV [2] and an acceptor level at $E_c - 0.38$ eV. Copper [85], oxygen [86], chromium [87], and carbon [88] have been evaluated in a similar approach.

DLTS studies by Bleka et al. [89] performed on p^+-n-n^+ silicon diode detectors produced from low-doped ([P] = 5.0×10^{12} cm^{-3}), high-purity FZ wafers. After irradiation with 6 MeV electrons to a dose of 5.0×10^{12} cm^{-2}, the well-known vacancy-oxygen (VO), double negatively charged vacancy-vacancy ($V_2^{(=/-)}$ or simply V_2 (=/-)) and single negatively charged vacancy-vacancy (V_2(-/0)) complexes are observed. There is the clear presence of a defect (X) with an energy level slightly shallower than that of V_2 (-/0) and a concentration about one-quarter that of V_2. Further analysis of the defect properties reveals an energy level 0.37 eV below the conduction band and an apparent capture cross-section of 1.0×10^{-14} cm^2.

6.8 Defect Decoration by Metals

Crystal defect decoration is a well-established method to facilitate the visualization of small crystal defects in silicon wafers that are otherwise scarcely revealed by etching alone. Metals such as copper or lithium precipitate preferentially at grown-in or process-induced crystal defects, such as dislocations under silicide formation (Cu_3Si, LiSi). This decoration promotes the visualization of the defects by preferential etching. Copper was the first metal used to explore this technique [90], followed by nickel [91]. Nickel was found to form haze defects only when introduced prior to heat treatment as an intentional contaminant. Copper was found to be a background contaminant-forming haze. An estimate of the amount of copper necessary to form this background yielded an upper limit of 5.0×10^{13} atoms/cm^3 [92]. Idrisi et al. [93] delineated the silicon-on-insulator (SOI) wafers and studied the influence of metal decoration (copper and lithium) on the preferential etching of crystal defects in silicon materials by measuring parameters such as removal rate, activation energy for the etching process, and selectivity. The effect of impurity doping on CZ-grown crystals with electrically active impurities (such as B, Ga, Sb, P, and As) and isoelectronic impurities (such as Sn) on the formation of swirl defects was investigated by means of preferential etching, Cu and Li decoration [31], x-ray transmission topography, and EBIC-mode SEM.

Kulkarni et al. [94] proposed a theoretical and experimental analysis of macrodecoration of defects in monocrystalline silicon by mainly emphasizing copper decoration. According to this model, the decorating etching creates pits at the microdefect sites as a result of the difference between the etching rate of the microdefect site and that of the surrounding defect-free silicon. The quality of the microdefect decoration improves by precipitation of copper as Cu_3Si particles around microdefects, known as copper precipitate colonies. The decorating efficiency of an etchant at the microdefect site, defined as the ratio of rate of increase in the microdefect–related etch-pit depth to the maximum possible rate of increase in the pit depth, can be given as a function of the effective kinetic and mass transport resistances at the site and on the perfect silicon surface. The propagation of etch pits is as important as their formation for an efficient microdefect decoration. Therefore, kinetic effects must be dominant over mass-transport effects to achieve an efficient microdefect decoration in silicon surfaces.

6.9　Electron Beam and High-Energy Electron Diffraction Studies

An electron beam is suitable for measuring the resistivity distribution of semiconductor crystals. Indeed, the beam is more convenient than an optical light for beam-scanning purposes because the beam can be electrostatically or magnetically deflected and controlled. The electron beam is more suitable for quantitative analysis because the current and energy are easily measured and modulated. Resistivity variations were measured in semiconductors using this method [95], and resistivity values were found to be quite reasonable.

High-energy electron diffraction (HEED) is used to study the surface condition after the silicon wafers have been treated with standard peroxide cleaning methods. Henderson [96] has reported that by using HEED patterns, one can look for the presence of SiC diffraction patterns to confirm the presence of carbon on the silicon surface. Different cleaning methods were used for different surface conditions, and the presence of carbon was evaluated. After using AES and HEED, the data was studied to assess the silicon surfaces. Similar studies on carbon were evaluated by Fontana et al. [97] using a photoemission electron microscope, and by Nakamura et al. [98] using photoluminescence measurements.

6.10　Flame Emission Spectrometry

In flame emission spectrometry, the sample under study is nebulized, or converted into a fine aerosol, and introduced into a flame where it is desolvated, vaporized, and atomized, all in rapid succession. Subsequently, the atoms and molecules are raised to the excited states via thermal collisions with the constituents of the partially burned flame gases. Upon their return to a lower or ground electronic state, the excited atoms and molecules emit radiation characteristic of the sample components. This emitted radiation passes through a monochromator that isolates the specific wavelength for analysis. A photodetector measures the radiant power of the selected radiation, which is then amplified and sent to a readout device or a microcomputer system to identify the characteristic wavelengths of those specific elements.

Analyzing and controlling sodium in the processing materials are important requirements for producing electrically stable metal-oxide semiconductor field-effect transistor (MOSFET) devices. Neutron activation and mass spectrography have been reported to provide accurate measurements of sodium in silicon and silicon dioxide. Yurash and Deal [99] reported on a new, simple method of determining sodium content at very low levels. With a sufficient amount of sample, it is possible to measure sodium concentrations down to ppb levels or less, a critical value necessary when fabricating electrically stable, reliable VLSI and ULSI devices. Though the presence of sodium is not recorded in the bulk silicon wafers, it is recorded on the process-bound wafers and on improperly cleaned polished silicon surfaces. Because so many chemicals are used in the fabrication process and come into contact with the silicon wafers, this is a strong source of sodium contamination. Knolle and Retaiczyk Jr. [77] systematically analyzed possible sodium contamination originating from different sources. Potassium is another contaminant. Knolle [100] has reported the details of such contamination and its effect on silicon surfaces. Chemical analysis using this method helps in identifying contamination sources and avoiding them if possible.

6.11 Four-Point Probe Technique for Resistivity Measurement and Mapping

Silicon wafer resistivity measurements are made on the flat ends of the crystal using the four-point probe technique. A current I is passed through the outer probes, and the voltage V is measured between the inner probes. The measured resistance (V/I) is converted to resistivity (Ω-cm) using the relationship

$$\rho = \left(\frac{V}{I}\right) 2\pi S$$

where S is the probe spacing in centimeters. The setup can measure resistivity of thin films and diffused/implanted regions. The four probes are arranged in a linear fashion such that two of the outer probes are connected to a current source and the inner probes are connected to a voltage meter, as shown in Figure 6.7. As current flows between the outer probes, the voltage drop across the inner probes is measured.

Sheet resistance of any material is defined as the electrical resistance of a square sheet of film. It is independent of the size of the square, but depends only on the film resistivity and the thickness of the film. The four-point probe works only on blank wafers with a continuous film on top and does not work on patterned wafers. In case if the samples are smaller in the size and are relatively compared to that of the physical length of 3S, geometric correction factors are to be included in the sheet resistance calculations. This method is destructive in nature, as the four points leave a dent on the films due to mechanical contact. Thin conducting films need to be isolated with an insulator beneath. Otherwise, the measured sheet resistance values will be erroneous and the real values will be missed during the analysis.

The four-point probe is set up in such a way that the DC current is adjusted to deliver a current value of 0.453 mA through the outer probes. This simplifies the sheet resistance

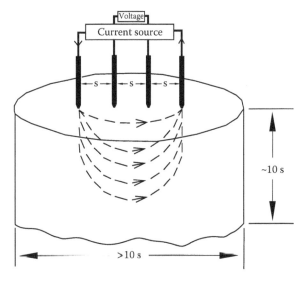

FIGURE 6.7
Typical four-point probe setup to measure wafer resistivity and sheet resistance measurements.

measurement by letting parameters arrive at sheet resistance values equal to 10 times the voltage value (in mV) measured between the two inner probe tips. The other way is to adjust the S value, with equal spacing between the linear arrangement, such that the $2\pi S$ value reaches unity. In this case, the equation becomes simpler, and the sheet resistance value is estimated by measuring the ratio directly between the voltage and current values.

The in-line array and square array configurations are shown in Figure 6.8. A current, I, is sent through probe points 1 and 4, and the voltage difference, V, is measured between points 2 and 3. From the sample geometry and the measured resistance, the resistivity for bulk material and the sheet resistance (ρ_s) for diffused layers can be calculated [101]. Probe loading, probe configuration, and tip radius also play a role in these measurements. Tong and Dupnock [102] have made certain recommendations for the measurement of resistivity of the wafers. Use of a broad probe tip is recommended, since a small tip radius might pierce through the diffusion and cause erratic readings. Probe points should be loaded with between 30 g and 60 g and lowered as slowly as possible to avoid accidentally breaking the wafers. Finally, the current levels should be adjusted so that the voltage difference between the two inner probes is less than 100 mV.

Probe material is an issue in view of the work function difference. Many options exist, including mercury. Severin and Bulle [103] reported the measurement details on n-type silicon on bulk and thin layers. Mercury contacts are extremely useful when assessing very thin, epitaxially grown, heterotype silicon layers or shallow ion-implanted samples, where conventional pointed probes cause substrate short-circuit or even penetrate to create permanent damage to the surface.

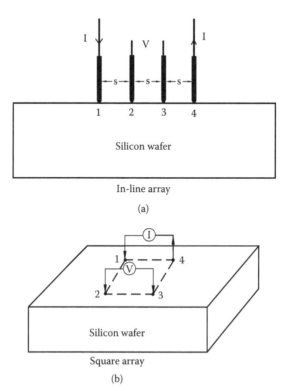

FIGURE 6.8
(a) In-line and (b) square array configuration of the four-point probe.

Recently, attention has been drawn to applying correction factors to large-diameter doped wafers to measure sheet resistance using a linear four-point probe. The homogeneity of doped wafers may be determined by measuring sheet resistance at a number of points using a conventional in-line four-point probe. Correction factors have been derived that take into account varying orientations of the probe array with respect to the wafer radius, making it possible to automate such measurements using standard *x-y* wafer probing equipment. Perloff [104] has covered this issue in depth, and it is strongly recommended for the researcher for mapping the wafers. Simplified expressions have also been obtained that relate the correction factor to a single dimensionless quantity when the probe separation is small (≤10%) in comparison to the wafer radius. Perloff et al. [105] has reported spatial distribution of dopant impurity with graphic display formats along with random measurements. Presently, commercial systems are available for this analysis.

Albers and Berkowitz [106] have reported on the relationship between the two-probe and linear four-probe resistance on nonuniform structures. Kopanski et al. [107] verified the use of these two techniques for silicon wafers. Sugawara et al. [108] applied finite difference analysis to determine the radial phosphorus dopant distributions in single crystals of silicon. The concentration was numerically simulated by solving the Navier-Stokes continuity and energy equations for flow and temperature fields in silicon melt and solving the dopant diffusion and segregation equations for the dopant distribution in the melt and in the grown crystal.

6.12   Fourier Transform Infrared Spectroscopy Measurements for Impurity Identification

Infrared spectroscopy has been a workhorse for material analysis for over several decades. One important aspect of this technique was the measurement of infrared absorption spectra by means of Fourier transform infrared spectroscopy (FTIR). FTIR is a technique that is used to obtain an infrared spectrum of radiation absorption, emission, photoconductivity, or Raman scattering of any material. When infrared radiation is passed through the sample, some of the radiation is absorbed and some of it is passed through. The resulting spectrum represents the molecular absorption and transmission, creating a molecular fingerprint of the sample. And like a fingerprint, no two unique molecular structures produce the same infrared spectrum. This technique makes infrared spectroscopy useful for several types of analysis. FTIR simultaneously collects spectral data in a wide range for impurity identification, and the technique is often used to evaluate silicon wafers for impurities.

As mentioned, carbon is one of the main impurities present in CZ-grown crystals. And the main source of this impurity is the graphite holder that supports the molten silicon in the quartz liner. Other sources are the heater assembly, the raw silicon material, and the purge gases used to maintain the internal environment of the chamber. Typical carbon concentration ([C]) in crystals grown using the CZ method will be on the order of 1.5×10^{16} to 1.0×10^{17} cm^{-3}. For the measurement of [C] the standard procedure is that the wafers should be polished on both the sides and properly cleaned. Unpolished wafer surface will be irregular and likely to retain unknown carbon residues and the recorded spectra may not reflect the true silicon bulk property. Measurements are generally carried out with a wavenumber resolution of 2.0 cm^{-1} at room temperature. Inoue et al. [109] were successful in detecting various complexes in low-carbon concentrations in CZ silicon. Electron irradiation experiments were carried out by them at an energy of 4.6 MeV with dosage values of 1.0×10^{15} and 1.0×10^{16} cm^{-2} for these studies. A low-pass filter was used to detect very weak absorption, with a good signal-to-noise ratio. Changes were implemented to avoid overlapping vibrations. As a result, [C] could be measured down to 1.0×10^{14} atoms/cm^3, one order of magnitude lower than the conventional detection limit. Very weak absorption lines were observed at 862 and 1108 cm^{-1} in low dosages with 1.0×10^{15} cm^{-2} samples. These were due to carbon interstitial (C_i)-oxygen interstitial (O_i) pairs. The team analyzed the samples and got some interesting results. Assuming that the oscillator strength of C_iO_i is nearly equal to that of C, O, N_i pairs and so on, $[C_iO_i]$ was estimated to be about 1.0×10^{14} cm^{-3}. This means that about 1% of carbon present in the silicon crystal has formed C_iO_i. In high-dose samples, about 1.0×10^{16} cm^{-2}, in addition to these absorption lines, lines at 545, 583, 831, 934, and 1020 cm^{-1} were observed. The lines at 545 and 583 cm^{-1} were known to be due to weaker absorption in addition to the C_iO_i. The 831 cm^{-1} line is due to VO (A-center), and the lines at 934 and 1020 cm^{-1} are due to $C_iO_iSi_i$. The absorbance of 862 and 1108 cm^{-1} lines increased by about one order with the one order increase in dosage. This means that the concentration is roughly proportional to the dosage, and about 10% of carbon formed C_iO_i in this case. The absorbance of the 831 cm^{-1} line was 0.002, meaning the VO concentration was about 1.0×10^{15} cm^{-3}. In contrast, the absorbance from $C_iO_iSi_i$ secondary defects was about 0.0002, and therefore, concentration was one order of magnitude lower, about 1.0×10^{14} cm^{-3}. Figure 6.9 shows the IR absorption spectra from the electron-irradiated samples. Individual wavenumbers are identified on the peaks. Careful study of the recorded spectra reveals a lot of information on the quality of silicon wafers. This is a nondestructive approach.

Nitrogen doping is becoming a key technology for growing defect-free CZ silicon crystals [16]. Because the nitrogen concentration is low and nitrogen has various configurations, it is difficult to quantify the dopant concentrations. In the IR absorption measurement, either the 963 cm^{-1} or 766 cm^{-1} absorption line has been used for both FZ and CZ silicon crystals for this purpose. These were assumed to be due to the N_2Si_2 interstitial N pair. There are several absorption lines in CZ silicon on the higher wavenumber sides of these lines that were attributed to the N-O complex. The absorption lines 963 cm^{-1} and 766 cm^{-1} are assigned to the modes of NSi$_3$ (a four-atom planar cluster) and Si-N-Si (a three-atom, nonlinear cluster) structures. As for the 963 cm^{-1} line group, the 996, 1018, and 1026 cm^{-1} line have been reported and are proposed to sum 963 (N_2), 996 (N_2O), and 1018 cm^{-1} (N_2O_2) lines to get a total N concentration. By using this approach, concentrations as low as 2.0×10^{14} cm^{-3} can be measured using in 10 mm thick samples, and above 5.0×10^{14} cm^{-3} using 2 mm thick samples.

It is believed that nitrogen-oxygen defects form stable nuclei at high temperatures. They have been taken as cluster group of stable configuration of nitrogen-vacancy-oxygen defects. Fujita et al. [17] investigated the binding energy, equilibrium concentration, local vibrational modes, and electrical activity of N_2O, N_2O_2, NO, and NO$_2$ defects in CZ silicon by means of local density functional theory. The binding energy of O with N_2 and the positions of four local vibrational modes of the N_2O center are in excellent agreement with their experiment. The N_2O_2 defect is the most common nitrogen-oxygen defect after N_2O, and it is suggested that the experimentally observed lines at 1018 and 810 cm^{-1} are due to this defect. In contrast with N_2O and N_2O_2, it was found that NO and NO$_2$ defects have very low concentrations, and this raises questions about their assignment to nitrogen-related shallow thermal donors.

Oxygen in silicon occupies both the interstitial and substitutional positions. Oxygen micro-segregation and swirl defect formation in CZ-grown silicon were studied by employing interface demarcation, spreading resistance measurements, IR absorption, high-resolution etching, and x-ray topography [21]. The concentration of dissolved oxygen at the interstitial position was determined from the intensity of the absorption line at 1107 cm^{-1}. The formation of oxygen precipitates was monitored by analyzing absorption spectra in the range of 1350–900 cm^{-1}. According to the most common interpretation, the absorption band in the range of 1250–1100 cm^{-1} may be attributed to small disc-shaped precipitates, while the band in the range of 1080–1020 cm^{-1} is due to large, mostly octahedral precipitates [22].

Oxygen distribution in the CZ single crystals of silicon is closely associated with melt convection occurring near the growth interface because oxygen atoms in the melt are transported by the fluid motion. Using micro-FTIR, Kawanishi et al. [20] measured oxygen concentration distribution in the growing crystal relating to melt convections. Axial oxygen distributions around the growth interface were measured at different radial distances R, 0, 27, 38, and 43 mm from the center ($R = 0$ mm), as shown in Figure 6.10. Arrows indicate the estimated position of the growth interface from both the shape of the meniscus and

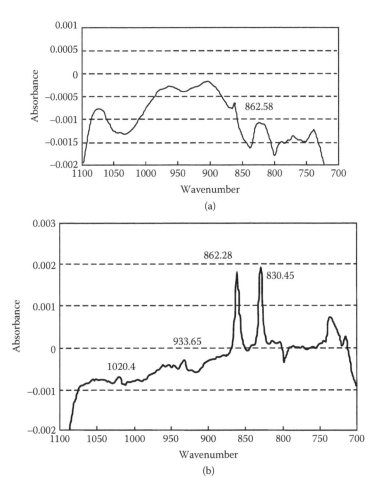

(a)

(b)

FIGURE 6.9

IR absorption spectrum from electron-irradiated sample, [C] = 1.5×10^{16} cm^{-3}. (a) The electron dosage at 4.6 MeV was about 1.0×10^{15} e/cm^2. (b) The electron dosage was about 1.0×10^{16} e/cm^2. (From N. Inoue, S. Yamazaki, Y. Goto, T. Kushida, and T. Sugiyama, "Infrared absorption spectroscopy of complexes in low carbon concentration, low electron dosage irradiated CZ silicon crystal," *210 ECE Meeting Cancun*, Mexico, Oct. 29–Nov. 3, 2006 [109].)

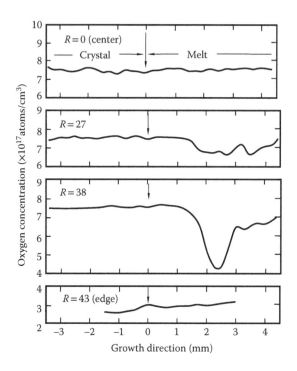

FIGURE 6.10

Oxygen concentration measured by micro-FTIR is plotted against the growth direction in the crystal and melt. Radial dependences are indicated at different radial distances: (a) $R = 0$ mm (center), (b) $R = 27$ mm, (c) $R = 38$ mm, and (d) $R = 43$ mm (edge). Arrows denote the estimated position of the growth interface. (From S. Kawanishi, S. Togawa, K. Izunome, K. Terashima, and S. Kimura, *Journal of Crystal Growth*, **152**, 266–273, 1995, and the references therein [20].)

etching observation. The axial oxygen distribution is plotted against the growth direction in the crystal and melts as follows: at the crystal center ($R = 0$ mm), the oxygen content was found to be 7.5×10^{17} atoms/cm^3 in both the crystal and the melt. At $R = 27$ mm, the oxygen concentration distribution changed, indicating that the oxygen content in the melt had begun to decrease. A further decrease was observed at $R = 38$ mm. At the crystal periphery ($R = 43$ mm), the oxygen content reached 3.0×10^{17} atoms/cm^3 in both the crystal and the melt. The results were supported by specific oxygen striations presenting a flow-like pattern with respect to melt convection.

Thermally induced microdefects in CZ-grown silicon crystals were investigated using an IR absorption technique, etching/optical microscopy, and transmission electron microscopy [37]. Thermal mapping of the grower was obtained via an IR thermovision system [110]. Thermally induced oxygen precipitation and redistribution phenomena in CZ-grown silicon crystals were investigated by means of transmission electron microscopy and differential-infrared absorption (DIR) by Shimura et al. [23]. The presence of oxygen [111–123], carbon [124–128], and nitrogen in silicon crystal has both advantages and disadvantages for silicon wafers used for VLSI and ULSI circuit fabrication. Their presence is continual, and many scientific teams are working on these trace materials in silicon lattice sites. Recently, Jiang et al. [129] reported that by using this method, a high concentration of Ge-doped, CZ single crystals of silicon were measured at room temperature and at 10 K using SEM-energy dispersive x-ray spectroscopy. A new peak appears at the wavenumber of 710 cm^{-1}. This peak becomes clearer as the concentration of germanium in silicon increases. The result indicated that FTIR technology is capable of determining the germanium concentration in a single crystal of CZ silicon.

6.13 Gas Fusion Analysis

In the last few decades, gas fusion analysis (GFA) has became the routine method of analyzing interstitial oxygen [Oi] in heavily doped silicon materials in which the free carrier absorption interferes with the Si-O absorption. GFA is a carrier gas heat extraction method for oxygen in pure materials, such as in high-purity silicon. With this process, a small portion of silicon is melted in a small pure-graphite double crucible under helium carrier gas flow at 1600°C [130]. The graphite reduces the silicon oxide in the melt and converts it to CO and CO_2. A catalyst oxidizes the CO to CO_2, and the oxygen concentration in the carrier gas can be determined by IR absorption in a gas cell. This method provides an inexpensive and simple way to determine [Oi] concentration values independent of dopant concentration.

GFA of oxygen present in silicon has been reported by Shaw et al. [131]. They have reported on the origins of oxygen evolution observed in silicon crystal lattice and strongly supported the presence of oxygen that originates in silicon. It has been shown that inert GFA of silicon yields three separate oxygen evolution peaks arising primarily from surface contamination, surface oxide, and bulk oxygen, respectively, as shown in Figure 6.11. The surface oxide is associated primarily with the second intermediate peak in the time sequence. Within a reasonable range of oxide thicknesses, it can be separated from the third peak by an appropriate program of power inputs. The third peak consists primarily of the bulk oxygen, but is found to increase as the surface is oxidized. For the present sample's geometry, this increase can be on the order of 10% of the typical oxygen level seen in CZ silicon. This contribution is tentatively attributed to a stable surface or near-surface oxygen on the oxidized samples. With nonoxidized samples, the range of oxygen concentration varies. This experiment has allowed us to gain a great deal of information on the presence of oxygen in single crystals of silicon. No information is available for other impurities present.

6.14 Hall Mobility

Carrier mobility is the ratio of the velocity to the electric field. High mobility means that the velocity of the charge is high. Mobility is an important parameter of any semiconductor materials used in devices. The Hall effect is the most common method to measure the mobility of charge carriers in semiconducting materials. It is used to determine the carrier concentration, carrier type, and, when coupled with a resistant measurement, the mobility of conducting materials. The traditional Hall measurement uses a DC magnetic field, but AC-field Hall measurements are popular for mobility estimations as well.

Single-crystal silicon mobility values for *n*-type and *p*-type do not change much, but in the case of polycrystalline silicon, the mobility values differ. The Hall method is often used to evaluate polycrystalline rods used for FZ silicon growth. Studies on zone refinement

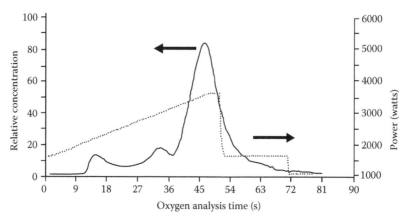

FIGURE 6.11
Power input (dotted line) and oxygen evolution vs. time for a sample with surface contamination and oxide. (From R. W. Shaw, R. Bredeweg, and P. Rossetto, *Journal of the Electrochemical Society*, **138**, 582–585, 1991 [131].)

will provide information on the quality of crystals when they undergo multiple passes for refinement. Olshefski et al. [132] have used direct polycrystalline rods to evaluate the average carrier densities for FZ silicon. Both *n*-type and *p*-type impurity carrier concentrations were evaluated with this method. Similarly, Hunt et al. [133] have adopted the method to determine the boron and phosphorus net concentration values for solar cell applications. This method is considered sufficient for making quality-control decisions with respect to boron and phosphorus concentration levels in metallurgical-grade silicon and the ingots grown from it. Hall measurements are considered standard and quite useful to evaluate silicon wafers [134] and ingot levels.

6.15 Mass Spectra Analysis

Mass spectrometry is an analytical tool used for measuring the molecular mass of a sample. This is a powerful technique for identifying unknowns, studying molecular structure, and probing the fundamental principles of chemistry. The unknown samples are introduced into the source region, and then neutral sample molecules are first ionized and then accelerated into the mass analyzer stage. The mass analyzer separates the ions, either in space or in time, according to their mass-to-charge ratio [135]. After the ions are separated, they are detected and the signal is transferred to a data system for analysis. This method can determine the unknowns present in the silicon, particularly those present as intentional and unintentional impurities.

By using this method, Kennicot [136] first reported the presence of nitrogen in silicon by using solid-source mass spectrometry. Mass spectra analysis of impurities and ion clusters, present in both amorphous and crystalline silicon films, were reported by Feldman and Satkiewicz [137]. With this approach, the relationship between cluster distribution and atomic neighbor environment was discussed. This process has some similarities with secondary ion mass spectroscopy. We shall discuss this method in more details in Section 6.22 of this chapter.

6.16 Minority Carrier Diffusion Length/Lifetime/Surface Photovoltage

Minority carrier lifetime is one of the most important parameters in the operation of many semiconductor devices. From a device operation point of view, a high carrier lifetime is desired, for example, to increase the refresh time in random-access memory (RAM) circuits and to achieve high efficiency for photovoltaic devices. Consequently, it is necessary to optimize the lifetime according to the device structure [1]. Minority carrier diffusion length and other techniques are also used to characterize the silicon crystals [134]. A novel carrier lifetime characterization approach uses IR radiation from a blackbody transmitted through the semiconductor wafer. A laser with $h\nu > E_G$ creates electron-hole pairs in the sample with and without the laser beam, one measures the "free carrier absorption" due to the excess carriers from which the lifetime is determined. In this case, no scanning is required, since both the blackbody and excitation laser are broad-area sources, covering the entire sample [76].

The effect of artificial mechanical damage in CZ silicon wafers was studied by Choi et al. [1] by measuring the structural and electrical characteristics of the wafers. Photoacoustic displacement, laser excitation, and microwave reflection photoconductance decay techniques were used to characterize the intensity of mechanical damage in the wafers. As the degree of mechanical damage increases, the density of oxygen-induced stacking faults increases, while the depth of their generation is almost independent of damage grade. The minority carrier lifetime is inversely proportional to the mechanical damage grade, while the photoacoustic displacement values are proportionally increased. Figure 6.12a and b show the results of minority carrier lifetime measurements for samples of reference and backside damaged wafers. A darker color on the lifetime map corresponds to a longer lifetime. The lifetime of the reference sample was around 900 µs and decreased to 200 µs as the grade of mechanical damage increased. This data clearly shows the relationship between the mechanical damage grade and the minority carrier lifetime [1]. By using this technique, one can assess the grade of mechanical damage that the wafer has undergone during crystal shaping methods.

Another method is described by Zimmermann et al. [3] that allows the fast determination of the concentration of vacancies and the visualization of their inhomogeneity in silicon wafers. Vacancy concentration wafer mapping has been suggested as a means to determine the quality of silicon wafers. The concentration of vacancies is determined via the measurement of the concentration of fast-diffusing elements, especially platinum or gold metals, which occupy silicon lattice vacancies. These occupied metals introduce deep energy levels into the bandgap, which act as recombination centers for the excess carriers; the presence of excess electrons or holes; and the associated excess carrier lifetime. Several important applications for this method have been suggested, such as the optimization of the crystal growth manufacturing process, wafer quality control, and the selection of wafers for the fabrication of specific devices.

Macdonald et al. [138] have reported minority carrier lifetime studies and their effect on phosphorus gettering on cast multicrystalline silicon solar cells. The largest increase in minority carrier lifetime after phosphorus gettering was found in samples with low dislocation densities and high concentrations of mobile impurities. The lifetimes of samples with dislocation densities above 10^6 cm^{-2} remained low after gettering, despite the extraction of a significant amount of mobile impurities. It is felt that the dislocations and microdefects in such samples are saturated with impurities, and those found out-diffusing are marked surplus. The magnitude of the lifetime response to phosphorus gettering was

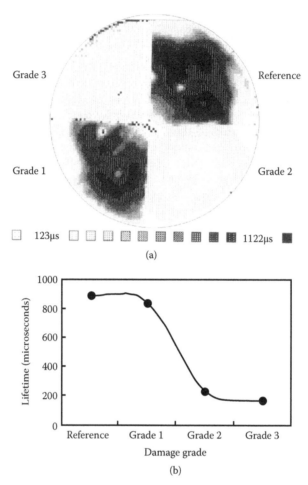

FIGURE 6.12
Relationship between mechanical damage grade and minority carrier recombination lifetime measured by microwave reflection photoconductance decay technique: (a) lifetime mapping and (b) average lifetime values; reference = 894.4 µs, grade 1 = 846 µs, grade 2 = 233.1 µs, and grade 3 = 158.8 µs. (From C.-Y. Choi, J.-H. Lee, and S.-H. Cho, *Solid-State Electronics*, **43**, 2011–2020, 1999, and the references therein [1].

also found to be ingot dependent. The team opined that gettering seems to be a viable option for improving solar cell efficiencies for substrates from such off-center regions. The final lifetimes after gettering are well correlated to the different crystallographic qualities of the various regions of both ingots. After gettering is complete, the minority carrier lifetime seems to be essentially determined by recombination centers associated with grain boundaries, dislocations, and other microdefects.

Horányi et al. [139] has reported that it is possible to evaluate metallic impurities present in silicon by carrier lifetime measurements. Iron [140] and copper [141] contamination studies have also been reported by using this method. This method finds more application in FZ-grown silicon crystals to evaluate the quality of the crystal and hence the wafers [142–145]. Studies have been extended to evaluate the quality of CCD imagers [146], CMOS circuits and how contamination affects their peformance [147], and their influence on reliability through surface voltage, DLTS, and other methods [148]. The evaluation is ongoing and expected to draw many researchers to study it further.

6.17 Optical Methods for Impurity Evaluation

IR absorption spectrophotometry has been used to determine the oxygen and carbon content of monocrystalline silicon. This method is particularly attractive because it has sensitivity in the ppba range. However, questions regarding the accuracy of this technique for measuring total oxygen and carbon content arise, because only interstitial oxygen and substitutional carbon [149], with absorption bands at 1130 cm^{-1} and 607 cm^{-1}, respectively, can be detected. Graff [150] proposes a precise evaluation of oxygen measurements by incorporating certain corrections to avoid measurement errors. By applying photoactivation analysis and IR absorption methods, Rath et al. [151] determined the oxygen content in silicon. According to this analysis, the precipitated oxygen in silicon does not contribute to the optical band spectra. Using charge particle activation analysis, conversion factors were suggested by Iizuka et al. [152] for the IR measurement between the absorption coefficient and the oxygen content actually present. Quantification of oxygen in silicon is still a major topic of research, and the presence of oxygen at interstitial, substitutional, and precipitations [153–160] in silicon is being carried out by many teams, but this is still a problem that needs a solution [158].

The carbon content in a single crystal of silicon has been investigated by several research teams using IR spectroscopy to show that carbon occupies substitutional sites in the silicon lattice. Regolini et al. [161] have reported that they could obtain a conversion factor for all the substitutional carbon that is typically recorded at 605 cm^{-1}. The impact of graphite furnace parts, the main source of carbon impurity in silicon, was analyzed by Gilmore et al. [162] in terms of the radial distribution of carbon in CZ-grown silicon crystals. They have also evaluated the graphite materials, both with and without SiC coatings, to confirm the radial distribution of carbon. Saito and Shirai [163] observed baseline variations near the standard of carbon in silicon, and it was found to be in the range of 630 cm^{-1} to 610 cm^{-1}. Alt et al. [164] proposed a new method to evaluate substitutional carbon in crystalline silicon, based on the second derivative of the IR absorption spectrum.

6.18 Photoluminescence Method for Determining Impurity Concentrations

Photoluminescence spectroscopy is a contactless, nondestructive method of probing the materials to extract information about the contents inside. This method is most commonly used at low temperatures for impurity identification and for determining low doping densities, particularly in silicon. Light is directed onto a sample, where it is absorbed and imparts excess energy into the material in a process called photoexcitation. One way this excess energy can be dissipated by the sample is through the emission of light, or luminescence. Excitation by absorbance of a photon leads to a major class of technically important luminescent species that can either fluoresce or phosphoresce. In general, fluorescence is a relatively fast process, whereas phosphorescence is a slow process. The energy of the emitted light relates to the difference in energy levels between the two electron states involved in the transition between the excited state and the equilibrium state. The quantity of the emitted light is affected by the relative contribution of the radiative process involved.

Photoluminescence spectroscopy has been widely used to characterize microdefects and impurities in semiconductor materials. Yamamoto and Nishihara [165] have examined the defects and defect-related deep-level traps in annealed CZ-grown wafers using this method. Depending on the interstitial oxygen reduction due to thermal annealing, the intensity values were found to differ, and these changes were recorded to the concentration of oxygen present in the wafers. By using the decay-time curve method, the band-edge photoluminescence intensity was correlated to the carrier lifetime.

Evaluation of oxygen present in the silicon was carried out by Kitagawara et al. [166] by using deep-level photoluminescence that corresponded to the D1 line observed at 0.81 eV. The measurements were performed at room temperature to obtain macroscopic and microscopic distributions of the D1 line intensity and the band-to-band intensity on the surfaces of the oxygen-precipitated silicon wafers. Figure 6.13a shows a typical room-temperature photoluminescence spectrum for the oxygen-precipitated CZ silicon in comparison with the corresponding spectrum at 4.2 K, as shown in Figure 6.13b. The 0.76 eV photoluminescence at room temperature corresponds to the D1 line at 4.2 K, while the 1.1 eV is the band-to-band emission. The 1.1 eV band-edge emissions are commonly observed in any silicon crystal. The D1 line at 0.76 eV was not detected in an as-grown crystal, but was sensitively detected in an oxygen-precipitated CZ silicon crystal after the two-step heat process involving 800°C heat treatment followed by the 1000°C heat treatment. The team indicated that a strong proportional relationship exists between the precipitated oxygen concentration and the relative D1 defect concentration, which implies that the D1 line defects in the evaluated sample crystals were induced by the oxygen precipitation formation. Leoni et al. [167] have demonstrated that a slight difference in the value, approximately 2.5 ppma, in the initial oxygen concentration is sufficient to give rise to the prevalence of a particular kind of oxygen nuclei after a treatment carried out at 650°C. In this way, one can provide information on the early stages of oxygen precipitation in CZ-grown silicon. With carbon impurity, the intensity peak is observed at 0.79 eV, and one can measure the concentration in the range of 1.0×10^{14} to 1.0×10^{15} atoms/cm^3. This

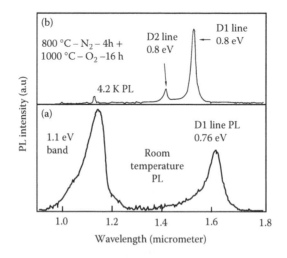

FIGURE 6.13
(a) A typical room-temperature photoluminescence spectrum for an oxygen-precipitated CZ silicon crystal. (b) The corresponding spectrum at 4.2 K. (From Y. Kitagawara, R. Hoshi, and T. Takenaka, *Journal of the Electrochemical Society*, 139, 2277–2281, 1992 [166].)

intensity does not interfere with the oxygen concentrations [168] and will show up independently in the intensity spectra.

The presence of copper can be measured in silicon crystals grown by both CZ and FZ methods. Nukamura [169] pointed out that copper intensity peaks do not shift much by the presence of secondary impurity peaks, indicating that the bonding character of copper around the atom was different and not entirely influenced by the secondary impurities. It was further pointed out that formation of the copper center was severely hindered by oxygen, but slightly enhanced by the presence of carbon. The intensities of the center were influenced by the boron and phosphorous dopants for CZ crystals. These observations supported the structural model of the copper center, with Cu collinearly bonded with Si atoms at the center of the Si-Si bond. The studies show good sensitivity up to the concentration values in the range of 1.0×10^{11} atoms/cm^3. Tajima et al. [170] have reported the data for calibrating arsenic and aluminum impurities in silicon by using this characterization tool. Recently, a photoluminescence-based commercial tool, SiPHER [76], has been introduced, operating at room temperature with two wavelengths: $\lambda = 532$ nm and 827 nm. It measures photoluminescence and the optical reflectance, and can detect defects such as dislocations, oxygen precipitates, and doping density striations in silicon wafers.

6.19 Gamma-Ray Diffractometry

Gamma-ray diffractometry has emerged as a powerful tool in structural and defect studies of crystals. Similar to x-rays and neutron diffraction in principle, the very short wavelength gives gamma-ray diffraction an edge over the other listed studies. The radiation does not get absorbed and is monoenergetic in nature [171]. The main advantages of gamma-ray diffraction—in particular, high sensitivity to lattice distortions—have been demonstrated in a study of the defect structure changes taking place during crystal growth, neutron transmutation doping, and device structure formation. Kurbakov [172] discusses in detail the role of nonequilibrium intrinsic point defects in defect formation during thermochemical treatment. Studies of plastic deformation of the CZ-grown crystals in the <110> orientation [25] was done by using an *in situ* gamma-ray diffractometry technique.

6.20 Scanning Electron Microscopy for Defect Analysis

The use of scanning electron microscopy (SEM) technology is an invaluable aid for detecting defects in silicon wafers and to analyze other specific properties. The SEM aims a focused beam of electrons across a small sample area. This focused electron beam generates several useful transitions, depending upon the electron beam energy and the sample material. Secondary electrons are a type of low-energy electron that is the primary transition used in SEM analysis. They are useful for imaging many important parameters. Contamination identification, metal decoration, defect analysis, and microcracks are only a few of the uses for SEM when performing wafer analysis. Other applications include the evaluation of wafer quality, surface finish, elemental analysis, and mechanical damages.

Cullis et al. [6] have used high-resolution SEM to study the dependence of maximum crystal growth velocity upon the crystallographic orientation of the growth surface. Discrete high-velocity areas of defective crystal growth have been identified, and defect formation and propagation processes characterized on the atomic scale. The effect of impurity doping with electrically active impurities (such as boron, gallium, antimony, phosphorus, and arsenic) and isoelectronic impurities (such as tin) on the formation of swirl defects were investigated using EBIC-mode SEM [31]. Defect formation and propagation processes are characterized on the atomic scale with the aid of high-resolution electron microscope images of growth breakdown interfaces.

Hackl et al. [173] analyzed the presence of iron and its detrimental impact on the nucleation and growth of oxygen precipitates during the internal gettering processes. Further reports estimate a threshold range for critical iron concentrations. The oxygen sample was precipitated efficiently with large precipitation areas, whereas almost no SiO_x precipitated in the other sample. Figure 6.14a and b show a significantly high difference in the precipitation amount and confirm that nucleation and growth of SiO_x in CZ-grown silicon are prevented in the presence of iron atoms. In this case, the SEM pictures provided compelling evidence that iron impurity retards oxygen precipitation. Similarly, the behavior of ultrafine particles on the silicon wafer surface was evaluated before and after the wafers were cleaned with standard cleaning techniques. Microroughness caused by the process was evaluated using SEM [174]. Such studies were also extended to FZ silicon crystals to analyze extended defects and local point defects [175] to elucidate the point defect reactions created by electron irradiation.

6.21   Scanning Optical Microscope

Optical evaluation of the polycrystalline silicon surface roughness was evaluated by Chiang et al. [176] using a nondestructive optical method. Spectrophotometer measurement of reflectance in the ultraviolet range was carried out. A single reading at one wavelength can be sufficient for this characterization. Structural perfection testing of films and wafers was carried out by Steigmeier and Auderset [177] using an optical scanner. This includes testing for silicon-on-sapphire, for low pressure CVD (LPCVD)-deposited amorphous silicon, and for the distribution of precipitated oxygen in silicon wafers treated for internal gettering. From the results, it appears that the scanners provide important information about wafer condition.

Kimura [178] used a phase-difference scanning optical microscope to observe oxygen precipitates in CZ-grown silicon wafers. The density distributions of the oxygen precipitates were measured in the vertical direction with a phase-differential scanning optical microscope that had a split detector. This microscope can produce images with information on phase differentials in only one direction, however. A quadrant detector is used to obtain two-dimensional phase-differential images, and the four signals are processed with the help of a computer. In experiments, the microscope was able to show decreases in defect density near wafer surfaces and to detect a difference in defect density distributions between wafers pulled at different growth rates. Minowa et al. [179] used an optical shallow defect analyzer to inspect wafers. With this, the team successfully found lattice strains and particles stuck on the wafer surface during the growth of silicon epitaxial films. In addition, the solid-melt interface fluctuations [180] in a silicon liquid bridge were observed.

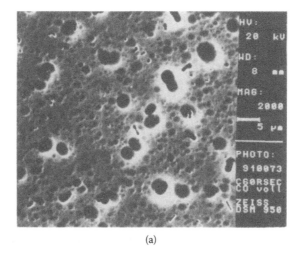

(a)

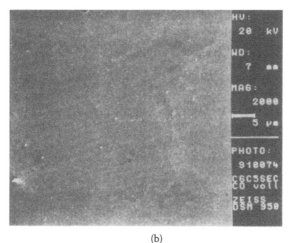

(b)

FIGURE 6.14
(a) SEM picture of the edge of the sample without iron contamination after the internal gettering temperature cycle, high density of oxygen precipitates (magnification 2000×). (b) SEM picture of the edge of the sample with 1.4×10^{14} iron atoms/cm^3 in the bulk after the internal gettering temperature cycle without any oxygen precipitates (magnification 2000×). (From B. Hackl, K.-J. Range, H. J. Gores, L. Fabry, and P. Stallhofer, *Journal of the Electrochemical Society*, **139**, 3250–3254, 1992 [173].)

6.22 Secondary Ion Mass Spectrometer for Impurity Distribution

Secondary ion mass spectrometry (SIMS) is a technique used in materials and surface sciences to analyze the composition of any solid surfaces and thin films by sputtering the surface of the specimen with a focused primary ion beam and collecting and analyzing any secondary ions ejected during this process. The mass/charge ratios of these ejected ions are measured with a mass spectrometer to determine the elemental, isotopic, or molecular composition of the surface to a depth of 10 to 20 Å. Due to the large variation in ionization probabilities among different materials to be analyzed, the SIMS technique is generally considered to be a qualitative technique, although quantitation is possible with the use of the standards and conversion factors. SIMS is the most sensitive surface analysis technique, with elemental detection limits ranging from parts per million to parts per billion, and is perfect for silicon wafer analysis. It can detect impurity contamination on the wafer surface, perform identification in bulk, and provide a map showing the complete distribution.

The technique has been successfully applied to determining in-depth concentration profiles of commonly used n-type and p-type dopants. Nitrogen doping is becoming a key technology for growing defect-free CZ silicon crystals [16]. Therefore, it is necessary to establish a method for measuring nitrogen concentration in these crystals. However, since the nitrogen concentration is low and nitrogen has various configurations, this is difficult. The proposed procedures for the American Society for Testing and Materials (ASTM) standard were adopted for SIMS. In addition, the raster change method was adopted to improve the detection limit. Levels as low as 4.0×10^{13} cm^{-3} can be successfully detected, and an acceptable accuracy is obtained for 2.0×10^{14} cm^{-3} with good reproducibility. A linear relationship was obtained between the IR absorption coefficients and SIMS. Voronkov and Falster [15] studied the out-diffusion profiles produced by annealing nitrogen-doped FZ and CZ wafers at 900°C. These were monitored to a depth of up to 120 µm using multiple SIMS runs. Fontana et al. [181] used this method effectively to analyze the silicon surface and evaluate the carbon residues on the surfaces of silicon integrated circuits.

Magee [182] measured the depth profiling of n-type dopants in silicon and gallium arsenide using Cs$^+$ ions. Depth profiles were obtained on phosphorus and arsenic in silicon, and, sulfur and silicon in gallium arsenide. With this technique, determining n-type dopants in silicon has become straightforward and as sensitive as determining p-type dopants. Figure 6.15 shows the depth profiles of two silicon samples implanted with 80 keV ^{31}P. One sample was implanted with a dose of 5.0×10^{15} atoms/cm^2 and another sample with 1.0×10^{14} atoms/cm^2. Both profiles were obtained using Cs$^+$ ion bombardment while monitoring ^{31}P$^-$ ions. Data points were taken by setting the mass spectrometer under computer control to mass number 31 and counting the arriving ions for 10 s durations. A matrix ion species (^{30}Si$^-$) was also counted for every 1 s interval to monitor the stability of the instrument during analysis. At a sputtering rate of 5 Å/s, data points were taken at approximately 50 Å depth intervals. The imaging secondary ion extraction lens was used to assure that only ions from the center of the rastered area were transmitted to the energy filter/quadrupole system. A detection limit of $<5.0 \times 10^{16}$ atoms/cm^3 was estimated from the profile of the low-dose sample. The ^{30}SiH background for both samples was <2 counts/sec, and the hydrogen concentration was $<1.0 \times 10^{18}$ atoms/cm^3. In this way, one can extract information regarding the nonsilicon atoms present in the silicon wafer. Both intentional and unintentional impurities can be extracted from the wafers using this method.

Due to stringent VLSI and ULSI production requirements, the metallic contamination of silicon surfaces must remain at very low concentration levels, often below 10^{10} atom/cm^2. Therefore, new alternatives to measure the presence of these contaminants and to quantify their amounts are very much needed. De Witte et al. [183] have proposed a time-of-flight SIMS (TOF-SIMS) approach and explored its capabilities in evaluating surface metallic contamination. This study compared and combined TOF-SIMS results with those from standard direct total reflectance x-ray fluorescence spectroscopy (TXRF). Both methods were quantified using relative sensitive factors according to two different approaches. In this case, the main metallic impurity was iron and it was distributed both on the surface and on the thin oxides present on the surface.

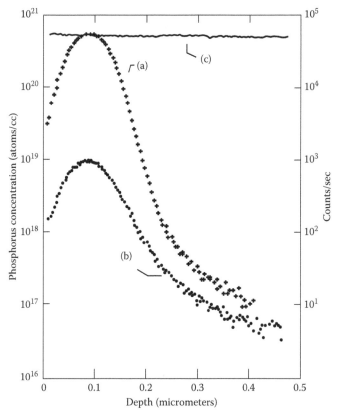

FIGURE 6.15

Depth profiles of 80 keV implants in silicon: (a) dose 5.0×10^{15} atoms/cm^2; (b) dose 1.0×10^{14} atoms/cm^2; (c) ^{30}Si$^-$ counts/sec, monitored to check instrumental stability. Bombarding species Cs$^+$; detected species ^{31}P$^-$. Sample chamber pressure: 4.0×10^{-10} Torr. No background subtraction. (From C. W. Magee, *Journal of the Electrochemical Society*, **126**, 660–663, 1979, and the references therein [182].)

6.23 Spreading Resistance and Two-Point Probe Measurement Technique

Spreading resistance profiling is a technique used to analyze resistivity versus depth in semiconductors. The electrically active carrier profile can be measured in the silicon bulk and in the processed wafer by proper beveling at small angles. This is a destructive approach, however, and sample preparation is key to obtaining the necessary information. It is strongly recommended that samples be beveled before probing to avoid native oxide formations. This method was first applied by Mazur and Dickey [184] to study the inhomogeneous silicon structures and determine local silicon resistivity variations. Correction factors were suggested [185, 186] for the measurements based on the multilayer step-junction theory. Two probes are used, typically made of tungsten carbide. The probe point is also a key issue in this measurement [187]. The probes are placed within a gap of 20 μm, and the probe tips are mounted on the end of a separate arm [188]. With minimum applied pressure, a small voltage of the order of 5 mV is applied to measure resistance between the tips. The details are shown in Figure 6.16. Ciszek [134] used this technique to characterize

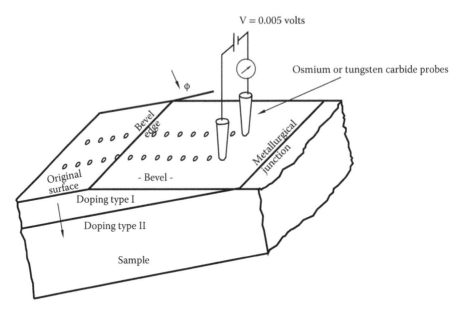

FIGURE 6.16
Schematic diagram of a spreading resistance measurement on a beveled sample. (From S. Loftis and D. Dickey, "An introduction to spreading resistance analysis and its application in the semiconductor industry," Seminar 09.pdf, Solecon Labs, 770 Trademark Dr, Reno, NV 89521, USA, https://www.solecon.com [188].)

the growth and properties of [100] and [111] dislocation-free crystals. De Kock et al. [10] reported the impurity doping striations on CZ- and FZ-grown crystals by spreading resistance measurements and preferential etching. Robertson, Jr. and O'Connor [189] conducted several experiments on strong magnetic field effects on FZ gallium-doped <100> oriented silicon crystals, and all were evaluated for dopant concentrations by spreading resistance measurements. The most uniform doping details were evaluated with this method to obtain the core regions of the crystal.

The method was modified to use mercury probes to avoid surface damage [190]. Different calibrations were also proposed for various *p*-type dopants in silicon [191]. Gupta et al. [192] used this method to detect inhomogeneities and strain areas in bulk and epitaxial silicon. Oxygen micro-segregation and swirl defect formation in CZ-grown silicon were studied by Murgai et al. [21] by employing interface demarcation and spreading resistance measurements. The difference in carrier concentrations, before and after oxygen gettering, was determined from the successive spreading resistance measurements and was equal to the thermally activated oxygen donor concentration. The latter measurements were converted to oxygen concentration through silicon samples from the same crystal, for which the thermal donor concentration was determined as earlier, as well as the oxygen concentration from IR absorption measurements. Lin and Stavola [193] also evaluated the oxygen segregation and microscopic-level inhomogeneity in CZ-grown single crystals. Through quenching experiments, variation in resistivity along the crystal growth direction was brought out.

The success of epitaxial silicon and the devices grown in this layer depend on the knowledge of the entire epitaxial impurity profile. The present spreading resistance technique is the only way to obtain the complete profile. Morris [194] reported on wide varieties of semiconductor devices and analyzed them in great detail. Iida et al. [195] proposed a new algorithm for this method by measuring the data generated from this analysis and applying it to thin epitaxial silicon wafers. Figure 6.17 shows an example of epitaxial film evaluations on measured resistivity and carrier concentration values for a typical high-resistivity epitaxial silicon wafer. The same can be extended to different multilayer structures in fabricated devices, or to any process-bound wafers [196–198].

6.24 Stress Measurements

The presence of residual stresses in single crystals of silicon greatly affects the performance and reliability of the integrated circuits fabricated using them. Residual stresses in both CZ and FZ methods occur in silicon wafers during the growth of bowls as well as in the subsequent thermal processing steps. These stresses can be determined in wafers by analyzing the out-of-plane deformation under transversally axisymmetrical loading. The out-of-plane deformation can be used in a model of a modified circular plate theory undergoing large defections. Vrinceanu and Danyluk [199] used a shadow interferometry approach to measure the surface-shape contour pattern of a wafer to extract the radial and tangential residual stresses.

The potential of nonlinear optical techniques for rapid online and nondestructive inspection and characterization of silicon wafers was discussed by Reif et al. [79]. As an example, the *in situ* detection of external stress on the wafer was reported, resulting from specific mounting conditions. The problem of radially nonuniform growth of the silicon crystal

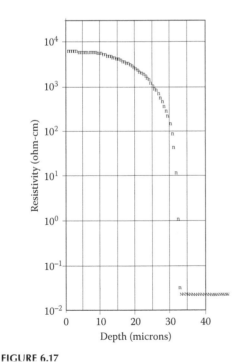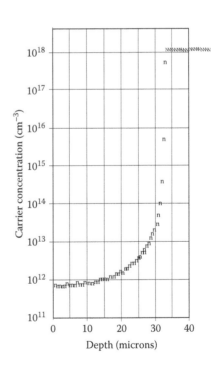

FIGURE 6.17

(a) Measured resistivity and (b) carrier concentration values for epitaxial wafers evaluated by spreading resistance method. (From S. Loftis and D. Dickey, "An introduction to spreading resistance analysis and its application in the semiconductor industry," Seminar09.pdf, Solecon Labs, 770 Trademark Dr, Reno, NV 89521, USA, https://www.solecon.com [188].)

when utilizing the CZ method was also addressed. A simple technique was proposed to differentiate those sections of the wafer that are ready for use in applications from those that are not useable for proper device fabrication, thus enabling the selection of appropriate material from as-grown crystals.

Stress is typically measured with Raman spectroscopy. A more recent method, polariscopy [200], mentioned earlier, produces two-dimensional stress maps. This is a nondestructive, highly sensitive, and quantitative mapping technique for residual strains in semiconductor wafers [76]. When a uniaxial stress is uniformly applied to an optically isotropic material with the stress direction in the plane of the sheet, a plane-polarized light ray is broken up into the "ordinary wave" and the "extraordinary wave." Since the two components propagate in the medium with different velocities, a phase difference is produced between the two that is proportional to the magnitude of the shear stress. A distribution of stresses produces bright and dark fringes in the form of contours of equal shear stress. Presently, this is a significant tool for wafer inspection.

Yang et al. [201] reported on the residual stress and fracture strength of crystalline silicon wafers by using a full-field, near-infrared polariscope [200]. Residual stress was analyzed in combination with observed surface defects, and the results were related to the measured fracture strength variation in the wafers. Measurements indicated that there is a sawing process-related residual stress in the as-cut wafers and that etch removal of ~5 μm from the wafer surface eliminates a damage layer that can significantly reduce the residual stress in the wafer, and therefore increases the observed fracture strength. Representative residual stress maps of multicrystalline-silicon (mc-Si) and monolike Si wafers in the as-cut condition and after 5 μm etching/damage layer removal are shown in Figure 6.18. The

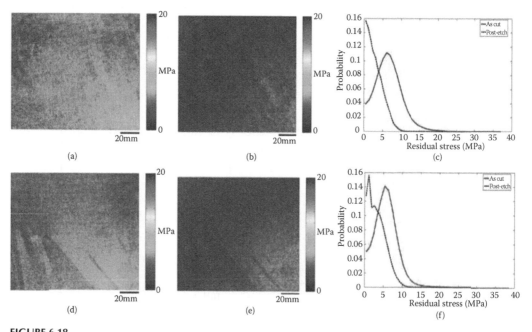

FIGURE 6.18
Residual stress maps for (a) as-cut cast mc-Si wafers, (b) post-etch cast mc-Si wafers, (d) as-cut cast mono like Si wafers, and (e) post-etch cast monolike Si wafers, and statistical distributions for (c) mc-Si wafers and (f) monolike Si wafers. Pixel size = 64 μm in the images. (From C. Yang, F. Mess, K. Skenes, S. Melkote, and S. Danyluk, *Applied Physics Letters*, **102**, 021909, 2013 [201].)

figure shows the residual stress maps for as-cut, post-etch cast, as-cut cast monolike, and post-etch cast monolike silicon wafers and the statistical distributions for the same. The results clearly show a detrimental effect of tensile residual stresses on the fracture strength of crystalline silicon wafers. Higher tensile residual stresses will lead to lower fracture strength. Therefore, it is critical to measure and quantify the residual stresses in a wafer and during the subsequent processing steps. The stress component is also measured by other methods, including the gravitational stress [202–206] and the induced dislocations generated.

6.25 Transmission Electron Microscopy

Transmission electron microscopy (TEM) is a technique in which a fine beam of high-energy electrons is transmitted through an ultra-thin specimen, interacting with the specimen as it passes through it. The beam behaves like a wave front, and when it passes through the specimen, the electrons are scattered. An image is formed from the interaction of the electrons transmitted through the specimen. TEMs are capable of imaging at a significantly higher resolution than ordinary optical beams due to the small de Broglie wavelength of electrons associated with them. This enables the investigator to examine fine details of the specimen under study. TEM provides a major method of evaluating many important crystal-related issues for the semiconductor industry.

Matsushita [37] used this method to investigate thermally induced microdefects in CZ-grown silicon crystals. In the TEM images of silicon crystals, extinction of fringes were observed by Hashimoto et al. [11] due to the periodic intensity distribution of electron waves. These extinction fringes, called "Pendellösung fringes," have appeared sometimes as "thickness fringes" and sometimes as "inclination fringes." Thermally induced oxygen precipitation and redistribution phenomena were investigated by means of TEM and DIR by Shimura et al. [23]. Swirl defects in quenched dislocation-free FZ silicon crystals also have been analyzed using this method. The TEM analysis shows that the Type A swirl defects are perfect extrinsic dislocation loops elongated along the <100> direction. The smaller Type B swirl defects remained undetectable by TEM, indicating that these defects cause very little lattice strain. It is concluded that the silicon self-interstitials are the dominant point defects during the FZ crystal growth process [32].

Shimura [207] used the TEM method to observe pyramidal hillocks bounded by four {111} planes of an annealed (001) wafer with oxidation-induced stacking faults. This showed that the pyramidal hillock is not an extraneous substance, such as SiO_2 precipitate formation, but an intrinsic part of the substrate silicon crystal. The possible mechanism of hillock formation is confirmed with the help of this tool. Figure 6.19 shows the various sizes of hillocks observed in both surface and inner parts of (001) wafers annealed at temperatures from 900°C to 1200°C. Similar studies on denuded zones in CZ wafers were reported by Wang et al. [208]. Oxygen precipitation induced by internal gettering was studied by Rivaud et al. [209], while chromium precipitation during intrinsic gettering was evaluated by Krieger-Kaddour et al. [210]. Grown-in defects in epitaxial silicon wafers were analyzed by Minowa et al. [211], Itsumi studied octahedral void defects [212], and Zeng et al. [213] analyzed dislocation motion in CZ crystals. More details are discussed in other chapters of this book.

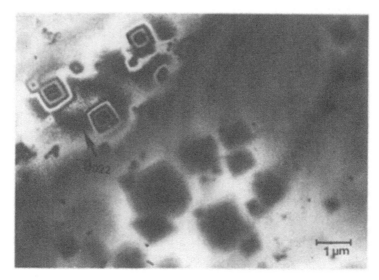

FIGURE 6.19
TEM image of pyramidal hillocks in (001) silicon wafer annealed at 1100°C. (From F. Shimura, *Journal of the Electrochemical Society*, **127**, 910–913, 1980 [207].)

6.26 van der Pauw Resistivity Measurement Technique for Irregular-Shaped Wafers

The van der Pauw method is a technique commonly used to measure the resistivity of any conducting sample. Its power lies in its ability to accurately measure the properties of a sample of any arbitrary shape, so long as the sample is approximately two-dimensional, without any openings, and the contacts are made at the perimeter for measurements. With this method, one can estimate the resistivity, dopant type, sheet carrier density, and mobility of the majority carriers. The method was first propounded by van der Pauw, which is why it is named after him.

Westbrook [214] studied the low mobility values in semiconductors and reported that this is generally construed to imply the presence of impurities or defects that adversely affect carrier transport. It is shown theoretically that low mobility values will also be obtained by van der Pauw measurements on samples containing certain types of radial inhomogeneities in the distribution of ionized impurities. The theoretical predictions have been confirmed by experiments on the slices cut from CZ-grown crystals. Bridge and quadrate-cross test structures were reported by Buehler et al. [215] for sheet resistance evaluations. The method is popular and widely used for wafer evaluations.

6.27 X-ray Technique for Crystal Perfection and Dislocation Density

An x-ray imaging technique based on the Bragg diffraction method is routinely used to evaluate crystal quality and perfection. Diffraction records the intensity profile of an x-ray beam diffracted by the crystal under study. The resulting topography represents a two-dimensional, spatial intensity mapping of reflected x-rays, and this intensity mapping reflects the distribution of scattering power inside the crystal. The mapping reveals the irregularities in a nonideal crystal lattice. It is mainly used to monitor crystal quality and visualize defects in the many different crystalline materials present. Ciszek [134] has used this method to study the growth and characterize the properties of the grown crystals. Carbon striations are analyzed using a special type of x-ray transmission section topography [10]. Micro-inhomogeneities, such as swirl defects, were evaluated by etching and x-ray topography [29] in FZ crystals. Oxygen micro-segregation and swirl defect formation in CZ-grown silicon were also analyzed using x-ray topography [21].

Convection of molten silicon during CZ single-crystal growth was directly observed by Kakimoto et al. [4] using x-ray radiography. The same team [27] used x-ray radiography to conduct *in situ* observation of the solid–liquid interface shape during CZ growth of single-crystal silicon. Simulation of the transmitted x-ray image by absorption calculation was carried out using absorption coefficients both for molten and solid silicon, and supports the fact that the solid–liquid interface can be observed.

An interesting problem is the possibility of revealing various kinds of small precipitates using x-ray diffraction topography, as shown by Wierzchowski et al. [22], which, in many cases, is very sensitive to small local strain changes. However, the x-ray methods seem to be less effective in the case of small oxygen precipitates, which seem to be imbedded coherently in crystals that are practically strain free. It is not yet clear how the efficiency of x-ray topographic methods will change with the increasing size of formed precipitates and with the use of more sensitive synchrotron methods. The considerable effect observed with all the topographic methods used was visible after relatively long annealing times (3 h at 1100°C). In this case, the individual precipitates were resolved in the topographs. The section topograph, shown in Figure 6.20, presents a region with significant variations in oxygen precipitate concentration. This concentration strongly increases along some bands, roughly corresponding to the striations, but is generally low on the left side of the reproduced area. In the left section, we can observe the distinct interference fringes previously mentioned. In contrast, in the right section of the picture, the fringes are not present and the topograph shows well-resolved precipitates. The representative transmission and double-crystal topographs are shown in the rest of the figure. Figure 6.21 shows the results of x-ray investigations for the sample annealed for 6 h at a temperature increasing from 600 to 1050°C and for a further 6 h at 1050°C. In this case, the interference fringes are not present and the oxygen precipitates can be resolved only in some areas. Also in this case, variation of precipitate concentration corresponds to the striations [22].

The effects of molten silicon flow on the distribution of impurity atoms, especially oxygen, in a single crystal of silicon grown using the CZ method have been studied by many researchers from a scientific point of view, as well as for application in semiconductor device technology. The molten silicon flow was observed using a double-beam x-ray visualization technique with solid tracer particles. The melt height was monitored using an x-ray radiograph image. Under these conditions, the temperature gradient over the crystal-melt interface was not exactly equal to that under crystal growth conditions. However, this was the only way to observe the flow as the melt height changed. If the flow observation was simultaneously carried out during crystal growth, tracer particles would be incorporated into the growing crystal. By observing the flow of molten silicon and the characteristics of the grown crystals from x-ray topographic images, Watanabe et al. [14] clarified that the crystal-melt interface shape changed from a convex to a gull-winged one due to the transition from a nonaxisymmetric flow with vortices to an axisymmetric flow caused by the baroclinic instability. Moreover, they also clarified by FTIR coupled with IR microscopy

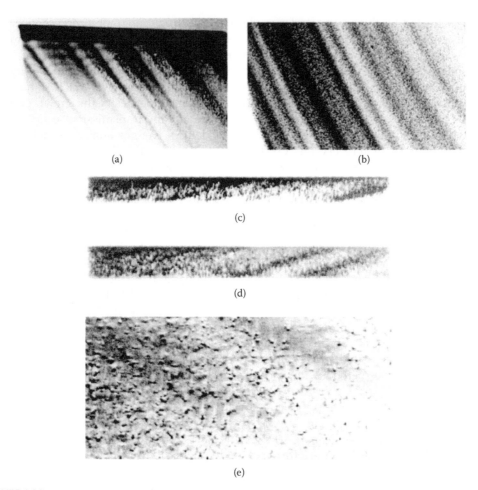

FIGURE 6.20
Representative topographic results for silicon CZ sample annealed for 3 h at 1100°C with a relatively small density of well-resolved oxygen precipitates. (a) 620 Bragg-case section topograph in 0.06 nm. In the left section of the reproduced area, the concentration of oxygen precipitates is smaller and the interference fringes caused by strain gradient are visible. (b) Projection back-reflection topograph in the same reflection as in (a). (c) Transmission section pattern in 553 reflection of 0.054 nm with the beam perpendicular to the sample. (d) Projection topograph for the same reflection as in (c). (e) Enlarged part of plane-wave synchrotron topograph in slightly asymmetric 511 reflection of 0.11 nm revealing interference fringes connected with some precipitates. (From W. Wierzchowski, K. Wieteska, W. Graeff, M. Pawłowska, B. Surma, and S. Strzelecka, *Journal of Alloys and Compounds*, **362**, 301–306, 2004, and the references therein [22].)

that the distribution of oxygen in the grown silicon crystal was modified by the flow-mode transition. Interstitial oxygen was distributed more homogeneously in the crystal grown under axisymmetric flow conditions than that grown under nonaxisymmetric flow conditions with vortices. The inhomogeneous distribution of oxygen in the crystal grown under nonaxisymmetric flow with vortices could be attributed to thermal asymmetry in the melt.

An instantaneous display of topographic images of individual dislocations was achieved by combining a high-power x-ray generator and a television unit with an x-ray sensing PbO-Vidiocon camera tube. Chikawa et al. [9] observed the silicon wafers that were deformed by tension (about 300 g/mm²) at about 1000°C in a vacuum furnace and, simultaneously,

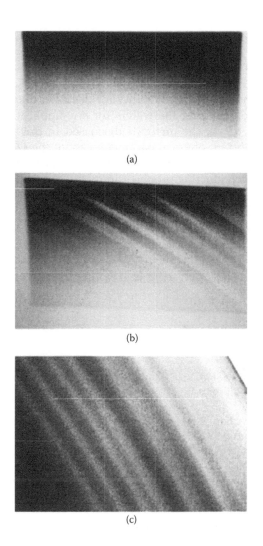

(a)

(b)

(c)

FIGURE 6.21
Representative topographic results for CZ silicon crystal annealed for 6 h at temperature increasing from 600 to 1050°C and for a further 6 h at 1050°C. (a) 620 Bragg-case section topograph in 0.06 nm for the central area of the sample with a uniform large concentration of oxygen precipitates. (b) 620 Bragg-case section topograph in 0.06 nm for the area closer to the edge showing regions with varying concentrations of oxygen precipitates. (c) Bragg-case projection topograph for the same region of the sample as in (b). (From W. Wierzchowski, K. Wieteska, W. Graeff, M. Pawłowska, B. Surma, and S. Strzelecka, *Journal of Alloys and Compounds*, **362**, 301–306, 2004, and the references therein [22].)

moving dislocations were observed continuously with the television unit. The observation of dislocations moving as fast as 0.3 mm/s may indicate the possibility that this technique can be a powerful tool in investigating the dynamic properties of dislocations. Faster dislocation motion was observed by a cine camera operated at 15 frames/sec (one TV frame on a movie frame). At present, the maximum observable velocity of dislocations is limited to a few mm/sec by the S/N of the images, but is an order of magnitude larger than that observed using electron microscopy. Parker and Porter [216] have used a high-speed x-ray topographic camera to evaluate silicon wafers. Many more interesting evaluations are possible using x-rays for wafer analysis [217–222].

6.28 Summary

In this chapter, we studied the different methods to evaluate silicon wafers to understand the detailed specifications and their suitability for device applications. It is important to have this information before using the wafers for any application, either for VLSI and ULSI or for MEMS structures. The methods discussed in this chapter allow investigators to extract many important physical parameters about the wafers. This includes electrical, physical, bulk, and compositional parameters. Optical methods are important to analyze the material semiconducting property, impurity composition, and intentional and unintentional dopant species present inside the silicon lattice. Four-point probe, spreading resistance method, SIMS, Hall mobility, and DLTS methods are useful to analzye the electrical parameters to ascertain the suitability of silicon wafers for any specific application. Chemical etching methods, metal decoration, x-ray, and gamma-ray diffractometry provide details of crystal perfection and any defects present inside the crystal lattice.

References

1. C.-Y. Choi, J.-H. Lee, and S.-H. Cho, "Characterization of mechanical damage on structural and electrical properties of silicon wafers," *Solid-State Electronics*, **43**, 2011–2020, 1999, and the references therein.
2. H. Lemke and K. Irmscher, "Proof of interstitial cobalt defects in silicon float zone crystals doped during crystal growth," *210 ECE Meeting Cancun*, Mexico, Oct. 29–Nov. 3, 2006.
3. H. Zimmermann, U. Gösele, M. Seilenthal, and P. Eichinger, "Vacancy concentration wafer mapping in silicon," *Journal of Crystal Growth*, **129**, 582–592, 1993.
4. K. Kakimoto, M. Eguchi, H. Watanabe, and T. Hibiya, "Direct observation by x-ray radiography of convection of molten silicon in the Czochralski growth method," *Journal of Crystal Growth*, **88**, 365–370, 1988.
5. T. Motooka, "Molecular-dynamics simulations of recrystallization processes in silicon: Nucleation and crystal growth in the solid-phase and melt," Department of Materials Science and Engineering, Kyushu University, Japan.
6. A. G. Cullis, N. G. Chew, H. C. Webber, and D. J. Smith, "Orientation dependence of high speed silicon crystal growth from the melt," *Journal of Crystal Growth*, **68**, 624–638, 1984.
7. M. Y. Ben-Sira and S. Bukshpan, "Spontaneous generation of dislocations during growth of silicon single crystals," *Journal of Crystal Growth*, **2**, 248–250, 1968.
8. D. G. Schimmel, "Defect etch for <100> silicon evaluation," *Journal of the Electrochemical Society*, **126**, 479–483, 1979.
9. J.-i. Chikawa, I. Fujimoto, and T. Abe, "X-ray topographic observation of moving dislocations in silicon crystals," *Applied Physics Letters*, **21**, 295–298, 1972.
10. A. J. R. De Kock, P. J. Roksnoer, and P. G. T. Boonen, "Formation and elimination of growth striations in dislocation-free silicon crystals," *Journal of Crystal Growth*, **28**, 125–137, 1975.
11. H. Hashimoto, S. Kozaki, and T. Ohkawa, "Observations of Pendellösung fringes and images of dislocations by x-ray shadow micrographs of Si crystals," *Applied Physics Letters*, **6**, 16–17, 1965.
12. I. Fusegawa, H. Yamagishi, T. Mori, H. Takayama, E. Iino, and K. Takano, "Characterization of interstitial oxygen striations in silicon single crystals," *Journal of Crystal Growth*, **128**, 293–297, 1993.

13. H. Sreedharamurthy, "Reducing iron in single crystal silicon grown using CZ process," *210 ECE Meeting Cancun*, Mexico, Oct. 29–Nov. 3, 2006.

14. M. Watanabe, M. Eguchi, K. Kakimoto, H. Ono, S. Kimura, and T. Hibiya, "Flow mode transition and its effects on crystal-melt interface shape and oxygen distribution for Czochralski-grown Si single crystals," *Journal of Crystal Growth*, **151**, 285–290, 1995, and the references therein.

15. V. V. Voronkov and R. Falster, "Nitrogen diffusion in silicon: A multi-species process," *210 ECE Meeting Cancun*, Mexico, Oct. 29–Nov. 3, 2006, and the references therein.

16. N. Inoue, K. Shingu, and K. Masumoto, "Measurement of nitrogen concentration in CZ silicon," presented at *Ninth International Symposium on Silicon Materials Science and Technology*, 201st Meeting, #621, Philadelphia, PA, May 12–17, 2002, and the references therein.

17. N. Fujita, R. Jones, S. Öberg, and P. R. Briddon, "Identification of nitrogen-oxygen defects in silicon," *210 ECE Meeting Cancun*, Mexico, Oct. 29–Nov. 3, 2006.

18. M. W. Jenkins, "A new preferential etch for defects in silicon crystals," *Journal of the Electrochemical Society*, **124**, 757–762, 1977, and the references therein.

19. A. G. Crouse, J. E. Huffman, C. S. Tindall, and M. L. W. Thewalt, "Float zone growth of high purity and high oxygen concentration silicon," *Journal of Crystal Growth*, **109**, 162–166, 1991.

20. S. Kawanishi, S. Togawa, K. Izunome, K. Terashima, and S. Kimura, "Melt quenching technique for direct observation of oxygen transport in the Czochralski-grown Si process," *Journal of Crystal Growth*, **152**, 266–273, 1995, and the references therein.

21. A. Murgai, H. C. Gatos, and W. A. Westdorp, "Effect of microscopic growth rate on oxygen microsegregation and swirl defect distribution in Czochralski-grown silicon," *Journal of the Electrochemical Society*, **126**, 2240–2245, 1979.

22. W. Wierzchowski, K. Wieteska, W. Graeff, M. Pawłowska, B. Surma, and S. Strzelecka, "X-ray topographic investigation of large oxygen precipitates in silicon," *Journal of Alloys and Compounds*, **362**, 301–306, 2004, and the references therein.

23. F. Shimura, H. Tsuya, and T. Kawamura, "Precipitation and redistribution of oxygen in Czochralski-grown silicon," *Applied Physics Letters*, **37**, 483–486, 1980.

24. A. Seidl and G. Müller, "Oxygen solubility in silicon melt measured in situ by an electrochemical solid ionic sensor," *Journal of the Electrochemical Society*, **144**, 3243–3245, 1997, and the references therein.

25. M. R. L. N. Murthy and J. J. Aubert, "Growth of dislocation-free silicon crystals in the <110> direction for use as neutron monochromators," *Journal of Crystal Growth*, **52**, 391–395, 1981.

26. G. K. Fraundorf and L. Shive, "Transmission electron microscope study of quartz crucibles used in growth of Czochralski silicon," *Journal of Crystal Growth*, **102**, 157–166, 1990.

27. K. Kakimoto, M. Eguchi, H. Watanabe, and T. Hibiya, "In-situ observation of solid-liquid interface shape by x-ray radiography during silicon single crystal growth," *Journal of Crystal Growth*, **91**, 509–514, 1988.

28. S. Kawado, T. Taishi, S. Iida, Y. Suzuki, Y. Chikaura, and K. Kajiwara, "Three-dimensional structure of dislocations in silicon determined by synchrotron white x-ray topography combined with a topo-tomographic technique," *Journal of Physics D: Applied Physics*, **38** (10A), A17–A22, 2005.

29. A. Eyer, B. O. Kolbesen, and R. Nitsche, "Floating zone growth of silicon single crystals in a double-ellipsoid mirror furnace," *Journal of Crystal Growth*, **57**, 145–154, 1982.

30. J.-i. Chikawa and S. Shirai, "Melting of silicon crystals and a possible origin of swirl defects," *Journal of Crystal Growth*, **39**, 328–340, 1977.

31. A. J. R. de Kock and W. M. van de Wijgert, "The effect of doping on the formation of swirl defects in dislocation-free Czochralski-grown silicon crystals," *Journal of Crystal Growth*, **49**, 718–734, 1980.

32. P. M. Petroff and A. J. R. De Kock, "The formation of interstitial swirl defects in dislocation-free floating-zone silicon crystals," *Journal of Crystal Growth*, **36**, 4–10, 1976.

33. S. Yasuami, M. Ogino, and S. Takasu, "The swirl formation of defects in Czochralski-grown silicon crystals," *Journal of Crystal Growth*, **39**, 227–230, 1977.

34. J. A. Stefani, J. K. Tien, K. S. Choe, and J. P. Wallace, "Eddy current monitoring system and data reduction protocol for Czochralski silicon crystal growth," *Journal of Crystal Growth*, **88**, 30–38, 1988.

35. K. S. Choe, J. A. Stefani, J. K. Tien, and J. P. Wallace, "Eddy current measurement of crystal axial thermal profiles during Czochralski silicon crystal growth," *Journal of Crystal Growth*, **88**, 39–52, 1988.

36. A. Virzi, "Computer modelling of heat transfer in Czochralski silicon crystal growth," *Journal of Crystal Growth*, **112**, 699–722, 1991.

37. Y. Matsushita, "Thermally induced microdefects in Czochralski-grown silicon crystals," *Journal of Crystal Growth*, **56**, 516–525, 1982.

38. F. Shimura, H. Tsuya, and T. Kawamura, "Thermally induced defect behavior and effective intrinsic gettering sink in silicon wafers," *Journal of the Electrochemical Society*, **128**, 1579–1583, 1981.

39. H. J. Ruiz, C. S. Williams, and F. A. Padovani, "Silicon slice analyzer using a He-Ne laser," *Journal of the Electrochemical Society*, **121**, 689–692, 1974.

40. B. R. Locke and R. P. Donovan, "Particle sizing uncertainties in laser scanning of silicon wafers: Calibration/evaluation of the Aeronca Wafer Inspection System 150," *Journal of the Electrochemical Society*, **134**, 1763–1771, 1987.

41. W. P. Lee and Y. L. Khong, "Laser microwave photoconductance studies of ultraviolet-irradiated silicon wafers: Effect of metallic contamination," *Journal of the Electrochemical Society*, **145**, 329–332, 1998.

42. K. Nakai, M. Hasebe, K. Ohta, and W. Ohashi, "Characterization of grown-in stacking faults and dislocations in CZ-Si crystals by bright field IR laser interferometer," *Journal of Crystal Growth*, **210**, 20–25, 2000.

43. K. Araki, H. Isogai, R. Takeda, K. Izunome, and X. Zhao, "Variation in Si (100) surface roughness caused by H-termination during high-temperature Ar annealing," *Journal of Crystal Growth*, **318**, 84–88, 2011.

44. K. Usuda and K. Yamada, "Atomic force microscopy observations of Si surfaces after rinsing in ultrapure water with low dissolved oxygen concentration," *Journal of the Electrochemical Society*, **144**, 3204-3207, 1997.

45. R. C. Henderson, "Silicon cleaning with hydrogen peroxide solutions: A high energy electron diffraction and Auger electron spectroscopy study," *Journal of the Electrochemical Society*, **119**, 772–775, 1972.

46. P. V. Fontana, J. P. Decosterd, and L. Wegmann, "Investigations of carbon residues on surfaces of silicon integrated circuits," *Journal of the Electrochemical Society*, **121**, 146–150, 1974.

47. M. G. Yang, K. M. Koliwad, and G. E. McGuire, "Auger electron spectroscopy of cleanup-related contamination on silicon surfaces," *Journal of the Electrochemical Society*, **122**, 675–678, 1975, and the references therein.

48. J. H. Thomas III, R. V. D'Aiello, and P. H. Robinson, "A scanning Auger electron spectroscopic study of particulate defects in metallurgical-grade silicon," *Journal of the Electrochemical Society*, **131**, 196–200, 1984.

49. Wet-chemical etching and cleaning of silicon, Virginia Semiconductor, Inc., 1501 Powhatan Street, Fredericksburg, VA 22401, USA, and the references therein. http://www.virginiasemi.com/pdf/siliconetchingandcleaning.pdf.

50. S. K. Ghandhi, *VLSI Fabrication Principles: Silicon and Gallium Arsenide*, John Wiley, New York, 2005, and the references therein.

51. E. Sirtl and A. Adler, "Chromsäure-Flussäure als spezifisches System zur Ätzgruben entwicklung auf Silizium," *Z. Metallkunde.* **52**, 529, 1961.

52. F. Secco d'Aragona, "Dislocation etch for (100) planes in silicon," *Journal of the Electrochemical Society*, **119**, 948–951, 1972.

53. D. E. Aspnes and A. A. Studna, Method of preparing semiconductor surfaces, U.S. 43,80,490A (1983); J. Heiss Jr., and J. Wylie, Removal of RTV silicon rubber encapsulants, U.S. Patent 4,089,704 (1978); *Quick Reference Manual for Silicon Integrated Circuit Technology*, W. E. Beadle, J. C. C. Tsai, R. D. Plummer (Eds.), Wiley-Interscience, 1985.

54. P. J. Whoriskey, "Two chemical stains for marking *p-n* junctions in silicon," *Journal of Applied Physics*, **29**, 867–868, 1958.
55. T. Iizuka and M. Kikuchi, "Anomalous etch patterns in heavily doped silicon," *Japanese Journal of Applied Physics*, **2**, 196, 1963.
56. Y. Yukimoto, A. Hirano, and Y. Sugioka, "X-ray observations of anomalous etch patterns in silicon crystals," *Japanese Journal of Applied Physics*, **6**, 420–421, 1967.
57. M. Kämper, "A new striation etch for silicon," *Journal of the Electrochemical Society*, **117**, 261–262, 1970.
58. D. G. Schimmel and M. J. Elkind, "An examination of the chemical staining of silicon," *Journal of the Electrochemical Society*, **125**, 152–155, 1978.
59. F. Shimura, "TEM observations of pyramidal hillocks formed on (100) silicon wafers during chemical etching," *Journal of the Electrochemical Society*, **127**, 910–913, 1980.
60. P. Rai-Choudhury and A. J. Noreika, "Hydrogen sulfide as an etchant for silicon," *Journal of the Electrochemical Society*, **116**, 539–541, 1969.
61. W. A. Porter, D. L. Parker, and L. G. Reed, "The effect of a prior Sirtl etch on subsequent thermally induced processing damage in silicon wafers," *Journal of the Electrochemical Society*, **123**, 146–147, 1976.
62. W. Wijaranakula, "Characterization of crystal originated defects in Czochralski silicon using nonagitated Secco etching," *Journal of the Electrochemical Society*, **141**, 3273–3277, 1994.
63. P. J. Holmes, "The use of etchants in assessment of semiconductor crystal properties," *Proceedings of Inst. Elec. Engrs*, **106B**, Suppl. 15 (May 1959). p. 861.
64. T. Iizuka, Y. Okada, and M. Kikuchi, "Generation of scratch-induced dislocations in silicon and its orientation dependence," *Japanese Journal of Applied Physics*, **4**, 237–238, 1965.
65. D. G. Schimmel, "A comparison of chemical etches for revealing <100> silicon crystal defects," *Journal of the Electrochemical Society*, **123**, 734–741, 1976.
66. T. Furuoya, "Dislocations in silicon single crystals," *Japanese Journal of Applied Physics*, **1**, 135–143, 1962.
67. T. Abe, T. Samizo, and S. Maruyama, "Etch pits observed in dislocation-free silicon crystals," *Japanese Journal of Applied Physics*, **5**, 458–459, 1966.
68. K. H. Yang, "An etch for delineation of defects in silicon," *Journal of the Electrochemical Society*, **131**, 1140–1145, 1984.
69. T. C. Chandler, "MEMC Etch – A chromium trioxide-free etchant for delineating dislocations and slip in silicon," *Journal of the Electrochemical Society*, **137**, 944–948, 1990.
70. K. Graff and P. Heim, "Chromium-free etch for revealing and distinguishing metal contamination defects in silicon," *Journal of the Electrochemical Society*, **141**, 2821–2825, 1994.
71. Y. Kashiwagi, R. Shimokawa, and M. Yamanaka, "Highly sensitive etchants for delineation of defects in single- and polycrystalline silicon materials," *Journal of the Electrochemical Society*, **143**, 4079–4087, 1996.
72. M. Akatsuka, K. Sueoka, and T. Yamamoto, "Classification of etch pits at silicon wafer surface using image-processing instrument," *Journal of Crystal Growth*, **210**, 366–369, 2000.
73. J. Lu and G. Rozgonyi, "Oxygen and carbon precipitation in crystalline sheet silicon: Depth profiling by infrared spectroscopy, and preferential defect etching," *Journal of the Electrochemical Society*, **153**, G986–G991, 2006.
74. I. A. Shah, B. M. A. van der Wolf, W. J. P. van Enckevort, and E. Vlieg, "Wet chemical etching of silicon {111}: Autocatalysis in pit formation," *Journal of the Electrochemical Society*, **155**, J79–J84, 2008.
75. I. A. Shah, B. M. A. van der Wolf, W. J. P. van Enckevort, and E. Vlieg, "Wet chemical etching of silicon {111}: Etch pit analysis by the Lichtfigur method," *Journal of Crystal Growth*, **311**, 1371–1377, 2009.
76. D. K. Schroder, "Some recent advances in contactless silicon characterization," 210 ECE Meeting Cancun, Mexico, Oct. 29–Nov. 3, 2006, and the references therein.
77. W. R. Knolle and T. F. Retajczyk, Jr., "Monitoring sodium contamination in silicon devices and processing materials by flame emission spectrometry," *Journal of the Electrochemical Society*, **120**, 1106–1111, 1973.

78. L. Zeni, R. Bernini, and R. Pierri, "Reconstruction of doping profiles in semiconductor materials using optical tomography," *Solid-State Electronics*, **43**, 761–769, 1999, and the references therein.

79. J. Reif, R. Schmid, T. Schneider, and D. Wolfframm, "Nonlinear optical characterization of the surface of silicon wafers: In-situ detection of external stress," *Solid-State Electronics*, **44**, 809–813, 2000, and the references therein.

80. A. Buczkowski, B. Orschel, S. Kim, S. Rouvimov, B. Snegirev, M. Fletcher, and F. Kirscht, "Photoluminescence intensity analysis in application to contactless characterization of silicon wafers," *Journal of the Electrochemical Society*, **150**, G436–G442, 2003.

81. L. Jastrzebski and J. Lagowski, "Deep levels study in float zone Si used for fabrication of CCD imagers," *Journal of the Electrochemical Society*, **128**, 1957–1963, 1981.

82. E. Simoen, C. Claeys, R. Job, A.G. Ulyashin, W.R. Fahrner, G. Tonelli, O. De Gryse, and P. Clauws, "Deep levels in oxygenated n-type high-resistivity FZ silicon before and after a low-temperature hydrogenation step," *Journal of the Electrochemical Society*, **150**, G520–G526, 2003.

83. K. Ryoo and W. E. Socha, "Correlation of surface photovoltaic technique with deep level transient spectroscopy for iron concentration measurement in *p*-type silicon wafers," *Journal of the Electrochemical Society*, **138**, 1424–1426, 1991.

84. K. Nauka and D. A. Gomez, "Surface photovoltage and deep level transient spectroscopy measurement of the Fe impurities in front-end operations of the IC CMOS process," *Journal of the Electrochemical Society*, **142**, L98–L99, 1995.

85. B. Hacki, K.-J. Range, P. Stallhofer, and L. Fabry, "Correlation between DLTS and TRXFA measurements of copper and iron contaminations in FZ and CZ silicon wafers: Application to gettering efficiencies," *Journal of the Electrochemical Society*, **139**, 1495–1498, 1992.

86. Y. Kitagawara, R. Hoshi, and T. Takenaka, "Evaluation of oxygen precipitated silicon crystals by deep-level photoluminescence at room temperature and its mapping," *Journal of the Electrochemical Society*, **139**, 2277–2281, 1992.

87. N-E. Chabane-Sari, S. Krieger-Kaddour, C. Vinante, M. Berenguer, and D. Barbier, "A deep level transient spectroscopy study of the internal gettering of Cr in Czochralski-grown silicon: Efficiency and reversibility upon lamp pulse annealing," *Journal of the Electrochemical Society*, **139**, 2900–2904, 1992.

88. D. Gilmore, T. Arahori, M. Ito, H. Murakami, and S.-i. Miki, "The impact of graphite furnace parts on radial impurity distribution in Czochralski-grown single-crystal silicon," *Journal of the Electrochemical Society*, **145**, 621–628, 1998.

89. J. H. Bleka, E. V. Monakhov, B. S. Avset, and B. G. Svensson, "DLTS study of room-temperature defect annealing in n-type high-purity FZ Si," *210 ECE Meeting Cancun*, Mexico, Oct. 29–Nov. 3, 2006, and the references therein.

90. T. Furuoya, "Dislocations in silicon single crystals," *Japanese Journal of Applied Physics*, **1**, 135–143, 1962.

91. M. Yoshida, Y. Yamaguchi, and H. Aoki, "Hexagonal precipitates on nickel in silicon," *Japanese Journal of Applied Physics*, **2**, 305–306, 1963.

92. W. T. Stacy, D. F. Allison, and T.-C. Wu, "Metal decorated defects in heat-treated silicon wafers," *Journal of the Electrochemical Society*, **129**, 1128–1133, 1982.

93. H. Idrisi, T. Sinke, V. Gerhardt, D. Ceglarek, and B. O. Kolbesen, "Decoration and preferential etching of crystal defects in silicon materials: Influence of metal decoration on the defect etching process," *Physica Status Solidi C*, **8**, 788–791, 2011, and the references therein.

94. M. S. Kulkarni, J. Libbert, S. Keltner, and L. Muléstagno, "A theoretical and experimental analysis of macrodecoration of defects in monocrystalline silicon," *Journal of the Electrochemical Society*, **149**, G153–G165, 2002.

95. C. Munakata, "An electron beam method of measuring resistivity distribution in semiconductors," *Japanese Journal of Applied Physics*, **6**, 963–971, 1967.

96. R. C. Henderson, "Silicon cleaning with hydrogen peroxide solutions: A high energy electron diffraction and Auger electron spectroscopy study," *Journal of the Electrochemical Society*, **119**, 772–775, 1972.

97. P. V. Fontana, J. P. Decosterd, and L. Wegmann, "Investigations of carbon residues on surfaces of silicon integrated circuits," *Journal of the Electrochemical Society*, **121**, 146–150, 1974.

98. M. Nakamura, E. Kitamura, Y. Misawa, T. Suzuki, S. Nagai, and H. Sunaga, "Photoluminescence measurement of carbon in silicon crystals irradiated with high energy electrons," *Journal of the Electrochemical Society*, **141**, 3576–3580, 1994.

99. B. Yurash and B. E. Deal, "A method for determining sodium content of semiconductor processing materials," *Journal of the Electrochemical Society*, **115**, 1191–1196, 1968, and the references therein.

100. W. R. Knolle, "Flame emission analysis of potassium contamination in silicon slice processing," *Journal of the Electrochemical Society*, **120**, 987–991, 1973.

101. G. Eranna and D. Kakati, "Limitations on the range of measurements of sheet resistivity of shallow diffused layers for profiling by the four-point probe technique," *Solid-State Electronics*, **25**, 611–614, 1982.

102. A. H. Tong and A. Dupnock, "The two-level effect in sheet resistance measurements made with a four-point probe," *Journal of the Electrochemical Society*, **118**, 390–394, 1971.

103. P. J. Severin and H. Bulle, "Four-point probe measurements on n-type silicon with mercury probes," *Journal of the Electrochemical Society*, **122**, 133–137, 1975.

104. D. S. Perloff, "Four-point probe correction factors for use in measuring large diameter doped semiconductor wafers," *Journal of the Electrochemical Society*, **123**, 1745–1750, 1976.

105. D. S. Perloff, F. E. Wahl, and J. Conragan, "Four-point sheet resistance measurements of semiconductor doping uniformity," *Journal of the Electrochemical Society*, **124**, 582–590, 1977.

106. J. Albers and H. L. Berkowitz, "The relation between two-probe and four-probe resistances on nonuniform structures," *Journal of the Electrochemical Society*, **131**, 392–398, 1984.

107. J. J. Kopanski, J. Albers, G. P. Carver, and J. R. Ehrstein, "Verification of the relation between two-probe and four-probe resistances as measured on silicon wafers," *Journal of the Electrochemical Society*, **137**, 3935–3941, 1990.

108. K. Sugawara, K. Ozeki, K. Fujioka, Y. Mamada, M. Igai, and H. Hirayama, "Finite difference analysis of radial phosphorus dopant distribution in Czochralski-grown silicon single crystals," *Journal of the Electrochemical Society*, **148**, G475–G480, 2001.

109. N. Inoue, S. Yamazaki, Y. Goto, T. Kushida, and T. Sugiyama, "Infrared absorption spectroscopy of complexes in low carbon concentration, low electron dosage irradiated CZ silicon crystal," *210 ECE Meeting Cancun*, Mexico, Oct. 29–Nov. 3, 2006.

110. G. Williams and R. E. Reusser, "Heat transfer in silicon Czochralski crystal growth," *Journal of Crystal Growth*, **64**, 448–460, 1983.

111. C. Gross, G. Gaetano, T. N. Tucker, and J. A. Baker, "Comparison of infrared and activation analysis results in determining the oxygen and carbon content in silicon," *Journal of the Electrochemical Society*, **119**, 926–929, 1972.

112. B. Pajot, "Characterization of oxygen in silicon by infrared absorption," *Analusis*, **5**, 293–303, 1977.

113. H. J. Rath, P. Stallhofer, D. Huber, and B. F. Schmitt, "Determination of oxygen in silicon by photon activation analysis for calibration of the infrared absorption," *Journal of the Electrochemical Society*, **131**, 1920–1923, 1984.

114. B. Pajot, H. J. Stein, B. Cales, and C. Naud, "Quantitative spectroscopy of interstitial oxygen in silicon," *Journal of the Electrochemical Society*, **132**, 3034–3037, 1985.

115. T. Iizuka, S. Takasu, M. Tajima, T. Arai, T. Nozaki, N. Inoue, and M. Watanabe, "Determination of conversion factor for infrared measurement of oxygen in silicon," *Journal of the Electrochemical Society*, **132**, 1707–1713, 1985.

116. K. G. Barraclough, R. W. Series, J. S. Hislop, and D. A. Wood, "Calibration of infrared absorption by gamma activation analysis for studies of oxygen in silicon," *Journal of the Electrochemical Society*, **133**, 187–191, 1986.

117. F. Schomann and K. Graff, "Correction factors for the determination of oxygen in silicon by IR spectrometry," *Journal of the Electrochemical Society*, **136**, 2025–2031, 1989.

118. D. E. Hill, "Determination of interstitial oxygen concentration in low-resistivity *n*-type silicon wafers by infrared absorption measurements," *Journal of the Electrochemical Society*, **137**, 3926–3928, 1990.
119. B. G. Rennex, J. R. Ehrstein, and R. I. Scace, "Methodology for the certification of reference specimens for determination of oxygen concentration in semiconductor silicon by infrared spectrophotometry," *Journal of the Electrochemical Society*, **143**, 258–263, 1996.
120. B. G. Rennex, J. R. Ehrstein, and R. I. Scace, "Methodology for the certification of reference specimens for determination of oxygen concentration in semiconductor silicon by infrared spectrophotometry," *Journal of the Electrochemical Society*, **143**, 258–263, 1996.
121. A. Sassella, A. Borghesi, and T. Abe, "Quantitative evaluation of precipitated oxygen in silicon by infrared spectroscopy: Still an open problem," *Journal of the Electrochemical Society*, **145**, 1715–1719, 1998.
122. O. De Gryse, P. Clauws, J. Vanhellemont, O. I. Lebedev, J. Van Landuyt, E. Simoen, and C. Claeys, "Characterization of oxide precipitates in heavily B-doped silicon by infrared spectroscopy," *Journal of the Electrochemical Society*, **151**, G598–G605, 2004.
123. S. Yang, Y. Li, Q. Ma, L. Liu, X. Xu, P. Niu, Y. Li, S. Niu, and H. Li, "Infrared absorption spectrum studies of the VO defect in fast-neutron-irradiated Czochralski silicon," *Journal of Crystal Growth*, **280**, 60–65, 2005.
124. J. L. Regolini, J. P. Stoquert, C. Ganter, and P. Siffert, "Determination of the conversion factor for infrared measurements of carbon in silicon," *Journal of the Electrochemical Society*, **133**, 2165–2168, 1986.
125. L. L. Hwang, J. Bucci, and R. McCormick, "Measurement of carbon concentration in polycrystalline silicon using FTIR," *Journal of the Electrochemical Society*, **138**, 576–581, 1991.
126. H. Saito and H. Shirai, "Baseline variations near the carbon impurity vibration band in infrared spectra of silicon," *Journal of the Electrochemical Society*, **147**, 1210–1212, 2000.
127. H. Ch. Alt, Y. Gomeniuk, B. Wiedemann, and H. Riemann, "Method to determine carbon in silicon by infrared absorption spectroscopy," *Journal of the Electrochemical Society*, **150**, G498–G501, 2003.
128. J. Lu and G. Rozgonyi, "Oxygen and carbon precipitation in crystalline sheet silicon: Depth profiling by infrared spectroscopy and preferential defect etching," *Journal of the Electrochemical Society*, **153**, G986–G991, 2006.
129. Z. Jiang, W. Zhang, X. Niu, and L. Yan, "Infrared measurement of Ge concentrations in CZ-Si," *Journal of Crystal Growth*, **279**, 65–69, 2005.
130. S. Pahlke and L. Fabry, "Determination of oxygen in semiconductor silicon by gas fusion analysis (GFA) – Historical and future trends," Wacker Siltronic AG, Central Research and Development, Central Analytical Laboratories, D-84479 Burghausen, Germany, www.electrochem.org/ma/203/pdfs/0725.pdf.
131. R. W. Shaw, R. Bredeweg, and P. Rossetto, "Gas fusion analysis of oxygen in silicon: Separation of components," *Journal of the Electrochemical Society*, **138**, 582–585, 1991.
132. P. J. Olshefski, D. J. Shombert, and I. R. Weingarten, "Use of Hall measurements in evaluating polycrystalline silicon," *Journal of the Electrochemical Society*, **108**, 362–365, 1961.
133. L. P. Hunt, R. W. Francis, and J. P. Dismukes, "Boron and phosphorus determination in low resistivity solar-grade silicon," *Journal of the Electrochemical Society*, **131**, 1888–1891, 1984.
134. T. F. Ciszek, "Growth and properties of [100] and [111] dislocation-free silicon crystals from a cold crucible," *Journal of Crystal Growth*, **70**, 324–329, 1984.
135. S. E. Van Bramer, "An introduction to mass spectrometry," Widener University, Department of Chemistry, One University Place, Chester, PA 19013, USA, September 1998.
136. P. R. Kennicott, "Determination of nitrogen in silicon by solid source mass spectrometry," *Journal of the Electrochemical Society*, **111**, 1101–1102, 1964.
137. C. Feldman and F. G. Satkiewicz, "Mass spectra analysis of impurities and ion clusters in amorphous and crystalline silicon films," *Journal of the Electrochemical Society*, **120**, 1111–1116, 1973.
138. D. Macdonald, A. Cuevas, and F. Ferrazza, "Response to phosphorus gettering of different regions of cast multicrystalline silicon ingots," *Solid-State Electronics*, **43**, 575–581, 1999.

139. T. S. Horányi, P. Tüttő, and Cs. Kovacsics, "Identification possibility of metallic impurities in *p*-type silicon by lifetime measurement," *Journal of the Electrochemical Society*, **143**, 216–220, 1996.

140. X. Gao and S. Yee, "Comparison of metal oxide semiconductor capacitance-time and surface photovoltage methods in investigating annealing behavior of iron contamination in boron-doped silicon," *Journal of the Electrochemical Society*, **140**, 2042–2046, 1993.

141. A. L. P. Rotondaro, T. Q. Hurd, A. Kaniava, J. Vanhellemont, E. Simoen, M. M. Heyns, C. Claeys, and G. Brown, "Impact of Fe and Cu contamination on the minority carrier lifetime of silicon substrates," *Journal of the Electrochemical Society*, **143**, 3014–3019, 1996.

142. A. Rohatgi and P. Rai-Choudhury, "Process-induced effects on carrier lifetime and defects in float zone silicon," *Journal of the Electrochemical Society*, **127**, 1136–1139, 1980.

143. T. F. Ciszek, T. Wang, T. Schuyler, and A. Rohatgi, "Some effects of crystal growth parameters on minority carrier lifetime in float-zoned silicon," *Journal of the Electrochemical Society*, **136**, 230–234, 1989.

144. S. K. Pang, A. Rohatgi, B. L. Sopori, and G. Fiegl, "A comparison of minority-carrier lifetime in as-grown and oxidized float-zone, magnetic Czochralski, and Czochralski silicon," *Journal of the Electrochemical Society*, **137**, 1977–1981, 1990.

145. M. Lozano, M. Ullán, C. Martínez, L. Fonseca, J.M. Rafí, F. Campabadal, E. Cabruja, C. Fleta, M. Key, and S. Bermejo, "Effect of combined oxygenation and gettering on minority carrier lifetime in high-resistivity FZ silicon," *Journal of the Electrochemical Society*, **151**, G652–G657, 2004.

146. L. Jastrzebski, R. Soydan, H. Elabd, W. Henry, and E. Savoye, "The effect of heavy metal contamination on defects in CCD imagers: Contamination monitoring by surface photovoltage," *Journal of the Electrochemical Society*, **137**, 242–249, 1990.

147. K. Nauka and D. A. Gomez, "Surface photovoltage and deep level transient spectroscopy measurement of the Fe impurities in front-end operations of the IC CMOS process," *Journal of the Electrochemical Society*, **142**, L98–L99, 1995.

148. T. Taishi, T. Hoshikawa, M. Yamatani, K. Shirasawa, X. Huang, S. Uda, and K. Hoshikawa, "Influence of crystalline defects in Czochralski-grown Si multicrystal on minority carrier lifetime," *Journal of Crystal Growth*, **306**, 452–457, 2007.

149. C. Gross, G. Gaetano, T. N. Tucker, and J. A. Baker, "Comparison of infrared and activation analysis results in determining the oxygen and carbon content in silicon," *Journal of the Electrochemical Society*, **119**, 926–929, 1972.

150. K. Graff, "Precise evaluation of oxygen measurements on CZ-silicon wafers," *Journal of the Electrochemical Society*, **130**, 1378–1381, 1983.

151. H. J. Rath, P. Stallhofer, D. Huber, and B. F. Schmitt, "Determination of oxygen in silicon by photon activation analysis for calibration of the infrared absorption," *Journal of the Electrochemical Society*, **131**, 1920–1923, 1984.

152. T. Iizuka, S. Takasu, M. Tajima, T. Arai, T. Nozaki, N. Inoue, and M. Watanabe, "Determination of conversion factor for infrared measurement of oxygen in silicon," *Journal of the Electrochemical Society*, **132**, 1707–1713, 1985.

153. B. Pajot, H. J. Stein, B. Cales, and C. Naud, "Quantitative spectroscopy of interstitial oxygen in silicon," *Journal of the Electrochemical Society*, **132**, 3034–3037, 1985.

154. A. Baghdadi, W. M. Bullis, M. C. Croarkin, Y.-z. Li, R.I. Scace, R.W. Series, P. Stallhofer, and M. Watanabe, "Interlaboratory determination of the calibration factor for the measurement of the interstitial oxygen content of silicon by infrared absorption," *Journal of the Electrochemical Society*, **136**, 2015–2026, 1989.

155. F. Schomann and K. Graff, "Correction factors for the determination of oxygen in silicon by IR spectrometry," *Journal of the Electrochemical Society*, **136**, 2025–2031, 1989.

156. D. E. Hill, "Determination of interstitial oxygen concentration in low-resistivity *n*-type silicon wafers by infrared absorption measurements," *Journal of the Electrochemical Society*, **137**, 3926–3928, 1990.

157. Y. Kitagawara, M. Tamatsuka, and T. Takenaka, "Accurate evaluation techniques of interstitial oxygen concentrations in medium-resistivity Si crystals," *Journal of the Electrochemical Society*, **141**, 1362–1364, 1994.

158. A. Sassella, A. Borghesi, and T. Abe, "Quantitative evaluation of precipitated oxygen in silicon by infrared spectroscopy: Still an open problem," *Journal of the Electrochemical Society*, **145**, 1715–1719, 1998.

159. O. De Gryse, P. Clauws, J. Vanhellemont, O.I. Lebedev, J. Van Landuyt, E. Simoen, and C. Claeys, "Characterization of oxide precipitates in heavily B-doped silicon by infrared spectroscopy," *Journal of the Electrochemical Society*, **151**, G598–G605, 2004.

160. J. Lu and G. Rozgonyi, "Oxygen and carbon precipitation in crystalline sheet silicon: Depth profiling by infrared spectroscopy, and preferential defect etching," *Journal of the Electrochemical Society*, **153**, G986–G991, 2006.

161. J. L. Regolini, J. P. Stoquert, C. Ganter, and P. Siffert, "Determination of the conversion factor for infrared measurements of carbon in silicon," *Journal of the Electrochemical Society*, **133**, 2165–2168, 1986.

162. D. Gilmore, T. Arahori, M. Ito, H. Murakami, and S.-i. Miki, "The impact of graphite furnace parts on radial impurity distribution in Czochralski-grown single-crystal silicon," *Journal of the Electrochemical Society*, **145**, 621–628, 1998.

163. H. Saito and H. Shirai, "Baseline variations near the carbon impurity vibration band in infrared spectra of silicon," *Journal of the Electrochemical Society*, **147**, 1210–1212, 2000.

164. H. Ch. Alt, Y. Gomeniuk, B. Wiedemann, and H. Riemann, "Method to determine carbon in silicon by infrared absorption spectroscopy," *Journal of the Electrochemical Society*, **150**, G498–G501, 2003.

165. T. Yamamoto and K. Nishihara, "Characterization of defects in annealed Czochralski-grown silicon wafers by photoluminescence method," *Journal of Crystal Growth*, **210**, 69–73, 2000.

166. Y. Kitagawara, R. Hoshi, and T. Takenaka, "Evaluation of oxygen precipitated silicon crystals by deep-level photoluminescence at room temperature and its mapping," *Journal of the Electrochemical Society*, **139**, 2277–2281, 1992.

167. E. Leoni, L. Martinelli, S. Binetti, G. Borionetti, and S. Pizzini, "The origin of photoluminescence from oxygen precipitates nucleated at low temperature in semiconductor," *Journal of the Electrochemical Society*, **151**, G866–G869, 2004.

168. M. Nakamura, E. Kitamura, Y. Misawa, T. Suzuki, S. Nagai, and H. Sunaga, "Photoluminescence measurement of carbon in silicon crystals irradiated with high energy electrons," *Journal of the Electrochemical Society*, **141**, 3576–3580, 1994.

169. M. Nukamura, "Influence of oxygen and carbon on the formation of the 1.014 eV photoluminescence copper center in silicon crystal," *Journal of the Electrochemical Society*, **147**, 796–798, 2000.

170. M. Tajima, T. Masui, D. Itoh, and T. Nishino, "Calibration of the photoluminescence method for determining As and Al concentrations in Si," *Journal of the Electrochemical Society*, **137**, 3544–3551, 1990.

171. D. B. Sirdeshmukh, "Gamma-ray diffraction — A powerful tool in crystal physics," *Current Science*, **72**, 631–640, 1997.

172. A. I. Kurbakov, "Gamma-ray diffraction in the study of silicon," *Materials Science and Engineering B*, **22**, 149–158, 1994.

173. B. Hackl, K.-J. Range, H. J. Gores, L. Fabry, and P. Stallhofer, "Iron and its detrimental impact on the nucleation and growth of oxygen precipitates during internal gettering processes," *Journal of the Electrochemical Society*, **139**, 3250–3254, 1992.

174. H. Morinaga, T. Futatsuki, T. Ohmi, E. Fuchita, M. Oda, and C. Hayashi, "Behavior of ultrafine metallic particles on silicon wafer surface," *Journal of the Electrochemical Society*, **142**, 966–970, 1995.

175. L. Fedina, A. Gutakovskii, and A. Aseev, "FZ-Si crystal growth and HREM study of new types of extended defects during in situ electron irradiation," *Journal of Crystal Growth*, **229**, 1–5, 2001.

176. K. L. Chiang, C. J. Dell'Oca, and F. N. Schwettmann, "Optical evaluation of polycrystalline silicon surface roughness," *Journal of the Electrochemical Society*, **126**, 2267–2269, 1979.

177. E. F. Steigmeier and H. Auderset, "Structural perfection testing of films and wafers by means of optical scanner," *Journal of the Electrochemical Society*, **131**, 1693–1699, 1984.

178. S. Kimura, "Observation of oxygen precipitates in CZ-grown Si wafers with a phase differential scanning optical microscope," *Journal of the Electrochemical Society*, **141**, L120–L122, 1994.
179. K. Minowa, K. Takeda, S. Tomimatsu, and K. Umemura, "TEM observation of grown-in defects in CZ and epitaxial silicon wafers detected with optical shallow defect analyzer," *Journal of Crystal Growth*, **210**, 15–19, 2000.
180. M. Sumiji, S. Nakamura, T. Azami, and T. Hibiya, "Optical observation of solid-melt interface fluctuation due to Marangoni flow in a silicon liquid bridge," *Journal of Crystal Growth*, **223**, 503–511, 2001.
181. P. V. Fontana, J. P. Decosterd, and L. Wegmann, "Investigations of carbon residues on surfaces of silicon integrated circuits," *Journal of the Electrochemical Society*, **121**, 146–150, 1974.
182. C. W. Magee, "Depth profiling of *n*-type dopants in Si and GaAs using Cs$^+$ bombardment negative secondary ion mass spectrometry in ultrahigh vacuum," *Journal of the Electrochemical Society*, **126**, 660–663, 1979, and the references therein.
183. H. De Witte, S. De Gendt, M. Douglas, T. Conard, K. Kenis, P. W. Mertens, W. Vandervorst, and R. Gijbels, "Evaluation of time-of-flight secondary ion mass spectrometry for metal contamination monitoring on Si wafer surfaces," *Journal of the Electrochemical Society*, **147**, 1915–1919, 2000.
184. R. G. Mazur and D. H. Dickey, "A spreading resistance technique for resistivity measurements in silicon," *Journal of the Electrochemical Society*, **113**, 255–259, 1966.
185. T. H. Yeh and K. H. Khokhani, "Multilayer theory of correction factors for spreading-resistance measurements," *Journal of the Electrochemical Society*, **116**, 1461–1464, 1969.
186. J. Albers, "Continuum formulation of spreading resistance correction factors," *Journal of the Electrochemical Society*, **127**, 2259–2263, 1980.
187. E. F. Gorey, C. P. Schneider and M. R. Poponiak, "Preparation and evaluation of spreading resistance probe tip," *Journal of the Electrochemical Society*, **117**, 721–725, 1970.
188. S. Loftis and D. Dickey, "An introduction to spreading resistance analysis and its application in the semiconductor industry," Seminar09.pdf, Solecon Labs, 770 Trademark Dr, Reno, NV 89521, USA.
189. G. D. Robertson, Jr. and D. J. O'Connor, "Magnetic field effects on float-zone Si crystal growth: III. Strong axial fields," *Journal of Crystal Growth*, **76**, 111–122, 1986.
190. P. J. Severin and H. Bulle, "Spreading resistance measurements on *n*-type silicon using mercury probes," *Journal of the Electrochemical Society*, **122**, 137–142, 1975.
191. J. R. Ehrstein, "Spreading resistance calibration for gallium- or aluminum-doped silicon," *Journal of the Electrochemical Society*, **127**, 1403–1404, 1980.
192. D. C. Gupta, J. Y. Chan, and P. Wang, "Observations on imperfections in silicon material using the spreading resistance probe," *Journal of the Electrochemical Society*, **117**, 1611–1613, 1970.
193. W. Lin and M. Stavola, "Oxygen segregation and microscopic inhomogeneity in Czochralski silicon," *Journal of the Electrochemical Society*, **132**, 1412–1416, 1985.
194. B. L. Morris, "Some device applications of spreading resistance measurements on epitaxial silicon," *Journal of the Electrochemical Society*, **121**, 422–426, 1974.
195. Y. Iida, H. Abe, and M. Kondo, "Impurity profile measurements of thin epitaxial silicon wafer by multilayer spreading resistance analysis," *Journal of the Electrochemical Society*, **124**, 1118–1122, 1977.
196. T. E. Hendrickson, "Ion implanted profiles from two point spreading resistance measurements," *Journal of the Electrochemical Society*, **122**, 1539–1541, 1975.
197. D. C. D'Avanzo, R. D. Rung, A. Gat, and R. W. Dutton, "High speed implementation and experimental evaluation of multilayer spreading resistance analysis," *Journal of the Electrochemical Society*, **125**, 1170–1176, 1978.
198. H. L. Berkowitz and R. A. Lux, "Errors in resistivities calculated by multilayer analysis of spreading resistances," *Journal of the Electrochemical Society*, **126**, 1479–1482, 1979.
199. I. D. Vrinceanu and S. Danyluk, "Measurement of residual stress in single crystal silicon wafers," *8th IEEE International Symposium on Advanced Packaging Materials*, 297–301, 2002.
200. M. Fukuzawa and M. Yamada, "Photoelastic characterization of Si wafers by scanning infrared polariscope," *Journal of Crystal Growth*, **229**, 22–25, 2001.
201. C. Yang, F. Mess, K. Skenes, S. Melkote, and S. Danyluk, "On the residual stress and fracture strength of crystalline silicon wafers," *Applied Physics Letters*, **102**, 021909, 2013.

202. R. Glang, R. A. Holmwood, and R. L. Rosenfeld, "Determination of stress in films on single crystal silicon substrates," *The Review of Scientific Instruments*, **36**, 7–10, 1965.

203. J. C. Hensel, H. Hasegawa, and M. Nakayama, "Cyclotron resonance in uniaxially stressed silicon: II. Nature of covalent bond," *Physical Review*, **138**, A225–A238, 1965.

204. H. Kotake and S. Takasu, "Quantitative measurement of stress in silicon by photoelasticity and its application," *Journal of the Electrochemical Society*, **127**, 179–184, 1980.

205. H. Shimizu, S. Isomae, K. Minowa, T. Satoh, and T. Suzuki, "Gravitational stress-induced dislocations in large-diameter silicon wafers studied by x-ray topography and computer simulation," *Journal of the Electrochemical Society*, **145**, 2523–2529, 1998.

206. K. Nishihira, S. Munetoh, and T. Motooka, "Uniaxial strain observed in solid/liquid interface during crystal growth from melted Si: A molecular dynamics study," *Journal of Crystal Growth*, **210**, 60–64, 2000.

207. F. Shimura, "TEM observations of pyramidal hillocks formed on (100) silicon wafers during chemical etching," *Journal of the Electrochemical Society*, **127**, 910–913, 1980.

208. P. Wang, L. Chang, L. J. Demer, and C. J. Varker, "Denuded zones in Czochralski silicon wafers," *Journal of the Electrochemical Society*, **131**, 1948–1952, 1984.

209. L. Rivaud, C. N. Anagnostopoulos, and G. R. Erikson, "A transmission electron microscopy (TEM) study of oxygen precipitation induced by internal gettering in low and high oxygen wafers," *Journal of the Electrochemical Society*, **135**, 437–442, 1988.

210. S. Krieger-Kaddour, N-E. Chabane-Sari, and D. Barbier, "Transmission electron microscopic study of the morphology of oxygen precipitates and of chromium precipitation during intrinsic gettering in Czochralski-grown silicon: Influence of lamp pulse annealing," *Journal of the Electrochemical Society*, **140**, 495–500, 1993.

211. K. Minowa, K. Takeda, S. Tomimatsu, and K. Umemura, "TEM observation of grown-in defects in CZ and epitaxial silicon wafers detected with optical shallow defect analyzer," *Journal of Crystal Growth*, **210**, 15–19, 2000.

212. M. Itsumi, "Octahedral void defects in Czochralski silicon," *Journal of Crystal Growth*, **237–239**, 1773–1778, 2002.

213. Z. Zeng, X. Ma, J. Chen, D. Yang, I. Ratschinski, F. Hevroth, and H.S. Leipner, "Effect of oxygen precipitates on dislocation motion in Czochralski silicon," *Journal of Crystal Growth*, **312**, 169–173, 2010.

214. R. D. Westbrook, "Effect of semiconductor inhomogeneities on carrier mobilities measured by the van der Pauw method," *Journal of the Electrochemical Society*, **121**, 1212–1215, 1974.

215. M. G. Buehler, S. D. Grant, and W. R. Thurber, "Bridge and van der Pauw sheet resistors for characterizing the line width of conducting layers," *Journal of the Electrochemical Society*, **125**, 650–654, 1978.

216. D. L. Parker and W. A. Porter, "A high speed x-ray topographic camera for semiconductor wafer evaluation," *Journal of the Electrochemical Society*, **123**, 407–409, 1976.

217. S. Kawado, "Large-area x-ray topography to observe 300-mm-diameter silicon crystals," Industial Applications, p89.pdf, X-ray Research Laboratory, Rigaku Corporation, Japan.

218. N. Parekh, C. Nieuwenhuizen, J. Borstrok, and O. Elgersma, "Analysis of thin films in silicon integrated circuit technology by x-ray fluorescence spectrometry," *Journal of the Electrochemical Society*, **138**, 1460–1465, 1991.

219. H. Bradaczek and G. Hildebrandt, "Real x-ray optics – A challenge for crystal growers," *Journal of Optoelectronics and Advanced Materials*, **1**, 3–8, 1999.

220. M. Akatsuka, K. Sueoka, H. Katahama, and N. Adachi, "Calculation of slip length in 300-mm silicon wafers during thermal processes," *Journal of the Electrochemical Society*, **146**, 2683–2688, 1999.

221. S. L. Morelhão, J. Härtwig, and D. L. Meier, "Dislocations in dendritic web silicon," *Journal of Crystal Growth*, **213**, 288–298, 2000.

222. G. Hildebrandt and H. Bradaczek, "High precision crystal orientation measurements with the x-ray Omega-Scan – A tool for the industrial use of quartz and other crystals," *Journal of Optoelectronics and Advanced Materials*, **6**, 5–21, 2004.

7

Resistivity and Impurity Concentration Mapping of Silicon Wafers

7.1 Introduction

Dopant impurities play a key role in semiconductor device operation, as they directly influence the electrical properties of different regions. The dopant concentration directly affects the parameters such as the threshold voltage in metal-oxide semiconductor (MOS) very large scale integration (VLSI) transistors and breakdown voltage of the devices fabricated. During silicon crystal growth, the impurity atoms with k_o value (i.e., equilibrium segregation coefficient) less than 1 are rejected by the advancing solid at a greater rate than they can diffuse into the bulk of the melt. Most values for the commonly used impurity dopant species for silicon are below this unity value, which means that during the crystal growth process, the dopant species are rejected into the silicon melt. With melt-crystal, at the beginning of the solidification process at a given crystal solid-melt interface, segregation takes place, and the rejected impurity atoms begin to accumulate in the melt layer near the growth interface and diffuse in the direction of the bulk of the melt [1]. An impurity concentration gradient develops ahead of advancing crystal.

Because of the equilibrium segregation coefficient, the impurity distribution will not be uniform with the axial distribution of the grown crystal. This variation is a function of crystal length, and the crystal tag end shows a continuous increase in the dopant concentration values. The impurity added at the beginning of the crystal growth, despite its homogeneous distribution in the liquid, may not reflect in the grown crystal. In addition to the axial variation, the grown crystals show radial variation, depending on the shape of the solidifying melt-solid interface, where the center of the wafer and the edges show variation in the resistivity values and the defects that generate during the process. With wafer thickness values around 1 mm or less, this variation is clearly distinguishable.

Therefore, to make semiconductor devices with controllable electrical properties, it is important to control the dopant concentration accurately, both along the growth axis of the grown silicon ingot and in the radial direction. The dopant concentration along the growth direction can be described by the normal freezing equation and is well understood, but factors controlling the dopant distribution in the radial direction are not well understood. Sugawara et al. [2] made a detailed, quantitative analysis of the radial dopant distribution in Czochralski (CZ) silicon crystals by solving the coupled Navier-Stokes, continuity, and energy equations for the silicon melt flow and temperature fields, and by solving the diffusion and segregation equations for the phosphorus impurity distribution in the melt and in the crystal. They reported good agreement between the measured and simulated radial dopant concentration in the CZ-grown crystals. Important components of the numerical simulation were carried out using the dopant concentration of the melt at the solid–liquid interface and the equilibrium segregation coefficient for dopant species, particularly for the phosphorus dopant.

Silicon crystals that have been doped with boron impurity are commercially available from 0.0005 to 50 Ω-cm. Arsenic- and phosphorus-doped crystals are available from 0.005 to 40 Ω-cm. Antimony-doped crystals are available from 0.01 Ω-cm and above. For lower resistivity ranges, arsenic dopant is preferred. Antimony-doped substrates are preferred as epitaxial substrates because of auto-doping effects, which are found to be at a minimum with this impurity compared to other dopant species. Resistivity values greater than few hundred Ω-cm are difficult to obtain with CZ crystals. Float-zone (FZ) crystals are the only other option. Though the intrinsic resistivity of the pure silicon is large, practically, it is difficult to achieve due to a large number of unintentional impurities getting into the growing crystal. These issues are discussed in Chapter-9.

Maintaining uniform dopant concentration is a major issue affecting quality silicon wafers. Many studies have been reported on the control of axial-specific resistivity distributions in bulk crystal growth. Various methods that have been proposed to avoid macroscopic axial and segregation can be classified into two main groups. In the first case, the key idea is to ensure that the composition of the liquid in the vicinity of the solid–liquid interface remains constant throughout the solidification process or the continuous liquid-feeding technique. A second approach consists of adapting the growth conditions to balance variations in bulk concentration. Wang et al. [3] have proposed a simple co-doping method for controlling the axial-specific resistivity distribution in single crystals of silicon by recombining the conduction electron and hole. Two sorts of impurities, boron and phosphorus, are doped simultaneously in the meltdown stage in the melt growth process, and the distributions are predictable, as shown in Figure 7.1. Numerical simulation and crystal growth experiments have demonstrated that the uniform axial-specific resistivity profile is possible to achieve using boron and phosphorus co-doping in the melt growth process of single crystals of silicon.

The concept of using the effective segregation coefficient and the thermal convection behavior of the silicon melt described so far can help explain the radial resistivity variation observed in the CZ-grown silicon wafers and explain the cause of microscopic resistivity variations measured in the ingot. The radial resistivity variations in the molten silicon arise because of the motion of the molten silicon due to the dominant thermal convection flow in large silicon melts held in the quartz crucible. The details are shown in Figure 7.2. This situation causes a thinner boundary layer to develop near the periphery of the growing ingot and a thicker one at the center of the growing ingot [4]. Figure 7.2 explains two situations for the boron impurity to the growing crystal about the radial impurity distribution in (100) and (111) crystal orientations. A similar situation takes place with the phosphorus impurity, but the features are not identical. If k_o is less than unity, more impurities will be incorporated near the center of the ingot and progressively fewer toward the edge of the grown crystal. When the wafers are produced with these ingots, the center part of the crystal differs in resistivity, forming a concentric circular pattern when the wafers are mapped. Practically achieving uniform resistivity in silicon crystals has been an important topic during the last 50 years and is likely to remain so for many researchers [5–7].

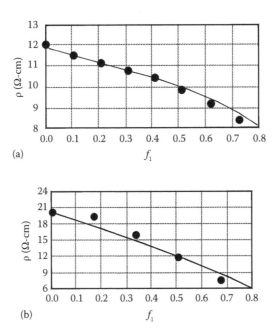

(a)

(b)

FIGURE 7.1
Axial resistivity distribution of (a) boron-doped and (b) phosphorus-doped silicon crystal as a function of solidified fraction. Experimental data and simulation results are indicated by the filled circles and solid line, respectively. (From J. H. Wang, *Journal of Crystal Growth*, **275**, e73–e78, 2005, and the references therein [3].)

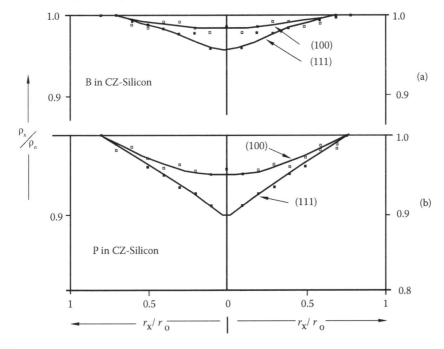

FIGURE 7.2
Radial resistivity variation in (a) boron-doped and (b) phosphorus-doped CZ-grown single crystals of silicon showing the influence of crystal orientation on the dopant concentration. (From W. Zuhlemer and D. Huber, "Czochralski-grown silicon," in *Crystals*, **8**, Springer-Verlag, Berlin, 1982 [4].)

7.2 Electrically Active and Inactive Impurities

Most of the dopant impurities present in the silicon lattice will occupy either interstitial positions or substitutional positions. The presence of a dopant impurity that belongs to Group III or Group V will be electrically active and contribute to the electrical conduction when it is occupying the substitutional locations. But impurities do not occupy all substitutional positions nor all interstitial positions. Hence, the presence of an impurity inside the lattice is no guarantee that it will participate in the electrical conduction. The presence of vacancies and the agglomeration of vacancy sites influence the dopant species by providing a physical location for them to precipitate. This brings out two types of impurities in the same species: those that are electrically active and those that are electrically inactive.

Dopant concentration and silicon electrical resistivity are two major issues for silicon integrated circuit fabrication, and boron and phosphorus are the two impurity species most widely used as common dopants for *p*-type and *n*-type semiconductors. With the introduction of Irvin's curves [8], the topic has changed stages and exact numerical equations are currently available [9] to estimate the concentration values from resistivity and vice versa [10]. Figure 7.3 shows the most popular resistivity and concentration graph used for the conversion of resistivity from impurity concentration for boron and phosphorus dopants. No such graphs are available for other impurity species, but slight differences are recorded for the deviations. Tables 7.1 and 7.2 show the conversion of electrical resistivity to dopant concentration for boron and phosphorus dopants. Tables 7.3 and 7.4 represent the conversion from the concentration values to resistivity for boron and phosphorus impurities. These tables can be used as a standard reference for any silicon wafer that has no defects in the crystal. However, the values are not applicable for polysilicon wafers and ingots because of grain boundaries. These grain boundaries accommodate several impurity species, and those trapped dopant species do not contribute to electrical conduction.

7.3 Surface Mapping and Concentration Contours

Surface mapping is carried out at the beginning of any integrated circuit fabrication. It is necessary to know the statistical range of resistivity and concentration values of silicon wafers. This will provide insight when estimating the variation in threshold voltage values (in MOS transistors) and the *p-n* junction breakdown values. If the tolerances are unacceptable, a decision, to use the silicon wafer for IC processing, needs to be made at this stage. Figure 7.4 shows a commercial system available to map a silicon wafer with an 8-inch (200 mm) diameter. The wafer resistivity values are measured using a standard four-point probe system with osmium tips. The values are measured at different locations and are subsequently mapped on the wafer. The number of mapping points depends on the degree of precision required. The results are displayed in Figure 7.5.

A fresh, commercially available, 6-inch silicon wafer that was polished on both sides was analyzed for the mapping of electrical resistivity to check the variations. This wafer was deliberately selected to highlight the variations. Figure 7.6a shows that the variations are too large to accept for wafer processing. The same results are redrawn as a contour to

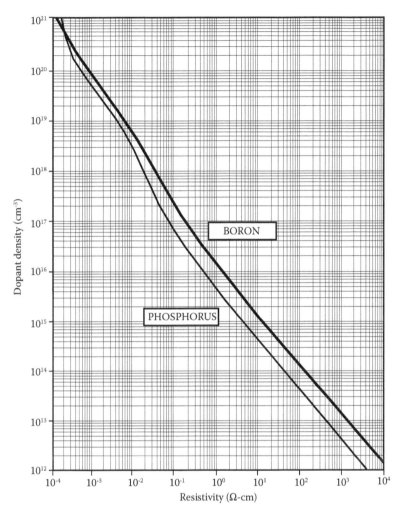

FIGURE 7.3
Conversion between resistivity and dopant density in silicon for the boron and phosphorus impurities at 23°C. (From W. E. Beadle, J. C. C. Tsai, and R. D. Plummer, *Quick Reference Manual for Silicon Integrated Circuit Technology*, John Wiley & Sons, New York, 1985 [10].)

check the specific locations on the wafer. In Figure 7.6b, the resistivity mapping shows that the wafer portion at the center is uniform, but the peripheries differ, and at certain points, it is just not acceptable. This example supports and corroborates the results that the wafer edges will not provide good circuits. This issue is highlighted further in the discussion on gate oxide integrity in the Chapters 8, 9, and 10.

7.4 Surface Roughness Mapping on a Complete Wafer

As discussed earlier, the surface roughness must be within the acceptable tolerance range (i.e., 100 Å) for these measurements. Rougher surfaces produce incomplete

TABLE 7.1

Resistivity to Dopant Concentration Density Conversion (at 23°C) for Boron-Doped Silicon [10]

	0	0.1	0.2	0.3	0.4	0.5	0.6	0.7	0.8	0.9
1.00E-04	1.20E+21	1.10E+21	1.00E+21	9.30E+20	8.60E+20	8.10E+20	7.60E+20	7.10E+20	6.70E+20	6.40E+20
2.00E-04	6.00E+20	5.80E+20	5.50E+20	5.20E+20	5.00E+20	4.80E+20	4.60E+20	4.50E+20	4.30E+20	4.20E+20
3.00E-04	4.00E+20	3.90E+20	3.80E+20	3.60E+20	3.50E+20	3.40E+20	3.30E+20	3.20E+20	3.20E+20	3.10E+20
4.00E-04	3.00E+20	2.90E+20	2.90E+20	2.80E+20	2.70E+20	2.70E+20	2.60E+20	2.50E+20	2.50E+20	2.40E+20
5.00E-04	2.40E+20	2.30E+20	2.30E+20	2.30E+20	2.20E+20	2.20E+20	2.10E+20	2.10E+20	2.10E+20	2.00E+20
6.00E-04	2.00E+20	2.00E+20	1.90E+20	1.90E+20	1.90E+20	1.80E+20	1.80E+20	1.80E+20	1.70E+20	1.70E+20
7.00E-04	1.70E+20	1.70E+20	1.60E+20	1.60E+20	1.60E+20	1.60E+20	1.60E+20	1.50E+20	1.50E+20	1.50E+20
8.00E-04	1.50E+20	1.50E+20	1.40E+20	1.40E+20	1.40E+20	1.40E+20	1.40E+20	1.40E+20	1.30E+20	1.30E+20
9.00E-04	1.30E+20	1.29E+20	1.28E+20	1.27E+20	1.25E+20	1.24E+20	1.22E+20	1.21E+20	1.20E+20	1.19E+20
1.00E-03	1.17E+20	1.06E+20	9.70E+19	8.92E+19	8.25E+19	7.67E+19	7.17E+19	6.72E+19	6.32E+19	5.97E+19
2.00E-03	5.65E+19	5.36E+19	5.09E+19	4.85E+19	4.63E+19	4.43E+19	4.24E+19	4.07E+19	3.91E+19	3.76E+19
3.00E-03	3.62E+19	3.49E+19	3.37E+19	3.25E+19	3.15E+19	3.04E+19	2.95E+19	2.86E+19	2.77E+19	2.69E+19
4.00E-03	2.61E+19	2.54E+19	2.47E+19	2.40E+19	2.34E+19	2.28E+19	2.22E+19	2.17E+19	2.11E+19	2.06E+19
5.00E-03	2.01E+19	1.97E+19	1.92E+19	1.88E+19	1.84E+19	1.80E+19	1.76E+19	1.72E+19	1.69E+19	1.65E+19
6.00E-03	1.62E+19	1.59E+19	1.56E+19	1.53E+19	1.50E+19	1.47E+19	1.44E+19	1.41E+19	1.39E+19	1.36E+19
7.00E-03	1.34E+19	1.32E+19	1.29E+19	1.27E+19	1.25E+19	1.23E+19	1.21E+19	1.19E+19	1.17E+19	1.15E+19
8.00E-03	1.13E+19	1.11E+19	1.10E+19	1.08E+19	1.06E+19	1.05E+19	1.03E+19	1.02E+19	1.00E+19	9.88E+18
9.00E-03	9.74E+18	9.60E+18	9.47E+18	9.33E+18	9.20E+18	9.08E+18	8.95E+18	8.83E+18	8.72E+18	8.60E+18
1.00E-02	8.49E+18	7.48E+18	6.65E+18	5.97E+18	5.39E+18	4.89E+18	4.47E+18	4.10E+18	3.78E+18	3.49E+18
2.00E-02	3.24E+18	3.02E+18	2.82E+18	2.64E+18	2.48E+18	2.33E+18	2.20E+18	2.07E+18	1.96E+18	1.86E+18
3.00E-02	1.77E+18	1.68E+18	1.60E+18	1.53E+18	1.46E+18	1.40E+18	1.34E+18	1.28E+18	1.23E+18	1.18E+18
4.00E-02	1.14E+18	1.09E+18	1.05E+18	1.02E+18	9.80E+17	9.46E+17	9.14E+17	8.84E+17	8.56E+17	8.29E+17
5.00E-02	8.03E+17	7.79E+17	7.55E+17	7.33E+17	7.12E+17	6.92E+17	6.73E+17	6.55E+17	6.37E+17	6.21E+17
6.00E-02	6.05E+17	5.89E+17	5.75E+17	5.61E+17	5.47E+17	5.34E+17	5.22E+17	5.10E+17	4.98E+17	4.87E+17
7.00E-02	4.76E+17	4.66E+17	4.56E+17	4.47E+17	4.37E+17	4.29E+17	4.20E+17	4.12E+17	4.03E+17	3.96E+17
8.00E-02	3.88E+17	3.81E+17	3.74E+17	3.67E+17	3.60E+17	3.54E+17	3.48E+17	3.42E+17	3.36E+17	3.30E+17
9.00E-02	3.24E+17	3.19E+17	3.14E+17	3.09E+17	3.04E+17	2.99E+17	2.94E+17	2.90E+17	2.85E+17	2.81E+17
1.00E-01	2.77E+17	2.40E+17	2.11E+17	1.88E+17	1.69E+17	1.53E+17	1.40E+17	1.28E+17	1.18E+17	1.10E+17
2.00E-01	1.02E+17	9.59E+16	9.01E+16	8.49E+16	8.02E+16	7.60E+16	7.22E+16	6.88E+16	6.56E+16	6.27E+16

3.00E-01	6.00E+16	5.76E+16	5.53E+16	5.32E+16	5.13E+16	4.94E+16	4.77E+16	4.61E+16	4.46E+16	4.32E+16
4.00E-01	4.19E+16	4.07E+16	3.95E+16	3.84E+16	3.73E+16	3.63E+16	3.54E+16	3.45E+16	3.36E+16	3.28E+16
5.00E-01	3.21E+16	3.13E+16	3.06E+16	2.99E+16	2.93E+16	2.87E+16	2.81E+16	2.75E+16	2.69E+16	2.64E+16
6.00E-01	2.59E+16	2.54E+16	2.49E+16	2.45E+16	2.40E+16	2.36E+16	2.32E+16	2.28E+16	2.24E+16	2.21E+16
7.00E-01	2.17E+16	2.14E+16	2.10E+16	2.07E+16	2.04E+16	2.01E+16	1.98E+16	1.95E+16	1.92E+16	1.89E+16
8.00E-01	1.87E+16	1.84E+16	1.82E+16	1.79E+16	1.77E+16	1.75E+16	1.72E+16	1.70E+16	1.68E+16	1.66E+16
9.00E-01	1.64E+16	1.62E+16	1.60E+16	1.58E+16	1.56E+16	1.54E+16	1.53E+16	1.51E+16	1.49E+16	1.47E+16
1.00E+00	1.46E+16	1.31E+16	1.20E+16	1.10E+16	1.01E+16	9.41E+15	8.79E+15	8.24E+15	7.76E+15	7.33E+15
2.00E+00	6.95E+15	6.60E+15	6.29E+15	6.01E+15	5.75E+15	5.51E+15	5.29E+15	5.08E+15	4.90E+15	4.72E+15
3.00E+00	4.56E+15	4.41E+15	4.27E+15	4.13E+15	4.01E+15	3.89E+15	3.78E+15	3.68E+15	3.58E+15	3.48E+15
4.00E+00	3.39E+15	3.31E+15	3.23E+15	3.15E+15	3.08E+15	3.01E+15	2.94E+15	2.88E+15	2.82E+15	2.76E+15
5.00E+00	2.70E+15	2.65E+15	2.60E+15	2.55E+15	2.50E+15	2.45E+15	2.41E+15	2.37E+15	2.32E+15	2.28E+15
6.00E+00	2.25E+15	2.21E+15	2.17E+15	2.14E+15	2.10E+15	2.07E+15	2.04E+15	2.01E+15	1.98E+15	1.95E+15
7.00E+00	1.92E+15	1.89E+15	1.87E+15	1.84E+15	1.82E+15	1.79E+15	1.77E+15	1.74E+15	1.72E+15	1.70E+15
8.00E+00	1.68E+15	1.66E+15	1.64E+15	1.62E+15	1.60E+15	1.58E+15	1.56E+15	1.54E+15	1.52E+15	1.51E+15
9.00E+00	1.49E+15	1.47E+15	1.46E+15	1.44E+15	1.43E+15	1.41E+15	1.40E+15	1.38E+15	1.37E+15	1.35E+15
1.00E+01	1.34E+15	1.22E+15	1.11E+15	1.03E+15	9.55E+14	8.91E+14	8.35E+14	7.85E+14	7.42E+14	7.02E+14
2.00E+01	6.67E+14	6.35E+14	6.06E+14	5.80E+14	5.56E+14	5.33E+14	5.13E+14	4.94E+14	4.76E+14	4.60E+14
3.00E+01	4.44E+14	4.30E+14	4.16E+14	4.04E+14	3.92E+14	3.81E+14	3.70E+14	3.60E+14	3.50E+14	3.41E+14
4.00E+01	3.33E+14	3.25E+14	3.17E+14	3.10E+14	3.03E+14	2.96E+14	2.89E+14	2.83E+14	2.77E+14	2.72E+14
5.00E+01	2.66E+14	2.61E+14	2.56E+14	2.51E+14	2.46E+14	2.42E+14	2.38E+14	2.34E+14	2.29E+14	2.26E+14
6.00E+01	2.22E+14	2.18E+14	2.15E+14	2.11E+14	2.08E+14	2.05E+14	2.02E+14	1.99E+14	1.96E+14	1.93E+14
7.00E+01	1.90E+14	1.87E+14	1.85E+14	1.82E+14	1.80E+14	1.77E+14	1.75E+14	1.73E+14	1.71E+14	1.68E+14
8.00E+01	1.66E+14	1.64E+14	1.62E+14	1.60E+14	1.58E+14	1.57E+14	1.55E+14	1.53E+14	1.51E+14	1.49E+14
9.00E+01	1.48E+14	1.46E+14	1.45E+14	1.43E+14	1.42E+14	1.40E+14	1.39E+14	1.37E+14	1.36E+14	1.34E+14
1.00E+02	1.07E+14	1.05E+14	1.03E+14	1.01E+14	9.85E+13	9.65E+13	9.50E+13	9.30E+13	9.15E+13	9.00E+13
2.00E+02	6.60E+13	6.30E+13	6.00E+13	5.80E+13	5.50E+13	5.30E+13	5.10E+13	4.90E+13	4.70E+13	4.60E+13
3.00E+02	4.40E+13	4.30E+13	4.20E+13	4.00E+13	3.90E+13	3.80E+13	3.70E+13	3.60E+13	3.50E+13	3.40E+13
4.00E+02	3.30E+13	3.20E+13	3.20E+13	3.10E+13	3.00E+13	3.00E+13	2.90E+13	2.80E+13	2.80E+13	2.70E+13
5.00E+02	2.70E+13	2.60E+13	2.60E+13	2.50E+13	2.50E+13	2.40E+13	2.40E+13	2.30E+13	2.30E+13	2.30E+13
6.00E+02	2.20E+13	2.20E+13	2.10E+13	2.10E+13	2.10E+13	2.00E+13	2.00E+13	2.00E+13	2.00E+13	1.90E+13
7.00E+02	1.90E+13	1.90E+13	1.80E+13	1.80E+13	1.80E+13	1.80E+13	1.70E+13	1.70E+13	1.70E+13	1.70E+13
8.00E+02	1.70E+13	1.60E+13	1.60E+13	1.60E+13	1.60E+13	1.60E+13	1.50E+13	1.50E+13	1.50E+13	1.50E+13

(Continued)

TABLE 7.1 (*Continued*)

	0	0.1	0.2	0.3	0.4	0.5	0.6	0.7	0.8	0.9
9.00E+02	1.50E+13	1.50E+13	1.40E+13	1.40E+13	1.40E+13	1.40E+13	1.40E+13	1.40E+13	1.40E+13	1.30E+13
1.00E+03	1.30E+13	1.20E+13	1.10E+13	1.00E+13	9.50E+12	8.90E+12	8.30E+12	7.80E+12	7.40E+12	7.00E+12
2.00E+03	6.60E+12	6.30E+12	6.00E+12	5.80E+12	5.50E+12	5.30E+12	5.10E+12	4.90E+12	4.70E+12	4.60E+12
3.00E+03	4.40E+12	4.30E+12	4.20E+12	4.00E+12	3.90E+12	3.80E+12	3.70E+12	3.60E+12	3.50E+12	3.40E+12
4.00E+03	3.30E+12	3.20E+12	3.20E+12	3.10E+12	3.00E+12	3.00E+12	2.90E+12	2.80E+12	2.80E+12	2.70E+12
5.00E+03	2.70E+12	2.60E+12	2.60E+12	2.50E+12	2.50E+12	2.40E+12	2.40E+12	2.30E+12	2.30E+12	2.30E+12
6.00E+03	2.20E+12	2.20E+12	2.10E+12	2.10E+12	2.10E+12	2.00E+12	2.00E+12	2.00E+12	2.00E+12	1.90E+12
7.00E+03	1.90E+12	1.90E+12	1.80E+12	1.80E+12	1.80E+12	1.80E+12	1.70E+12	1.70E+12	1.70E+12	1.70E+12
8.00E+03	1.70E+12	1.60E+12	1.60E+12	1.60E+12	1.60E+12	1.60E+12	1.50E+12	1.50E+12	1.50E+12	1.50E+12
9.00E+03	1.50E+12	1.50E+12	1.40E+12	1.40E+12	1.40E+12	1.40E+12	1.40E+12	1.40E+12	1.40E+12	1.30E+12
1.00E+04	1.30E+12	1.20E+12	1.10E+12	1.00E+12	9.50E+11	8.90E+11	8.30E+11	7.80E+11	7.40E+11	7.00E+11
2.00E+04	6.60E+11	6.30E+11	6.00E+11	5.80E+11	5.50E+11	5.30E+11	5.10E+11	4.90E+11	4.70E+11	4.60E+11
3.00E+04	4.40E+11	4.30E+11	4.20E+11	4.00E+11	3.90E+11	3.80E+11	3.70E+11	3.60E+11	3.50E+11	3.40E+11
4.00E+04	3.30E+11	3.20E+11	3.20E+11	3.10E+11	3.00E+11	3.00E+11	2.90E+11	2.80E+11	2.80E+11	2.70E+11
5.00E+04	2.70E+11	2.60E+11	2.60E+11	2.50E+11	2.50E+11	2.40E+11	2.40E+11	2.30E+11	2.30E+11	2.30E+11
6.00E+04	2.20E+11	2.20E+11	2.10E+11	2.10E+11	2.10E+11	2.00E+11	2.00E+11	2.00E+11	2.00E+11	1.90E+11
7.00E+04	1.90E+11	1.90E+11	1.80E+11	1.80E+11	1.80E+11	1.80E+11	1.70E+11	1.70E+11	1.70E+11	1.70E+11
8.00E+04	1.70E+11	1.60E+11	1.60E+11	1.60E+11	1.60E+11	1.60E+11	1.50E+11	1.50E+11	1.50E+11	1.50E+11
9.00E+04	1.50E+11	1.50E+11	1.40E+11	1.40E+11	1.40E+11	1.40E+11	1.40E+11	1.40E+11	1.40E+11	1.30E+11

Source: W. E. Beadle, J. C. C. Tsai, and R. D. Plummer, *Quick Reference Manual for Silicon Integrated Circuit Technology*, John Wiley & Sons, New York, 1985 [10].

TABLE 7.2

Resistivity to Dopant Concentration Density Conversion (at 23°C) for Phosphorus-Doped Silicon [10]

	0	0.1	0.2	0.3	0.4	0.5	0.6	0.7	0.8	0.9
1.00E-04	1.60E+21	1.40E+21	1.20E+21	1.10E+21	9.40E+20	8.50E+20	7.70E+20	7.00E+20	6.40E+20	5.90E+20
2.00E-04	5.50E+20	5.10E+20	4.80E+20	4.50E+20	4.30E+20	4.00E+20	3.80E+20	3.60E+20	3.50E+20	3.30E+20
3.00E-04	3.15E+20	3.02E+20	2.89E+20	2.78E+20	2.67E+20	2.58E+20	2.48E+20	2.40E+20	2.32E+20	2.24E+20
4.00E-04	2.17E+20	2.10E+20	2.04E+20	1.98E+20	1.93E+20	1.87E+20	1.82E+20	1.78E+20	1.73E+20	1.69E+20
5.00E-04	1.65E+20	1.61E+20	1.57E+20	1.53E+20	1.50E+20	1.47E+20	1.44E+20	1.41E+20	1.38E+20	1.35E+20
6.00E-04	1.32E+20	1.30E+20	1.27E+20	1.25E+20	1.23E+20	1.20E+20	1.18E+20	1.16E+20	1.14E+20	1.12E+20
7.00E-04	1.10E+20	1.09E+20	1.07E+20	1.05E+20	1.04E+20	1.02E+20	1.00E+20	9.90E+19	9.75E+19	9.61E+19
8.00E-04	9.48E+19	9.34E+19	9.21E+19	9.09E+19	8.97E+19	8.85E+19	8.73E+19	8.62E+19	8.51E+19	8.40E+19
9.00E-04	8.30E+19	8.19E+19	8.09E+19	8.00E+19	7.90E+19	7.81E+19	7.72E+19	7.63E+19	7.54E+19	7.46E+19
1.00E-03	7.38E+19	6.64E+19	6.04E+19	5.53E+19	5.11E+19	4.74E+19	4.42E+19	4.14E+19	3.90E+19	3.68E+19
2.00E-03	3.48E+19	3.30E+19	3.14E+19	2.99E+19	2.86E+19	2.73E+19	2.62E+19	2.51E+19	2.41E+19	2.32E+19
3.00E-03	2.24E+19	2.16E+19	2.08E+19	2.01E+19	1.94E+19	1.88E+19	1.82E+19	1.76E+19	1.71E+19	1.66E+19
4.00E-03	1.61E+19	1.56E+19	1.52E+19	1.47E+19	1.43E+19	1.40E+19	1.36E+19	1.32E+19	1.29E+19	1.26E+19
5.00E-03	1.22E+19	1.19E+19	1.16E+19	1.14E+19	1.11E+19	1.08E+19	1.06E+19	1.03E+19	1.01E+19	9.87E+18
6.00E-03	9.65E+18	9.44E+18	9.24E+18	9.04E+18	8.84E+18	8.66E+18	8.47E+18	8.30E+18	8.13E+18	7.96E+18
7.00E-03	7.80E+18	7.65E+18	7.49E+18	7.35E+18	7.20E+18	7.06E+18	6.93E+18	6.80E+18	6.67E+18	6.54E+18
8.00E-03	6.42E+18	6.30E+18	6.19E+18	6.07E+18	5.96E+18	5.86E+18	5.75E+18	5.65E+18	5.55E+18	5.46E+18
9.00E-03	5.36E+18	5.27E+18	5.18E+18	5.09E+18	5.01E+18	4.92E+18	4.84E+18	4.76E+18	4.68E+18	4.61E+18
1.00E-02	4.53E+18	3.87E+18	3.34E+18	2.91E+18	2.55E+18	2.25E+18	2.00E+18	1.78E+18	1.60E+18	1.45E+18
2.00E-02	1.31E+18	1.20E+18	1.09E+18	1.00E+18	9.26E+17	8.56E+17	7.94E+17	7.39E+17	6.89E+17	6.44E+17
3.00E-02	6.04E+17	5.68E+17	5.35E+17	5.05E+17	4.77E+17	4.52E+17	4.29E+17	4.08E+17	3.88E+17	3.70E+17
4.00E-02	3.53E+17	3.38E+17	3.23E+17	3.10E+17	2.98E+17	2.86E+17	2.75E+17	2.65E+17	2.55E+17	2.46E+17
5.00E-02	2.37E+17	2.29E+17	2.22E+17	2.15E+17	2.08E+17	2.02E+17	1.95E+17	1.90E+17	1.84E+17	1.79E+17
6.00E-02	1.74E+17	1.69E+17	1.65E+17	1.61E+17	1.56E+17	1.53E+17	1.49E+17	1.45E+17	1.42E+17	1.38E+17
7.00E-02	1.35E+17	1.32E+17	1.29E+17	1.27E+17	1.24E+17	1.21E+17	1.19E+17	1.16E+17	1.14E+17	1.12E+17
8.00E-02	1.10E+17	1.08E+17	1.06E+17	1.04E+17	1.02E+17	9.99E+16	9.81E+16	9.64E+16	9.48E+16	9.32E+16
9.00E-02	9.16E+16	9.01E+16	8.87E+16	8.72E+16	8.59E+16	8.45E+16	8.32E+16	8.20E+16	8.07E+16	7.95E+16
1.00E-01	7.84E+16	6.83E+16	6.04E+16	5.40E+16	4.88E+16	4.45E+16	4.09E+16	3.78E+16	3.51E+16	3.27E+16
2.00E-01	3.07E+16	2.88E+16	2.72E+16	2.58E+16	2.45E+16	2.33E+16	2.22E+16	2.12E+16	2.03E+16	1.95E+16

(Continued)

TABLE 7.2 (*Continued*)

	0	0.1	0.2	0.3	0.4	0.5	0.6	0.7	0.8	0.9
3.00E−01	1.87E+16	1.80E+16	1.73E+16	1.67E+16	1.61E+16	1.56E+16	1.51E+16	1.46E+16	1.42E+16	1.38E+16
4.00E−01	1.34E+16	1.30E+16	1.26E+16	1.23E+16	1.20E+16	1.17E+16	1.14E+16	1.11E+16	1.09E+16	1.06E+16
5.00E−01	1.04E+16	1.02E+16	9.94E+15	9.73E+15	9.53E+15	9.34E+15	9.15E+15	8.97E+15	8.80E+15	8.64E+15
6.00E−01	8.48E+15	8.32E+15	8.18E+15	8.03E+15	7.89E+15	7.76E+15	7.63E+15	7.51E+15	7.39E+15	7.27E+15
7.00E−01	7.16E+15	7.05E+15	6.94E+15	6.83E+15	6.73E+15	6.64E+15	6.54E+15	6.45E+15	6.36E+15	6.27E+15
8.00E−01	6.19E+15	6.10E+15	6.02E+15	5.95E+15	5.87E+15	5.79E+15	5.72E+15	5.65E+15	5.58E+15	5.51E+15
9.00E−01	5.45E+15	5.38E+15	5.32E+15	5.26E+15	5.20E+15	5.14E+15	5.08E+15	5.03E+15	4.97E+15	4.92E+15
1.00E+00	4.86E+15	4.39E+15	4.00E+15	3.68E+15	3.40E+15	3.16E+15	2.95E+15	2.77E+15	2.61E+15	2.47E+15
2.00E+00	2.34E+15	2.22E+15	2.12E+15	2.02E+15	1.93E+15	1.85E+15	1.78E+15	1.71E+15	1.65E+15	1.59E+15
3.00E+00	1.53E+15	1.48E+15	1.43E+15	1.39E+15	1.35E+15	1.31E+15	1.27E+15	1.23E+15	1.20E+15	1.17E+15
4.00E+00	1.14E+15	1.11E+15	1.08E+15	1.06E+15	1.03E+15	1.01E+15	9.86E+14	9.64E+14	9.43E+14	9.24E+14
5.00E+00	9.05E+14	8.86E+14	8.69E+14	8.52E+14	8.36E+14	8.20E+14	8.05E+14	7.91E+14	7.77E+14	7.63E+14
6.00E+00	7.50E+14	7.37E+14	7.25E+14	7.13E+14	7.02E+14	6.91E+14	6.80E+14	6.70E+14	6.60E+14	6.50E+14
7.00E+00	6.40E+14	6.31E+14	6.22E+14	6.13E+14	6.05E+14	5.97E+14	5.89E+14	5.81E+14	5.73E+14	5.66E+14
8.00E+00	5.58E+14	5.51E+14	5.45E+14	5.38E+14	5.31E+14	5.25E+14	5.19E+14	5.13E+14	5.07E+14	5.01E+14
9.00E+00	4.95E+14	4.89E+14	4.84E+14	4.79E+14	4.74E+14	4.68E+14	4.63E+14	4.59E+14	4.54E+14	4.49E+14
1.00E+01	4.45E+14	4.03E+14	3.69E+14	3.40E+14	3.15E+14	2.94E+14	2.75E+14	2.59E+14	2.44E+14	2.31E+14
2.00E+01	2.19E+14	2.09E+14	1.99E+14	1.90E+14	1.82E+14	1.75E+14	1.68E+14	1.62E+14	1.56E+14	1.50E+14
3.00E+01	1.45E+14	1.40E+14	1.36E+14	1.32E+14	1.28E+14	1.24E+14	1.21E+14	1.17E+14	1.14E+14	1.11E+14
4.00E+01	1.08E+14	1.06E+14	1.03E+14	1.01E+14	9.83E+13	9.61E+13	9.40E+13	9.20E+13	9.00E+13	8.81E+13
5.00E+01	8.64E+13	8.46E+13	8.30E+13	8.14E+13	7.99E+13	7.84E+13	7.70E+13	7.56E+13	7.43E+13	7.30E+13
6.00E+01	7.18E+13	7.06E+13	6.94E+13	6.83E+13	6.72E+13	6.62E+13	6.51E+13	6.42E+13	6.32E+13	6.23E+13
7.00E+01	6.14E+13	6.05E+13	5.96E+13	5.88E+13	5.80E+13	5.72E+13	5.65E+13	5.57E+13	5.50E+13	5.43E+13
8.00E+01	5.36E+13	5.29E+13	5.23E+13	5.16E+13	5.10E+13	5.04E+13	4.98E+13	4.92E+13	4.87E+13	4.81E+13
9.00E+01	4.76E+13	4.70E+13	4.65E+13	4.60E+13	4.55E+13	4.50E+13	4.45E+13	4.41E+13	4.36E+13	4.32E+13
1.00E+02	4.27E+13	3.88E+13	3.55E+13	3.28E+13	3.04E+13	2.83E+13	2.65E+13	2.50E+13	2.36E+13	2.23E+13
2.00E+02	2.10E+13	2.00E+13	1.90E+13	1.80E+13	1.80E+13	1.70E+13	1.60E+13	1.60E+13	1.50E+13	1.50E+13
3.00E+02	1.40E+13	1.40E+13	1.30E+13	1.30E+13	1.20E+13	1.20E+13	1.20E+13	1.10E+13	1.10E+13	1.10E+13
4.00E+02	1.00E+13	1.00E+13	1.00E+13	9.80E+12	9.50E+12	9.30E+12	9.10E+12	8.90E+12	8.70E+12	8.50E+12
5.00E+02	8.40E+12	8.20E+12	8.00E+12	7.90E+12	7.70E+12	7.60E+12	7.50E+12	7.30E+12	7.20E+12	7.10E+12

6.00E+02	7.00E+12	6.80E+12	6.70E+12	6.60E+12	6.50E+12	6.40E+12	6.30E+12	6.20E+12	6.10E+12	6.00E+12
7.00E+02	6.00E+12	5.90E+12	5.80E+12	5.70E+12	5.60E+12	5.60E+12	5.50E+12	5.40E+12	5.30E+12	5.30E+12
8.00E+02	5.20E+12	5.10E+12	5.10E+12	5.00E+12	5.00E+12	4.90E+12	4.80E+12	4.80E+12	4.70E+12	4.70E+12
9.00E+02	4.60E+12	4.60E+12	4.50E+12	4.50E+12	4.40E+12	4.40E+12	4.30E+12	4.30E+12	4.20E+12	4.20E+12
1.00E+03	4.20E+12	3.80E+12	3.50E+12	3.20E+12	3.00E+12	2.80E+12	2.60E+12	2.40E+12	2.30E+12	2.20E+12
2.00E+03	2.10E+12	2.00E+12	1.90E+12	1.80E+12	1.70E+12	1.60E+12	1.60E+12	1.50E+12	1.50E+12	1.40E+12
3.00E+03	1.40E+12	1.30E+12	1.30E+12	1.20E+12	1.20E+12	1.20E+12	1.10E+12	1.10E+12	1.10E+12	1.00E+12
4.00E+03	1.00E+12	1.00E+12	9.70E+11	9.50E+11	9.30E+11	9.10E+11	8.90E+11	8.70E+11	8.50E+11	8.30E+11
5.00E+03	8.20E+11	8.00E+11	7.80E+11	7.70E+11	7.50E+11	7.40E+11	7.30E+11	7.10E+11	7.00E+11	6.90E+11
6.00E+03	6.80E+11	6.70E+11	6.60E+11	6.50E+11	6.40E+11	6.30E+11	6.20E+11	6.10E+11	6.00E+11	5.90E+11
7.00E+03	5.80E+11	5.70E+11	5.60E+11	5.60E+11	5.50E+11	5.40E+11	5.30E+11	5.30E+11	5.20E+11	5.10E+11
8.00E+03	5.10E+11	5.00E+11	4.90E+11	4.90E+11	4.80E+11	4.80E+11	4.70E+11	4.70E+11	4.60E+11	4.60E+11
9.00E+03	4.50E+11	4.50E+11	4.40E+11	4.40E+11	4.30E+11	4.30E+11	4.20E+11	4.20E+11	4.10E+11	4.10E+11
1.00E+04	4.00E+11	3.70E+11	3.40E+11	3.10E+11	2.90E+11	2.70E+11	2.50E+11	2.40E+11	2.20E+11	2.10E+11
2.00E+04	2.00E+11	1.90E+11	1.80E+11	1.70E+11	1.70E+11	1.60E+11	1.50E+11	1.50E+11	1.40E+11	1.40E+11
3.00E+04	1.30E+11	1.30E+11	1.30E+11	1.20E+11	1.20E+11	1.10E+11	1.10E+11	1.10E+11	1.10E+11	1.00E+11
4.00E+04	1.00E+11	9.60E+10	9.50E+10	9.30E+10	9.10E+10	8.90E+10	8.70E+10	8.50E+10	8.30E+10	8.10E+10
5.00E+04	8.00E+10	7.80E+10	7.70E+10	7.50E+10	7.40E+10	7.20E+10	7.10E+10	7.00E+10	6.90E+10	6.70E+10
6.00E+04	6.60E+10	6.50E+10	6.40E+10	6.30E+10	6.20E+10	6.10E+10	6.00E+10	5.90E+10	5.80E+10	5.80E+10
7.00E+04	5.70E+10	5.60E+10	5.50E+10	5.40E+10	5.40E+10	5.30E+10	5.20E+10	5.20E+10	5.10E+10	5.00E+10
8.00E+04	5.00E+10	4.90E+10	4.80E+10	4.80E+10	4.70E+10	4.70E+10	4.60E+10	4.60E+10	4.50E+10	4.50E+10
9.00E+04	4.40E+10	4.40E+10	4.30E+10	4.30E+10	4.20E+10	4.20E+10	4.10E+10	4.10E+10	4.00E+10	4.00E+10

Source: W. E. Beadle, J. C. C. Tsai, and R. D. Plummer, *Quick Reference Manual for Silicon Integrated Circuit Technology*, John Wiley & Sons, New York, 1985 [10].

TABLE 7.3
Dopant Concentration Density to Resistivity Conversion (at 23°C) for Boron-Doped Silicon [10]

	0	0.1	0.2	0.3	0.4	0.5	0.6	0.7	0.8	0.9
1.00E+12	1.30E+04	1.20E+04	1.10E+04	1.00E+04	9.30E+03	8.70E+03	8.20E+03	7.70E+03	7.20E+03	6.90E+03
2.00E+12	6.50E+03	6.20E+03	5.90E+03	5.70E+03	5.40E+03	5.20E+03	5.00E+03	4.80E+03	4.70E+03	4.50E+03
3.00E+12	4.30E+03	4.20E+03	4.10E+03	4.00E+03	3.80E+03	3.70E+03	3.60E+03	3.50E+03	3.40E+03	3.30E+03
4.00E+12	3.30E+03	3.20E+03	3.10E+03	3.00E+03	3.00E+03	2.90E+03	2.80E+03	2.80E+03	2.70E+03	2.70E+03
5.00E+12	2.60E+03	2.60E+03	2.50E+03	2.50E+03	2.40E+03	2.40E+03	2.30E+03	2.30E+03	2.30E+03	2.20E+03
6.00E+12	2.20E+03	2.10E+03	2.10E+03	2.10E+03	2.00E+03	2.00E+03	2.00E+03	1.90E+03	1.90E+03	1.90E+03
7.00E+12	1.90E+03	1.80E+03	1.80E+03	1.80E+03	1.80E+03	1.70E+03	1.70E+03	1.70E+03	1.70E+03	1.70E+03
8.00E+12	1.60E+03	1.60E+03	1.60E+03	1.60E+03	1.60E+03	1.50E+03	1.50E+03	1.50E+03	1.50E+03	1.50E+03
9.00E+12	1.50E+03	1.40E+03	1.40E+03	1.40E+03	1.40E+03	1.40E+03	1.40E+03	1.30E+03	1.30E+03	1.30E+03
1.00E+13	1.30E+03	1.20E+03	1.10E+03	1.00E+03	9.30E+02	8.70E+02	8.20E+02	7.70E+02	7.30E+02	6.90E+02
2.00E+13	6.50E+02	6.20E+02	5.90E+02	5.70E+02	5.40E+02	5.20E+02	5.00E+02	4.80E+02	4.70E+02	4.50E+02
3.00E+13	4.40E+02	4.20E+02	4.10E+02	4.00E+02	3.80E+02	3.70E+02	3.60E+02	3.50E+02	3.40E+02	3.40E+02
4.00E+13	3.30E+02	3.20E+02	3.10E+02	3.00E+02	3.00E+02	2.90E+02	2.80E+02	2.80E+02	2.70E+02	2.70E+02
5.00E+13	2.60E+02	2.60E+02	2.50E+02	2.50E+02	2.40E+02	2.40E+02	2.30E+02	2.30E+02	2.30E+02	2.20E+02
6.00E+13	2.20E+02	2.10E+02	2.10E+02	2.10E+02	2.00E+02	2.00E+02	2.00E+02	2.00E+02	1.90E+02	1.90E+02
7.00E+13	1.90E+02	1.80E+02	1.80E+02	1.80E+02	1.80E+02	1.70E+02	1.70E+02	1.70E+02	1.70E+02	1.70E+02
8.00E+13	1.60E+02	1.60E+02	1.60E+02	1.60E+02	1.60E+02	1.50E+02	1.50E+02	1.50E+02	1.50E+02	1.50E+02
9.00E+13	1.50E+02	1.40E+02	1.40E+02	1.40E+02	1.40E+02	1.40E+02	1.40E+02	1.30E+02	1.30E+02	1.30E+02
1.00E+14	1.30E+02	1.20E+02	1.10E+02	1.00E+02	9.40E+01	8.70E+01	8.20E+01	7.70E+01	7.30E+01	6.90E+01
2.00E+14	6.56E+01	6.25E+01	5.97E+01	5.71E+01	5.47E+01	5.26E+01	5.05E+01	4.87E+01	4.69E+01	4.53E+01
3.00E+14	4.38E+01	4.24E+01	4.11E+01	3.99E+01	3.87E+01	3.76E+01	3.66E+01	3.56E+01	3.47E+01	3.38E+01
4.00E+14	3.29E+01	3.21E+01	3.14E+01	3.07E+01	3.00E+01	2.93E+01	2.87E+01	2.81E+01	2.75E+01	2.69E+01
5.00E+14	2.64E+01	2.59E+01	2.54E+01	2.49E+01	2.45E+01	2.40E+01	2.36E+01	2.32E+01	2.28E+01	2.24E+01
6.00E+14	2.20E+01	2.17E+01	2.13E+01	2.10E+01	2.07E+01	2.04E+01	2.01E+01	1.98E+01	1.95E+01	1.92E+01
7.00E+14	1.89E+01	1.87E+01	1.84E+01	1.81E+01	1.79E+01	1.77E+01	1.74E+01	1.72E+01	1.70E+01	1.68E+01
8.00E+14	1.66E+01	1.64E+01	1.62E+01	1.60E+01	1.58E+01	1.56E+01	1.54E+01	1.53E+01	1.51E+01	1.49E+01
9.00E+14	1.48E+01	1.46E+01	1.44E+01	1.43E+01	1.41E+01	1.40E+01	1.38E+01	1.37E+01	1.36E+01	1.34E+01
1.00E+15	1.33E+01	1.21E+01	1.11E+01	1.03E+01	9.55E+00	8.93E+00	8.38E+00	7.90E+00	7.47E+00	7.08E+00
2.00E+15	6.74E+00	6.42E+00	6.14E+00	5.88E+00	5.64E+00	5.42E+00	5.22E+00	5.03E+00	4.85E+00	4.69E+00

3.00E+15	4.54E+00	4.40E+00	4.26E+00	4.14E+00	4.02E+00	3.91E+00	3.81E+00	3.71E+00	3.61E+00	3.52E+00
4.00E+15	3.44E+00	3.36E+00	3.28E+00	3.21E+00	3.14E+00	3.07E+00	3.01E+00	2.95E+00	2.89E+00	2.83E+00
5.00E+15	2.78E+00	2.72E+00	2.67E+00	2.63E+00	2.58E+00	2.53E+00	2.49E+00	2.45E+00	2.41E+00	2.37E+00
6.00E+15	2.33E+00	2.30E+00	2.26E+00	2.23E+00	2.19E+00	2.16E+00	2.13E+00	2.10E+00	2.07E+00	2.04E+00
7.00E+15	2.02E+00	1.99E+00	1.96E+00	1.94E+00	1.91E+00	1.89E+00	1.87E+00	1.84E+00	1.82E+00	1.80E+00
8.00E+15	1.78E+00	1.76E+00	1.74E+00	1.72E+00	1.70E+00	1.68E+00	1.66E+00	1.64E+00	1.63E+00	1.61E+00
9.00E+15	1.59E+00	1.58E+00	1.56E+00	1.54E+00	1.53E+00	1.51E+00	1.50E+00	1.48E+00	1.47E+00	1.46E+00
1.00E+16	1.44E+00	1.32E+00	1.22E+00	1.13E+00	1.06E+00	9.93E-01	9.37E-01	8.87E-01	8.42E-01	8.03E-01
2.00E+16	7.67E-01	7.34E-01	7.04E-01	6.77E-01	6.52E-01	6.29E-01	6.08E-01	5.88E-01	5.70E-01	5.53E-01
3.00E+16	5.37E-01	5.22E-01	5.08E-01	4.95E-01	4.82E-01	4.70E-01	4.59E-01	4.49E-01	4.39E-01	4.29E-01
4.00E+16	4.20E-01	4.12E-01	4.03E-01	3.95E-01	3.88E-01	3.81E-01	3.74E-01	3.67E-01	3.61E-01	3.55E-01
5.00E+16	3.49E-01	3.43E-01	3.38E-01	3.33E-01	3.28E-01	3.23E-01	3.18E-01	3.14E-01	3.09E-01	3.05E-01
6.00E+16	3.01E-01	2.97E-01	2.93E-01	2.89E-01	2.86E-01	2.82E-01	2.79E-01	2.76E-01	2.72E-01	2.69E-01
7.00E+16	2.66E-01	2.63E-01	2.60E-01	2.58E-01	2.55E-01	2.52E-01	2.50E-01	2.47E-01	2.44E-01	2.42E-01
8.00E+16	2.40E-01	2.37E-01	2.35E-01	2.33E-01	2.31E-01	2.29E-01	2.27E-01	2.25E-01	2.23E-01	2.21E-01
9.00E+16	2.19E-01	2.17E-01	2.15E-01	2.13E-01	2.12E-01	2.10E-01	2.08E-01	2.07E-01	2.05E-01	2.04E-01
1.00E+17	2.02E-01	1.88E-01	1.76E-01	1.66E-01	1.58E-01	1.50E-01	1.43E-01	1.37E-01	1.32E-01	1.27E-01
2.00E+17	1.22E-01	1.18E-01	1.15E-01	1.11E-01	1.08E-01	1.05E-01	1.02E-01	9.99E-02	9.75E-02	9.53E-02
3.00E+17	9.32E-02	9.12E-02	8.93E-02	8.76E-02	8.59E-02	8.43E-02	8.27E-02	8.13E-02	7.99E-02	7.86E-02
4.00E+17	7.73E-02	7.61E-02	7.49E-02	7.38E-02	7.27E-02	7.17E-02	7.07E-02	6.98E-02	6.88E-02	6.79E-02
5.00E+17	6.71E-02	6.63E-02	6.54E-02	6.47E-02	6.39E-02	6.32E-02	6.25E-02	6.18E-02	6.11E-02	6.05E-02
6.00E+17	5.98E-02	5.92E-02	5.86E-02	5.80E-02	5.75E-02	5.69E-02	5.64E-02	5.58E-02	5.53E-02	5.48E-02
7.00E+17	5.43E-02	5.39E-02	5.34E-02	5.29E-02	5.25E-02	5.21E-02	5.16E-02	5.12E-02	5.08E-02	5.04E-02
8.00E+17	5.00E-02	4.96E-02	4.92E-02	4.89E-02	4.85E-02	4.82E-02	4.78E-02	4.75E-02	4.71E-02	4.68E-02
9.00E+17	4.65E-02	4.62E-02	4.58E-02	4.55E-02	4.52E-02	4.49E-02	4.47E-02	4.44E-02	4.41E-02	4.38E-02
1.00E+18	4.35E-02	4.10E-02	3.88E-02	3.69E-02	3.53E-02	3.38E-02	3.24E-02	3.12E-02	3.00E-02	2.90E-02
2.00E+18	2.81E-02	2.72E-02	2.64E-02	2.56E-02	2.49E-02	2.42E-02	2.36E-02	2.30E-02	2.25E-02	2.19E-02
3.00E+18	2.14E-02	2.10E-02	2.05E-02	2.01E-02	1.97E-02	1.93E-02	1.89E-02	1.86E-02	1.82E-02	1.79E-02
4.00E+18	1.76E-02	1.73E-02	1.70E-02	1.67E-02	1.64E-02	1.62E-02	1.59E-02	1.57E-02	1.54E-02	1.52E-02
5.00E+18	1.50E-02	1.48E-02	1.46E-02	1.44E-02	1.42E-02	1.40E-02	1.38E-02	1.36E-02	1.35E-02	1.33E-02
6.00E+18	1.31E-02	1.30E-02	1.28E-02	1.27E-02	1.25E-02	1.24E-02	1.22E-02	1.21E-02	1.19E-02	1.18E-02
7.00E+18	1.17E-02	1.16E-02	1.14E-02	1.13E-02	1.12E-02	1.11E-02	1.10E-02	1.09E-02	1.08E-02	1.07E-02
8.00E+18	1.06E-02	1.05E-02	1.04E-02	1.03E-02	1.02E-02	1.01E-02	9.98E-03	9.89E-03	9.80E-03	9.72E-03

(Continued)

TABLE 7.3 (*Continued*)

	0	0.1	0.2	0.3	0.4	0.5	0.6	0.7	0.8	0.9
9.00E+18	9.63E-03	9.55E-03	9.47E-03	9.39E-03	9.31E-03	9.23E-03	9.16E-03	9.08E-03	9.01E-03	8.94E-03
1.00E+19	8.87E-03	8.22E-03	7.67E-03	7.18E-03	6.76E-03	6.39E-03	6.05E-03	5.75E-03	5.48E-03	5.24E-03
2.00E+19	5.01E-03	4.81E-03	4.62E-03	4.45E-03	4.29E-03	4.14E-03	4.00E-03	3.87E-03	3.75E-03	3.63E-03
3.00E+19	3.53E-03	3.43E-03	3.33E-03	3.24E-03	3.16E-03	3.08E-03	3.00E-03	2.93E-03	2.86E-03	2.79E-03
4.00E+19	2.73E-03	2.67E-03	2.61E-03	2.56E-03	2.50E-03	2.45E-03	2.40E-03	2.36E-03	2.31E-03	2.27E-03
5.00E+19	2.23E-03	2.19E-03	2.15E-03	2.11E-03	2.08E-03	2.04E-03	2.01E-03	1.98E-03	1.94E-03	1.91E-03
6.00E+19	1.88E-03	1.86E-03	1.83E-03	1.80E-03	1.77E-03	1.75E-03	1.72E-03	1.70E-03	1.68E-03	1.66E-03
7.00E+19	1.63E-03	1.61E-03	1.59E-03	1.57E-03	1.55E-03	1.53E-03	1.51E-03	1.49E-03	1.48E-03	1.46E-03
8.00E+19	1.44E-03	1.43E-03	1.41E-03	1.39E-03	1.38E-03	1.36E-03	1.35E-03	1.33E-03	1.32E-03	1.30E-03
9.00E+19	1.29E-03	1.28E-03	1.26E-03	1.25E-03	1.24E-03	1.23E-03	1.21E-03	1.20E-03	1.19E-03	1.18E-03
1.00E+20	1.17E-03	1.07E-03	9.83E-04	9.11E-04	8.49E-04	7.94E-04	7.47E-04	7.04E-04	6.67E-04	6.33E-04
2.00E+20	6.00E-04	5.70E-04	5.50E-04	5.30E-04	5.00E-04	4.90E-04	4.70E-04	4.50E-04	4.30E-04	4.20E-04
3.00E+20	4.10E-04	3.90E-04	3.80E-04	3.70E-04	3.60E-04	3.50E-04	3.40E-04	3.30E-04	3.20E-04	3.10E-04
4.00E+20	3.10E-04	3.00E-04	2.90E-04	2.90E-04	2.80E-04	2.70E-04	2.70E-04	2.60E-04	2.60E-04	2.50E-04
5.00E+20	2.50E-04	2.40E-04	2.40E-04	2.30E-04	2.30E-04	2.20E-04	2.20E-04	2.20E-04	2.10E-04	2.10E-04
6.00E+20	2.10E-04	2.00E-04	2.00E-04	2.00E-04	1.90E-04	1.90E-04	1.90E-04	1.80E-04	1.80E-04	1.80E-04
7.00E+20	1.80E-04	1.70E-04	1.70E-04	1.70E-04	1.70E-04	1.70E-04	1.60E-04	1.60E-04	1.60E-04	1.60E-04
8.00E+20	1.60E-04	1.50E-04	1.50E-04	1.50E-04	1.50E-04	1.50E-04	1.40E-04	1.40E-04	1.40E-04	1.40E-04
9.00E+20	1.40E-04	1.40E-04	1.40E-04	1.30E-04	1.30E-04	1.30E-04	1.30E-04	1.30E-04	1.30E-04	1.30E-04

Source: W. E. Beadle, J. C. C. Tsai, and R. D. Plummer, *Quick Reference Manual for Silicon Integrated Circuit Technology*, John Wiley & Sons, New York, 1985 [10].

TABLE 7.4

Dopant Concentration Density to Resistivity Conversion (at 23°C) for Phosphorus-Doped Silicon [10]

	0	0.1	0.2	0.3	0.4	0.5	0.6	0.7	0.8	0.9
1.00E+12	4.30E+03	3.90E+03	3.60E+03	3.30E+03	3.00E+03	2.80E+03	2.70E+03	2.50E+03	2.40E+03	2.20E+03
2.00E+12	2.10E+03	2.00E+03	1.90E+03	1.90E+03	1.80E+03	1.70E+03	1.60E+03	1.60E+03	1.50E+03	1.50E+03
3.00E+12	1.40E+03	1.40E+03	1.30E+03	1.30E+03	1.30E+03	1.20E+03	1.20E+03	1.10E+03	1.10E+03	1.10E+03
4.00E+12	1.10E+03	1.00E+03	1.00E+03	9.90E+02	9.70E+02	9.40E+02	9.20E+02	9.00E+02	8.90E+02	8.70E+02
5.00E+12	8.50E+02	8.30E+02	8.20E+02	8.00E+02	7.90E+02	7.70E+02	7.60E+02	7.50E+02	7.30E+02	7.20E+02
6.00E+12	7.10E+02	7.00E+02	6.90E+02	6.70E+02	6.60E+02	6.50E+02	6.40E+02	6.30E+02	6.30E+02	6.20E+02
7.00E+12	6.10E+02	6.00E+02	5.90E+02	5.80E+02	5.70E+02	5.70E+02	5.60E+02	5.50E+02	5.40E+02	5.40E+02
8.00E+12	5.30E+02	5.20E+02	5.20E+02	5.10E+02	5.10E+02	5.00E+02	4.90E+02	4.90E+02	4.80E+02	4.80E+02
9.00E+12	4.70E+02	4.70E+02	4.60E+02	4.60E+02	4.50E+02	4.50E+02	4.40E+02	4.40E+02	4.30E+02	4.30E+02
1.00E+13	4.30E+02	3.90E+02	3.50E+02	3.30E+02	3.00E+02	2.80E+02	2.70E+02	2.50E+02	2.40E+02	2.20E+02
2.00E+13	2.10E+02	2.00E+02	1.90E+02	1.80E+02	1.80E+02	1.70E+02	1.60E+02	1.60E+02	1.50E+02	1.50E+02
3.00E+13	1.42E+02	1.37E+02	1.33E+02	1.29E+02	1.25E+02	1.22E+02	1.18E+02	1.15E+02	1.12E+02	1.09E+02
4.00E+13	1.07E+02	1.04E+02	1.02E+02	9.92E+01	9.69E+01	9.48E+01	9.27E+01	9.08E+01	8.89E+01	8.71E+01
5.00E+13	8.53E+01	8.37E+01	8.21E+01	8.05E+01	7.90E+01	7.76E+01	7.62E+01	7.49E+01	7.36E+01	7.24E+01
6.00E+13	7.12E+01	7.00E+01	6.89E+01	6.78E+01	6.68E+01	6.57E+01	6.48E+01	6.38E+01	6.29E+01	6.20E+01
7.00E+13	6.11E+01	6.02E+01	5.94E+01	5.86E+01	5.78E+01	5.70E+01	5.63E+01	5.56E+01	5.49E+01	5.42E+01
8.00E+13	5.35E+01	5.28E+01	5.22E+01	5.16E+01	5.10E+01	5.04E+01	4.98E+01	4.92E+01	4.87E+01	4.81E+01
9.00E+13	4.76E+01	4.71E+01	4.66E+01	4.61E+01	4.56E+01	4.51E+01	4.46E+01	4.42E+01	4.37E+01	4.33E+01
1.00E+14	4.29E+01	3.90E+01	3.58E+01	3.31E+01	3.07E+01	2.87E+01	2.69E+01	2.54E+01	2.40E+01	2.27E+01
2.00E+14	2.16E+01	2.06E+01	1.97E+01	1.88E+01	1.80E+01	1.73E+01	1.67E+01	1.61E+01	1.55E+01	1.50E+01
3.00E+14	1.45E+01	1.40E+01	1.36E+01	1.32E+01	1.28E+01	1.25E+01	1.21E+01	1.18E+01	1.15E+01	1.12E+01
4.00E+14	1.09E+01	1.07E+01	1.04E+01	1.02E+01	9.96E+00	9.74E+00	9.53E+00	9.33E+00	9.14E+00	8.96E+00
5.00E+14	8.79E+00	8.62E+00	8.46E+00	8.30E+00	8.15E+00	8.01E+00	7.87E+00	7.73E+00	7.60E+00	7.48E+00
6.00E+14	7.36E+00	7.24E+00	7.12E+00	7.01E+00	6.91E+00	6.80E+00	6.70E+00	6.61E+00	6.51E+00	6.42E+00
7.00E+14	6.33E+00	6.24E+00	6.16E+00	6.08E+00	6.00E+00	5.92E+00	5.85E+00	5.77E+00	5.70E+00	5.63E+00
8.00E+14	5.56E+00	5.50E+00	5.43E+00	5.37E+00	5.30E+00	5.24E+00	5.19E+00	5.13E+00	5.07E+00	5.02E+00
9.00E+14	4.96E+00	4.91E+00	4.86E+00	4.81E+00	4.76E+00	4.71E+00	4.66E+00	4.62E+00	4.57E+00	4.53E+00
1.00E+15	4.48E+00	4.09E+00	3.76E+00	3.48E+00	3.24E+00	3.03E+00	2.85E+00	2.69E+00	2.55E+00	2.42E+00
2.00E+15	2.31E+00	2.20E+00	2.11E+00	2.02E+00	1.94E+00	1.87E+00	1.80E+00	1.74E+00	1.68E+00	1.62E+00

(Continued)

TABLE 7.4 (*Continued*)

	0	0.1	0.2	0.3	0.4	0.5	0.6	0.7	0.8	0.9
3.00E+15	1.57E+00	1.53E+00	1.48E+00	1.44E+00	1.40E+00	1.36E+00	1.33E+00	1.29E+00	1.26E+00	1.23E+00
4.00E+15	1.20E+00	1.18E+00	1.15E+00	1.12E+00	1.10E+00	1.08E+00	1.06E+00	1.04E+00	1.02E+00	9.97E−01
5.00E+15	9.78E−01	9.61E−01	9.44E−01	9.27E−01	9.12E−01	8.96E−01	8.82E−01	8.68E−01	8.54E−01	8.41E−01
6.00E+15	8.28E−01	8.16E−01	8.04E−01	7.92E−01	7.81E−01	7.70E−01	7.59E−01	7.49E−01	7.39E−01	7.29E−01
7.00E+15	7.20E−01	7.11E−01	7.02E−01	6.93E−01	6.85E−01	6.76E−01	6.68E−01	6.60E−01	6.53E−01	6.45E−01
8.00E+15	6.38E−01	6.31E−01	6.24E−01	6.17E−01	6.11E−01	6.04E−01	5.98E−01	5.92E−01	5.86E−01	5.80E−01
9.00E+15	5.74E−01	5.69E−01	5.63E−01	5.58E−01	5.52E−01	5.47E−01	5.42E−01	5.37E−01	5.32E−01	5.28E−01
1.00E+16	5.23E−01	4.81E−01	4.45E−01	4.15E−01	3.89E−01	3.67E−01	3.47E−01	3.30E−01	3.14E−01	3.00E−01
2.00E+16	2.87E−01	2.76E−01	2.65E−01	2.56E−01	2.47E−01	2.39E−01	2.31E−01	2.24E−01	2.17E−01	2.11E−01
3.00E+16	2.06E−01	2.00E−01	1.95E−01	1.90E−01	1.86E−01	1.82E−01	1.78E−01	1.74E−01	1.70E−01	1.67E−01
4.00E+16	1.64E−01	1.61E−01	1.58E−01	1.55E−01	1.52E−01	1.49E−01	1.47E−01	1.44E−01	1.42E−01	1.40E−01
5.00E+16	1.38E−01	1.36E−01	1.34E−01	1.32E−01	1.30E−01	1.28E−01	1.27E−01	1.25E−01	1.23E−01	1.22E−01
6.00E+16	1.20E−01	1.19E−01	1.17E−01	1.16E−01	1.15E−01	1.13E−01	1.12E−01	1.11E−01	1.10E−01	1.09E−01
7.00E+16	1.08E−01	1.06E−01	1.05E−01	1.04E−01	1.03E−01	1.02E−01	1.01E−01	1.00E−01	9.95E−02	9.87E−02
8.00E+16	9.78E−02	9.69E−02	9.61E−02	9.53E−02	9.45E−02	9.37E−02	9.30E−02	9.22E−02	9.15E−02	9.08E−02
9.00E+16	9.01E−02	8.94E−02	8.87E−02	8.81E−02	8.74E−02	8.68E−02	8.62E−02	8.56E−02	8.50E−02	8.44E−02
1.00E+17	8.38E−02	7.86E−02	7.43E−02	7.05E−02	6.72E−02	6.44E−02	6.19E−02	5.96E−02	5.76E−02	5.57E−02
2.00E+17	5.41E−02	5.25E−02	5.11E−02	4.98E−02	4.87E−02	4.75E−02	4.65E−02	4.55E−02	4.46E−02	4.38E−02
3.00E+17	4.30E−02	4.22E−02	4.15E−02	4.08E−02	4.02E−02	3.96E−02	3.90E−02	3.84E−02	3.79E−02	3.74E−02
4.00E+17	3.69E−02	3.65E−02	3.60E−02	3.56E−02	3.52E−02	3.48E−02	3.44E−02	3.40E−02	3.37E−02	3.33E−02
5.00E+17	3.30E−02	3.27E−02	3.23E−02	3.20E−02	3.17E−02	3.15E−02	3.12E−02	3.09E−02	3.07E−02	3.04E−02
6.00E+17	3.01E−02	2.99E−02	2.97E−02	2.94E−02	2.92E−02	2.90E−02	2.88E−02	2.86E−02	2.84E−02	2.82E−02
7.00E+17	2.80E−02	2.78E−02	2.76E−02	2.74E−02	2.72E−02	2.71E−02	2.69E−02	2.67E−02	2.66E−02	2.64E−02
8.00E+17	2.62E−02	2.61E−02	2.59E−02	2.58E−02	2.56E−02	2.55E−02	2.53E−02	2.52E−02	2.51E−02	2.49E−02
9.00E+17	2.48E−02	2.47E−02	2.45E−02	2.44E−02	2.43E−02	2.41E−02	2.40E−02	2.39E−02	2.38E−02	2.37E−02
1.00E+18	2.36E−02	2.25E−02	2.16E−02	2.07E−02	2.00E−02	1.93E−02	1.87E−02	1.81E−02	1.76E−02	1.71E−02
2.00E+18	1.66E−02	1.62E−02	1.58E−02	1.54E−02	1.51E−02	1.47E−02	1.44E−02	1.41E−02	1.38E−02	1.35E−02
3.00E+18	1.32E−02	1.30E−02	1.27E−02	1.25E−02	1.23E−02	1.21E−02	1.19E−02	1.17E−02	1.15E−02	1.13E−02
4.00E+18	1.11E−02	1.09E−02	1.08E−02	1.06E−02	1.04E−02	1.03E−02	1.01E−02	1.00E−02	9.86E−03	9.73E−03

5.00E+18	9.60E-03	9.47E-03	9.34E-03	9.22E-03	9.11E-03	8.99E-03	8.88E-03	8.77E-03	8.66E-03	8.56E-03
6.00E+18	8.46E-03	8.36E-03	8.27E-03	8.17E-03	8.08E-03	7.99E-03	7.90E-03	7.82E-03	7.74E-03	7.65E-03
7.00E+18	7.57E-03	7.50E-03	7.42E-03	7.34E-03	7.27E-03	7.20E-03	7.13E-03	7.06E-03	6.99E-03	6.92E-03
8.00E+18	6.86E-03	6.79E-03	6.73E-03	6.67E-03	6.61E-03	6.55E-03	6.49E-03	6.44E-03	6.38E-03	6.32E-03
9.00E+18	6.27E-03	6.22E-03	6.16E-03	6.11E-03	6.06E-03	6.01E-03	5.96E-03	5.91E-03	5.87E-03	5.82E-03
1.00E+19	5.78E-03	5.35E-03	4.99E-03	4.68E-03	4.40E-03	4.15E-03	3.93E-03	3.73E-03	3.56E-03	3.39E-03
2.00E+19	3.25E-03	3.11E-03	2.99E-03	2.87E-03	2.77E-03	2.67E-03	2.58E-03	2.49E-03	2.41E-03	2.34E-03
3.00E+19	2.27E-03	2.20E-03	2.14E-03	2.08E-03	2.03E-03	1.98E-03	1.93E-03	1.88E-03	1.84E-03	1.79E-03
4.00E+19	1.75E-03	1.71E-03	1.68E-03	1.64E-03	1.61E-03	1.58E-03	1.54E-03	1.51E-03	1.49E-03	1.46E-03
5.00E+19	1.43E-03	1.41E-03	1.38E-03	1.36E-03	1.34E-03	1.31E-03	1.29E-03	1.27E-03	1.25E-03	1.23E-03
6.00E+19	1.21E-03	1.20E-03	1.18E-03	1.16E-03	1.15E-03	1.13E-03	1.11E-03	1.10E-03	1.09E-03	1.07E-03
7.00E+19	1.06E-03	1.04E-03	1.03E-03	1.02E-03	1.01E-03	9.94E-04	9.82E-04	9.71E-04	9.60E-04	9.49E-04
8.00E+19	9.38E-04	9.28E-04	9.18E-04	9.08E-04	8.98E-04	8.89E-04	8.80E-04	8.71E-04	8.62E-04	8.53E-04
9.00E+19	8.45E-04	8.37E-04	8.29E-04	8.21E-04	8.13E-04	8.06E-04	7.98E-04	7.91E-04	7.84E-04	7.77E-04
1.00E+20	7.70E-04	7.08E-04	6.57E-04	6.13E-04	5.76E-04	5.43E-04	5.15E-04	4.89E-04	4.67E-04	4.47E-04
2.00E+20	4.29E-04	4.12E-04	3.97E-04	3.84E-04	3.71E-04	3.60E-04	3.49E-04	3.39E-04	3.30E-04	3.21E-04
3.00E+20	3.10E-04	3.10E-04	3.00E-04	2.90E-04	2.90E-04	2.80E-04	2.80E-04	2.70E-04	2.70E-04	2.60E-04
4.00E+20	2.60E-04	2.50E-04	2.50E-04	2.40E-04	2.40E-04	2.40E-04	2.30E-04	2.30E-04	2.30E-04	2.20E-04
5.00E+20	2.20E-04	2.20E-04	2.20E-04	2.10E-04	2.10E-04	2.10E-04	2.10E-04	2.10E-04	2.00E-04	2.00E-04
6.00E+20	2.00E-04	2.00E-04	2.00E-04	1.90E-04	1.90E-04	1.90E-04	1.90E-04	1.90E-04	1.90E-04	1.90E-04
7.00E+20	1.80E-04	1.80E-04	1.80E-04	1.80E-04	1.80E-04	1.80E-04	1.80E-04	1.80E-04	1.80E-04	1.70E-04
8.00E+20	1.70E-04	1.70E-04	1.70E-04	1.70E-04	1.70E-04	1.70E-04	1.70E-04	1.70E-04	1.70E-04	1.70E-04
9.00E+20	1.70E-04	1.70E-04	1.60E-04	1.60E-04	1.60E-04	1.60E-04	1.60E-04	1.60E-04	1.60E-04	1.60E-04

Source: W. E. Beadle, J. C. C. Tsai, and R. D. Plummer, *Quick Reference Manual for Silicon Integrated Circuit Technology*, John Wiley & Sons, New York, 1985 [10].

FIGURE 7.4
Four-point probe resistivity mapping system used for evaluating silicon wafers prior to fabrication to confirm the range of resistivity and doping uniformity.

(a)

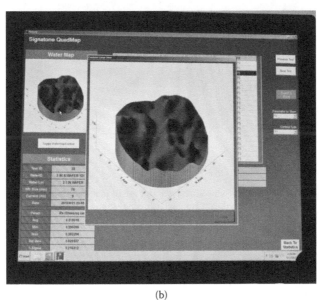

(b)

FIGURE 7.5
(a) Command menu for the mapping system to extract the wafer data. (b) Results obtained from the mapping system and other parameter evaluation details.

contacts with the wafer surface, and resistivity measurements with the four-point probe may produce contact resistance issues. The application of pressure during the measurement should be monitored during wafer mapping. A good starting wafer enhances the chances of getting the best VLSI and ultra large scale integration (ULSI) circuits with excellent yield.

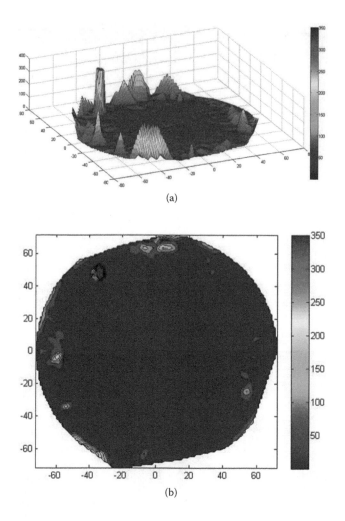

FIGURE 7.6

(a) Resistivity contour map for a 6-inch silicon wafer using the four-point probe mapping system. (b) Nonuniformity in the wafer and concentration variations near the wafer periphery.

7.5 Summary

In this chapter, we analyzed silicon wafers for resistivity and impurity concentration mapping. These are two key parameters for any silicon wafer, as they determine threshold voltage and the gate oxide integrity of metal-oxide semiconductor field-effect transistor (MOSFET) devices. Through wafer mapping, one can estimate the quality of the silicon wafer selected for processing. The four-point probe approach is the only nondestructive method and is often used for wafer resistivity evaluations.

References

1. J. H. Wang, "Resistivity distribution of silicon single crystals using codoping," *Journal of Crystal Growth*, **280**, 408–412, 2005.
2. K. Sugawara, K. Ozeki, K. Fujioka, Y. Mamada, M. Igai, and H. Hirayama, "Finite difference analysis of radial phosphorus dopant distribution in Czochralski-grown silicon single crystals," *Journal of the Electrochemical Society*, **148**, G475–G480, 2001.
3. J. H. Wang, "Resistivity distribution in bulk growth of silicon single crystals," *Journal of Crystal Growth*, **275**, e73–e78, 2005, and the references therein.
4. W. Zuhlemer and D. Huber, "Czochralski-grown silicon," in *Crystals* **8**, Springer-Verlag, Berlin, 1982.
5. M. Tanenbaum and A. D. Mills, "Preparation of uniform resistivity *n*-type silicon by nuclear transmutation," *Journal of the Electrochemical Society*, **108**, 171–176, 1961.
6. R. D. Westbrook, "Effect of semiconductor inhomogeneities on carrier mobilities measured by the van der Pauw method," *Journal of the Electrochemical Society*, **121**, 1212–1215, 1974.
7. H.-J. Schulze, A. Lüdge, and H. Riemann, "High-resolution measurement of resistivity variations in power devices by the photoscanning method," *Journal of the Electrochemical Society*, **143**, 4105–4108, 1996.
8. J. C. Irvin, "Resistivity of bulk silicon and of diffused layers in silicon," *The Bell System Technical Journal*, **41**, 387–410, 1962.
9. N. Sclar, "Resistivity and deep impurity levels in silicon at 300 K," *IEEE Transactions on Electron Devices*, **ED-24**, 709–712, 1977.
10. W. E. Beadle, J. C. C. Tsai, and R. D. Plummer, *Quick Reference Manual for Silicon Integrated Circuit Technology*, John Wiley & Sons, New York, 1985.

8

Impurities in Silicon Wafers

8.1 Effect of Intentional and Unintentional Impurities and Their Influence on Silicon Devices

There are two types of impurities in silicon wafers. Any nonsilicon atoms are considered impurities. Some of them are added intentionally, whereas others are present despite the best efforts to drive them away from the silicon lattice. As was discussed earlier, the silicon lattice is not very dense, and its filling factor is much less when compared to other close-packed structures, so there is a lot of scope to occupy the interstitial positions in the lattice. This accommodative property of silicon traps many metallic ions, which are relatively small, and it is difficult to drive them from these locations. These ions also get accommodated if vacancy sites are available in the regular silicon lattice sites participating in chemical bonding. The presence of these nonsilicon atoms disturbs the electrical properties of the silicon to a great extent and, ultimately, may affect many electrical and reliability issues of devices fabricated using these wafers. The atoms/ions present in the interstitial positions move freely when the silicon wafer undergoes high-temperature thermal cycles, and this movement agglomerates many moving species, leading to precipitations. This may, in turn, lead to unwanted electrical paths and may contribute to device leakage paths.

The grown silicon wafers are expected to provide uniform doping concentration and resistivity values. Radial resistivity uniformity for n-doped material ranges from 10% to 50%, depending on the diameter, dopant impurity, crystal orientation, and process conditions [1]. The demand for better wafer uniformity is a key issue as the wafer experiences identical process conditions, and dopant nonuniformity may lead to dissimilar electrical properties of the devices and circuits.

Efforts have been made to study the soft reverse characteristics of many junctions where excess reverse current flows at moderate to lower applied voltages greatly below the junction breakdown voltages. This high leakage current is believed to be due to the localized high electric fields generated because of metallic ion precipitates [2]. Particulate defects in metallurgical-grade silicon were first reported by Thomas III et al. [3]. They reported that the silicon contained mainly iron, nickel, titanium, vanadium, and aluminum, which was determined through Auger electron spectroscopy analysis. It was also reported that the impurities detected in bulk segregate were silicide-like particles that formed during the crystal solidification process. Other impurities reported are oxygen, carbon, and sulfur.

Metal impurities degrade the performance of devices fabricated from silicon wafers. Therefore, the concentration of metal impurities in wafers must be reduced when increasing the integration density of these devices. The presence of metals introduces defect levels in the bandgap, leading to enhanced carrier recombination that affects the lifetime of charge carriers. The metals may also precipitate and contribute to the formation of shunts and junction leakage, or even junction breakdowns. It is well known that metal impurities are introduced primarily during the cleaning of silicon wafers, but also during device fabrication processes, while the contamination from bulk silicon is ignored. However, metal impurities introduced during the growth of single-crystal silicon affect the formation of defects such as oxidation-induced stacking fault (OISF) and oxygen precipitates, which also degrade the device performance [4]. Metal precipitates enhance the injection rate of electrons into the SiO_2 layer, and interstitial metallic impurities increase the thermally activated component of the leakage current through *p-n* junctions. For next-generation devices with even higher integration density, which must have fewer metal impurities and fewer defects, Czochralski (CZ) single-crystal silicon with even greater purity is required. In order to purify CZ silicon crystals, we need to know where the metal impurities originate from and how they get into the growing silicon crystal. It is difficult, however, to directly to determine the metal concentrations in any single-crystal CZ silicon because they are too low in concentration to be detectable by conventional quantitative analyses. Possible sources of metal impurities in single-crystal CZ silicon are shown in Figure 8.1: a, the poly-Si feed; b, the dopant added to the charge and its purity; c, the quartz crucible holding the charge; d, the carbon parts of the heating element; e, the purity of argon gas; and f, the crystal puller and seed holder [4]. Metal impurities from these sources will be incorporated into the single-crystal CZ silicon through the melt silicon. Among these sources, the amount of metal impurities from the poly-Si and dopant depend on their initial weight, but those from the other sources change as the single-crystal CZ silicon grows. Their incorporation into the growing crystal changes, depending upon the prevailing conditions at that point of crystal growth. These points are shown in Figure 8.2.

Control of fast-diffusing metal impurities is of considerable importance in the production of 256-K and 1-Mbit dynamic random access memory (DRAM) and in the development of 4-Mbit DRAM. Alpern et al. [5] reported the detection of fast-diffusing metal impurities in

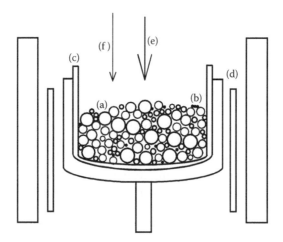

FIGURE 8.1
Possible sources of metal impurities contained in single-crystal CZ silicon: (a) polysilicon, (b) dopant, (c) quartz crucible, (d) carbon parts, (e) Ar gas, and (f) puller. (From K. Harada, H. Tanaka, J. Matsubara, Y. Shimanuki, and H. Furuya, *Journal of Crystal Growth*, **154**, 47–53, 1995, and the references therein [4].)

silicon by using a haze test and also by modulating the optical reflectance method to detect the critical elements, particularly iron, nickel, and copper. The method is an alternative way to detect metallic impurity precipitates in the vicinity of the surface. The stringent demands for advanced processing technologies used for the fabrication of 64-Mbit DRAM and 16-Mbit static access random memory devices require the removal of particles larger than one-tenth of the minimum feature size, with a goal of metallic concentrations lower than 10^{10} cm^{-2} [6].

The role of heavy metal contamination on defect formation in charge-coupled device (CCD) imagers fabricated in silicon has been presented by Jastrzebski et al. [7]. Their evaluations show that the crystallographic defects, such as dislocations, stacking faults, and oxygen precipitates, form localized dark-current generation sites when present in the space charge regions of junctions or metal-oxide semiconductor (MOS) capacitors. The dark-current generation rate is a function of decoration of these defects by heavy metals present in the device processing line. They have demonstrated that the dark currents in CCD imagers and yield loss caused by white spots are a strong function of the heavy metal contamination level (diffusion length) measured in the wafers after the completion of the fabrication sequence. The analysis identified iron as one of the main contaminants. By depositing poly-Si using the low pressure CVD (LPCVD) technique, Morin et al. [8] evaluated the solid-state metallic contamination through the minority carrier lifetime evaluations and confirmed the presence of iron in silane gas that affects the poly-Si gate properties.

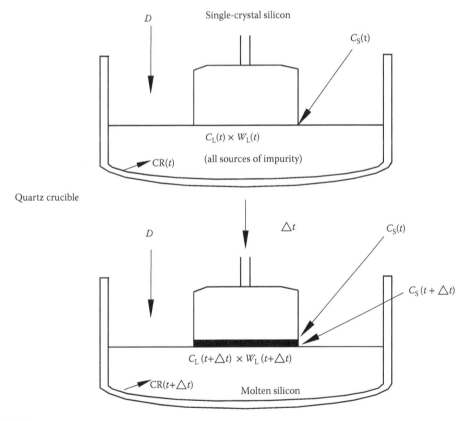

FIGURE 8.2
Schematic diagram of how metal impurities are incorporated into single-crystal CZ silicon. (From K. Harada, H. Tanaka, J. Matsubara, Y. Shimanuki, and H. Furuya, *Journal of Crystal Growth*, **154**, 47–53, 1995, and the references therein [4].)

New experimental data by de Kock and van de Wijgert [9] strongly suggests that the formation of different types of swirl defects in silicon crystals are due to the parallel condensation process involving silicon interstitial atoms and the vacancies. It is proposed that during CZ crystal growth three types of swirl defects—type A, type B, and type C—can form. Doping with donor concentration species (such as P, As, and Sb) greater than 1.0×10^{17} cm^{-3} suppresses the formation of type A swirl defects. Doping with acceptors (B and Ga) with concentration values greater than 1.0×10^{17} cm^{-3} eliminates the formation of type B and type C swirl defects. The observed doping effects are explained in terms of complex formation as a result of Coulomb attraction between dopant species and the charged thermal point defects in the lattice. The effect of impurity doping with electrically active impurities (such as B, Ga, Sb, P, and As in substitutional locations) and isoelectronic impurities (such as Sn) on the formation of swirl defects was investigated by means of preferential chemical etching, copper and lithium decoration, x-ray transmission topography, and electron-beam induced current (EBIC)-mode scanning electron microscopy (SEM).

8.2   Intentional Dopant Impurities in Silicon Wafers

8.2.1   Aluminum

Aluminum is a typical dopant used to produce *p*-type silicon. With three electrons in its outer shell, its behavior is typical to that of other dopant species normally used for producing *p*-type material. It is also an unintentional impurity. Baur et al. [10] used x-ray Raman scattering in total reflection x-ray fluorescence (TXRF) spectroscopy to evaluate aluminum impurities in silicon. Using the angular dependence of the Raman scattering background and the aluminum signal intensity, the minimum detection limit is calculated as a function of the incident angle. The results show that the given minimum detection limit continuously decreases with increasing angles of incidence, up to the critical angle, resulting in a best limit of 2.8×10^{10} atoms/cm^2 right at the critical angle. Their calculations further demonstrated a minimum detection limit of aluminum in silicon at a value of 6.0×10^9 atoms/cm^2 for a 10,000 s count time. Tajima et al. [11] have described the first preparation of standard samples for round-robin analysis to explain the experimental procedures for the photoluminescence and resistivity measurements. The results reported are based on their analysis, and highly reliable calibration curves were presented for this dopant. However, aluminum is not a popular dopant for very large scale integration (VLSI) and ultra large scale integration (ULSI) circuits.

8.2.2   Antimony

Antimony (Sb) is an intentional dopant used to produce *n*-type silicon. Sb-doped CZ silicon has attracted increased interest in advanced complementary metal-oxide semiconductor (CMOS) fabrication in recent years. It has been established that heavily Sb-doped CZ silicon contains less oxygen than lightly doped crystals. These Sb-doped substrates are preferred as epitaxial substrates because autodoping effects are minimized in them.

Single crystals of silicon doped with a large amount of antimony are typically grown with the CZ method, and the dopant is added in the crucible before initiating the crystal growth process. Kodera [12] reported that when the concentration of antimony exceeded a certain value, corrugation appeared on the grown crystal surface and the crystal became polycrystalline. This critical concentration in the crystal was experimentally determined and was found to be 2.9×10^{19} cm^{-3} under the growth condition, where the pull rate was 0.2–1.0 mm/min and rotation rate was about 5 rpm [12]. The crystals were further analyzed perpendicular to the growth direction. Cross-sections near the point where the crystal became polycrystalline were investigated, and several precipitates were present in the silicon matrix. These precipitates were short along the growth direction and were of various shapes and sizes. However, most of their edges lay in the <110> direction. These precipitates had a metallic brilliancy.

In order to study the mechanism of reduced oxygen content in antimony-doped silicon crystals, weight analysis experiments have been performed by Liu and Carlberg [13]. They reported that a reduction in oxygen is due to an increased evaporation rate of SiO species from the silicon melt surface. The reason for this increased evaporation is a shift in the free energy in the reaction by which SiO is formed. Through thermodynamic calculations and SEM studies, the evaporated deposits from antimony-doped silicon melts were analyzed. The deposits were mainly silicon oxide, silicon, and antimony. No antimony oxide was identified. The effects of antimony doping on the oxygen equilibrium solubility and on the viscosity of silicon melts have also been calculated. Significant improvements in oxygen precipitation and in density of bulk defects show that the nucleation-growth process does improve the effectiveness of oxygen precipitation [14]. This, in turn, greatly improves internal gettering in epitaxial wafers. Through thermodynamic calculation, Liu et al. [15] further supported that the oxygen loss by evaporation from the free surface of the melt is only due to the formation of SiO, and Sb_2O_3 evaporation can be neglected. The basic reason for a reduction in oxygen concentration in heavily Sb-doped CZ silicon was that oxygen solubility decreased when antimony with a larger radius was doped degenerately into the silicon crystal. By using x-ray diffraction (XRD) and x-ray photoelectron spectroscopy (XPS) analysis, Yang et al. [16] indicated that during heavily Sb-doped CZ silicon growth, antimony interacts with SiO to enhance the evaporation of SiO from the silicon melt, which results in a reduction in oxygen concentration in the heavily Sb-doped CZ silicon.

8.2.3 Arsenic

Arsenic is a preferred dopant for lower resistivity ranges and is an intentional dopant in CZ and float-zone (FZ)-grown crystals. Many researchers have explored this impurity using electrical and luminescence studies. The diffusion coefficient of this impurity is slow, but fits well in the silicon lattice. Tajima et al. [11] described the first preparation of standard samples for round-robin analysis and then explained the experimental procedures for the photoluminescence and resistivity measurements. The results reported are based on their analysis, and highly reliable calibration curves are presented. A small amount of residual arsenic impurity is frequently detected in highly pure materials using the photoluminescence method; therefore, accurate calibration curves for arsenic impurity are eagerly sought.

The hypothesis of equilibrium between point defects leading to the formation of an Si-As complex was tested by Nobili et al. [17] using single crystals of silicon doped in a broad range of concentrations. The amount of precipitation that is consistent with that of the electrically inactive dopant was confirmed by small-angle x-ray scattering, which also provided details on the size and shape of the particles. The latter are in the form of thin platelets and match the structure of the silicon lattice. These results clearly contradict the models developed to account for the difference between carrier and arsenic concentration, which hypothesize the formation of a high-equilibrium concentration of complex point defects, making the excess dopant electrically inactive.

Oxygen precipitation behavior in heavily arsenic-doped CZ silicon wafers, with and without prior rapid thermal processing (RTP) at 1100°C to 1250°C, and subjected to subsequent two-step annealing at various lower temperatures, was reported by Wang et al. [18]. They proposed that the As-O-V complexes formed due to the heavy arsenic doping provide an additional path for the heterogeneous oxygen precipitate nucleation at 650°C. Moreover, the As-O complexes are likely to play important role in the heterogeneous oxygen precipitate nucleation at 450°C. However, the previously mentioned arsenic-related complexes are not available at 800°C for the heterogeneous oxide precipitate nucleation. In the case with the RTP at high temperatures, such as 1250°C, O_2V complexes facilitate heterogeneous oxide precipitate nucleation at 800°C.

8.2.4 Boron

Boron is a popular dopant for both CZ and FZ crystals. High boron doping has been reported to influence the overall crystal-originated particle (COP) density of crystal defects on silicon wafers. The studies of Suhren et al. [19] have presented the impact of boron concentration as an important additional parameter influencing the radial defect distribution in the silicon crystals. According to them, silicon wafers with boron concentrations up to 2.0×10^{19} cm^{-3} were characterized by delineating defects and analyzed with respect to COPs. No OISF ring was noticed, and the entire wafer displays a homogeneous COP density for low-boron-doped ingots in a resistivity range of several Ω-cm and appropriate pulling conditions. By keeping the identical conditions but increasing boron concentration, changes in the radial COP distribution were observed. The area with a high COP density shrinks and vanishes in the center of the wafer when the boron concentration approaches a level of about 1.0×10^{19} cm^{-3}. It is assumed that boron doping at a sufficiently high level modifies the balance of vacancies and interstitials generated in the crystal pulling process and changes the radial defect distribution in the silicon crystals as well.

Kamiura et al. [20] reported on the effects of heat treatments on the electrical properties of boron-doped crystals. They presented the results of electrical resistivity changes induced by high-temperature heat treatments and subsequent annealing and discussed these results in connection with the roles of iron, iron-related defects, and other defects. Further, they have investigated the effects of heat treatment conditions such as temperature and atmosphere. When applying iron gettering in the device fabrication process, it is proposed that annealing around 300°C is most suitable as the final heat treatment step to remove iron and related defects from active regions of the devices.

Ono et al. [21, 22] reported on the effect of heavy boron doping on oxide precipitate growth in CZ silicon wafers with resistivities ranging from 6 to 40 mΩ-cm. These were studied with prolonged annealing conditions from 800°C to 1000°C. Transmission electron microscopy (TEM) revealed that the growth rate of oxide platelet precipitates is proportional to the square root of time in 40 mΩ-cm samples, and the precipitate morphology changed from plate to polyhedral. The strain around the precipitate also decreased during the annealing cycle at 900°C in 6, 9, and 18 mΩ-cm samples. It is concluded that these changes are taking place due to oxygen precipitation and that the physical size effects of the boron atoms are enhanced by decreasing the boron concentration levels. Further, the strain around the precipitate was estimated by considering the effect of the high interstitial silicon supersaturation due to the high density of oxide precipitate growth.

Yonenaga et al. [23] investigated dislocation-impurity interaction in the growth of CZ silicon crystals that were heavily doped with boron and germanium impurities by using x-ray topography and etch-pit techniques. In terms of crystal growth, doping of boron and co-doping of germanium and boron impurities is effective in suppressing the generation of slip dislocations, mainly due to the preferential segregation on them. However, the dislocation velocity in motion under a high stress in boron-doped silicon is higher than that in undoped silicon. The difference in the lattice parameter between the seed and the crystal induced by impurity doping leads to the generation of misfit dislocations at the seed/crystal interface in the growing crystal.

8.2.5 Gallium

Gallium-doped CZ silicon was recently identified as a promising substrate material for solar cells, since it shows a high minority carrier lifetime with no lifetime degradation effect. It has been reported that the minority carrier lifetime increased as the germanium concentration increased [24]. In addition, the higher minority carrier lifetime was associated with a decrease in interstitial oxygen related to D-defects in the Si crystal. Although the gallium-doped CZ silicon shows a high minority carrier lifetime, different types of grown-in microdefects may exist in dislocation-free crystals and limit the minority carrier lifetime by acting as recombination centers. Arivanandhan et al. [25] have experimentally shown that the dissolved oxygen interstitial concentration decreased and the number of grown-in oxygen precipitates increased as the germanium concentration increased in gallium and germanium co-doped CZ single-crystal silicon. To further improve the minority carrier lifetime of the gallium-doped CZ silicon, and thus the conversion efficiency of the solar cell, it is essential to understand the formation mechanism of grown-in microdefects in CZ silicon and their effects on minority carrier lifetime.

8.2.6 Phosphorus

Phosphorus is another common dopant for CZ-grown silicon wafers. The accurate relationship between the electrical resistivity and the phosphorus concentration in silicon has been reported by Mousty et al. [26]. Electrical and analytical measurements allowed the authors to determine a relationship between resistivity and phosphorus concentration in doped silicon. This relationship was compared to that obtained in the case of n-type silicon, and the differences due to the various doping agents were critically analyzed. Differences between concentrations determined by using analytical techniques and by using Hall effect measurements were reported for doped silicon. Electron mobility values were empirically fitted for phosphorus concentration in silicon by Baccarani and Ostoja [27], and the deviations from the standard known curves were recorded. Masetti et al. [28] also reported on the carrier mobility values of phosphorus-doped silicon.

Using a phosphorus impurity in silicon has an added advantage of trapping gold in the crystal. Gold is a fast diffusant and acts as an effective recombination center in silicon. Phosphorus acts as an impurity gettering point [29] and is an effective way to reduce the activity of gold in silicon. An investigation of the macroscopic radial distribution of phosphorus dopant found it to vary by a factor of 40% in the (100) case and by 70% in the (111) case [1]. Series [30] made a systematic study of an axial magnetic field on the incorporation of oxygen, carbon, and phosphorus into CZ silicon crystals.

8.3 Unintentional Dopant Impurities in Silicon Wafers

Metallurgical-grade silicon is typically 98% to 99% pure and mainly consists of aluminum (1570 ppm), boron (44 ppm), iron (2070 ppm), phosphorus (28 ppm), chromium (137 ppm), manganese (70 ppm), nickel (4 ppm), titanium (163 ppm), vanadium (100 ppm), and carbon (80 ppm) as impurities [31]. Through the chemical vapor deposition (CVD) epitaxial route, the purity of silicon is improved to electronic-grade silicon. After acid leaching and gas blowing through the silicon melt, the purity will reach nearly the level of 99.99%. Once the silicon is converted into the electronic-grade form, most of the impurities listed earlier are eliminated by a chemical process that converts the metals to their respective volatile compounds. In this way, aluminum, calcium, carbon, magnesium, iron, boron, phosphorus, and titanium metals can be removed [32]. Despite all these efforts, however, the presence of some metallic species will still be felt, either during the device fabrication processes or at the time of device/circuit usage. Some of the segregation coefficients of both intentional and unintentional impurities are listed in Table 8.1.

Istrato et al. [32] presented a clear picture of these metallic impurities at different stages of silicon refinement. Figure 8.3a shows the impurities present in the metallurgical-grade silicon. It is apparent that the impurity content and the levels are high, but the values from

TABLE 8.1

Typical Impurities Present in Molten Silicon and Their Effective Segregation Coefficients

Impurity	Segregation Coefficient (k_o)
Aluminum (Al)	$2.0 - 2.8 \times 10^{-3}$
	0.0018
Antimony (Sb)	0.023
	0.02
Arsenic (As)	0.3
	0.27
Bismuth (Bi)	7.0×10^{-4}
Boron (B)	0.8
	0.72
Carbon (C)	0.07
Cobalt (Co)	8.0×10^{-6}
Copper (Cu)	4.0×10^{-4}
Gallium (Ga)	8.0×10^{-3}
	0.0072
Gold (Au)	2.5×10^{-5}
Iron (Fe)	8.0×10^{-6}
Lithium (Li)	1.0×10^{-4}
Manganese (Mn)	1.0×10^{-5}
Nickel (Ni)	3.0×10^{-5}
Oxygen (O)	0.25–0.5
	1.4
Phosphorus (P)	0.35
	0.32
Silver (Ag)	1.0×10^{-6}
Titanium (Ti)	9.0×10^{-6}

the different sources rarely vary by more than an order of magnitude. The impurity content in the commercial microcrystalline silicon is shown in Figure 8.3c, and is two to five orders of magnitude lower than in metallurgical-grade silicon, and some elements such as titanium, manganese, vanadium, zirconium, and manganese are below the detection limit of neutron activation analysis. The authors pointed out that the variation in metal concentration by one to two orders of magnitude reflects the differences in growth conditions and in the quality of feedstock used by different manufacturers. In Figure 8.3b, the metal concentration in solar-grade silicon is shown. In this case, the variation in metallic impurities is much less than that of metallurgical-grade silicon, but higher than that of microcrystalline silicon. Probably one or more purification steps are being carried out to arrive at this concentration range. The metal content in solar-grade silicon is much higher than that in microcrystalline polysilicon feed used in the production of integrated circuits. At present, solar-grade silicon contains practically every transition metal in concentrations between 3.0×10^{13} and 1.0×10^{16} atoms/cm^3. In comparison to the actual silicon concentration of 5.0×10^{22} atoms/cm^3, these metallic impurities are smaller in number, but they have a large influence on the performance of active devices. These impurities severely affect the minority carrier diffusion length and solar cell efficiency. Figure 8.4 presents data for interstitial iron and FeB pairs for copper and for nickel [32]. The shaded area on the plot indicates the range of minority carrier diffusion lengths typically found in microcrystalline solar cells. The threshold concentration of interstitial iron acceptable for a solar cell is around 2.0×10^{12} cm^{-3}. Copper and nickel can be acceptable up to values of 10^{14}–10^{16} cm^{-3}. Copper is the least detrimental metal impurity, while heavier metals such as titanium can degrade performance in concentrations as low as 10^{11}–10^{12} cm^{-3}.

8.3.1 Carbon

Along with oxygen, carbon is a major unintentionally introduced impurity in CZ or FZ silicon crystals. It is found in concentrations between 1.0×10^{16} and 1.0×10^{17} cm^{-3}. Polycrystalline sheets grown by a mold-shaping technique had an oxygen content of 1.7 ppm, a carbon content of 0.33 ppm, and lower levels of other contaminants [33]. At high concentrations, carbon can create precipitates and stacking faults during high-temperature processing steps. The effects of carbon are, however, not well understood. As with oxygen, infrared (IR) spectroscopy is used to measure the concentration of carbon, which exhibits strong absorption at ≈16.6 µm [34]. Carbon comes from the crucible holding molten silicon. It occupies a substitutional lattice site in the silicon and can be conveniently measured. A more rapid, precise, and sensitive analytical method was reported by Vidrine [35] to measure the carbon concentration in silicon at room temperature. Typical seed-end concentrations of carbon range from 1.0×10^{16} and lower. At the butt ends, its value will reach 5.0×10^{17} atoms/cm^3. The main source of high carbon levels in silicon crystals grown from melt under reduced pressures and contained in silica crucibles supported by a graphite retainer/susceptor has been identified through thermodynamic analysis [36]. The calculations have been verified by experimental results, and the carbon level can be reduced by approximately 50% with the use of molybdenum retainers. At these concentration levels, carbon does not precipitate like oxygen and is electrically inactive. It only aids in the formation of defects in the crystal lattice.

IR absorption spectrophotometry has been used to determine the carbon content of monocrystalline silicon. This method is particularly attractive because it has sensitivity in the parts per billion atomic (ppba) ranges, and it is an inexpensive, quick, and nondestructive method. Gross et al. [37] have reported on ^{3}He activation analysis and compared the results. The results were found to be in good agreement with those obtained by IR

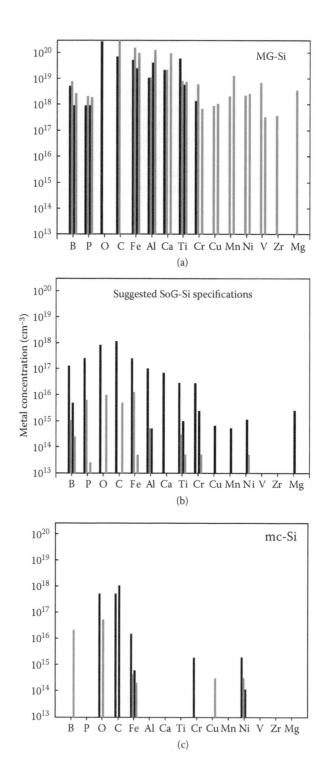

FIGURE 8.3
Typical concentrations of impurities in (a) metallurgical-grade, (b) solar-grade, and (c) multicrystalline silicon solar cells that are currently in production. The data points are from limited sources and are indicative only, not absolute. (From A. A. Istratov, T. Buonassisi, M. D. Pickett, M. Heuer, and E. R. Weber, *Materials Science and Engineering B*, **134**, 282–286, 2006, and the references therein [32].)

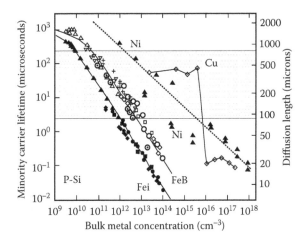

FIGURE 8.4
Impact of iron, copper, and nickel on minority carrier diffusion length in single-crystalline silicon. The shaded area represents a typical range of minority carrier diffusion lengths in multicrystalline silicon solar cells. (From A. A. Istratov, T. Buonassisi, M. D. Pickett, M. Heuer, and E. R. Weber, *Materials Science and Engineering B*, **134**, 282–286, 2006, and the references therein [32].)

spectrophotometry and those obtained by the activation analysis method within a 40% range in determining the carbon impurity concentration in silicon. The carbon content of semiconductor-grade silicon was found to be on the order of a few ppma, which is in disagreement with chemical analysis results. Quantification of carbon in silicon is still a major topic of research. Many groups are actively working on this key material present in silicon crystals [38–47], as is evident from the publications.

Three types of microdefects, formed by the agglomeration of either self-interstitials or vacancies, have been observed in dislocation-free single crystals grown using the pedestal pulling technique (crucibleless growth) [48]. It is suggested that a fraction of the carbon atoms in the crystal is incorporated at interstitial lattice sites and that the interaction of interstitial carbon with the vacancies is responsible for the change in the dominant type of point defect. Inoue et al. [49] were able to detect the carbon concentration down to 1.0×10^{14} cm^{-3} using IR absorption spectroscopy. Typical carbon concentration in samples of CZ crystals will be on the order of 1.5×10^{16} to 1.0×10^{17} cm^{-3}. When measuring carbon concentration, it is standard for the wafers to be polished on both sides. Electron irradiation experiments were carried out at an energy of 4.6 MeV with dosage values of 1.0×10^{15} and 1.0×10^{16} cm^{-2}. Measurements were carried out using Fourier transform infrared spectroscopy (FTIR) with a wavenumber resolution of 2.0 cm^{-1} at room temperature. They were successful in detecting various complexes in low carbon concentrations in CZ silicon at lower concentrations.

Series [30] made a systematic study of an axial magnetic field on the incorporation of carbon into CZ-grown silicon crystals. The radial dopant distribution was correlated with the interface shape and its radius of curvature. No correlation was found between the interface shape and radial distribution of oxygen or carbon [50]. The detailed measurements by Series and Barraclough [51] on carbon contamination grown in a low-pressure CZ system were measured as 0.1 μg/s, in the absence of other sources of contamination. This would limit the carbon content of the crystals to about 2.0×10^{15} atoms/cm^3 near the seed and 2.0×10^{16} atoms/cm^3 at a fraction solid of 0.7. The measured carbon content of the crystals was generally higher than, indicating that contamination of the crucible, starting charge, and melt prior to growth were the dominant sources of carbon. It is further reported [52] that the rate of carbon contamination of a silicon melt using a low-pressure CZ puller technique by repeated dips. They also studied the effect of a gas baffle placed between the heater and the melt and found that the contamination is mainly from the charge.

Ekhult and Carlberg [53] proposed a new technique to control the diameter of the growing silicon crystal. This technique involves the use of fast, adjustable IR heaters directed toward the melt surface and facilitates growth at constant pull rate. Using this method, two crystals were grown and compared to a reference crystal grown with the normal CZ technique. The axial carbon distributions showed higher levels in the crystal grown using IR heating. Radial carbon and oxygen measurements did not reveal any significant differences between the two techniques.

The concentration of carbon in semiconductor silicon and the behavior of carbon in the fusion and crystallization of silicon have been studied by Nozaki et al. [54] by using charged particle activation analysis. The details of the phase diagram are shown in Figure 8.5 at an extremely low carbon concentration range. The presence of carbon can lead to the formation of SiC clusters and may lead to many unwanted issues in terms of VLSI and ULSI circuits.

Carbon impurity was reported to cause a change from the vacancy growth mode to the interstitial growth mode at a doping level of 1.7×10^{17} cm^{-3}. The interstitial carbon diffusivity is known at very low T, and thus $D_i^*(T)$ can be approximated using an Arrhenius plot in the entire range, from low T up to T_m, the crystal melting point:

$$D_i^* = 4.4 \times 10^{-4} \exp\left(\frac{-0.7 \text{ eV}}{kT}\right) \text{cm}^2/\text{s}$$

In modern CZ crystals, the carbon content is low, about 3.0×10^{15} cm^{-3}. The kick-out temperature, 770°C, is well below the nucleation temperature, which is reported to be over 900°C. Yet the fraction of carbon interstitials at T_n, the nucleation temperature of interstitial microdefects, is still appreciable: 10% at 930°C. It is likely that the interstitial agglomeration in CZ crystals is also controlled by the carbon impurity [55]. The carbon-containing globular clusters may be identified by the so-called B-defects, which are present along with the interstitial-type loops (A-defects) and form a separate B-band at the edge of an interstitial region. The B-defects may possibly be the precursors of the A-defects and the B → A conversion might occur if the B-defects are large enough [55].

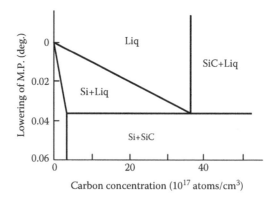

FIGURE 8.5
Phase diagram of Si-C system in extremely low carbon concentrations. (From T. Nozaki, Y. Yatsurugi, and N. Akiyama, *Journal of the Electrochemical Society*, **117**, 1566–1568, 1970 [54].)

Atomic configuration around the carbon atoms, in silicon network, change after silicon crystallization takes place. Figure 8.6 shows the crystal growth processes of liquid silicon, including three carbon atoms near the solid–liquid interface under the same growth conditions. Although almost no impurity effect was observed up to two carbon impurity atoms, a remarkable difference was found in the local atomic configurations around the three carbon atoms between [001] and [111] crystal growth, as shown in the figure. This can be attributed to the difference in the crystal growth mechanism between [001] and [111] described earlier. In addition, disturbance due to carbon atoms tends to be localized in the former case, while it tends to be transmitted in the latter [56]. It is also important to investigate the effects of oxygen on crystal growth and defect formation, since ~10^{18} cm^{-3} oxygen is included in CZ-grown silicon.

8.3.2 Chromium

Chromium concentration values in silicon will be present at levels of 10^{12} to 10^{13} per cm^3. This is an unintentional impurity, but does influence the performance of solar cells.

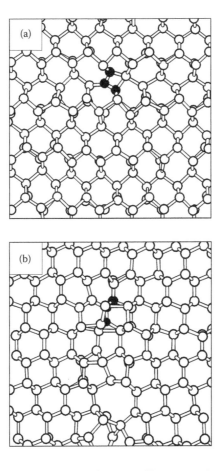

FIGURE 8.6
Atomic configurations around the carbon atoms after crystallization: (a) [001] and (b) [111] crystal growth. (From T. Motooka, "Molecular dynamics simulations of crystal growth from melted silicon," *Proceedings of the First Symposium on Atomic-scale Surface and Interface Dynamics*, March 13–14, 1997, Tokyo, and the references therein [56].)

A systematic study of this impurity has been carried out by Pizzini et al. [57] on the diffusion length of minority carriers as a function of the impurity concentration and microstructural features, such as grain boundaries, dislocations, twins, and stacking faults. The influence is systematically recorded. By using deep-level transient spectroscopy analysis, Chabane-Sari et al. [58] analyzed the internal gettering of chromium about the electrically active metal profiles after precipitation. By use of rapid thermal lamp pulse annealing for the metal diffusion before gettering, it is found to impede the internal gettering mechanism in silicon wafers. This strange behavior was correlated to oxygen nucleation and precipitation. The authors used boron-doped silicon substrates for different gettering cycles. Further, they reported that the intrinsic gettering efficiency strongly depends on the oxygen precipitate morphology. The team concluded [59] that the polyhedral morphology, which occurs only after an extended high-temperature treatment for low-carbon silicon wafers, improves the intrinsic gettering efficiency.

8.3.3 Copper

Copper is an unintentional impurity present in the lattice. Its presence causes performance failures of devices and drastically reduces the manufacturing yield. Its detrimental impact on leakage current, gate oxide quality, and defect generation is a major issue for VLSI and ULSI devices [60]. It has been shown that copper was gettered onto substitutional sites by phosphorus diffusion, while these impurities were gettered onto nonsubstitutional sites by boron diffusion. This pointed to an association mechanism—namely, that copper, which is an acceptor on substitutional sites in silicon, was paired with phosphorus donors. However, in the case of boron diffusion, the solubility of copper donors would be enhanced so that it would be on nonsubstitutional sites. This complementary behavior was experimentally verified by Meek et al. [61]. These results gave a qualitative understanding and formed the basis for detailed solubility calculations. When the solubility is exceeded, new phases are formed in silicon crystals. The formation of precipitates for boron diffusions has been studied by electron microscopy to support the behavior.

The minority carrier diffusion length has been found to decrease monotonically in n-type silicon and to exhibit a step-like behavior in p-type silicon at copper concentrations above a certain critical level. Sachdeva et al. [62] suggested that the impact of copper on the minority carrier diffusion length is determined by the formation of copper precipitates. This process is retarded in perfect silicon due to the large lattice mismatch between Cu_3Si and the silicon lattice, and is even more retarded in p-type silicon due to electrostatic repulsion effects between the positively charged copper precipitates and interstitial copper ions present in the silicon lattice.

Schwuttke [63, 64] reported that the precipitation behavior of copper is strongly influenced by the presence of imperfections and impurities in silicon. The precipitation occurs entirely during the cooling period. The amount of precipitation depends on the temperature and cooling rate. This influence of the oxygen concentration on the shape of the precipitate suggests the formation of a copper-oxygen complex. Two types of precipitates have been reported consisting of needle-like and plate-like shapes. The precipitates have a definite preferred orientation with respect to the silicon crystal lattice. They prefer to segregate on {100} planes. The needles grow in <110> directions, while the plates are oriented in such a way that their edges are parallel to the <100> directions. Composite structures appear as crosses when observed in the {100} planes. If the plane of observation is a {111} plane, their projection appears as a star pattern. The needles always bisect the angle between the plates.

Effects of copper contamination in silicon, particularly to MOS thin silicon dioxide gate breakdown, have been reported by Ramappa and Henley [65]. For advanced ULSI process technologies, the most critical yield-limiting aspect of heavy metal contamination, such as copper, is the detrimental effect of heavy metal silicide precipitates on thin gate oxide dielectric breakdown. Metallic precipitates formed at the Si/SiO_2 interface could lead to localized thinning of oxides. A secondary, and probably a more detrimental, effect is the formation of a conductive geometric singularity in or near the interface. Electrostatic analysis has shown that such conductive singularities in a dielectric medium can enhance the local electrical field in the dielectric. The problem is exacerbated as technology (and design rule) scaling reduces feature size and film thickness such that the effects of precipitate defects are more pronounced. This was further supported by Lin et al. [66] for the metal-oxide semiconductor field-effect transistor (MOSFET) thin oxide gates in the thickness ranges of 30 Å to 50 Å. There are excellent review articles that provide more detail on copper [67] in VLSI and ULSI production lines. Copper precipitations and intrinsic gettering of this material is an important topic, and many research teams are endeavoring [68–72] to understand the behavior of copper in silicon.

8.3.4 Germanium

The tetravalent germanium atoms exist at substitutional sites in silicon crystal after incorporating into the crystal during growth. Germanium doping of CZ silicon has attracted attention in recent years. Yang [73] pointed out that germanium doping can result in smaller, denser voids, which can be eliminated easily by annealing the crystals at high temperature, and is expected to improve the gate oxide integrity of MOSFET devices. Oxygen precipitation in CZ silicon can be enhanced by germanium doping so that the internal gettering ability of the wafers is improved substantially. In addition, germanium can suppress the formation of thermal donors. Chen et al. [74] reported that COPs with high density but small sizes were inclined to generate in germanium-doped CZ silicon wafers. They have proposed that the combination of germanium atom and vacancy could reduce the free vacancy concentration and the onset temperature for void generation, thus forming denser but smaller voids. While the stress compensation induced by boron and germanium atoms could increase the vacancy fluxes in heavily boron-doped germanium-doped CZ silicon crystal, the presence of oxygen atoms in germanium-doped silicon would tend to benefit the formation of inner oxide walls of void, in particular, ones that are small. Furthermore, the thinner oxide walls within the silicon crystal void are charged for easy annihilation by the germanium doping.

Co-doping is another approach that many scientific groups are studying as a means to achieve better silicon crystals. Dislocation and impurity interactions and improved fracture strength of silicon [75–78] are some of the advantages that have been reported with germanium doping in silicon. In view of these advantages, it is believed that germanium-doped CZ silicon can be used as a better substrate for VLSI and ULSI circuits, which have a high-quality subsurface layer, high internal gettering (IG) ability, and a stable electrical property, in order to achieve better circuits and better performances.

8.3.5 Gold

Gold is an unintentional impurity present in the lattice. Considerable attention has been paid to the properties of gold in silicon because of its use in controlling minority carrier lifetime. At room temperature, this element introduces a donor level 0.35 eV above the valence band

and an acceptor level 0.54 eV below the conduction band. Both of these levels are believed to be due to substitutionally occupied gold atoms in the silicon lattice sites [79]. Gold is a double-level impurity in silicon, introducing both an acceptor level and a donor level in the bandgap between the conduction band and the valence bands [80]. The impurity center can change charge states in the semiconductor only by the emission of carriers, either to the conduction band or to the valence band, and the capture of carriers from one of the bands. Gold is also introduced into the silicon bulk for specific applications such as IR detection applications.

It has been shown that gold was gettered onto substitutional sites by phosphorus diffusion, while these impurities were gettered onto nonsubstitutional sites by boron impurity diffusion. This pointed to an association mechanism—namely, that gold, which is an acceptor on substitutional sites in silicon, was paired with phosphorus donors. However, in the case of boron diffusion, the solubility of gold donors was enhanced so that it would be on nonsubstitutional sites. This complementary behavior involving gold on substitutional sites under a diffused donor profile and on nonsubstitutional sites under a diffused acceptor profile was experimentally verified by Meek et al. [61]. These observations gave a qualitative understanding of this behavior and formed the basis for detailed solubility calculations. When the solubility is exceeded, new phases are formed in silicon crystals. Electron microscopy studies are documented the formation of these precipitates for boron diffusion.

8.3.6   Helium

Not much information regarding the Si:He system is known. Helium manifests no chemical activity toward silicon. It is known to dissolve into the silicon at interstitial sites and moves in a zig-zag path passing through hexagonal sites. No appreciable lattice rearrangement has been recorded during the movement in the silicon crystal lattice, but there is a tendency toward a clustering mechanism [81]. A helium atom located initially in a vacancy diffuses out to an interstitial site, and a single vacancy does not act as a trapping center for these helium atoms. They exist in close proximity to the vacancy and in a divacancy state. More details are not available on their effect on the silicon crystal lattice. Helium does not behave as a catalyst, but rather as a solvent and participates in the thermodynamics.

8.3.7   Hydrogen

Hydrogen is an unintentional impurity that is quite unstable, but shows its presence in *p*-type silicon wafers doped with boron. There is a lot of interest in the properties of hydrogen in crystalline silicon. This is mainly due to the ability of hydrogen to passivate the electrical activity of shallow acceptors, as well as the many deep impurity and defect levels present. The determination of the electrical properties, such as energy levels and charge states, of isolated hydrogen in silicon remains a challenge for both experimentalists and theorists. Atomic hydrogen is expected to be incorporated into silicon during a variety of processing and device operation steps. In particular, near the surface, close to 1 μm depth range hydrogen passivation of boron acceptor impurities has been demonstrated by Chantre et al. [82] following reactive ion etching, wafer polishing, or during avalanche injection of electrons into the silicon dioxide. According to them, the possibility of atomic hydrogen getting into the *p*-type material occurs during the polishing stage.

The effect of hydrogen annealing on oxygen precipitation behavior and on gate oxide integrity has been reported by Abe et al. [83]. The annealing, with a specific ramp-up rate, has a strong dependence on the bulk microdefect density in relation to DRAM devices and the gate oxide integrity improves after heat treatment. Further, it was found that oxygen precipitation in hydrogen annealing was the same as in nonannealed wafers and strongly depended on the ramp-up rate. The effect of hydrogen annealing is limited to the near-surface regions, and it is in this part of the wafer where gate oxides are generally defined.

Hydrogen was the first impurity reported to affect formation of grown-in microdefects. It was found that the critical growth rate for the disappearance of swirl defects (loops) was strongly reduced by a factor of 2 in the presence of hydrogen, in a concentration of 1.0×10^{15} cm^{-3}. No further details of vacancy trapping by hydrogen can be given due to the lack of information on the vacancy-hydrogen complexes [55]. Hydrogen impurity was found to suppress defect growth after solidification of the droplets [84]. Sueoka et al. [85] have concluded that hydrogen-annealed wafers have no mechanical strength problems during actual low-temperature processes, whereas the existence of oxide precipitates degrades the mechanical strength of CZ wafers. Simoen et al. [86] reported that the role of hydrogen is more complex than acting merely as a catalyst that accelerates the oxygen transport in silicon.

8.3.8 Iron

Iron is another unintentional impurity present in the silicon lattice. Its presence can cause performance failures of devices and drastically reduces the manufacturing yield. Its detrimental impact on leakage current, gate oxide quality, and defect generation are some of the major issues for VLSI and ULSI devices and are related to their circuit failures [60]. Unintentional iron impurity in silicon exists in two forms: as an interstitial and as a complex in the presence of other defects. At the interstitial position, iron introduces a donor level at 0.375 eV [87]. Because of near-mid gap energy and a large capture cross-section, it is expected that iron will produce high recombination or low minority carrier lifetime. The interstitial iron in p-Si is positively charged at room temperature and at slightly elevated temperatures. As a result, it tends to form pairs with negatively charged defects, such as shallow acceptors. This has resulted in more than 30 complexes that can be formed between iron and other defects present in the silicon lattice, and about 20 deep levels are associated with these complexes. In boron-doped silicon crystals, iron is known to form a B–Fe (or Fe–B or FeB) complex.

Metals such as iron tend to precipitate at extended crystal defects such as grain boundaries and dislocations. Metal dissolution from these precipitates is held responsible for a decrease in recombination lifetime after high-temperature processing steps, such as the thermal oxidation steps, in the range of 830°C to 1050°C. Imaging of metal impurities, particularly about the iron in silicon, has been demonstrated by Schubert et al. [88] by using luminescence spectroscopy and synchrotron techniques. The spatial distribution of iron, which is the most frequently found metal contamination in silicon, has been specifically addressed including imaging and mapping measurement techniques. Photoluminescence and x-ray fluorescence spectra were detected to measure the recombination lifetime and distribution of metal precipitates in the silicon. Numerical simulations of carrier diffusion were applied to show evidence of impurity dissolution from grain boundaries after oxidation. These methods were applied to clarify the distribution of iron after high-temperature processes and to identify breakdown sites. Advanced two-dimensional simulations of the iron distribution in multicrystalline silicon blocks after crystallization were also presented. It has been shown that, in addition to reducing the recombination lifetime, iron in the form of precipitates may introduce breakdown sites.

Harada et al. [4] examined the origins of metal impurities in CZ single-crystal silicon in order to develop improved crystal purification techniques. They calculated the mass of metal impurities incorporated into molten silicon from each origin during crystal growth and estimated the concentration of impurities in the molten silicon and single-crystal silicon. To confirm these results, they analyzed a sample of the residual melt after crystal growth. By comparing the calculated and experimental results, it was shown that the main origins of aluminum and nickel are probably the quartz crucible and poly-Si, respectively, while iron originates from both. The calculated concentration of Fe and Ni in single-crystal silicon is 10^8–10^9 cm^3, and that of Al is 10^{11}–10^{12} cm^3. We confirm that the concentrations of Fe and Ni can be reduced by using high-purity silicon as a raw material.

In the calculation, Harada et al. [4] used the following equation:

$$\frac{dC_s(t)}{dt} = \frac{1}{W_L(t)}\left((k_0 - 1)C_s(t)\frac{dW_L(t)}{dt} + k_0 CR(t) + k_0 D\right)$$

where $C_s(t)$ is the impurity concentration at the freezing interface in the crystal at time t, $W_L(t)$ is the weight of the melt, $CR(t)$ is the contamination rate from the crucible, D is the contamination rate from the ambience in the puller, and k_0 is a distribution coefficient. In the calculation, D is assumed to be constant, and $CR(t)$ is variable because the area of the interface between the silicon melt and the crucible changes during crystal growth. The Fe concentration in the single-crystal CZ silicon is shown in Figure 8.7 as an example of the calculation results that were obtained using the previous equation. The growth conditions of the crystal and the analyzed values are discussed elsewhere. The horizontal axis indicates the solidified fraction, and the vertical axis is the Fe concentration in the crystal. Figure 8.7 shows two cases: (a) Fe is incorporated into the molten silicon from all origins (poly-Si, quartz crucible, and ambience in the puller), as shown by the solid line, and (b) Fe is incorporated into the molten silicon only from the poly-Si, as shown by the dashed line. The difference between curve "a" and curve "b" indicates the contribution of the quartz crucible and the ambience in the puller, from which the ratio increases relative to the solidified fraction (g). It is reported that the concentration of nickel was one order of magnitude smaller in the purified crystal than in the reference. This result supports the assertion that

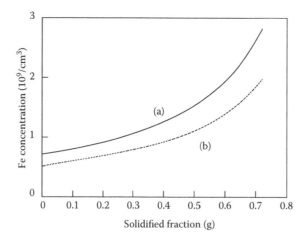

FIGURE 8.7
Iron concentration in single-crystal Czochralski silicon. (From K. Harada, H. Tanaka, J. Matsubara, Y. Shimanuki, and H. Furuya, *Journal of Crystal Growth*, **154**, 47–53, 1995, and the references therein [4].)

the main origin of nickel is the poly-Si charge. The decrease in iron concentration is less than that of nickel, showing that one of the main origins of iron is again poly-Si, but the effect of iron in the quartz crucible is comparatively large.

It is reported that iron concentration values in the silicon will be present at concentration levels of 10^{13} per cm^3. This is an unintentional impurity, but it does influence the performance of solar cells. A systematic study of this impurity has been carried out by Pizzini et al. [57] on the diffusion length of minority carriers as a function of the impurity concentration and micro-structural features like grain boundaries, dislocations, twins, and stacking faults. The influence has been systematically recorded. Similarly, Laades et al. [89] reported on the impact of iron contamination on the performance of industrial CZ-silicon solar cells. There is presently a lot of active research [90–94] on this key impurity in silicon crystal.

8.3.9 Nickel

Nickel is also an unintentional impurity present in the silicon lattice. Its presence causes performance failures of devices and drastically reduces the manufacturing yield of integrated circuits, but it shows good solid solubility in silicon [95]. Its detrimental impact on leakage current, gate oxide quality, and defect generation is a major issue for VLSI and ULSI devices and is related to their circuit failures [60]. Iizuka and Kikuchi [96] reported that nickel impurities in silicon precipitate in a lump shape. In addition, hexagonal platelets were found near the decorated <110> dislocation lines in CZ-grown crystals. These hexagons lie on the {111} plane, and the sides of them lie in the <110> directions. The physical size may be very large. With FZ-grown crystals, no hexagons are found, implying that oxygen concentration in the lattice plays a role in the formation of hexagons as an impurity for the pinning of dislocations. It is felt that the hexagons are probably nickel platelets coherent with stacking faults surrounded by partial dislocations.

Harada et al. [4] examined the origins of nickel impurities in CZ single-crystal silicon in order to develop improved crystal purification techniques. They estimated the mass of impurities incorporated into molten silicon from each origin during crystal growth and estimated the concentration of impurities. The main origin of nickel was felt to be probably the quartz crucible and poly-Si, respectively. The calculated concentration of nickel in single-crystal silicon is 10^8–10^9 cm^3. They confirmed that the concentrations of nickel can be reduced by using high-purity poly-Si as a raw material. By using pure poly-Si, the concentration of nickel was re-estimated, and their results support the assertion that the main origin of nickel is poly-Si only. However, the detection and removal of bulk nickel impurities are major issues for device engineers because of the possible interaction of dopants, particularly with boron-doped *p*-type silicon [97–99], and the formation of nickel silicides affecting normal device performance.

8.3.10 Nitrogen

Nitrogen doping is becoming key to growing defect-free CZ silicon crystals [100]. Therefore, it is indispensable to establish a method for measuring nitrogen concentration in CZ silicon crystals. However, since the nitrogen concentration is low and nitrogen has various configurations, this is a difficult task.

For a proper understanding of the beneficial effects of nitrogen, it is necessary to have a clear idea of the nitrogen states in the lattice and their contribution to the diffusion. A conventional notion is that the major form of nitrogen is dimeric interstitial N_2; under certain conditions, a small amount of substitutional nitrogen N_s also can be present. The data on the evolution of implanted nitrogen profiles by a rapid thermal annealing shows that the diffusivity of the major nitrogen form is very low; the profile evolves by dissociation of the major species into minor but highly mobile species, within the conventional notion, into monomers: $N_2 \rightarrow 2\,N_I$. However, this simple treatment of nitrogen species turns out to be oversimplified. Studies by Voronkov and Falster [101] on the out-diffusion profiles produced by annealing nitrogen-doped FZ and CZ wafers at 900°C were monitored to a depth of up to 120 µm using multiple secondary ion mass spectrometry (SIMS) runs. As the probe depth was increased, astounding features were revealed, as shown in Figure 8.8. The profile slope increased abruptly at a characteristic depth, about 70 µm, which results in a stepwise overall shape. In material with a higher nitrogen content, about 2.25×10^{15} cm^{-3}, three characteristic regions can be distinguished: the near-surface region (A), the deeper portion (B), and the intermediate part (C), as shown in Figure 8.8a. In material with a lower nitrogen concentration, about 4.5×10^{14} cm^{-3}, as shown in Figure 8.8b, only the B portion is

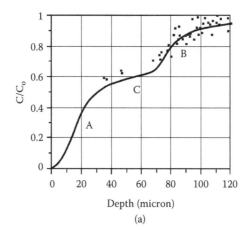

(a)

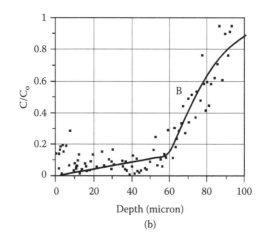

(b)

FIGURE 8.8
(a) and (b): Nitrogen out-diffusion profiles normalized by the bulk concentration C_o (FZ samples, 900°C). (a) $C_o = 2.25 \times 10^{15}$ cm^{-3}, diffusion time 15 min, (b) $C_o = 4.50 \times 10^{14}$ cm^{-3}, diffusion time 10 min. (From V. V. Voronkov and R. Falster, "Nitrogen diffusion in silicon: A multi-species process," *210 ECE Meeting Cancun, Mexico*, Oct. 29–Nov. 3, 2006, and the references therein [101].)

manifested. A simple stepwise shape found for a lower nitrogen content can be accounted for, assuming that the nitrogen exists mostly in a precipitated form, as numerous small clusters, with a low solubility C_{Be}. Due to out-diffusion of the dissolved species, their concentration C_B becomes smaller than C_{Be} in a near-surface region, which results in dissolution of the clusters. Eventually, the clusters disappear within some depth L that increases with time by a normal square law $t^{1/2}$. In this denuded zone $z < L$, the nitrogen concentration is remarkably low, less than C_{Be}. Deeper in the bulk, the total nitrogen concentration is mostly due to the clusters, and a sharp rise in $C(z)$ corresponds to an intermediate region of dissolving clusters. The S-like shape of part (A) clearly corresponds to the dissociation mechanism of diffusion of the A-component—a dominant, almost immobile, A-species dissociates into two minor mobile species. The simplest assignment of the A-component is to a larger species N_6, the B-component to N_2 transported through dissociation into N_1, and for the C-component N_4 or N_5.

Nitrogen in silicon strongly enhances oxygen precipitation, which improves the gettering potential of CZ silicon. It is believed that nitrogen-oxygen defects form stable nuclei at high temperatures where oxygen-only nuclei are unstable. It has been suggested that these are nitrogen-vacancy-oxygen defects. Fujita et al. [102] investigated the binding energy, equilibrium concentration, local vibrational modes, and electrical activity of N_2O, N_2O_2, NO, and NO_2 defects in CZ silicon by means of local density functional theory. The binding energy of oxygen with N_2 and the positions of four local vibrational modes of the N_2O center are in excellent agreement with the experiment. The N_2O_2 defect is the most common nitrogen-oxygen defect after N_2O, and it is suggested that the experimentally observed lines at 1018 and 810 cm^{-1} are due to this defect. The concentrations of these defects are greater than those of oxygen dimers at temperatures around 650°C and hence, these defects, and not nitrogen-oxygen-vacancy centers, could be nuclei for oxygen precipitates in nitrogen-doped CZ silicon. In contrast with N_2O and N_2O_2, it was found that NO and NO_2 defects have very low concentrations, and this raises questions about their assignment to nitrogen-related shallow thermal donors.

The nitrogen impurity is thought to be present in the di-interstitial (molecular) state N_2 at room temperature. Though it dissociates on raising T, producing atomic interstitials N_1, the first-principle calculations suggest that the di-interstitial state prevails even at T_m, the melting point of silicon. The calculated energies of various nitrogen/vacancy species—substitutional nitrogen (VN, a complex of a vacancy and one interstitial nitrogen atom), substitutional molecule (VN$_2$), and the V$_2$N$_2$ complex—are all produced from the basic molecular state N_2. The species concentrations, in equilibrium with the solid solution of nitrogen molecules, can be expressed by a mass-action law using the reported energies [55]. A change in the vibrational entropy upon formation of these species will be neglected, and the resulting numbers should be considered as rough estimates. The square brackets will be used to denote the atomic fraction of the species (the ratio of the concentration to the site density ρ).

For atomic nitrogen interstitials:

$$[N_1] = [N_2]^{1/2} \exp\left(\frac{-2.15 \text{ eV}}{kT}\right)$$

Formation of substitutional nitrogen VN for the equilibrium case ($C_v = C_{vc}$) can be performed without assisting vacancies (by removing a silicon atom to the crystal surface), and the mass-action law reads quite similarly to the expression with the formation energy of 1.74 eV. However, when this species is produced by trapping a vacancy, the concentration of VN is proportional to C_v and, therefore, includes the supersaturation factor C_v/C_{vc}:

$$[VN] = \left(\frac{C_V}{C_{VC}}\right)[N_2]^{1/2}\exp\left(\frac{-1.74\ eV}{kT}\right)$$

The concentrations of the other vacancy/nitrogen species are expressed similarly:

$$[VN_2] = \left(\frac{C_V}{C_{VC}}\right)[N_2]\exp\left(\frac{-2.33\ eV}{kT}\right)$$

$$[V_2N_2] = \left(\frac{C_V}{C_{VC}}\right)^2[N_2]\exp\left(\frac{-1.26\ eV}{kT}\right)$$

Assuming that for the total nitrogen concentration C a typical value of 1.0×10^{15} cm^{-3}, the calculated concentrations at the crystal melting point (when $C_v/C_{vc} = 1$) are all much lower than the concentration of the molecular form. The latter is thus close to $C/2 = 5.0 \times 10^{14}$ cm^{-3}. The concentrations of the other species, as defined by the mathematical expressions, are as follows (in units of cm^{-3}): atomic interstitials N_1: 1.9×10^{12}; substitutional species VN: 3.1×10^{13}; VN_2: 5.5×10^7; V_2N_2: 8.6×10^{10}. The formation energy is the lowest for V_2N_2.

The equilibrium concentration of the vacancy species at $C = 1.0 \times 10^{15}$ cm^{-3} is plotted in Figure 8.9. The binding temperature is close to 1115°C if the contribution of the V_2N_2 species is neglected [55]. Nitrogen in silicon is another important topic for engineers and is a major field of research at present [103–107]. The properties of crystals doped with nitrogen are still under investigation.

8.3.11 Oxygen

Oxygen is an important impurity occurring in silicon crystals, especially those grown by the CZ method. It is one of the main impurities that are unintentionally introduced

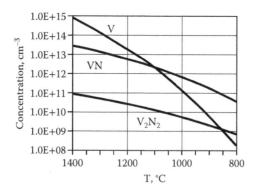

FIGURE 8.9
Equilibrium concentrations of vacancy species V, VN, and V_2N_2 in nitrogen-doped silicon, at the nitrogen concentration $C = 1.0 \times 10^{15}$ cm^{-3}. The VN_2 species is of too low a concentration to show in the figure. (From V. V. Voronkov and R. Falster, *Journal of the Electrochemical Society*, **149**, G167–G174, 2002, and the references therein [55].)

in silicon during crystal growth in silica (quartz) crucibles, and it enters the silicon melt via dissolution of the crucible. Most dissolved oxygen (over 95%) escapes from the free melt surface as silicon monoxide, and only a small amount (<5%) is incorporated into the growing crystal [108]. The oxygen level of silicon wafers currently being processed in the semiconductor industry ranges from 24–33 ppma. Oxygen concentration in silicon crystal is determined by the oxygen distribution in the melt, at the crystal-melt interface, and in the bulk melt. Those sources are affected, in turn, by the dissolution rate, transfer in the melt, via convection and diffusion, and the ratio (m) of the free melt surface to the melt in contact with the crucible wall. Dissolution of the SiO_2 crucible into the silicon melt is influenced by crucible wall temperature, crucible rotation rate, and fluid-flow instabilities in the melt. In the oxygen transportation mechanism, oxygen enters the silicon melt mainly from a silica crucible. The dissolution rate of the crucible by the melt depends on the temperature, adjacent melt flow rate, and surface condition of the crucible, and many other factors [109]. It is transported by diffusion, natural convection, and forced convection driven by crucible and crystal rotation. Most of the oxygen present in the melt reaches the free silicon liquid surface and evaporates as SiO. The rest is incorporated into the growing crystal according to a segregation mechanism. Figure 8.10 explains the incorporation of oxygen species dissolution into the silicon melt and its incorporation into the growing crystal. As discussed earlier, not all of the dissolved oxygen will enter into the crystal, and most of it will escape to the chamber environment. The escaped oxygen will be carried away by the high-purity argon purging gas.

The typical concentration of oxygen in silicon crystal varies from 1.0×10^{17} to 1.0×10^{18} cm^{-3}. Oxygen becomes a major device detractor when present at high concentrations in silicon, creating dislocations and sites for heavy metal precipitation. At concentrations above 1.0×10^{17} cm^{-3}, it causes precipitates of silicon-oxygen complexes when the crystal is subjected to high-temperature cycles. This is accompanied by the generation of mobile dislocations around the precipitates. In an interstitial position, oxygen forms a silicon-oxygen bond that exhibits a strong IR absorption band at 9.1 μm. This allows routine measurements of the oxygen content in silicon by IR spectroscopy [34]. Oxygen is of supreme importance in CZ silicon and is dealt with in greater detail in [111]. In many cases, this impurity plays a negative role because of its tendency to form precipitates in thermal process cycles used in the IC processing technology. The positive interest in oxygen precipitates arises from the possibility of using them as an easily introduced gettering factor when CZ-grown crystals are used for manufacturing VLSI devices [112]. The fundamental aspects of oxygen incorporation during the CZ technique are discussed by Benson and Lin [113], including the recent data obtained on the segregation coefficient. Oxygen precipitation in CZ silicon has garnered attention because of its importance in crystal and device technology. The precipitate nucleation in lightly and heavily doped CZ silicon was investigated by Sueoka [114] by using epitaxial wafers, and the data was recorded on the nature of oxygen species in silicon lattice sites. The radial dopant distribution was correlated with the interface shape and its radius of curvature [50]. No correlation was found between the interface shape and radial distribution of oxygen or carbon. As mentioned, the main source of oxygen is the poly-Si material. The polycrystalline sheets grown using the mold-shaping technique had an oxygen content of 1.7 ppm, a carbon content of 0.33 ppm, and lower levels of other contaminants. The surface area of the poly-Si may be a possible source for the larger concentration values of oxygen here.

Conceptually, the presence of oxygen in silicon wafers can be both beneficial and harmful. Interstitial oxygen is the main impurity in single crystals of silicon grown using the conventional CZ method. Oxygen in excess of the solubility limit can precipitate, with the subsequent formation of bulk stacking faults bounded by partial dislocations. Such defects, if penetrating the active device area, may prove harmful to junction characteristics

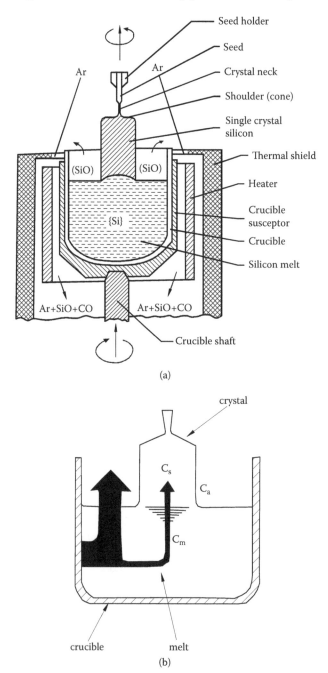

FIGURE 8.10
Oxygen gets into silicon crystals from (a) polysilicon (which is made out of silica, SiO_2) and the (b) crucible (CZ growth). (From R. Falster and V.V. Voronkov, "Lattice Defects in Silicon", Monsanto Electronic Materials Company (MEMC) website, 2006, and the references therein [110].)

and for the operation of devices. On the other hand, proper control over the positioning of the oxygen-induced defects can be potentially beneficial, since they can act as gettering sites [115] and may become a sink for many metallic impurities. Ideally, one might prefer to have the surface, and perhaps the first 20–30 μm of depth be void of bulk oxygen defects, while confining the bulk oxygen defects to the central depth position of the wafer. These can be controlled by proper processing. Properly positioned bulk oxygen defects can provide a beneficial gettering effect. Hence, when using silicon with a high oxygen content, an understanding and control over the distribution of oxygen in the depth is important. In addition, control must be exercised as to whether the oxygen is present in interstitial form or in a precipitated form [115]. Figure 8.11 shows the position of interstitial oxygen in the silicon lattice. The behavior is somewhat similar to the typical resonance structure we come across in some key bulk molecules. We shall discuss this more in Chapter 9. Measuring oxygen in silicon has been a major research topic over the last 30 years, and still draws attention of many scientific groups who are evaluating silicon wafers for the quantification of oxygen primarily by using IR absorption, photoactivation analysis, and spectroscopy measurements [116–130]. The behavior of oxygen and its position in the silicon crystal lattice is a subject of continued interest.

The formation of oxygen precipitates and their transformation in different thermal processes have been the subject of many research studies. In view of the small diameters of the formed microdefects, it is important for their structural characterization to use TEM, which is often accompanied by other methods. The subtle selective etching method and Fourier IR spectroscopy are effective in revealing small oxygen precipitates [112]. Recently, the effective use of laser scanning tomography has been introduced, which could reveal small precipitates. Thanks to the scattering of IR radiation.

Models for homogeneous nucleation of oxide precipitates in silicon are based on nucleation of incoherent SiO_x particles. However, these models provide feasible results only if the strain of the growing nuclei is relieved by the emission of interstitials or the absorption of vacancies. These models are in agreement with the well-known precipitation enhancement by vacancy supersaturation and the amorphous or cristobalite structure of precipitates

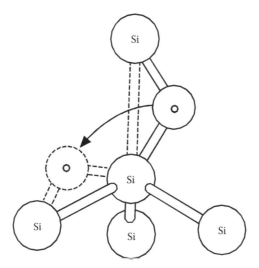

FIGURE 8.11
Position of interstitial oxygen in the silicon lattice. (From R. Falster and V.V. Voronkov, "Lattice Defects in Silicon", Monsanto Electronic Materials Company (MEMC) website, 2006, and the references therein [110].)

observed in various TEM investigations. A more advanced model treats the incorporation of vacancies and oxygen into incoherent precipitates independently. Another more advanced model of incoherent nucleation also incorporates the segregation of interstitials to the precipitate nuclei. It is assumed that part of the generated self-interstitials is incorporated into the growing oxide precipitate instead of being emitted into the host matrix. Nucleation of oxide precipitates in the silicon lattice usually takes place below 1000°C. This is a typical temperature range where the majority of vacancies are bound to oxygen by forming VO_2 complexes, as shown in Figure 8.12. Thus, the sequence of oxide precipitation according to the proposed advanced model by Kissinger and Dabrowski [131] would start with mono-layered seed-SiO. However, SiO is just a metastable phase, and in thermal equilibrium, SiO_2 exists in a matrix of silicon saturated with interstitial oxygen. The heterogeneous nucleation of amorphous SiO_2 or cristobalite precipitates at the seed-SiO plates is regarded as a possible path to reach the equilibrium. The heterogeneous nucleation rate would depend on the size of the seed-SiO plates. This model explains the nucleation of oxygen species in association with the vacancies and is supported by TEM observations. The large oxide precipitates are square-shaped platelets with about 500 Å to 2600 Å half-diagonal length.

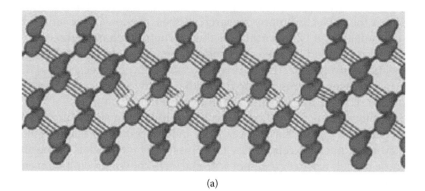

(a)

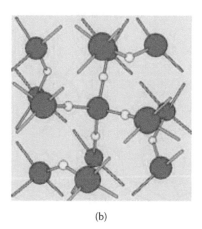

(b)

FIGURE 8.12
(a) Ball-and-stick models of seed-SiO and seed-SiO_2 shown here as a four-VO_2 cluster. (b) The larger spheres represent silicon atoms, and the smaller spheres are oxygen atoms. (From G. Kissinger and J. Dabrowski, "Oxide precipitate nucleation via coherent 'seed'-oxide phases," *210 ECE Meeting Cancun*, Mexico, Oct. 29–Nov. 3, 2006, and the references therein [131].)

The observed precipitate density is considerably higher than that expected for homogeneous nucleation at the temperature range of 1000 to 1250°C [132]. The super-cooling degree for the nucleation is about 50°C to 300°C and is smaller than that for homogeneous nucleation.

Thermally induced oxygen precipitation and redistribution phenomena in CZ-grown silicon crystals were investigated by means of TEM and differential-infrared absorption (DIR) by Shimura et al. [133]. CZ-grown silicon crystals generally contain interstitial oxygen in concentrations on the order of 1.0×10^{18} cm^{-3}. The radial distribution of the oxygen content was observed to be uniform overall, except for a rim of 5 mm in the wafers. When such crystals are subjected to heat treatments, a variety of microdefects are generated as a result of the precipitation of oxygen. Although oxygen precipitation in silicon has been the subject of considerable investigation modeling, it still is a subject of debate. In order to investigate the role of thermal history and the oxygen supersaturated condition for oxygen precipitation, a two-step annealing process needs to be adopted. Evident existence of SiO_2, in cristobalite form, was observed in as-grown crystals by DIR at 1225 cm^{-1}. As a result, it was ascertained that oxygen precipitation does not occur by homogeneous nucleation depending on the ratio of supersaturated oxygen, but occurs by heterogeneous nucleation.

Oxygen precipitates in samples of CZ-grown silicon crystal annealed at various temperatures from 750°C to 1100°C were studied by Wierzchowski et al. [112] with x-ray methods both with white and monochromatic synchrotron radiation, particularly Bragg-case section topography. The samples were additionally controlled by observation of a selective etching pattern, IR, Fourier spectroscopy, and laser scattering tomography (LST). The visibility of oxygen precipitates in synchrotron topographic methods was strongly dependent on annealing temperature and time. For annealing performed at 750°C, the effect was relatively weak and consisted of small irregularities in Pendellösung fringes. The back-reflection section topographs usually revealed fringes characteristic of bent crystals, with some irregularities caused by oxygen precipitates. The significant effect was observed with x-ray diffraction methods for 3 h annealing at 1100°C, when the individual precipitates were resolved in the topographs. In the case of longer combined annealing cycles, including annealing at 1050°C, an irresolvable quantity of defects and complete dampening of interference fringes were observed. The concentration of precipitates strongly increased along some striations, while in other parts of the crystal, the interference fringes were present.

The oxygen impurity in silicon is well known to exist in the interstitial state. The substitutional (off-center) state of oxygen or the vacancy/oxygen complex (VO, also known as A-center) is produced in irradiation experiments. By annealing the irradiated samples, the A-center disappears at $T > 350$°C, presumably by attaching more oxygen atoms and not by dissociation. The binding energy of a vacancy to an oxygen atom may be large enough for the VO species to be present in an appreciable amount, even at high T, in equilibrium with free vacancies. Evidence of high-temperature vacancy trapping by oxygen comes from the fact that, in the course of crystal growth, some vacancies escape being consumed by voids; these "residual vacancies" control subsequent oxygen precipitation, giving rise to a complicated pattern. This effect can be accounted for if the free vacancies become bound (trapped) by oxygen below some characteristic binding temperature T_b (estimated to be around 1050°C); otherwise, the vacancies would be completely consumed by the voids. A vacancy, due to its four dangling bonds, can absorb either one or two oxygen atoms, giving rise to VO and VO_2 species. It is assumed that the transformation reactions of V into VO (by vacancy diffusion) and of VO into VO_2 (by oxygen diffusion) are fast enough to support the equilibrium ratio between species concentrations C_v, C_1, and C_2 (of V, VO, and VO$_2$, respectively). It is also assumed that the free energy change, upon attachment of the first and the second oxygen atom to a vacancy, is the same. A qualitative argument in favor of such a simplification is that an oxygen atom in the A-center is close to two of the silicon neighbors, leaving the two other silicon neighbors in the same state as that in the free vacancy. With this assumption, the mass-action law for the previously mentioned reaction

$$\frac{C_1}{C_v} = \frac{C_2}{C_1} = AC \exp\left(\frac{E_b}{kT}\right)$$

where C is the oxygen concentration, the pre-factor A includes a change in the vibrational entropy due to the dangling bonds on attaching an oxygen atom to a vacancy. The binding temperature may be defined by the condition $C_1/C_v = C_2/C_1 = 1$. Below T_b, this concentration ratio becomes high, and then $C_2 \gg C_1 \gg C_v$, which means that the prevailing vacancy state is VO$_2$. At temperatures greater than T_b, this ratio becomes small, and $C_2 \ll C_1 \ll C_v$. This means that free vacancies prevail with some small fraction of VO species and a much smaller fraction of VO$_2$ species [55].

Figure 8.13 shows the typical equilibrium concentrations of vacancy species V, VO, and VO$_2$. They should be considered only a rough description of the actual relationship between the V, VO, and VO$_2$ species based on the assumptions made. And yet, the resulting value of the binding temperature is in reasonable agreement with a rough estimate of T_b based on the residual vacancy consideration. If E_v were assumed to be 4.6 eV, a too high value of T_b (about 1100°C) would be obtained. This is an argument in favor of a lower formation energy ($E_v = 4.0$ eV), which is adopted from now on [55].

A reduction in the amount of oxygen assimilated by the crystal when a magnetic field is applied during the crystal growth can be forecast [134]. Many attempts have been made to reduce the presence of oxygen species in the growing silicon. CZ silicon crystal growth in the presence of an axially symmetric cusp magnetic field was reported by Hirata and Hoshikawa [135]. The free surface of the melt is centered between two superconducting coils, and in this way, the oxygen concentration was successfully controlled from 1.0×10^{18} to 2.0×10^{17} atoms/cm^3 by increasing the cusp magnetic field strength up to 3500 Oe at the

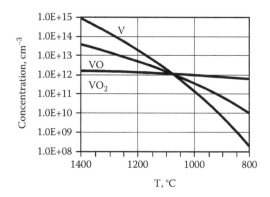

FIGURE 8.13
Equilibrium concentrations of vacancy species V, VO, and VO$_2$, at the oxygen concentration $C = 8.0 \times 10^{17}$ cm^{-3}. (From V. V. Voronkov and R. Falster, *Journal of the Electrochemical Society*, **149**, G167–G174, 2002, and the references therein [55].)

center of the bottom melt-silica crucible interface, while keeping the crystal rotation constant at 30 rpm and crucible rotation at −10 rpm. Both the oxygen and dopant concentrations were homogenized by the presence of a magnetic field. Series [136] further reported, through a series of experiments, on the effect of a shaped magnetic field on the CZ growth of silicon. By choosing the appropriate growth conditions, crystals can be grown with oxygen contents that vary over a wide range. The radial uniformity of both dopant and oxygen is comparable to that of crystals grown under the corresponding zero-field conditions. The magnetic field configuration is amenable to scaling the large diameters. The oxygen concentration in CZ crystals was increased and homogenized along the growth axes by applying an axially symmetric cusp magnetic field. The orthogonal components of the cusp magnetic field are independently changed on the melt-free surface and at the melt-crucible interface where oxygen evaporation and dissolution takes place, from the silicon melt. The concentration at a large solidified fraction was increased by slightly reducing the oxygen evaporation rate to 14.1×10^{17} atoms/cm^3, 35% more than that with no magnetic field [137].

Oxygen transport in magnetic CZ growth of silicon was studied by Kobayashi [138] to interpret the different kinds of behavior between the transverse and axial magnetic CZ on the oxygen content in grown silicon crystals. It was found that they were low in transverse-field CZ, but very high in axial-field grown crystals. The effect of a configured magnetic field on the uptake of oxygen by a CZ-grown silicon crystal was studied by Hicks et al. [139]. According to them, the amount of oxygen assimilated by the crystal is reduced as the magnetic-field strength increases, and a more uniform distribution within the crystal than what can be achieved with a uniform magnetic field was realized. According to Hjellming and Walker [140], the problem of oxygen concentration in the silicon crystal is intrinsically unsteady in magnetic CZ-grown silicon, and the oxygen concentration of a crystallizing particle depends on its trajectory since the beginning of crystal growth. The results provide a base solution for the mass transport problem with much weaker magnetic fields.

A new technique was proposed by Ekhult and Carlberg [53] to control the diameter of the growing crystal. This technique involved the use of fast, adjustable IR heaters directed toward the melt surface and facilitated growth at a constant pull rate. Using this method, two crystals were grown and compared to a reference crystal grown with the normal CZ technique. The axial oxygen distribution in the reference crystal could be understood reasonably well by applying a steady-state model of the dynamic oxygen equilibrium in the melt. The other two crystals showed a more even oxygen concentration than predicted, and it was clear that adjusting the power of the IR heaters was followed by variations that could not be accounted for by the steady-state model. By applying a new time-dependent model that emphasized the kinetics at the interfaces toward the crucible and the atmosphere, it became evident that dynamic oxygen equilibrium in the melt was never reached. The deviations from the steady-state model could be explained by this, as could the short-range oxygen variations. Kobayashi [141] simulated the oxygen transport under an axial magnetic field. The magnetic field not only damps the convection in the melt, but also acts to make the flow in the melt axially two-dimensional. There are two flow procedures, depending on the convection in the melt: a thermal convection dominant procedure, and a forced convection dominant procedure. When the thermal convection due to temperature inhomogeneity is dominant in the melt, the oxygen concentration incorporated into the crystal decreases as the magnetic field increases. When the forced convection due to counter-rotation between the crystal and the crucible is dominant, the oxygen concentration increases. The forced convection due to a single rotation of the crystal decreases the oxygen concentration in the growing crystal.

The amounts and distribution of oxygen, dopants, and other contaminants in CZ-grown silicon depend critically on motions in the melt during the crystal growth. The motions in the melt, induced mainly by temperature gradients and by crystal and crucible rotations, can be suppressed and the crystal quality can be improved by applying a magnetic field. The flow fields indicated that the oxygen incorporation into the crystal would be less for the growth in a cusp magnetic field than in an axial magnetic field [142]. Silicon with high concentrations of interstitial oxygen has been proposed using the FZ technique by Crouse et al. [143]. In this material, interstitial oxygen concentration [O] approximately equal to 9.0×10^{17} cm^{-3} was achieved while maintaining good crystalline structure and a ratio of oxygen to shallow impurities greater than 10^5. This method provides efficient oxygen "doping" by exposing only the melted zone during boule growth. This method is of fundamental importance in the investigation of thermal donors in silicon, where contamination from Group III and Group V impurities, as well as carbon, complicates the role of oxygen in the formation of thermal donors. Issues relating to thermal donors are discussed in Chapter 9.

Salnick [144] conducted an experimental study on magnetic field effects, including the effect of a weak field, on oxygen in 105–155 mm diameter crystals grown in a conventional CZ puller. The oxygen concentration and distribution in the crystal depend upon the conditions of oxygen-rich melt transport from the crucible wall to the melt-crystal interface. An area of low-oxygen (down to 2.0×10^{17} cm^{-3}) large-diameter crystal growth was found. The features of oxygen distribution in the crystals were qualitatively explained by the author. Oxygen precipitation occurs along growth striations in CZ-grown single crystals of silicon [145]. Interstitial oxygen (O_i) striations in various CZ and horizontal magnetic field CZ-grown (HMCZ) single crystals of silicon were studied with a micro-FTIR mapping system. These striations were found in most crystals. The O_i profiles were irregular, but their periods and heights were about 0.8 mm and 0.2×10^{17} atoms/cm^3, except for several cases. Subsequently, oxygen micro-precipitation was investigated by Fusegawa et al. [145] for various conditions of thermal treatment. It was found that homogeneous nucleation was inferred to be operative in both CZ and HMCZ methods.

The presence of oxygen precipitates (SiO_2) in a swirl-like pattern in CZ-grown silicon is often referred to as swirl defects, while no such SiO_2 precipitation is observed for FZ silicon. This leads to the question as to whether or not the swirls in CZ silicon act as nucleation centers for SiO_2 precipitates. The oxygen concentration distribution in silicon can be determined by spreading resistance measurements after the oxygen in the crystal is activated to donor levels by heat treatment at 450°C for 50 h [146]. It was found that for forced convection conditions of growth, rates of oxygen segregation are microscopically controlled. Maxima in oxygen concentration occurred at minima in the microscopic rates, indicating that the segregation coefficient is greater than unity. The distribution of swirl defects was found to be critically dependent on the microscopic growth-rate fluctuations; this dependence made it possible to show that the critical impurity nuclei are not related to oxygen, but to some impurity, most likely carbon, that has a segregation coefficient less than unity. It was finally concluded that the segregation coefficient of oxygen in silicon is greater than 1, which is consistent with earlier reports. A closer examination of the swirl defect distributions show that the swirl defect striations do not coincide with the dopant striations; instead, the swirl microdefects are located at the seed side of the dopant striations and thus, near microscopic growth-rate maxima.

Ma et al. [147] found that the density of oxygen precipitates in nitrogen-doped CZ (NCZ) silicon was slightly higher than that in CZ silicon during a multistep thermal process.

When pre-annealed by a thermal process, such as a rapid thermal process at 1250°C for 60 s, the density of oxygen precipitates in NCZ silicon was higher than that in CZ silicon, and the denuded zone was formed near the surface of the NCZ silicon, though it was not as wide as that of CZ silicon. In addition, when pre-annealed by sequential steps of 1250°C (for 60 s), 800°C (for 4 h), and 1000°C (for 16 h), the concentration of oxygen in both CZ silicon and NCZ silicon reached the solid solubility of oxygen.

In CZ single crystals of silicon, antimony doping decreases the oxygen concentration by enhancing the oxygen species evaporation from the melt surface. Argon ambient pressures of around 100 Torr over the silicon melt were found to suppress this, however. To clarify the effect of the growth chamber ambient pressure on oxygen concentration in grown crystals, heavily Sb-doped CZ silicon crystals were grown by Izunome et al. [148] under argon pressures of 30, 60, and 100 Torr. Increasing the argon pressure increases the oxygen and antimony concentrations at the melt surface. The oxygen concentration under argon pressure of 100 Torr was 1.2 times higher than that under 30 and 60 Torr when the solidified fractions were 0.5 or larger. The oxygen evaporation rate was controllable by gas-phase transport of Sb_2O at high argon pressures. Figure 8.14 shows the variation of the oxygen concentration along the growth direction. The oxygen concentration decreases in response to an increase in the solidified fraction due to the enhancement of oxide evaporation from the melt surface with increasing antimony concentration. The oxygen concentration under argon pressure of 100 Torr is 1.2 times as large as that under 30 and 60 Torr when the solidified fractions are 0.5 or larger. Thus, the oxygen concentrations in the crystals can clearly be controlled by adjusting the argon pressure in the chamber. Other reports [149–152] support the reduction of oxygen concentrations in antimony-doped crystals.

Oxygen distribution in the CZ single crystal of silicon has been reported to be closely associated with melt convection occurring near the growth interface because oxygen atoms in the melt are transported by fluid motion. By using the melt-quenching technique developed utilizing the double-layered CZ process, Kawanishi et al. [153] studied

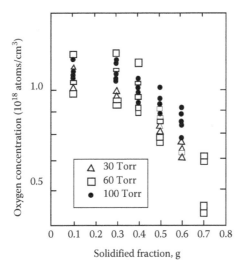

FIGURE 8.14
Variation of the oxygen concentration along the growth direction under various argon pressures. (From K. Izunome, X. Huang, S. Togawa, K. Terashima, and S. Kimura, *Journal of Crystal Growth*, **151**, 291–294, 1995, and the references therein [148].)

the oxygen transport phenomena around the growth interface. Micro-FTIR measurements showed that the oxygen concentration distribution in the growing crystal was related to melt convections, such as the Cochran flow and fluid motion from the melt surface. The details are shown in Figure 8.15. Here, the oxygen content measured at a depth of 2 mm from the growth interface was chosen for comparison with the typical oxygen concentration distribution in the crystal. The oxygen content in the melt was found to decrease gradually from 7.5×10^{17} to 3.0×10^{17} atoms/cm^3 as the radial distance approached the crystal edge. On the other hand, in the crystal itself, the radial oxygen distribution is uniform except at the edge. The radial oxygen distribution in the crystal is thus not necessarily consistent with that in the melt. This agrees with characteristic oxygen striation in both the crystal and melt observed using micro-FTIR and Wright etching. The authors concluded that melt quenching using the double-layer CZ process is effective in directly observing the oxygen concentration distribution in the melt during crystal growth.

As mentioned, the control of the behavior of oxygen in silicon is one of the most important challenges in VLSI and ULSI processing. Control of oxygen precipitation is essential for the development of internal gettering. Such gettering schemes play an important role in yield management. Since the discovery of the gettering effect, many scientists and engineers have struggled with the problem of precisely and reliably controlling the precipitation of oxygen in silicon. This has met with only partial success, in the sense that the defect engineering of conventional silicon wafers is still, by and large, an empirical exercise. It consists largely of careful, empirical tailoring of wafer type, oxygen concentration, crystal growth method, and details of pre-heat treatments to match the specific process details of the application in order to achieve good, reliable performance. Reliable and efficient gettering requires the robust formation of oxygen-precipitate-free surface regions and a bulk defective layer consisting of a minimum density (at least about 10^8 cm^{-3}) of oxygen precipitates during the wafer processing [154]. Uncontrolled precipitation of oxygen in the near-surface region of the wafer represents a risk to device yield. The basis

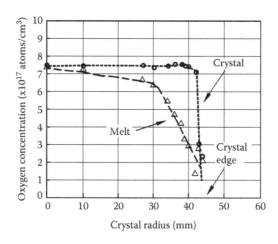

FIGURE 8.15
Typical radial oxygen distribution in the crystal and melt. The typical oxygen distribution is shown for the crystal, and the oxygen content measured 2 mm deep from the growth interface is indicated for the melt. (From S. Kawanishi, S. Togawa, K. Izunome, K. Terashima, and S. Kimura, *Journal of Crystal Growth*, **152**, 266–273, 1995, and the references therein [153].)

of the conventional method for dealing with the creation of such a layered structure has been to ensure sufficient out-diffusion of oxygen from the near-surface region. In recent years, due to radical reductions in the total thermal budgets of processes, this is no longer possible.

Ever since it was recognized that oxygen precipitates could act as gettering sites for fast diffusing and harmful transition metal contamination, their use has played an important role in contamination management schemes throughout the integrated circuit industry. It turned out to be quite a difficult problem from a practical perspective. The gettering part is largely easy, but the difficult part lies in the precise control of oxygen precipitation behavior without harmful side effects [155]. Many approaches have been developed over the years to manage and achieve the desired gettering effects. These include wafer pretreatments. Other approaches have included attempts to narrowly specify oxygen concentration in the crystal growth process or even the segments of the crystal from which wafers should be taken for specific application.

Changing crucible and crystal rotation rates is commonly used to control oxygen concentration, resulting in the following: (1) oxygen concentration increases as the crucible rotation rate increases; (2) oxygen concentration increases as the crystal rotation rate increases (in the absence of a horizontal magnetic field); and (3) as the crystal rotation increases, radial distribution of oxygen concentration becomes flatter [109]. Figure 8.16 shows the oxygen concentration in the crystal measured by FTIR using the old American Society for Testing and Materials (ASTM) conversion-coefficient value. Although the oxygen concentration decreased at 300 mm because of a change in crucible rotational speed from 8 to 6 rpm, the fluctuation of oxygen was 5.2% while the raw materials were charged. Before and after the material feeding, the oxygen concentration became lower because the contact area between the molten silicon and the crucible narrowed as the melt was decreased. According to Shiraishi et al. [156], oxygen in the silicon crystal can be easily controlled by adjusting the amount of silicon initially charged in the crucible.

Kobayashi [157] proposed a model to describe oxygen precipitation in CZ silicon. The model includes the nucleation process of oxygen precipitates and the interaction of intrinsic point defects in silicon with the precipitates. This is the first approach to model oxygen precipitation in CZ silicon. It is assumed that the shape of the precipitate is spherical and that the growth process is diffusion limited. This is a major impurity, with a typical concentration on the order of 10^{18} at/cm^3, and it is precipitated as SiO_2 during the heat

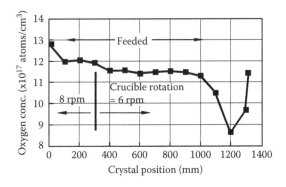

FIGURE 8.16
Oxygen concentration along the growth direction of a 6-inch crystal. (From Y. Shiraishi, S. Kurosaka, and M. Imai, *Journal of Crystal Growth*, **166**, 685–688, 1996, and the references therein [156].)

treatments. Oxygen will be precipitated during the crystal growth, and is supersaturated in the growing crystal, except for in the vicinity of the melt-crystal interface. The effects of the pulling rate on oxygen precipitation in a growing crystal are discussed on the basis of the proposed model. The calculated results agree with the observed data quantitatively in size, but only qualitatively in the density of oxygen precipitate. Figure 8.17 shows (a) the average size and (b) total density of precipitates for R (radius of precipitate) ≥ 1 nm for two levels of the pulling rates. As the pulling rate increases, the size decreases but the density increases. The calculated sizes are about 100 nm for 0.5 mm/min and 80 nm for 1.0 mm/min. The calculated densities are reported to be on the order of 10^8 cm^{-3}. Except for the size, the calculated results of oxygen precipitation show large discrepancies from the observed data on grown-in defects. It is also shown that voids scarcely form from bare intrinsic point defects above 1200 K [158]. Although the discrepancies between the oxygen precipitation model and the grown-in defects can be reduced by selecting data on the interface energy between SiO and SiO$_2$, as the literature has shown, the relief mechanism of stress energy of oxygen precipitates has not yet been clarified. Three possibilities for a stress relief mechanism have been discussed: (1) direct void formation from oxygen precipitate, (2) void formation through intrinsic stacking fault, and (3) composition change of oxygen precipitate between SiO and SiO$_2$. It is suggested that (1) the grown-in defects will be different in composition from oxygen precipitates grown by usual wafer annealing, even if those are oxygen precipitates; and (2) the mechanism of void formation will be related to oxygen precipitation, but is not a simple process even if the grown-in defects are voids. The kinetics of composition change, as well as the formation path of voids, is subject to future study.

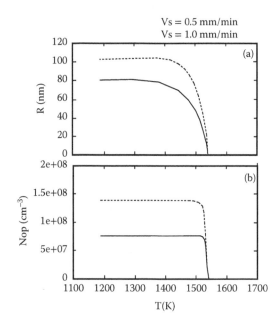

FIGURE 8.17
The effects of the pulling rate on (a) the average size of oxygen precipitates ($R \geq 1$ nm) and (b) the density of oxygen precipitates ($R \geq 1$ nm). (From S. Kobayashi, *Journal of Crystal Growth*, **174**, 163–169, 1997, and the references therein [157].)

With the introduction of larger-diameter CZ-grown silicon crystals, the oxygen content becomes an increasing concern, due to its greater incorporation into the crystal during growth and subsequent modification of defect formation during processing. The studies include saucer or shallow etch pits, OISFs, bulk stacking faults, epitaxial stacking faults, and precipitates. The interstitial oxygen content can be controlled simply by selecting those crystals or wafers with the lowest oxygen content and then heat-treating the wafers to lower the interstitial oxygen content near the wafer surface, or through crystal growth modifications. Katz and Hill [115] have observed that with large-diameter CZ-grown silicon crystals, there is an increase in the oxygen content. Oxygen in excess of the solubility limit will precipitate, resulting in bulk crystal defects that can modify the gettering of impurities known to degrade the device's electrical characteristics. They investigated different crystal growth parameters and related them to the properties of the crystals duly processed under various conditions. Defect formation was evaluated as a function of processing and related to the crystal properties. By growing crystals at high seed-rotation rates so as to maximize the oxygen content, they were able to suppress etch pit formation and epitaxial stacking fault formation. The crystals with a high oxygen content behaved in a manner similar to other size silicon crystals with respect to defect formation.

Many techniques have been developed over the years to control the axial and radial gradients of oxygen in CZ silicon. Several approaches apply a magnetic field during the basic CZ process. Various magnetic CZ (MCZ) techniques have been developed and are being used effectively to control oxygen in 200 mm silicon. Control of the oxygen level, uniformity, and precipitation behavior in CZ silicon has been and will continue to be a challenge in the face of ever-increasing crystal diameters, from 200 mm today to 300–400 mm in the near future. Optimization of crystal growth parameters using such approaches as the MCZ technique has been largely effective in meeting the challenge so far; but, with the larger crystal-growth systems (with 150–600 kg melts in 24–36-inch crucibles) required for larger wafers, the problems will persist [108]. Efforts to control oxygen segregation have led to remarkable production capabilities in the silicon wafer industry. With today's 200 mm silicon, typical specifications at ppma can be routinely met, with a radial gradient <5%. Lower ranges are becoming possible, though usually at great difficulty and/or cost. With 300–400 mm crystals, it is expected that similar specifications will eventually be met by the crystal growing industry.

Thus, oxygen precipitation will continue to be a major topic of research, as it has been for the last 35 years. Many scientific teams are evaluating key parameters in this field, and many interesting and important observations [159–182] on the dual role oxygen plays in the silicon lattice have been made. Many models have been proposed to analyze the importance of oxygen precipitation for MOS devices, particularly with respect to memory cells [183–186]. The affinity of oxygen to silicon is so important and at the same time plays a crucial role in device performance; thus, it will remain a focus of investigations in the coming years.

8.3.12 Tin

Doping silicon with a Group IV isovalent impurity like tin offers a unique flexibility in terms of modifying the mechanical lattice properties of the material without affecting the doping density. Tin has a covalent radius of 1.40 Å, compared with 1.17 Å for silicon, and can be incorporated into substitutional lattice sites at a large concentration. The solid solubility in silicon is ~5.0 × 10^{19} cm^{-3} at 1000°C [187]. Consequently, tin acts as a selective vacancy trap, while, in contrast, not affecting interstitial reactions. This leads to a reduced formation of oxygen thermal donors in *n*-type silicon and lowers the concentration of vacancy-oxygen and divacancy centers in irradiated material. Tin-doped silicon is, in many respects, an interesting tool for studying the impact of native point defects on impurity diffusion and defect formation processes. However, the occurrence of specific deep-level centers could prohibit its practical usefulness as a selective vacancy trap. It is anticipated that its major field of application lies in mechanical stress compensation and associated bandgap engineering. The performance parameters of devices fabricated with tin-doped silicon remain to be seen.

8.4   Other Metallic Impurities

We could cover only a few metallic impurities present in the silicon single crystals. There are many metallic impurities, which form as a part of crystals grown. With silicon atomic density at 5.0 × 10^{22} atoms/cm^3 there are some impurities, which are at ppb levels and it is difficult to quantify them in silicon. Very sophisticated instrumentations are required for this purpose. Some may get into the silicon depending on the selection of raw material used for getting metallurgical grade silicon, from quartz crucible container, graphite supporter and the gases, one uses to purge the environment in the crystal chamber during the growth process.

The effect of impurity doping, in the CZ-grown crystals, with electrically active impurities (such as boron, gallium, antimony, phosphorus, and arsenic) and isoelectronic impurities, such as tin, on the formation of swirl defects were investigated by means of preferential etching, copper and lithium decoration, X-ray transmission topography, and EBIC-mode SEM [9].

It is reported that vanadium, titanium, and zirconium concentration values in the silicon will be present at a concentration levels of 10^{12} per cm^3. These are unintentional impurities, but influence the performance in solar cells. A systematic study of this impurity has been carried out by Pizzini et al. [57] on the diffusion length of minority carriers as a function of the impurity concentration and micro structural features like grain boundaries (GB), dislocations, twins, and stacking faults. The influence is systematically recorded in each case.

Impurities can never be removed completely; their presence is always there. However, gettering methods can be used to either nullify their activity or drive them away from the active regions where VLSI and ULSI circuits are fabricated [188–194]. We shall discuss this issue in more detail in Chapters 9 and 10.

8.5   Summary

In this chapter, we studied the effect of intentional and unintentional impurities on silicon crystals. Details regarding aluminum, antimony, arsenic, boron, gallium, and phosphorus were studied, as these are the intentional impurities used to alter the electrical properties of silicon. These impurities are added to the molten silicon during the growth process. However, other impurities are present in the silicon, and they either occupy the interstitial positions or will try to occupy the lattice defects and become a part of the crystal, disturbing the very nature of the crystal properties. With large defects or at defect cluster points, they precipitate and form an even larger defect and provide extra charge-leakage paths between different circuit locations. This disturbs the circuit functionality and will lead to various reliability issues. We discussed the unintentional impurities, such as carbon, chromium, copper, germanium, gold, helium, hydrogen, iron, nickel, nitrogen, oxygen, and tin. The composition of these impurities is difficult to guess, as it depends on the basic raw material selected to prepare metallurgical-grade silicon.

References

1. W. Keller, "Experimental influence of some growth parameters upon the shape of the melt interfaces and the radial phosphorus distribution during float-zone growth of silicon single crystals," *Journal of Crystal Growth*, **36**, 215–231, 1976.
2. M. Nakamura, T. Kato, and N. Oi, "A study of the gettering effect of metallic impurities in silicon," *Japanese Journal of Applied Physics*, **7**, 512–519, 1968.
3. J. H. Thomas III, R. V. D'Aiello, and P. H. Robinson, "A scanning Auger electron spectroscopic study of particulate defects in metallurgical-grade silicon," *Journal of the Electrochemical Society*, **131**, 196–200, 1984.
4. K. Harada, H. Tanaka, J. Matsubara, Y. Shimanuki, and H. Furuya, "Origins of metal impurities in single-crystal Czochralski silicon," *Journal of Crystal Growth*, **154**, 47–53, 1995, and the references therein.
5. P. Alpern, W. Bergholz, and R. Kakoschke, "Detection of fast diffusing metal impurities in silicon by haze test and by modulated optical reflectance: A comparison," *Journal of the Electrochemical Society*, **136**, 3841–3848, 1989.
6. O. J. Anttila, M. V. Tilli, M. Schaekers, and C. L. Claeya, "Effect of chemicals on metal contamination on silicon wafers," *Journal of the Electrochemical Society*, **139**, 1180–1185, 1992.
7. L. Jastrzebski, R. Soydan, H. Elabd, W. Henry, and E. Savoye, "The effect of heavy metal contamination on defects in CCD imagers: Contamination monitoring by surface photovoltage," *Journal of the Electrochemical Society*, **137**, 242–249, 1990.
8. M. Morin, M. Koyanagi, J. M. Friedt, and M. Hirose, "Metal impurity evaluation in silane gas from the qualification of poly-Si layers," *Journal of the Electrochemical Society*, **141**, 274–277, 1994.
9. A. J. R. de Kock and W. M. van de Wijgert, "The effect of doping on the formation of swirl defects in dislocation-free Czochralski-grown silicon crystals," *Journal of Crystal Growth*, **49**, 718–734, 1980.
10. K. Baur, J. Kerner, S. Brennan, A. Singh, and P. Pianetta, "Aluminum impurities in silicon: Investigation of x-ray Raman scattering in total reflection x-ray fluorescence spectroscopy," Stanford Synchrotron Radiation Laboratory, Stanford University, Stanford CA 94309, USA, www-project.slac.stanford.edu/ssrltxrf/publications/al.pdf.
11. M. Tajima, T. Masui, D. Itoh, and T. Nishino, "Calibration of the photoluminescence method for determining As and Al concentrations in Si," *Journal of the Electrochemical Society*, **137**, 3544–3551, 1990.
12. I I. Kodera, "Precipitation of antimony in heavily doped silicon identified by x-ray microanalyzer," *Japanese Journal of Applied Physics*, **2**, 193–194, 1963.
13. Z. Liu and T. Carlberg, "On the mechanism of oxygen content reduction by antimony doping of Czochralski silicon melts," *Journal of the Electrochemical Society*, **138**, 1488–1492, 1991.

14. A. R. Comeau, "Improvement to bulk oxygen precipitate density necessary for internal gettering in antimony doped n/n+ epitaxial wafers using the 3-step technique and ramped nucleation," *Journal of the Electrochemical Society*, **139**, 1455–1463, 1992.

15. C. Liu, H. Wang, Y. Li, Q. Wang, B. Ren, Y. Xu, and D. Que, "Study on the oxygen concentration reduction in heavily Sb-doped silicon," *Journal of Crystal Growth*, **196**, 111–114, 1999.

16. D. Yang, C. Li, M. Luo, J. Xu, and D. Que, "Reduction of oxygen during the crystal growth in heavily antimony-doped Czochralski silicon," *Journal of Crystal Growth*, **256**, 261–265, 2003.

17. D. Nobili, A. Carabelas, G. Celotti, and S. Solmi, "Precipitation as the phenomenon responsible for the electrically inactive arsenic in silicon," *Journal of the Electrochemical Society*, **130**, 922–928, 1983.

18. B. Wang, X. Zhang, X. Ma, and D. Yang, "Effects of high temperature rapid thermal processing on oxygen precipitation in heavily arsenic-doped Czochralski silicon," *Journal of Crystal Growth*, **318**, 183–186, 2011.

19. M. Suhren, D. Gräf, U. Lambert, and P. Wagner, "Crystal defects in highly boron-doped silicon," *Journal of the Electrochemical Society*, **144**, 4041–4044, 1997.

20. Y. Kamiura, F. Hashimoto, and M. Yoneta, "Effects of heat treatments on electrical properties of boron-doped silicon crystals," *Journal of the Electrochemical Society*, **137**, 3642–3647, 1990.

21. T. Ono, E. Asayama, H. Horie, M. Hourai, K. Sueoka, H. Tsuya, and G.A. Rozgonyi, "Effect of heavy boron doping on oxide precipitate growth in Czochralski silicon," *Journal of the Electrochemical Society*, **146**, 2239–2244, 1999.

22. T. Ono, A. Romanowski, E. Asayama, H. Horie, K. Sueoka, H. Tsuya, and G.A. Rozgonyi, "Oxide precipitate-induced dislocation generation in heavily boron-doped Czochralski silicon," *Journal of the Electrochemical Society*, **146**, 3461–3465, 1999.

23. I. Yonenaga, T. Taishi, X. Huang, and K. Hoshikawa, "Dislocation-impurity interaction in Czochralski-grown Si heavily doped with B and Ge," *Journal of Crystal Growth*, **275**, e501–e505, 2005.

24. R. Gotoh, M. Arivanandhan, K. Fujiwara, and S. Uda, "Ga segregation during Czochralski-Si crystal growth with Ge codoping," *Journal of Crystal Growth*, **312**, 2865–2870, 2010.

25. M. Arivanandhan, R. Gotoh, K. Fujiwara, T. Ozawa, Y. Hayakawa, and S. Uda, "The impact of Ge co-doping on grown-in O precipitates in Ga-doped Czochralski-silicon," *Journal of Crystal Growth*, **321**, 24–28, 2011.

26. F. Mousty, P. Ostoja, and L. Passari, "Relationship between resistivity and phosphorus concentration in silicon," *Journal of Applied Physics*, **45**, 4576–4580, 1974.

27. G. Baccarani and P. Ostoja, "Electron mobility empirically related to the phosphorus concentration in silicon," *Solid-State Electronics*, **18**, 579–580, 1975.

28. G. Masetti, M. Severi, and S. Solmi, "Modeling of carrier mobility against carrier concentration in arsenic-, phosphorus-, and boron-doped silicon," *IEEE Transactions on Electron Devices*, **ED-30**, 764–765, 1983.

29. S. P. Murarka, "A study of the phosphorus gettering of gold in silicon by use of neutron activation analysis," *Journal of the Electrochemical Society*, **123**, 765–767, 1976.

30. R. W. Series, "Czochralski growth of silicon under an axial magnetic field," *Journal of Crystal Growth*, **97**, 85–91, 1989.

31. C. W. Pearce, "Crystal growth and wafer preparation" in *VLSI Technology*, edited by S. M. Sze, McGraw-Hill, New York, 1988, pp. 9–54, and the references therein.

32. A. A. Istratov, T. Buonassisi, M. D. Pickett, M. Heuer, and E. R. Weber, "Control of metal impurities in 'dirty' multicrystalline silicon for solar cells," *Materials Science and Engineering B*, **134**, 282–286, 2006, and the references therein.

33. I. Hide, T. Yokoyama, T. Matsuyama, K. Sawaya, M. Suzuki, M. Sasaki, and Y. Maeda, "Mould-shaping silicon crystal growth with a mould coating material by the spinning method," *Journal of Crystal Growth*, **79**, 583–589, 1986.

34. B. El-Kareh, *Fundamentals of Semiconductor Processing Technology*, Kluwer Academic Publishers, Boston, 1995.

35. D. W. Vidrine, "Room temperature carbon and oxygen determination in single-crystal silicon," *Analytical Chemistry*, **52**, 92–96, 1980.

36. F. Schmid, C. P. Khattak, T. G. Digges, Jr., and L. Kaufman, "Origin of SiC impurities in silicon crystals grown from the melt in vacuum," *Journal of the Electrochemical Society*, **126**, 935–938, 1979.
37. C. Gross, G. Gaetano, T. N. Tucker, and J. A. Baker, "Comparison of infrared and activation analysis results in determining the oxygen and carbon content in silicon," *Journal of the Electrochemical Society*, **119**, 926–929, 1972.
38. J. L. Regolini, J. P. Stoquert, C. Ganter, and P. Siffert, "Determination of the conversion factor for infrared measurements of carbon in silicon," *Journal of the Electrochemical Society*, **133**, 2165–2168, 1986.
39. L. L. Hwang, J. Bucci, and R. McCormick, "Measurement of carbon concentration in polycrystalline silicon using FTIR," *Journal of the Electrochemical Society*, **138**, 576–581, 1991.
40. H. Saito and H. Shirai, "Baseline variations near the carbon impurity vibration band in infrared spectra of silicon," *Journal of the Electrochemical Society*, **147**, 1210–1212, 2000.
41. H. Ch. Alt, Y. Gomeniuk, B. Wiedemann, and H. Riemann, "Method to determine carbon in silicon by infrared absorption spectroscopy," *Journal of the Electrochemical Society*, **150**, G498–G501, 2003.
42. J. Lu and G. Rozgonyi, "Oxygen and carbon precipitation in crystalline sheet silicon: Depth profiling by infrared spectroscopy and preferential defect etching," *Journal of the Electrochemical Society*, **153**, G986–G991, 2006.
43. B. Gao and K. Kakimoto, "Global simulation of coupled carbon and oxygen transport in a Czochralski furnace for silicon crystal growth," *Journal of Crystal Growth*, **312**, 2972–2976, 2010.
44. L. Raabe, O. Pätzold, I. Kupka, J. Ehrig, S. Würzner, and M. Stelter, "The effect of graphite components and crucible coating on the behavior of carbon and oxygen in multicrystalline silicon," *Journal of Crystal Growth*, **318**, 234–238, 2011.
45. M. Nakamura, E. Kitamura, Y. Misawa, T. Suzuki, S. Nagai, and H. Sunaga, "Photoluminescence measurement of carbon in silicon crystals irradiated with high energy electrons," *Journal of the Electrochemical Society*, **141**, 3576–3580, 1994.
46. M. Nukamura, "Influence of oxygen and carbon on the formation of the 1.014 eV photoluminescence copper center in silicon crystal," *Journal of the Electrochemical Society*, **147**, 796–798, 2000.
47. S. Lazanu and I. Lazanu, "Role of oxygen and carbon impurities in the radiation resistance of silicon detectors," *Journal of Optoelectronics and Advanced Materials*, **5**, 647–652, 2003.
48. P. J. Roksnoer, "The mechanism of formation of microdefects in silicon," *Journal of Crystal Growth*, **68**, 596–612, 1984.
49. N. Inoue, S. Yamazaki, Y. Goto, T. Kushida, and T. Sugiyama, "Infrared absorption spectroscopy of complexes in low carbon concentration, low electron dosage irradiated CZ silicon crystal," *210 ECE Meeting Cancun*, Mexico, Oct. 29–Nov. 3, 2006.
50. H. M. Liaw, "Interface shape and radial distribution of impurities in <111> silicon crystals," *Journal of Crystal Growth*, **67**, 261–270, 1984.
51. R. W. Series and K. G. Barraclough, "Carbon contamination during growth of Czochralski silicon," *Journal of Crystal Growth*, **60**, 212–218, 1982.
52. R. W. Series and K. G. Barraclough, "Control of carbon in Czochralski silicon crystals," *Journal of Crystal Growth*, **63**, 219–221, 1983.
53. U. Ekhult and T. Carlberg, "Oxygen and carbon incorporation during infrared assisted Czochralski growth of silicon crystals," *Journal of Crystal Growth*, **98**, 801–809, 1989.
54. T. Nozaki, Y. Yatsurugi, and N. Akiyama, "Concentration and behavior of carbon in semiconductor silicon," *Journal of the Electrochemical Society*, **117**, 1566–1568, 1970.
55. V. V. Voronkov and R. Falster, "Intrinsic point defects and impurities in silicon crystal growth," *Journal of the Electrochemical Society*, **149**, G167–G174, 2002, and the references therein.
56. T. Motooka, "Molecular dynamics simulations of crystal growth from melted silicon," *Proceedings of the First Symposium on Atomic-scale Surface and Interface Dynamics*, March 13–14, 1997, Tokyo, and the references therein.
57. S. Pizzini, L. Bigoni, M. Beghi, and C. Chemelli, "On the effect of impurities on the photovoltaic behavior of solar grade silicon: II. Influence of titanium, vanadium, chromium, iron, and zirconium on photovoltaic behavior of polycrystalline solar cells," *Journal of the Electrochemical Society*, **133**, 2363–2373, 1986.

58. N-E. Chabane-Sari, S. Krieger-Kaddour, C. Vinante, M. Berenguer, and D. Barbier, "A deep-level transient spectroscopy study of the internal gettering of Cr in Czochralski-grown silicon: Efficiency and reversibility upon lamp pulse annealing," *Journal of the Electrochemical Society*, **139**, 2900–2904, 1992.

59. S. Krieger-Kaddour, N-E. Chabane-Sari, and D. Barbier, "Transmission electron microscopic study of the morphology of oxygen precipitates and of chromium precipitation during intrinsic gettering in Czochralski-grown silicon: Influence of lamp pulse annealing," *Journal of the Electrochemical Society*, **140**, 495–500, 1993.

60. B. Hacki, K.-J. Range, P. Stallhofer, and L. Fabry, "Correlation between DLTS and TRXFA measurements of copper and iron contaminations in FZ and CZ silicon wafers; Application to gettering efficiencies," *Journal of the Electrochemical Society*, **139**, 1495–1498, 1992.

61. R. L. Meek, T. E. Seidel, and A. G. Cullis, "Diffusion gettering of Au and Cu in silicon," *Journal of the Electrochemical Society*, **122**, 786–796, 1975.

62. R. Sachdeva, A. A. Istratov, and E. R. Weber, "Recombination activity of copper in silicon," *Applied Physics Letters*, **79**, 2937–2939, 2001.

63. T. G. Digges, Jr. and R. Shima, "The effect of growth rate, diameter and impurity concentration on structure in Czochralski silicon crystal growth," *Journal of Crystal Growth*, **50**, 865–869, 1980.

64. G. H. Schwuttke, "Study of copper precipitation behavior in silicon single crystals," *Journal of the Electrochemical Society*, **108**, 163–167, 1961.

65. D. A. Ramappa and W. B. Henley, "Effects of copper contamination in silicon on thin oxide breakdown," *Journal of the Electrochemical Society*, **146**, 2258–2260, 1999, and the references therein.

66. Y. H. Lin, Y. C. Chen, K. T. Chan, F. M. Pan, I. J. Hsieh, and A. Chin, "The strong degradation of 30 Å gate oxide integrity contaminated by copper," *Journal of the Electrochemical Society*, **148**, F73–F76, 2001.

67. A. A. Istratov and E. R. Weber, "Physics of copper in silicon," *Journal of the Electrochemical Society*, **149**, G21–G30, 2002.

68. S. Isomae, H. Ishida, T. Itoga, and K. Hozawa, "Intrinsic gettering of copper in silicon wafers," *Journal of the Electrochemical Society*, **149**, G343–G347, 2002.

69. K. Sueoka, "Modeling of internal gettering of nickel and copper by oxide precipitates in Czochralski-Si wafers," *Journal of the Electrochemical Society*, **152**, G731–G735, 2005.

70. L. Fabry, R. Hoelzl, A. Andrukhiv, K. Matsumoto, J. Qiu, S. Koveshnikov, M. Goldstein, A. Grabau, H. Horie, and R. Takeda, "Test methods for measuring bulk copper and nickel in heavily doped p-type silicon wafers," *Journal of the Electrochemical Society*, **153**, G566–G571, 2006.

71. K.-S. Kim, S.-W. Lee, H.-B. Kang, B.-Y. Lee, and S.-M. Park, "Quantitative evaluation of gettering efficiencies below 1×10^{12} atoms/cm^3 in p-type silicon using a ^{65}Cu tracer," *Journal of the Electrochemical Society*, **155**, H912–H917, 2008.

72. W. Wang, D. Yang, X. Ma, and D. Que, "Copper precipitation in nitrogen-doped Czochralski silicon," *Journal of Applied Physics*, **104**, 013508, 2008.

73. D. Yang, "Defects in germanium-doped Czochralski silicon," *Physica Status Solidi (a)*, **202**, 931–938, 2005.

74. J. Chen, D. Yang, H. Li, X. Ma, D. Tian, L. Li, and D. Que, "Crystal-originated particles in germanium-doped Czochralski silicon crystal," *Journal of Crystal Growth*, **306**, 262–268, 2007.

75. I. Yonenaga, T. Taishi, X. Huang, and K. Hoshikawa, "Dislocation-impurity interaction in Czochralski-grown Si heavily doped with B and Ge," *Journal of Crystal Growth*, **275**, e501–e505, 2005.

76. R. Gotoh, M. Arivanandhan, K. Fujiwara, and S. Uda, "Ga segregation during Czochralski-Si crystal growth with Ge codoping," *Journal of Crystal Growth*, **312**, 2865–2870, 2010.

77. M. Arivanandhan, R. Gotoh, K. Fujiwara, T. Ozawa, Y. Hayakawa, and S. Uda, "The impact of Ge codoping on grown-in O precipitates in Ga-doped Czochralski-silicon," *Journal of Crystal Growth*, **321**, 24–28, 2011.

78. P. Wang, X. Yu, Z. Li, and D. Yang, "Improved fracture strength of multicrystalline silicon by germanium doping," *Journal of Crystal Growth*, **318**, 230–233, 2011.

79. R. C. Dorward and J. S. Kirkaldy, "Solubility of gold in p-type silicon," *Journal of the Electrochemical Society*, **116**, 1284–1285, 1969.

80. L. Forbes, "Gold in silicon: Characterisation and infrared detector applications," Department of Electrical Engineering, University of California at Davis, Davis, California, USA, http://download.springer.com/static/pdf/982/art%253A10.1007%252FBF03215429.pdf?auth66 = 1395127873_7d04c 738856f231 ee99ebcc12d536be1&ext =.pdf.

81. G. F. Cerofolini, F. Corni, S. Frabboni, C. Nobili, G. Ottaviani, and R. Tonini, "Hydrogen and helium bubbles in silicon," *Materials Science and Engineering*, **R27**, 1–52, 2000, and the references therein.

82. A. Chantre, L. Bouchet, and E. Andre, "On the hydrogen content of commercial silicon wafers," *Journal of the Electrochemical Society*, **135**, 2867–2869, 1988.

83. H. Abe, I. Suzuki, and H. Koya, "The effect of hydrogen annealing on oxygen precipitation behavior and gate oxide integrity in Czochralski Si wafers," *Journal of the Electrochemical Society*, **144**, 306–311, 1997.

84. J.-i. Chikawa and S. Shirai, "Melting of silicon crystals and a possible origin of swirl defects," *Journal of Crystal Growth*, **39**, 328–340, 1977.

85. K. Sueoka, M. Akatsuka, H. Katahama, and N. Adachi, "Investigation of the mechanical strength of hydrogen-annealed Czochralski silicon wafers," *Journal of the Electrochemical Society*, **146**, 364–366, 1999.

86. E. Simoen, Y. L. Huang, Y. Ma, J. Lauwaert, P. Clauws, J.M. Rafi, A. Ulyashin, and C. Claeys, "What do we know about hydrogen-induced thermal donors in silicon?," *Journal of the Electrochemical Society*, **156**, H434–H442, 2009.

87. B. Sopori, "Impurities and defects in photovoltaic Si devices: A review," National Renewable Energy Laboratory, 1617 Cole Boulevard, Golden, Colorado 80401, NREL/CP-520–27524, November 1999.

88. M. C. Schubert, J. Schön, P. Gundel, H. Habenicht, W. Kwapil, and W. Warta, "Imaging of metal impurities in silicon by luminescence spectroscopy and synchrotron techniques," *Journal of Electronic Materials*, **39**, 787–793, 2010.

89. A. Laades, K. Lauer, M. Bähr, C. Maier, A. Lawerenz, D. Alber, J. Nutsch, J. Lossen, C. Koitzsch, and R. Kibizov, "Impact of iron contamination on CZ-silicon solar cells," http://web.cismst.org/uploads/documents/1–1298537127.pdf.

90. K. Graff and H. Pieper, "The properties of iron in silicon," *Journal of the Electrochemical Society*, **128**, 669–674, 1981.

91. P. J. Ward, "A survey of iron contamination in silicon substrates and its impact on circuit yield," *Journal of the Electrochemical Society*, **129**, 2573–2576, 1982.

92. B. Hackl, K.-J. Range, H. J. Gores, L. Fabry, and P. Stallhofer, "Iron and its detrimental impact on the nucleation and growth of oxygen precipitates during internal gettering processes," *Journal of the Electrochemical Society*, **139**, 3250–3254, 1992.

93. K. Nauka and D. A. Gomez, "Surface photovoltage and deep level transient spectroscopy measurement of the Fe impurities in front-end operations of the IC CMOS process," *Journal of the Electrochemical Society*, **142**, L98–L99, 1995.

94. D. Caputo, P. Bacciaglia, C. Carpanese, M.L. Polignano, P. Lazzeri, M. Bersani, L. Vanzetti, P. Pianetta, and L. Moro, "Quantitative evaluation of iron at the silicon surface after wet cleaning treatments," *Journal of the Electrochemical Society*, **151**, G289–G296, 2004.

95. Y. Yamaguchi, M. Yoshida, and H. Aoki, "Solid solubility of nickel in silicon determined by use of ^{63}Ni as a tracer," *Japanese Journal of Applied Physics*, **2**, 714–718, 1963.

96. T. Iizuka and M. Kikuchi, "Hexagonal platelets observed in nickel diffused silicon," *Japanese Journal of Applied Physics*, **2**, 309–310, 1963.

97. K. Sueoka, S. Sadamitsu, Y. Koike, T. Kihara, and H. Katahama, "Internal gettering for Ni contamination in Czochralski silicon wafers," *Journal of the Electrochemical Society*, **147**, 3074–3077, 2000.

98. K. Sueoka, "Modeling of internal gettering of nickel and copper by oxide precipitates in Czochralski-Si wafers," *Journal of the Electrochemical Society*, **152**, G731–G735, 2005.

99. L. Fabry, R. Hoelzl, A. Andrukhiv, K. Matsumoto, J. Qiu, S. Koveshnikov, M. Goldstein, A. Grabau, H. Horie, and R. Takeda, "Test methods for measuring bulk copper and nickel in heavily doped *p*-type silicon wafers," *Journal of the Electrochemical Society*, **153**, G566–G571, 2006.

100. N. Inoue, K. Shingu, and K. Masumoto, "Measurement of nitrogen concentration in CZ silicon," presented at *N1-Ninth International Symposium on Silicon Materials Science and Technology*, 201st Meeting, #621, Philadelphia, PA, May 12–17, 2002, and the references therein.

101. V. V. Voronkov and R. Falster, "Nitrogen diffusion in silicon: A multi-species process," *210 ECE Meeting Cancun*, Mexico, Oct. 29–Nov. 3, 2006, and the references therein.

102. N. Fujita, R. Jones, S. Öberg, and P. R. Briddon, "Identification of nitrogen-oxygen defects in silicon," *210 ECE Meeting Cancun*, Mexico, Oct. 29–Nov. 3, 2006.

103. L. Jastrzebski, G. W. Cullen, R. Soydan, G. Harbeke, J. Lagowski, S. Vecrumba, and W. N. Henry, "The effect of nitrogen on the mechanical properties of float-zone silicon and on CCD device performance," *Journal of the Electrochemical Society*, **134**, 466–470, 1987.

104. B. M. Park, G. H. Seo, and G. Kim, "Nitrogen-doping effect in a fast-pulled CZ-Si single crystal," *Journal of Crystal Growth*, **222**, 74–81, 2001.

105. M. S. Kulkarni, "Defect dynamics in the presence of nitrogen in growing Czochralski silicon crystals," *Journal of Crystal Growth*, **310**, 324–335, 2008.

106. M. Trempa, C. Reimann, J. Friedrich, and G. Müller, "The influence of growth rate on the formation and avoidance of C and N related precipitates during directional solidification of multi crystalline silicon," *Journal of Crystal Growth*, **312**, 1517–1524, 2010.

107. X. Yu, D. Yang, and K. Hoshikawa, "Investigation of nitrogen behaviors during Czochralski silicon crystal growth," *Journal of Crystal Growth*, **318**, 178–182, 2011.

108. K.-M. Kim, "Growing improved silicon crystals for VLSI/ULSI," *Solid State Technology*, **39**, p. 70, 1996, and the references therein.

109. I. Kanda, T. Suzuki, and K. Kojima, "Influence of crucible and crystal rotation on oxygen-concentration distribution in large-diameter silicon single crystals," *Journal of Crystal Growth*, **166**, 669–674, 1996, and the references therein.

110. R. Falster and V.V. Voronkov, "Lattice defects in silicon", 123–5.5&5.6.ppt, from Monsanto Electronic Materials Company (MEMC) website, 2006.

111. W. Zulehner, "Czochralski growth of silicon," *Journal of Crystal Growth*, **65**, 189–213, 1983.

112. W. Wierzchowski, K. Wieteska, W. Graeff, M. Pawłowska, B. Surma, and S. Strzelecka, "X-ray topographic investigation of large oxygen precipitates in silicon," *Journal of Alloys and Compounds*, **362**, 301–306, 2004, and the references therein.

113. K. E. Benson and W. Lin, "The role of oxygen in silicon for VLSI," *Journal of Crystal Growth*, **70**, 602–608, 1984.

114. K. Sueoka, "Oxygen precipitation in lightly and heavily doped Czochralski silicon," *210 ECE Meeting Cancun*, Mexico, Oct. 29–Nov. 3, 2006.

115. L. E. Katz and D. W. Hill, "High oxygen Czochralski silicon crystal growth relationship to epitaxial stacking faults," *Journal of the Electrochemical Society*, **125**, 1151–1155, 1978.

116. B. Pajot, "Characterization of oxygen in silicon by infrared absorption," *Analusis*, **5**, 293–303, 1977.

117. K. Graff, "Precise evaluation of oxygen measurements on CZ-silicon wafers," *Journal of the Electrochemical Society*, **130**, 1378–1381, 1983.

118. H. J. Rath, P. Stallhofer, D. Huber, and B. F. Schmitt, "Determination of oxygen in silicon by photon activation analysis for calibration of the infrared absorption," *Journal of the Electrochemical Society*, **131**, 1920–1923, 1984.

119. B. Pajot, H. J. Stein, B. Cales, and C. Naud, "Quantitative spectroscopy of interstitial oxygen in silicon," *Journal of the Electrochemical Society*, **132**, 3034–3037, 1985.

120. T. Iizuka, S. Takasu, M. Tajima, T. Arai, T. Nozaki, N. Inoue, and M. Watanabe, "Determination of conversion factor for infrared measurement of oxygen in silicon," *Journal of the Electrochemical Society*, **132**, 1707–1713, 1985.

121. K. G. Barraclough, R. W. Series, J. S. Hislop, and D. A. Wood, "Calibration of infrared absorption by gamma activation analysis for studies of oxygen in silicon," *Journal of the Electrochemical Society*, **133**, 187–191, 1986.

122. A. Baghdadi, W. M. Bullis, M. C. Croarkin, Y.-z. Li, R. I. Scace, R. W. Series, P. Stallhofer, and M. Watanabe, "Interlaboratory determination of the calibration factor for the measurement of the interstitial oxygen content of silicon by infrared absorption," *Journal of the Electrochemical Society*, **136**, 2015–2026, 1989.

123. F. Schomann and K. Graff, "Correction factors for the determination of oxygen in silicon by IR spectrometry," *Journal of the Electrochemical Society*, **136**, 2025–2031, 1989.

124. D. E. Hill, "Determination of interstitial oxygen concentration in low-resistivity *n*-type silicon wafers by infrared absorption measurements," *Journal of the Electrochemical Society*, **137**, 3926–3928, 1990.

125. Y. Kitagawara, M. Tamatsuka, and T. Takenaka, "Accurate evaluation techniques of interstitial oxygen concentrations in medium-resistivity Si crystals," *Journal of the Electrochemical Society*, **141**, 1362–1364, 1994.

126. C. Maddalon-Vinante, J. P. Vallard, and D. Barbier, "Infrared study of the effect of rapid thermal annealing, thermal donor formation, and hydrogen on the precipitation of oxygen," *Journal of the Electrochemical Society*, **142**, 2071–2076, 1995.

127. B. G. Rennex, J. R. Ehrstein, and R. I. Scace, "Methodology for the certification of reference specimens for determination of oxygen concentration in semiconductor silicon by infrared spectrophotometry," *Journal of the Electrochemical Society*, **143**, 258–263, 1996.

128. A. Sassella, A. Borghesi, and T. Abe, "Quantitative evaluation of precipitated oxygen in silicon by infrared spectroscopy: Still an open problem," *Journal of the Electrochemical Society*, **145**, 1715–1719, 1998.

129. O. De Gryse, P. Clauws, J. Vanhellemont, O. I. Lebedev, J. Van Landuyt, E. Simoen, and C. Claeys, "Characterization of oxide precipitates in heavily B-doped silicon by infrared spectroscopy," *Journal of the Electrochemical Society*, **151**, G598–G605, 2004.

130. M. Furuhashi and K. Taniguchi, "Formation mechanisms of divacancy-oxygen complex in silicon," *Journal of the Electrochemical Society*, **155**, H160–H163, 2008.

131. G. Kissinger and J. Dabrowski, "Oxide precipitate nucleation via coherent 'seed'-oxide phases," *210 ECE Meeting Cancun*, Mexico, Oct. 29–Nov. 3, 2006, and the references therein.

132. K. Wada, H. Nakanishi, H. Takaoka, and N. Inoue, "Nucleation temperature of large oxide precipitates in as-grown Czochralski silicon crystal," *Journal of Crystal Growth*, **57**, 535–540, 1982.

133. F. Shimura, H. Tsuya, and T. Kawamura, "Precipitation and redistribution of oxygen in Czochralski-grown silicon," *Applied Physics Letters*, **37**, 483–486, 1980.

134. A. E. Organ and N. Riley, "Oxygen transport in magnetic Czochralski growth of silicon," *Journal of Crystal Growth*, **82**, 465–476, 1987.

135. H. Hirata and K. Hoshikawa, "Silicon crystal growth in a cusp magnetic field," *Journal of Crystal Growth*, **96**, 747–755, 1989.

136. R. W. Series, "Effect of a shaped magnetic field on Czochralski silicon growth," *Journal of Crystal Growth*, **97**, 92–98, 1989.

137. H. Hirata and K. Hoshikawa, "Homogeneous increase in oxygen concentration in Czochralski silicon crystals by a cusp magnetic field," *Journal of Crystal Growth*, **98**, 777–781, 1989.

138. S. Kobayashi, "Numerical analysis of oxygen transport in magnetic Czochralski growth of silicon," *Journal of Crystal Growth*, **85**, 69–74, 1987.

139. T. W. Hicks, A. E. Organ, and N. Riley, "Oxygen transport in magnetic Czochralski growth of silicon with a nonuniform magnetic field," *Journal of Crystal Growth*, **94**, 213–228, 1989.

140. L. N. Hjellming and J. S. Walker, "Mass transport in a Czochralski puller with a strong magnetic field," *Journal of Crystal Growth*, **85**, 25–31, 1987.

141. N. Kobayashi, "Oxygen transport under an axial magnetic field in Czochralski silicon growth," *Journal of Crystal Growth*, **108**, 240–246, 1991.

142. P. Sabhapathy and M. E. Salcudean, "Numerical study of Czochralski growth of silicon in an axisymmetric magnetic field," *Journal of Crystal Growth*, **113**, 164–180, 1991.

143. A. G. Crouse, J. E. Huffman, C. S. Tindall, and M. L. W. Thewalt, "Float-zone growth of high purity and high oxygen concentration silicon," *Journal of Crystal Growth*, **109**, 162–166, 1991.

144. Z. A. Salnick, "Oxygen in Czochralski silicon crystals grown under an axial magnetic field," *Journal of Crystal Growth*, **121**, 775–780, 1992.
145. I. Fusegawa, H. Yamagishi, T. Mori, H. Takayama, E. Iino, and K. Takano, "Characterization of interstitial oxygen striations in silicon single crystals," *Journal of Crystal Growth*, **128**, 293–297, 1993.
146. A. Murgai, H. C. Gatos, and W. A. Westdorp, "Effect of microscopic growth rate on oxygen microsegregation and swirl defect distribution in Czochralski-grown silicon," *Journal of the Electrochemical Society*, **126**, 2240–2245, 1979.
147. Q. Ma, D. Yang, X. Ma, and D. Que, "Oxygen precipitation of nitrogen-doped Czochralski silicon subjected to multi-step thermal process," *210 ECE Meeting Cancun*, Mexico, Oct. 29–Nov. 3, 2006.
148. K. Izunome, X. Huang, S. Togawa, K. Terashima, and S. Kimura, "Control of oxygen concentration in heavily antimony-doped Czochralski Si crystals by ambient argon pressure," *Journal of Crystal Growth*, **151**, 291–294, 1995, and the references therein.
149. Z. Liu and T. Carlberg, "On the mechanism of oxygen content reduction by antimony doping of Czochralski silicon melts," *Journal of the Electrochemical Society*, **138**, 1488–1492, 1991.
150. A. R. Comeau, "Improvement to bulk oxygen precipitate density necessary for internal gettering in antimony doped n/n+ epitaxial wafers using the 3-step technique and ramped nucleation," *Journal of the Electrochemical Society*, **139**, 1455–1463, 1992.
151. D. Yang, C. Li, M. Luo, J. Xu, and D. Que, "Reduction of oxygen during the crystal growth in heavily antimony-doped Czochralski silicon," *Journal of Crystal Growth*, **256**, 261–265, 2003.
152. C. Liu, H. Wang, Y. Li, Q. Wang, B. Ren, Y. Xu, and D. Que, "Study on the oxygen concentration reduction in heavily Sb-doped silicon," *Journal of Crystal Growth*, **196**, 111–114, 1999.
153. S. Kawanishi, S. Togawa, K. Izunome, K. Terashima, and S. Kimura, "Melt quenching technique for direct observation of oxygen transport in the Czochralski-grown Si process," *Journal of Crystal Growth*, **152**, 266–273, 1995, and the references therein.
154. R. Falster and V. V. Voronkov, "Intrinsic point defects and their control in silicon crystal growth and wafer processing," *MRS Bulletin*, **25**, pp 28–32, June 2000, and the references therein.
155. R. Falster, "Defect control in silicon crystal growth and wafer processing," MEMC Electronic Materials SpA, Novara 28100, Italy, www.suneditionsemi.com/assests/file/technology/papers, and the references therein.
156. Y. Shiraishi, S. Kurosaka, and M. Imai, "Silicon crystal growth using a liquid-feeding Czochralski method," *Journal of Crystal Growth*, **166**, 685–688, 1996, and the references therein.
157. S. Kobayashi, "A model for oxygen precipitation in Czochralski silicon during crystal growth," *Journal of Crystal Growth*, **174**, 163–169, 1997, and the references therein.
158. S. Kobayashi, "Mathematical modeling of grown-in defects formation in Czochralski silicon," *Journal of Crystal Growth*, **180**, 334–342, 1997.
159. P. E. Freeland, "Oxygen precipitation in silicon at 650°C," *Journal of the Electrochemical Society*, **127**, 754–756, 1980.
160. F. Shimura and H. Tsuya, "Oxygen precipitation factors in silicon," *Journal of the Electrochemical Society*, **129**, 1062–1066, 1982.
161. N. Inoue, J. Osaka, and K. Wada, "Oxide micro-precipitates in as-grown CZ silicon," *Journal of the Electrochemical Society*, **129**, 2780–2788, 1982.
162. L. Jastrzebski, P. Zanzucchi, D. Thebault, and J. Lagowski, "Method to measure the precipitated and total oxygen concentration in silicon," *Journal of the Electrochemical Society*, **129**, 1638–1641, 1982.
163. H. R. Huff, H. F. Schaake, J. T. Robinson, S. C. Baber, and D. Wong, "Some observations on oxygen precipitation/gettering in device processed Czochralski silicon," *Journal of the Electrochemical Society*, **130**, 1551–1555, 1983.
164. S. K. Bains, D. P. Griffiths, J. G. Wilkes, R. W. Series, and K. G. Barraclough, "Oxygen precipitation in heavily doped silicon," *Journal of the Electrochemical Society*, **137**, 647–652, 1990.
165. W. Huber and M. Pagani, "The behavior of oxygen precipitates in silicon at high process temperature," *Journal of the Electrochemical Society*, **137**, 3210–3213, 1990.

166. Y. Kitagawara, R. Hoshi, and T. Takenaka, "Evaluation of oxygen precipitated silicon crystals by deep-level photoluminescence at room temperature and its mapping," *Journal of the Electrochemical Society*, **139**, 2277–2281, 1992.

167. B. Hackl, K.-J. Range, H. J. Gores, L. Fabry, and P. Stallhofer, "Iron and its detrimental impact on the nucleation and growth of oxygen precipitates during internal gettering processes," *Journal of the Electrochemical Society*, **139**, 3250–3254, 1992.

168. S. Kimura, "Observation of oxygen precipitates in CZ-grown Si wafers with a phase differential scanning optical microscope," *Journal of the Electrochemical Society*, **141**, L120–L122, 1994.

169. H.-D. Chiou, Y. Chen, R. W. Carpenter, and J. Jeong, "Warpage and oxide precipitate distributions in CZ silicon wafers," *Journal of the Electrochemical Society*, **141**, 1856–1862, 1994.

170. H. Fujimori, "Dependence on morphology of oxygen precipitates upon oxygen supersaturation in Czochralski silicon crystals," *Journal of the Electrochemical Society*, **144**, 3180–3184, 1997.

171. T. Ono, G. A. Rozgonyi, C. Au, T. Messina, R. K. Goodall, and H. R. Huff, "Oxygen precipitation behavior in 300-mm polished Czochralski silicon wafers," *Journal of the Electrochemical Society*, **146**, 3807–3811, 1999.

172. L. B. Xu, "A statistical thermodynamic model for oxygen segregation during Czochralski growth of silicon single crystals," *Journal of Crystal Growth*, **200**, 414–420, 1999.

173. K. Nakashima, Y. Watanabe, T. Yoshida, and Y. Mitsushima, "A method to detect oxygen precipitates in silicon wafers by highly selective reactive ion etching," *Journal of the Electrochemical Society*, **147**, 4294–4296, 2000.

174. C. Claeys, E. Simoen, V. B. Neimash, A. Kraitchinskii, M. Kras'ko, O. Puzenko, A. Blondeel, and P. Clauws, "Tin doping of silicon for controlling oxygen precipitation and radiation hardness," *Journal of the Electrochemical Society*, **148**, G738–G745, 2001.

175. Y. X. Li, H. Y. Guo, B. D. Liu, T. J. Liu, Q. Y. Hao, C. C. Liu, D. R. Yang, and D. L. Que, "The effects of neutron irradiation on the oxygen precipitation in Czochralski-silicon," *Journal of Crystal Growth*, **253**, 6–9, 2003.

176. K. Sueoka, M. Akatsuka, M. Okui, and H. Katahama, "Computer simulation for morphology, size, and density of oxide precipitates in CZ silicon," *Journal of the Electrochemical Society*, **150**, G469–G475, 2003.

177. K. Nakashima, T. Yoshida, and Y. Mitsushima, "Measurements of size, morphology, and spatial distribution of oxygen precipitates in Si wafers using RIE," *Journal of the Electrochemical Society*, **152**, G339–G344, 2005.

178. Y. Zeng, J. Chen, M. Ma, W. Wang, and D. Yang, "Enhancement effect of nitrogen co-doping on oxygen precipitation in heavily phosphorus-doped Czochralski silicon during high-temperature annealing," *Journal of Crystal Growth*, **311**, 3273–3277, 2009.

179. Z. Zeng, X. Ma, J. Chen, D. Yang, I. Ratschinski, F. Hevroth, and H. S. Leipner, "Effect of oxygen precipitates on dislocation motion in Czochralski silicon," *Journal of Crystal Growth*, **312**, 169–173, 2010.

180. Y.-Y. Teng, J.-C. Chen, C.-W. Lu, H.-I Chen, C. Hsu, and C.-Y. Chen, "Effects of the furnace pressure on oxygen and silicon oxide distributions during the growth of multicrystalline silicon ingots by the directional solidification process," *Journal of Crystal Growth*, **318**, 224–229, 2011.

181. A. Sarikov, V. Litovchenko, I. Lisovskyy, M. Voitovich, S. Zlobin, V. Kladko, N. Slobodyan, V. Machulin, and C. Claeys, "Mechanisms of oxygen precipitation in CZ-Si wafers subjected to rapid thermal anneals," *Journal of the Electrochemical Society*, **158**, H772–H777, 2011.

182. P. K. Kulshreshtha, Y. Yoon, K. M. Youssef, E. A. Good, and G. Rozgonyi, "Oxygen precipitation related stress-modified crack propagation in high growth rate Czochralski silicon wafers," *Journal of the Electrochemical Society*, **159**, H125–H129, 2012.

183. W. K. Tice, "Oxygen precipitation and formation of diffusion shorts in FET circuits," *Journal of the Electrochemical Society*, **136**, 1572–1573, 1989.

184. T. Tuomi, M. Tuominen, E. Prieur, J. Partanen, J. Lahtinen, and J. Laakkonen, "Synchrotron section topographic study of Czochralski-grown silicon wafers for advanced memory circuits," *Journal of the Electrochemical Society*, **142**, 1699–1701, 1995.

185. H. Fujimori, Y. Ushiku, T. Ihnuma, Y. Kirino, and Y. Matsushita, "The interrelation between the morphology of oxide precipitates and the junction leakage current in Czochralski silicon crystals," *Journal of the Electrochemical Society*, **146**, 702–706, 1999.

186. J.-C. Chen, Y.-Y. Teng, W.-T. Wun, C.-W. Lu, H. I Chen, C.-Y. Chen, and W.-C. Lan, "Numerical simulation of oxygen transport during the CZ silicon crystal growth process," *Journal of Crystal Growth*, **318**, 318–323, 2011.

187. C. Claeys, E. Simoen, V. B. Neimash, A. Kraitchinskii, M. Kras'ko, O. Puzenko, A. Blondeel, and P. Clauws, "Tin doping of silicon for controlling oxygen precipitation and radiation hardness," *Journal of the Electrochemical Society*, **148**, G738–G745, 2001.

188. M. Nakamura, T. Kato, and N. Oi, "A study of gettering effect of metallic impurities in silicon," *Japanese Journal of Applied Physics*, **7**, 512–519, 1968.

189. K. P. Lisiak and A. G. Milnes, "A comparison of the process-induced gettering of atomic platinum and gold in silicon," *Journal of the Electrochemical Society*, **123**, 305–308, 1976.

190. L. Baldi, G. Cerofolini, and G. Ferla, "Heavy metal gettering in silicon-device processing," *Journal of the Electrochemical Society*, **127**, 164–169, 1980.

191. C. J. Werkhoven, C. W. T. Bulle-Lieuwma, B. J. H. Leunissen, and M. P. A. Viegers, "Characterization of metallic precipitates in epitaxial Si by means of preferential etching and TEM," *Journal of the Electrochemical Society*, **131**, 1388–1391, 1984.

192. H. Morinaga, T. Futatsuki, T. Ohmi, E. Fuchita, M. Oda, and C. Hayashi, "Behavior of ultrafine metallic particles on silicon wafer surface," *Journal of the Electrochemical Society*, **142**, 966–970, 1995.

193. W. P. Lee and Y. L. Khong, "Laser microwave photoconductance studies of ultraviolet-irradiated silicon wafers: Effect of metallic contamination," *Journal of the Electrochemical Society*, **145**, 329–332, 1998.

194. H. Lemke and K. Irmscher, "Proof of interstitial cobalt defects in silicon float zone crystals doped during crystal growth," *210 ECE Meeting Cancun*, Mexico, Oct. 29–Nov. 3, 2006.

9

Defects in Silicon Wafers

9.1 Introduction

Single-crystal silicon grown by both the Czochralski (CZ) process and the float-zone (FZ) process inherently contains many crystallographic imperfections popularly known as microdefects or grown-in defects. Prior to the 1960s, crystals contained dislocations induced by the thermomechanical stresses near the vicinity of the melt-crystal interface. A major breakthrough by Dash [1] allowed crystal growth without the thermomechanically induced dislocations, and this brought a revolution in growing single crystals of silicon with minimum defects. However, even though the crystals are free of thermomechanically induced dislocations, various other microdefects can form through an agglomeration of point defects that exist as solutes in the silicon crystal matrix. There are two basic types of point defects in silicon: vacancies, which are formed when a silicon atom is missing from the regular lattice site, and self-interstitials (or simply, interstitials), which are interstitial silicon atoms not bonded with the other silicon atoms forming the lattice. Microdefect dynamics, or simply defect dynamics [2], is a collective term that describes the interplay between the point defects transport, the Frenkel reaction, the nucleation of the point defects, and the growth of the microdefects. The two-dimensional figures in Figures 9.1 and 9.2 show various defects that are generally present in the silicon crystal lattice.

The onus to produce high-quality silicon substrates for ultra large scale integration (ULSI) applications falls to the wafer makers, who must acquire a greater understanding of the mechanisms that are responsible for device failure in order to facilitate improvements. Of particular importance is the control of as-grown silicon crystals and wafers and process-induced crystallographic defects that are known to have adverse effects on the electrical characteristics of devices. The thermal history of CZ silicon crystals has a dominating influence on the formation of microdefects. Most of these defects (e.g., clusters of point defects), such as vacancies, self-interstitials, complexes with oxygen, and other impurities, are formed during crystal growth. Thus, by adjusting the temperature in the crystal through hot-zone modifications such as heat shields, these defects can be effectively controlled. The results on thermal simulations by Dornberger et al. [3] demonstrate that advanced numerical models can be applied to calculate the temperature distribution in CZ silicon crystals and in the furnace parts. These models can be used to predict the temperature distribution of growing silicon crystals, which is important to optimize growth processes and the engineering of as-grown microdefects. The density and type of defects in single-crystal silicon wafers are functions of the growth rate and the temperature gradient at the liquid–solid interface [4]. Different types of defects dominate different ranges of the input variables, and it is possible to identify a process range that minimizes defects.

9.2 Impact of Defects in Silicon Devices and Structures

Crystalline defects in a single-crystal silicon wafer, whether present before or introduced during processing, often have a deleterious effect on device performance. The most common microdefects in CZ silicon, voids and dislocation loops, are formed by an agglomeration of point defects, vacancies, and self-interstitials, respectively. The ever-shrinking size of microelectronic devices places challenging restrictions on the quality of the silicon substrate used in the manufacture of very large scale integration (VLSI) and ULSI devices. The quality of the substrate silicon is essentially determined by the type and distribution of microdefects. In the CZ process, a crystalline ingot is continuously pulled from the melt in a quartz crucible, and in the FZ process, a molten zone traverses through a polycrystalline ingot to form monocrystalline silicon. Both point defects and secondary defects become increasingly important as we march toward the era of giga and terabit memory devices.

Today, silicon device manufacturers face two main challenges when it comes to defects, neither of which can be solved simply by increasing the level of crystal purity. So important are these problems that they in fact challenge the very status of the traditional, polished, CZ-grown silicon wafer as a suitable material for the future generations of advanced VLSI and ULSI processing. One of these challenges arises from the intrinsic, or native, point defects, lattice vacancies, and silicon self-interstitials, and the other from the most important extrinsic point defect in CZ-grown silicon: i.e., oxygen [5]. Large advances have been made recently in these areas, producing solutions to these engineering problems that are separate yet related. In particular, studies of point defects from the point of view of crystal growth and oxygen-precipitation control can give new insights into the elusive properties of these defects.

Microdefects in silicon related to intrinsic point defects were first observed in the early 1960s. Eliminating dislocations from silicon crystals simultaneously eliminated an important distributed sink for grown-in intrinsic point defects, thus allowing them to homogeneously agglomerate. The study of these microdefect agglomerates has yielded a rich array of information on intrinsic point defects, especially their properties at high temperatures. The incorporation was primarily studied in terms of thermal history and how the crystal was solidified and cooled after it was pulled from the crucible. It is very important to

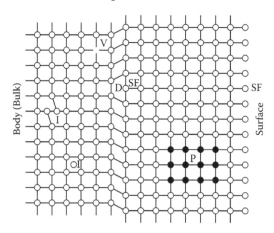

FIGURE 9.1
Various possible crystal defects in a hypothetical silicon lattice. V, vacancy; I, interstitial; D, defect; P, precipitation; and SF, stacking faults.

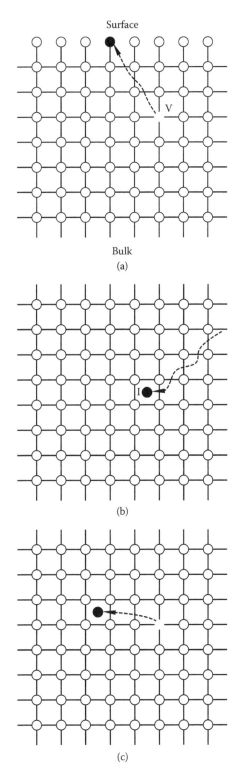

FIGURE 9.2
Various defects in bulk silicon crystal lattice: (a) Schottky, (b) interstitial, and (c) Frenkel defects.

understand the thermal environment of the CZ furnace in order to improve the crystal quality and the process yield. However, direct temperature measurement with thermocouples is difficult [6]. Interstitial- and vacancy-type microdefects occur in two clearly defined modes of crystal growth. Grown-in microdefects are controlled simply by controlling the ratio of pull rate to the thermal gradient. Swirl defects will form if the ratio is below a certain measurement; D-defects will form otherwise [5]. This simple and important rule holds both for the FZ- and CZ-grown crystals. At a low interstitial concentration, only B-defects will form. At still-lower concentrations, no detectable defects will form. Accordingly, the main interstitial region of a crystal (the region containing A-defects) is surrounded by a band of B-defects.

Lawrence [7] provided early reports on the behavior of dislocations in silicon. It was proved experimentally that unstable dislocations are slightly more electrically active than are low-energy stable dislocations. The precipitation of metallic contaminants such as copper can greatly alter the electrical properties of the dislocations present in the lattice. Devices formed using material with few dislocations are more likely to degrade due to metal precipitate formation in the crystal lattice. The degradations attributed to metal precipitates are large *p-n* junction reverse currents, emitter-to-collector short-circuits, and wide variations in the transistor gain parameters. Similarly, Katz [8] reported on the relationship between process-induced defects and soft *p-n* junctions in silicon. The defects were evaluated by chemical etching. Osburn and Ormond [9] reported on the effect of silicon wafer imperfections on minority carrier generation and dielectric breakdown phenomenon in metal-oxide semiconductor (MOS) structures. They confirmed that the thermal oxidation increases the defect density, making the front and back surfaces more alike. The amount of oxidation-induced damage also depends on the quality of the starting silicon wafer. Kissinger et al. [10] provided a clear picture of the grown-in oxide precipitate in CZ silicon wafers and its role in complementary metal-oxide semiconductor (CMOS) device processing at different steps, corroborating thermal history–related defect density spectra.

Accurate control of the defects in silicon wafers is a subject of immense importance to both the silicon and VLSI industries. The exploding costs of wafer development and production, as well as the processing of 300 mm and larger wafers means that predictive defect engineering is a requirement for both industries more than ever [11]. Intrinsic point defect concentration and reaction control in the growth crystals, including effects of impurities and the use of vacancy concentration profiles installed in the silicon wafer, are necessary to achieve ideal oxygen precipitation performance. The importance of accurate modeling of defect dynamics is necessary. In addition, significantly higher levels of mechanical stress in the processing of large wafers has led to a new appreciation of the role played by oxygen species in the locking of dislocations and the dynamics of wafer hardening during processing. Metrology and sampling issues are only part of the problem. The second major problem lies in the fact that there is a strong coupling between the various defect formation mechanisms and the ultimate performance of the material in specific applications and their usage.

With the introduction of ULSI devices, severe demands were placed on the quality and specifications of CZ silicon wafers, beginning with the raw material. The relationship between gate oxide integrity (GOI) for MOS devices and grown-in defects has been studied in order to improve wafer quality. It is now well known that the GOI of a MOS device depends on the CZ crystal pulling rate in any of the grown silicon wafers [12, 13]. Oxygen precipitation in hydrogen annealing was the same as in nonannealed wafers, but GOI strongly depended on the increase in temperature ramp rate of the CZ crystal pulling rate during the growth process. A single-bit failure rate in some dynamic random access memory (DRAM) reliability tests [14] turned out to correlate with GOI.

The sensitivity to defects in charge-coupled device (CCD) imagers is about two orders of magnitude higher than in the most sophisticated DRAM circuits. This makes CCD imagers one of the most difficult and challenging VLSI circuits to manufacture. This is particularly important for photosensitive memories, such as CCD imagers intended for use as the image sensor in a television camera. In the case of large-area CCD imagers for TV applications, on the order of 10^5 elements, a defect may affect the generation and recombination current and reflect in the image. Thus, high-quality wafers are essential for these CCD devices. By using deep-level transient spectroscopy (DLTS) and derivative surface photovoltage spectroscopy (DSPS), Jastrzebski and Lagowski [15] were able to evaluate FZ-grown silicon wafers. The DSPS measurements revealed the presence of deep centers, in the range of 10^{11}–10^{12} cm^{-3}, around the middle of the energy gap in the as-grown and the heat-treated wafers. The effect of these deep centers on the quantum efficiency and the fixed pattern noise were evaluated. The authors believed that the deep levels were introduced by point defect complexes or their clusters with impurities rather than by the heavy metals or any other chemical impurities. Jastrzebski et al. [16] reported that the best results are possible with magnetic CZ (MCZ)–grown silicon wafers, and epitaxial layers grown on internally gettered substrates are good for CCD imagers. The role of heavy metal contamination on defect formation has been studied [17], and a certain threshold contamination level exists beyond which the probability of crystallographic defect decoration and formation of stacking faults increases in the wafer. This level varies according to the type of defect as well as the type of wafer. Epitaxial layers showed the lowest resistance to contamination-induced defect generation and decoration, and internally gettered wafers showed the highest resistance.

The leakage of stored charge is one of the major causes of failure in DRAM devices. The capacitors lose their charge due to various sources of leakage. Kim and Wijaranakula [18] reported on the effect of crystal grown-in defects on the pause-tail characteristics of megabit DRAM devices. The refresh time failure in the memory devices is strongly affected by the crystallographic defects originating from the CZ substrate. The degradation of device performance is caused by residual crystal defects located within several microns beneath the epitaxial layer. This implies that the crystal grown-in defects affecting the refresh time of the trench storage capacitors are confined mainly within the CZ silicon substrate and the defects present in it. It was further reported [19] that the refresh time degradation in trench-type memory devices is directly related to the grown-in oxide polyhedra located in the device's active region. Based upon this analysis, the formation of the oxide precipitate ring in the bottom section of the crystal is hypothesized to be caused by radiative heating during and after the crystal tailing-off process.

Dornberger et al. [20] provided details on the silicon crystal defects and their impact on DRAM characteristics. According to them, the grown-in defect formation in silicon crystals is dominated by the aggregation of vacancies and silicon self-interstitials. Depending on the crystal pulling conditions, one of the two species will prevail. This is illustrated in Figure 9.3 with an axially cut crystal, which was grown with a pull rate decrease from top to bottom. At the top of the crystal, the pull rate was high and only vacancy-type defects, the so-called D-defects or crystal-oriented particles (COPs), appear. When the pull rate is reduced, the vacancy-rich region shrinks and interstitial-type defects are formed near the crystal edges. These are networks of dislocation loops as large pits of 20 μm. Wafers with large pit defect (or L-pits) are detrimental for devices because of their large physical size. A high density of oxidation-induced stacking faults (OISF) can be observed at the boundary between vacancy and interstitial regions after wet oxidation at 1100°C. According to Voronkov's theory, the boundary between vacancy and interstitial excess is located where

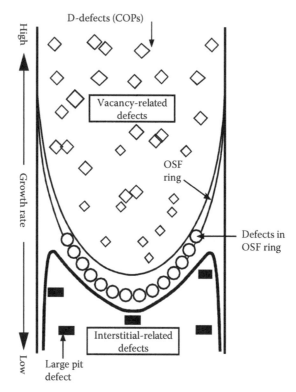

FIGURE 9.3
Grown-in defects in a silicon crystal grown with a varying pulling rate. (From E. Dornberger, D. Temmler, and W. von Ammon, *Journal of the Electrochemical Society*, **149**, G226–G231, 2002, and the references therein [20].)

the ratio of the pull rate and the temperature gradient at the solidification interface ratio is equal to a critical value C_{crit} of 1.4×10^{-3} cm^2/K min, as shown in Figure 9.4. Crystals become fully interstitial rich at low pull rates, where C_{crit} is lower than 1.4×10^{-3} cm^2/K min over the complete radius. In a relatively narrow window of approximately $\pm 10\%$ of the critical ratio, neither vacancy nor interstitial reaches the critical supersaturation for the nucleation of larger defects. This explains most of the defects generated during crystal growth, since they depend on the ratio between pull rate and temperature gradient. State-of-the-art ULSI DRAM failure analysis down to 0.14 µm feature size has been reported by Ruprecht et al. [21] based on defects present in the silicon wafers. Hirano et al. [22] reported on the impact of defects in flash memory characteristics. They found that the residual misalignment of photolithography and electrical properties of flash memory are closely related to defects in silicon wafers.

9.3   Point Defects and Vacancies

Point defects are incorporated into silicon when it crystallizes. As a growing crystal cools, vacancies and self-interstitials may diffuse, recombine, and agglomerate to form various microdefects such as swirl defects and oxygen precipitation centers. There have been two models qualitatively describing the formation of microdefects in silicon crystals. Grown-in microdefects in dislocation-free silicon crystals have been studied extensively over the past 40 years. This interest is due, in large part, to their influence on the parameters of silicon-based devices. With the control of dopants and oxygen now fairly well in hand, crystal growers are devoting increasing attention to understanding and controlling the distribution of native point defects, particularly those of the vacancies and self-interstitials [23]. Silicon crystals grown using FZ techniques generally contain two types of vacancy clusters [24] and fewer of them than in crystals grown using the CZ technique. Defect formation and propagation processes are characterized on the atomic scale with the aid of high-resolution electron microscope images [25] of growth breakdown interfaces. Thermal oxidation of silicon induces growth of stacking faults [26]. The growth of oxidation stacking faults is caused by the supersaturation of silicon interstitials in bulk silicon during oxidation.

The structural properties of as-grown silicon are controlled by the type and concentration of intrinsic point defects incorporated into the growing crystal. An incorporation model assumes a fast recombination of intrinsic point defects in the vicinity of the crystal-melt interface. The annihilation stage is effectively complete when the temperature is below the melting point: about 100°C. Only one kind of defect, either vacancy or self-interstitial, is present in a supersaturated concentration, while the competing defects rapidly disappear. Due to radial and axial nonuniformity in the gradient temperature and crystal pull rate, the same crystal may contain both vacancy and interstitial regions separated by a well-defined boundary. Voronkov and Falster [27] have explained the exact relationship between the intrinsic point defects and the presence of impurities in silicon in a more systematic fashion that uses the ratio between the temperature gradient and the crystal pull rate. Any surviving intrinsic defects will agglomerate into various structural microdefects as the temperature lowers. At a low vacancy concentration, oxide particles are formed instead of voids by joint agglomeration of vacancies and oxygen species; this effect is responsible for the formation of OISFs in a ring-like distribution. At low self-interstitial concentration values, B-defects (globular interstitial aggregates) are formed instead of loops. The critical ratio V/G is of fundamental importance to the growth of dislocation-free silicon crystals with controlled microdefect properties. The incorporation of intrinsic point defects into a

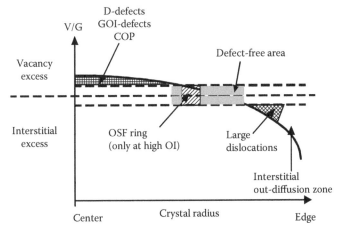

FIGURE 9.4
Radial variation of *V/G* ratio in CZ silicon crystals grown with a transition from vacancy rich to interstitial rich, from center to rim. (From E. Dornberger, D. Temmler, and W. von Ammon, *Journal of the Electrochemical Society*, **149**, G226–G231, 2002, and the references therein [20].)

growing crystal is affected by the presence of impurities that can react with vacancies and interstitials. Some impurities, such as oxygen, nitrogen, and hydrogen, trap vacancies and cause a downward shift in the critical ratio. Other impurities, like carbon, trap self-interstitials and cause an upward shift in the critical V/G ratio. The impurities affect both the incorporation and agglomeration stages of microdefect production. For more details, readers are urged to refer to [27] to better understand the governing relationship between intrinsic point defects and impurities. Budil et al. [28] reported on the physical properties of point defects from back-side oxidation experiments, particularly with the silicon self-interstitials.

Puzanov et al. [23] have used the theory and values of Voronkov and Falster [27], axial temperature gradient, and cooling rate measured during crystal growth. The numerical calculations quantitatively reproduce experimental data, such as the transition between the vacancy and interstitial microdefects when the pulling rate passes through a critical value, the size and shape of the regions of interstitial (A and B) defects and vacancy-related A' defects, and the position and shape of the defect-free zone in silicon crystals 8–16 cm in diameter. By fitting the calculations to the observed microdefect patterns, the activation energy for point defect migration was estimated to be 1.3 eV for temperatures ranging from the melting point to 1273 K. Further, the authors have shown how to evaluate the radial distribution of the axial temperature gradient near the growth interface from the actual microdefect patterns revealed in silicon crystals. Figure 9.5 shows the axial temperature gradient (G) versus the distance (z) from the central and peripheral regions of the growth interface at different crystal regions.

As mentioned, the quality of single crystals of silicon grown using CZ and FZ methods depends on the distribution of microdefects formed by silicon vacancies and interstitials and by impurities such as oxygen and carbon. Brown et al. [29] attempted to model the formation of these defects by combining an atomistic-level simulation of the equilibrium, transport, and kinetics of point defects and impurities in silicon with continuum modeling of defect transport and their possible reactions. The continuum models were written in terms of classical equilibrium, transport, and kinetic coefficients, which were estimated using atomistic simulations based on the Stillinger-Weber interatomic potential to describe the interactions. The simulations were reported for the equilibrium and transport properties of interstitials and vacancies in pure silicon. The calculations predicted that interstitials prefer to form <110> dumbbells in the diamond lattice and that these point defects become delocalized at elevated temperatures. The authors proposed a model for the recombination of vacancies and interstitials that leads to a high entropic energy barrier at high temperatures due to this delocalization within the crystal lattice.

The temperature inside a growing crystal sharply drops as the distance from the melt-crystal interface increases, which decreases the equilibrium concentrations of the point defects. Under such conditions, the Frenkel reaction decreases the concentrations of both the point defects. The point defect concentration gradients drive the diffusion of the point defects into the crystal. The physical displacement or convection of the crystal also contributes to the flux of the point defects relative to a fixed coordinate system. The net flux of the point defects, defined as the difference between the flux of, say, vacancies and interstitials, very close to the interface determines the difference between the vacancy concentration and the interstitial concentration in the crystal, which is a short distance from the interface, termed the recombination length. In the absence of external sources and sinks, such as the crystal surface or the thermomechanically induced dislocations, the established point defect concentration difference remains constant. The process of establishing the concentrations of the surviving point defect species beyond the recombination length is termed the initial incorporation. After the initial incorporation, beyond the recombination

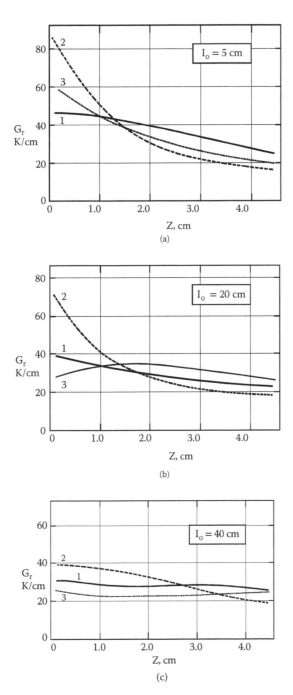

FIGURE 9.5
Axial temperature gradient vs. distance from the crystal-melt interface for different crystal lengths: (a) $I_o = 5$, (b) $I_o = 20$, (c) $I_o = 40$ cm: "1" in the center of the crystal for $V = 0.5$ and 1.5 mm/min; "2" near the cylindrical surface for $V = 0.5$ mm/min; "3" near the cylindrical surface for $V = 1.5$ mm/min. (From N. I. Puzanov, A. M. Eidenzon, and D. N. Puzanov, *Journal of Crystal Growth*, **178**, 468–478, 1997, and the references therein [23].)

length, the established concentration difference determines the type of the prevailing point defect species at a distance far away from the interface. As the pull rate, denoted by V (cm/s), of the crystal increases, the convection of the point defects dominates their diffusion from the interface, leading to vacancy-rich conditions because the equilibrium concentration of vacancies at the interface is higher than the equilibrium concentration of interstitials. As the magnitude of the axial temperature gradient near the interface, denoted by G (K/cm), increases, the sharp temperature drop near the interface dramatically increases the concentration gradients of the point defects, thus allowing the diffusion to dominate the convection. Typically, in a growing crystal, G increases along the radial position, which leads to the incorporation of excess vacancies in the central region and of excess interstitials in the peripheral region. The two regions are separated by a boundary. Accordingly, there is a central region on the vacancy agglomerates and a peripheral region on the interstitial agglomerates separated by a microdefect-free region [2]. Self-interstitials diffuse faster than vacancies and lead to interstitial-rich conditions. Thus, vacancies remain in excess at a higher V/G ratio, self-interstitials remain in excess at a lower V/G ratio, and no point defect species dominates at a V/G ratio closer to its critical value. The surviving point defects precipitate at a lower temperature to form microdefects. This quantification of defect dynamics explains most of the defect formations in growing crystals. Also, the model predictions agree very well with the experimental results.

Point defects incorporate during crystallization. A computer simulation estimation, by Kim [30], on the silicon crystal point defects and their thermodynamic equilibrium concentrations at ~1.0×10^{15}/cm^3 for vacancies, and about 20% less for the interstitials, at the melting point of silicon: 1412°C. As the crystal cools continually during CZ crystal growth, the solubility of the point defects decreases, resulting in supersaturation. Intense recombination of vacancies and interstitials takes place at a high temperature range (from the melting point to ~1200°C). Also, some out-diffusion of interstitials occurs near the periphery of the crystal surface. The simulation also shows ~10% of the original point defects in both vacancy and interstitial-rich regions survive after the intense recombination, but they are still in a high degree of supersaturation below about 1200°C and form agglomerates or clusters. Survival of minority point defects in the vacancy and interstitial-rich regions is about 1%. Vacancies agglomerate into octahedral voids in the 1150–1080°C temperature range.

Matsushita [31] investigated the thermally induced microdefects in CZ-grown silicon crystals. It was found that the microdefect nature varies with the temperature and the defect density increases exponentially as the annealing temperature decreases. Microdefect formation strongly depends on the thermal history during crystal growth. When the annealing temperature is 800°C or higher, the induced microdefects are caused by the growth of micro-precipitates present in the as-grown crystal, but not the nucleation of oxygen precipitates. When the annealing temperature is as low as 650°C, nuclei of oxygen precipitates can form during annealing. Furthermore, it is concluded that the micro-precipitates or nuclei are heterogeneously formed at sites closely correlated with carbon impurities in the silicon crystal. The size distribution of micro-precipitates strongly depends on the thermal history of the entire crystal growth process and its termination. One has to pay attention to the final stages of crystal cooling, as they directly affect defect density.

Nakanishi et al. [32] studied the influence of annealing during the growth of silicon crystals. They used two silicon crystals that were grown based upon the measured temperature profile and with different cooling processes. One was quenched directly from about 1000°C to room temperature, and the other was not quenched but gradually cooled to room temperature, as are crystals grown using the standard CZ method. The quenched crystal with super-saturated oxygen differed greatly from the standard grown crystals.

In the quenched crystal, the density of defects (silicon oxide precipitates) detected after heat treatment at 1000°C for 12 h was reduced by about three orders of magnitude more than the standard grown crystal when the crystals contain about 1×10^{18} oxygen atoms per cm^3.

Three types of microdefects, formed by the agglomeration of either self-interstitials or vacancies, have been observed in dislocation-free single crystals grown using the pedestal pulling technique (crucible-less growth). Based on the microdefect distribution in quenched crystals, Roksnoer [33] concluded that the dominant type of point defect at the melting point of silicon is the vacancy, whereas an excess of self-interstitials is usually present at lower temperatures.

The interstitial and substitutional defects of transition metals generate deep energy levels in the forbidden gap of silicon. These defects are particularly found in crystals grown using the FZ method [34]. In cobalt-doped crystals, it is generally assumed that the energy levels observed with DLTS are caused mainly by the substitutional defects, with a donor level at $E_v + 0.4$ eV and an acceptor level at $E_c - 0.38$ eV. DLTS studies by Bleka et al. [35] were performed on p^+-n-n^+ silicon diode detectors produced from low-doped ([P] = 5.0×10^{12} cm^{-3}), high-purity FZ wafers. After irradiation with 6 MeV electrons to a dose of 5.0×10^{12} cm^{-2}, the well-known vacancy-oxygen (VO), doubly negatively charged vacancy – vacancy (V_2 $^{(=/-)}$ or simply $V_2(=/-)$), and singly negatively charged vacancy-vacancy ($V_2(-/0)$) complexes were observed. There was the clear presence of a defect (X) with an energy level slightly shallower than that of $V_2(-/0)$ and the concentration was about one-quarter of that of V_2. Further analysis of the defect properties reveals an energy level 0.37 eV below the conduction band and an apparent capture cross-section of 1.0×10^{-14} cm^2.

CZ crystal growth processes that provide vacancy dominance in the crystal are usually preferred and pursued because oxygen precipitation and extended defect formation are more easily controlled during subsequent wafer processing [36]. Point defects in FZ silicon, which has oxygen levels about two to three orders of magnitude lower than CZ silicon, form some distinct secondary defects. Interstitials form A-defects, and vacancies form D-defects as a result of agglomeration due to supersaturation *in situ* during crystal growth. Point defects (and especially their secondary defects) in CZ silicon are somewhat different from those in FZ silicon because of the higher oxygen concentration in the CZ silicon.

Nakamura et al. [37] analyzed the formation process of grown-in defects in CZ-grown silicon crystals and the simulation of point defect diffusions. They explored the mechanism of the grown-in defect formation in CZ silicon. They also determined a set of diffusion coefficients (D_v: vacancies, D_i: self-interstitials) and equilibrium concentrations (C_v^{eq}: vacancies, C_v^{eq}: self-interstitials) of point defects that has been satisfied with the dependence of two-dimensional defect patterns on growth rate and axial temperature gradient and with the reported product values of D_vC and $D_iC_i^{eq}$. Based on the TEM observations that the grown-in defects in the vacancy-dominant region are voids with an octahedral shape, the idea that the grown-in defects are the voids formed by the vacancy aggregation has been examined by the simulation model. This model describes well the behaviors of grown-in defect formation during the crystal growth and shows that it is possible for a void to form as the grown-in defect in the silicon.

Three kinds of defects have been reported as grown-in defects related to the GOI yield. These are named according to their observation techniques, such as infrared (IR) scattering centers detected through the use of laser scattering tomography (LSTDs), the small COPs revealed by SC-1 cleaning, and flow pattern defects (FPD) consisting of etch pits revealed by Secco etching. It was reported that the GOI yield in every CZ wafer was in proportion to the density of each of the three types of defects. According to Nakamura et al. [37], it is presumed that the conditions that controlled formation of the three types of defects are the same, or that the defects are of the same origin.

Wijaranakula [38] performed a real-time simulation using the numerical method in order to study the aggregation process of the intrinsic point defects during CZ crystal growth. The point defect aggregation was dominated by the recombination process between silicon interstitials and the vacancies. During the initial cone growth, the crystal experienced both a large fluctuation in the growth rate and a significant radial temperature gradient during the transition period at the solid-melt interface from a convex to a concave shape. These conditions initiated an aggregation of point defects. Figure 9.6a shows the x-ray topography of the cone section. The transient behavior of these point defect distributions was analyzed by combining the thermal history of the crystal at various positions. Figure 9.6b and c show the temperature distribution in the crystal at t = 0 and 3 h, respectively. The isotherms are plotted in 100°C increments. Here, the isotherms show radial temperature gradients near the cone surface, while the isotherms become flatter as they approach the solid-melt interface. In silicon saturated with interstitial oxygen the formation of vacancy-oxygen pairs contributes to silicon interstitial saturation, which in turn gives rise to a formation of interstitial-type defects. Therefore, in addition to the crystal growth rate and the temperature gradient, the concentration of dissolved interstitial oxygen species plays a significant role in the aggregation process of point defects. Grown-in point defects and microscopic defect formations were theoretically analyzed by Tiller et al. [39] in the one-dimensional, steady-state, and neutral species-only approximation limit. Many research groups are actively engaged in analyzing these point defect(s) by different models [40–43], thermal stress issues [44–46], crystal rotation effect on defect generation [47], magnetic effect [48], in the presence of interstitial oxygen species [49, 50], and their properties. The issues are many, but our goal is achieving perfect single-crystal silicon.

Larger crystal defects, or simply defects, in single-crystal silicon is a major concern for silicon semiconductor devices, as they affect how the device functions and for how long, and leakage current. Early experiments [51–60] in growing a perfect crystal provided insight into the bonding behavior among silicon atoms and the nonsilicon atoms present in the crystal lattice. Defect evaluation was carried out using different techniques.

Lägel et al. [61] used a scanning Kelvin probe to study the defects and chemical contamination present in silicon wafers. This is a noncontact, no-wafer-preparation detection system for defects and chemical contamination on the surfaces. They have shown that the dark work function indicates changes in surface charge due to contamination, whereas measurement of the saturation surface photovoltage allows the total surface charge to be determined. The results of the measurements show that iron contamination on silicon wafers induces a negative charge in the surface layer. With this method, one can evaluate the presence of iron at the defect locations. Other evaluation methods include IR light-scattering tomography [62–64], high-resolution electron microscopy (HREM) with *in situ* electron irradiation [65], minority carrier lifetime [66], and photoluminescence [67] methods.

9.4 Line Defects

Line defects are larger and are considered a major crystal defect. The presence of such defects is detrimental for wafer processing. They may not be easy to identify initially, but the defect will grow very fast, and with each thermal cycle, and the wafer may not

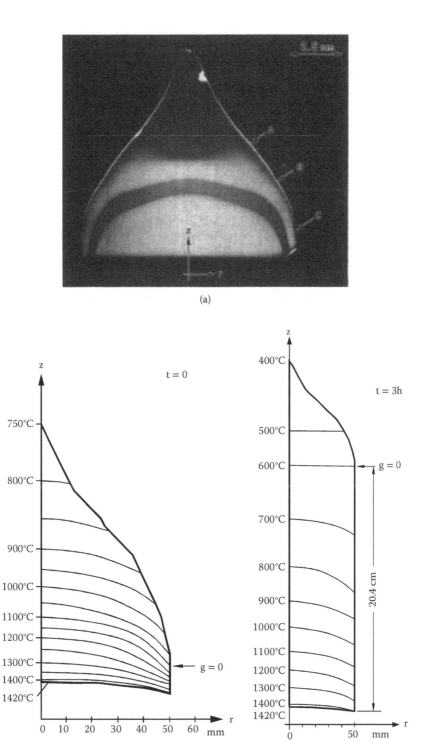

FIGURE 9.6
(a) X-ray topography of the cone section after 100 min at 1100°C. Points A, B, and C refer to different growth rate points. Isotherms in the crystal (b) at $t = 0$ h, and (c) at $t = 3$h. (From W. Wijaranakula, *Journal of the Electrochemical Society*, **139**, 604–616, 1992 [38].)

reach the end stage of processing. These defects can be detected with chemical etching. Mechanical as well as thermal shocks and/or pressure, either due to the weight of the large wafers or due to handling pressure, are possible reasons for these defects. Maruyama and Okada [68] have reported that a pressure on the order of 2.99×10^{10} dynes/cm^2 from the pressure of 400 g and a point contact area of 1.31×10^{-5} cm^2 is enough to create such defects in a crystal.

9.5 Bulk Defects and Voids

The performance of MOSFET devices, especially the GOI and the refresh times of high-density DRAMs, is degraded by grown-in secondary defects or microdefects. As mentioned, these secondary defects evolve from an agglomeration of the vacancies, interstitials, and/or oxygen, which become supersaturated *in situ* during the crystal growth batch process. Secondary defects or microdefects that degrade the GOI include FPDs, COPs, and other grown-in defects. In general, the grown-in defects are referred to as LSTDs, after the detection technique. LSTDs are affected by the *in situ* dwell time during crystal growth at a temperature of about 1050°C [36]. The formation of LSTDs is believed to be facilitated by the vacancies at that high temperature value. Oxygen supersaturation alone, however, cannot provide a sufficient driving force for homogeneous nucleation at high temperatures.

Ben-Sira and S. Bukshpan [69] initially reported on the spontaneous generation of dislocations in the silicon crystals and a large concentration of etch pits at the peripheries of the wafers. Kalaev et al. [70] presented 3D unsteady analysis of the silicon melt turbulent convection coupled with heat transfer in the crystal and crucible during 300 mm CZ silicon crystal growth to calculate bulk defects. The impact of melt turbulence fluctuations on the formation of defects was analyzed.

Yang et al. [71] reported that germanium doping in silicon crystals can effectively suppress void defects. During the silicon crystal growth, the combination of vacancies present in the lattice and doped germanium atoms prior to the vacancy aggregation form voids, which relaxes the strain originating from the mismatch of germanium atoms in the silicon crystal lattice and reduces the concentration of free vacancies quite significantly. This reduction of free vacancies, on one hand, leads to a suppression of grown-in voids in the lattice; on the other hand, it decreases the formation temperature and thus, the size of the voids, leading to poorer thermal stability as verified by the efficient elimination of flow pattern defects in germanium-doped CZ silicon at relatively lower temperatures. Figure 9.7a and b show the optical microscope photographs of FPDs in the head samples in the as-grown CZ and Ge-doped CZ silicon samples. The difference is apparent. Figure 9.7c shows the density variation when these crystals are subjected to annealing in argon ambient. The defects reduced significantly when compared to the as-grown silicon crystals.

Voronkov and Falster [72] developed a model of void formation in silicon crystals from a supersaturated vacancy solution in the presence of nitrogen and oxygen. At the void formation stage, the presence of nitrogen is represented mostly by dimeric species, but the major vacancy traps are single nitrogen interstitials. The trapping is remarkably strong, and yet the voids are easily produced at a lower temperature and in increased density. These results are in quantitative agreement with the reported data. The role of oxygen impurity here is to reduce the void surface energy and to enhance the pairing/dissociation rate of nitrogen in the silicon lattice.

Akatsuka et al. [73] developed a simulation model that calculates the nucleation and growth of void defects during CZ silicon crystal growth. The growth of inner oxide films of voids is also incorporated in this model. It predicts the void size and its density in the

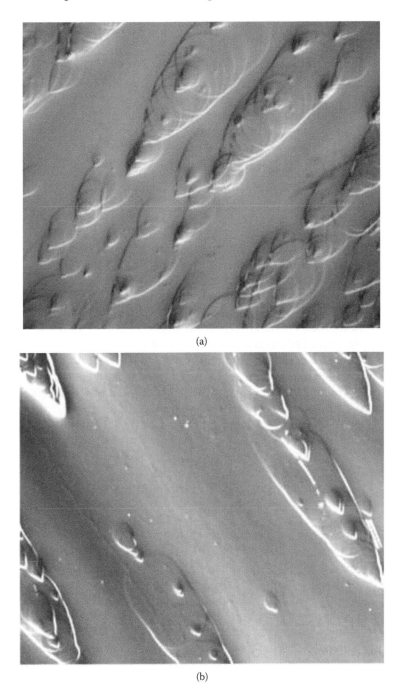

(a)

(b)

FIGURE 9.7 *(Continued)*

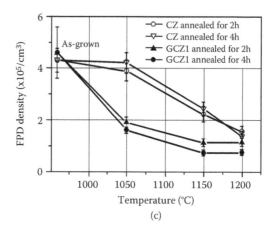

FIGURE 9.7
Optical microscope photographs of flow pattern defects in the head samples of (a) CZ and (b) Ge-doped CZ silicon crystals. (c) Flow pattern defect densities in the CZ and Ge-doped CZ silicon as a function of annealing temperature and time. (From D. Yang, X. Yu, X. Ma, J. Xu, L. Li, and D. Que, *Journal of Crystal Growth*, **243**, 371–374, 2002 [71].)

grown crystals. The void strongly affects the GOI, which actually relates to the cooling rate of the crystal and the concentration of nitrogen present in the lattice sites. This model could simulate not only the behavior of void defects in ordinary crystal growth, but detaching or halting crystal growth methods as well. Both the size distribution of the voids and the thickness of the inner oxide film agreed with the experimental results with high accuracy. Figure 9.8 shows the experimental and calculated results of void size distribution in 300 mm diameter crystal. The oxygen concentration [Oi] of these crystals was in the range of $11.0–14.0 \times 10^{17}$ atoms/cm^3. The calculated results agreed well with experimental results found in 300 mm diameter crystals. The calculated peak sizes of voids were approximately 240 and 170 nm for crystal D and E, respectively [73] in the samples studied by the authors. These results imply that the current model developed by this team can simulate void size distribution in both 200 mm and 300 mm diameter crystals. Figure 9.9 shows the experimentally obtained void size and density in the lightly boron-doped crystals, with a diameter of 150 mm, that were detached or detached and followed by a halt for 5 h at various temperatures during the crystal growth. The void density shows no difference between detaching and detaching followed by a halt. The model and experimental results strongly support that the proposed model for the void size and density to the CZ-grown silicon wafers are highly accurate.

Itsumi [74] reported on the dual-type octahedral void defects in CZ silicon. The formation of these defects is closely related to the agglomeration of lattice vacancies during the crystal growth process. As discussed earlier, the point defects and their relative prevalence is determined by growth conditions. It is widely believed that the voids are vacancy-related defects or the agglomeration of (supersaturated) vacancies resulting in the formation of voids. One such dual-type octahedral void defect is shown in Figure 9.10a, and its schematic illustration in Figure 9.10b. Here it is truncated by two (001) subplanes on the top and bottom. The TEM observation of the void shows that the bigger void is truncated by two (001) subplanes on the top and bottom. The smaller void is also truncated by two (001) subplanes on the top and bottom. Considering that the (001) subplane occurs during the void shrinkage process in the thin sample, the author believes that a factor

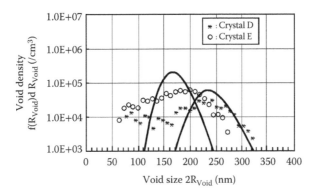

FIGURE 9.8
Size distribution of void defect in 300 mm wafers with different growth rates. (From M. Akatsuka, M. Okui, S. Umeno, and K. Sueoka, *Journal of the Electrochemical Society*, **150**, G587–G590, 2003 [73].)

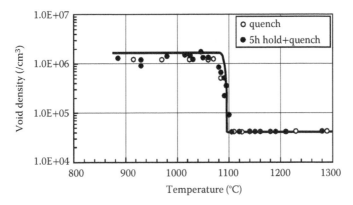

FIGURE 9.9
Void density in quenched or quenched followed by halted for 5 h. (From M. Akatsuka, M. Okui, S. Umeno, and K. Sueoka, *Journal of the Electrochemical Society*, **150**, G587–G590, 2003 [73].)

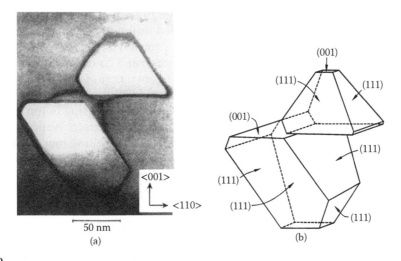

FIGURE 9.10
(a) TEM image and result of grown-in void defect in CZ-grown silicon wafer and (b) schematic illustration. (From M. Itsumi, *Journal of Crystal Growth*, **237–239**, 1773–1778, 2002 [74].)

causing the occurrence of the (001) subplane remains in the thin CZ silicon sample. The most probable assumption is that some local fluctuations in point defect concentrations or thermal stress in the crystal growth direction are responsible for this.

9.6   Dislocations and Screw Dislocations

Joshi [75] reported on the effect of fast cooling on imperfections in silicon. Studies on misfit dislocations in phosphorus-diffused silicon show that all the atoms suffer rearrangement on fast cooling. These dislocations extend in specific directions and act as centers for phosphorus precipitations. Different chemical etchants [76–78] are popular to study the formation of defects and defect-related issues.

Kawado et al. [79] studied the propagation and elimination of dislocations generated at the early stage of CZ silicon crystal growth using synchrotron white x-ray topography combined with a topotomographic technique. Two silicon crystals with [001] growth axes were examined for this purpose. The first one was intentionally grown without enlarging its diameter to easily observe the features of the dislocation propagation, and the other was grown with Dash necking, followed by a 2-inch enlargement of its diameter in order to observe the elimination of the dislocations. The three-dimensional structure of the individual dislocation (i.e., the direction of the dislocation line), its Burgers vector, and the glide plane were subsequently determined. These investigations revealed that the dislocation half-loops, which were generated from tangled dislocations, were expanded on their glide planes and were often deformed by their interaction, cross-slip, and collision with the crystal surface, followed by a gradual decrease in their density. The dislocation-elimination effect of the Dash necking was caused by the expansion of the dislocation half-loops being terminated within the crystal and by their pinning on the crystal surface of silicon.

Dislocations of the 60° type can cause an increase of several orders of magnitude in the generation rate [80]. In *n*-type silicon, this behavior can be described by the Shockley-Read recombination-generation center with an energy level at a distance of 0.06 eV from the center of the energy gap. If this level is situated in the upper half of the energy gap, it probably acts as a donor; if in the lower half, it probably acts as an acceptor in silicon. From the viewpoint of crystal growth, doping of boron and co-doping of germanium and boron impurities are effective in suppressing the generation of slip dislocations due to the preferential segregation on them [81]. It is reported [82] that the dislocation propagation velocity in boron-doped silicon seed decreased with increasing boron concentration value in the seed. Dislocations also affect the crystal-melt interface. Wang and Kakimoto [83] reported that in the case of low dislocation density, the melting took place uniformly, and the shape of the crystal-melt interface was flat. In the case of high dislocation density, inhomogeneous melting was observed and the shape of the crystal-melt interface was not flat. The inhomogeneous melting was related to a large amount of dislocations induced by thermal stress.

The thermal stress caused by transient temperature variation is known to induce a collective motion of dislocations at surface irregularities, residual micro-structural defects in wafer edges, and at oxide precipitates, accompanying punched-out dislocation loops in bulk silicon during thermal heat treatments. In a vertical furnace, the circular wafers are set horizontally on supporting jigs, often on the point-contact types. At the fulcrum of the jigs, the wafers bow elastically due to the gravitational stress from the weight of the wafers. This stress potentially creates nucleation sites and promotes the collective motion of slip dislocations in the wafer. Shimizu et al. [84] studied these gravitational stress-induced dislocations in large-diameter silicon wafers. They described an initiation and a collective motion of dislocations under the gravitational stress in 200 mm diameter silicon wafers during high-temperature device processing, and characters of induced dislocations were investigated using x-ray diffraction topography. By using a three-point jig, in the case of a single wafer, the evaluation was carried out to show that the dislocations multiply. Figure 9.11a shows how the dislocations multiply in three cases under different single-wafer heat treatments using a three-point jig. This figure shows an x-ray topograph where the wafer was subject to direct heat up to 1200°C and was held there for just 1 min. This case did not give rise to heavy slip, indicating that thermal stress and gravitational stress were not high enough to cause a collective motion of slip. On the contrary, after preheat treatments at 400°C and 600°C for 30 min, the single-wafer heat treatments up to 1200°C for 30 min led to multiplied dislocations from the supporting jigs. These multiplications are shown in Figure 9.11b and c. Three types of half-loop dislocations were identified to be 60° terminating at the surface with screw-type characteristics on two different (111) planes. It is further reported that these were dislocations with Burgers vectors of [011] in $(\bar{1}\,\bar{1}\,1)$, [101] in $(\bar{1}\,\bar{1}\,1)$, and $[01\bar{1}]$ in (111) planes. Based on the gravitational stresses estimated by the finite-element method (FEM), the authors computed the gravitational stress in a 300 mm diameter wafer. By comparing the gravitational stress with the critical stress to multiply slip dislocations, conditions for suppressing slip bands were predicted for silicon wafers larger than 300 mm diameter.

In the TEM images of silicon crystals, extinction of fringes were observed by Hashimoto et al. [85] due to the periodic intensity distribution of electron waves. These extinction fringes, called Pendellösung fringes, have appeared sometimes as "thickness fringes" and sometimes as "inclination fringes." Pendellösung fringes were observed by Chen [86] in reflection-section topographs of bent single crystals of silicon. The number of fringes increases while the distance between fringes decreases with the decreasing radius of curvature. Excessive bending of the lattice planes adjacent to a dislocation produces finer fringes. These fringes may be employed as a means to detect minute lattice strains.

Ciszek [87] produced experimental evidence that the <115> crystallographic directions are viable orientations for dislocation-free silicon float-zoning, and <115> crystals have several properties of interest for MOS applications, as well as other applications where

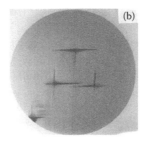

FIGURE 9.11
Dislocation multiplication in 200 mm diameter silicon wafer under different single-wafer heat treatments using a three-point jig: (a) 1 min at 1200°C, (b) 30–36 min at 400°C–1200°C, and (c) 30–30 min at 600°C–1200°C. (From H. Shimizu, S. Isomae, K. Minowa, T. Satoh, and T. Suzuki, *Journal of the Electrochemical Society*, **145**, 2523–2529, 1998 [84].)

uniform doping is of importance. The relatively high (115) surface-free energy allows more uniform growth, particularly for dopant incorporation, than is characteristic of low-surface, free-energy planes such as (111). The difference between [111] and [115] FZ crystals are expected to hold true in CZ crystals also.

9.7 Swirl Defects

Swirl defects are found in large dislocation-free silicon crystals. They occur in two variants, A-swirl and B-swirl defects, and are named for the typical spiral, or swirl-like, pattern they present. During crystal pulling, the growth interface is convex at the start and becomes concave at the end of crystal growth due to heat flow conditions. Since silicon grows layer by layer in absence of dislocations, when a wafer is cut, circular defects are present in the wafer corresponding to each layer grown. These defects are cluster-of-point defects. The following photograph explains these swirl defects. A preferentially etched and delineated silicon wafer with these defects can be visualized under high-magnification microscope. As the figure shows, there are many small defects, called the B-swirls (white dots), and a much smaller number of larger defects, the A-swirls (black-white contrasts) [88]. The details are shown in Figure 9.12. Quantitative evaluation of the micrograph shows that B-swirls are delineated as small and shallow pits, whereas the A-swirls are delineated as hillocks. There is a good correspondence between the striated pattern consisting of hillocks and the buried microdefects. These defects form a perfect dislocation loop cluster and tetrahedral precipitate, respectively. In addition, a kind of tiny microdefect is found to be distributed preferentially in the vicinity of the swirl pattern, although there is no detectable correspondence between the hillocks and the microdefects. The experimental results of Fan et al. [89] show some indications of the existence of oxygen and carbon in the core of the precipitate and suggest that oxygen and carbon may play important roles in the formation of these swirl defects. These swirl defects are mainly generated by the agglomeration of point defects while the crystal is cooling.

Föll and Bolbesen [90] reported that a single swirl defect consists of a dislocation loop or a cluster of dislocation loops. Contrast experiments have shown that these loops are formed by the agglomeration of self-interstitial silicon atoms. Generally, the loops have a/2 <110> Burgers vectors, but in specimens with high concentrations of carbon ($\sim 10^{17}$ cm^{-3}) and oxygen ($\sim 10^{16}$ cm^{-3}), dislocation loops that included a stacking fault were observed. In crystals grown at growth rates higher than V = 4 mm/min, no swirls are observed; lower growth rates do not markedly affect the size and shape of these dislocation loops. With decreasing oxygen and carbon content, the swirl density decreases, whereas the dislocation loop clusters become larger and more complex. The authors presented a model to describe the formation of swirls in terms of agglomeration of silicon self-interstitials and impurity atoms present in the crystal lattice.

The occurrence of swirl defects and the amplitude of dopant striations are strongly influenced by the cooling rate during the crystal growth process. Roksnoer et al. [91] reported that the formation of swirls can be prevented if the cooling rate does not exceed 5°C/min. At this cooling rate, the concentration of thermal defects decreases below the critical value necessary for swirl formation by diffusion to the crystal surface, permitting homogeneous formation of silicon crystals. Explaining the mechanism by which swirl defects form, Voronkov [92] confirmed that the influence of growth rate (V)

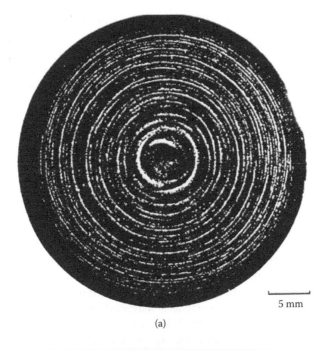

(a)

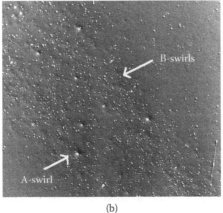

(b)

FIGURE 9.12
(a) Typical spiral or swirl-like pattern seen in grown crystals. (b) A- and B-swirl defects in grown crystals. (From www.tf.uni-kiel.de/matwis/amat/def_en/kap_1/illustr/t1_3_5.html. Prof. Helmut Föll, University of Kiel, Faculty of Engineering, Germany, 1976 [88].)

and the temperature gradient (G), the first stage of defect formation, is due to recombination and diffusion of vacancies and self-interstitials in the vicinity of the crystallization front.

Yasuami et al. [93] analyzed swirl defects in dislocation-free CZ silicon using x-ray topography with two annealing processes. The first annealing was done in nitrogen ambient for 2 h, and Pendellösung fringes were clearly observed. The second annealing in steam at 1200°C decorated the defects so as to reveal the configuration in the stage of the first annealing. When the temperature of the first annealing was sufficiently high (above 800°C), some of the grown-in defects migrated and aligned along the growth layers.

When the temperature was below 800°C, they could not migrate as much and remained along the crystal pulling directions.

Eyer et al. [94] used a closed double-ellipsoid mirror heating facility in conjunction with the FZ technique to grow dislocation-free single crystals with a relatively small diameter. Micro-inhomogeneities such as swirl defects were evaluated by etching and x-ray topography.

Oxygen is the most predominant impurity in CZ-grown silicon and has been found to affect various stages of thermal device processing. Oxygen precipitates (SiO_2) in a swirl-like pattern in CZ-grown silicon are often referred to as swirls, while no SiO_2 precipitation is observed in FZ silicon. This leads to the question as to whether or not the swirls in CZ silicon act as nucleation centers for SiO_2 precipitates. Murgai et al. [95] reported that for forced convection conditions of growth, the growth rate of oxygen segregation is controlled microscopically. Maxima in oxygen concentration occurred at minima in the microscopic rates, indicating that the segregation coefficient is greater than unity. The distribution of swirl defects was found to be critically dependent on the microscopic growth rate fluctuations; this dependence made it possible to show that the critical impurity nuclei are not related to oxygen, but to some impurity, most likely carbon, with a segregation coefficient less than unity. Further, it was concluded that the segregation coefficient of oxygen in silicon is greater than 1, consistent with earlier reports available in the open literature. A closer examination of the swirl defect distributions indicates that the striations do not coincide with the dopant striations; instead, the swirl defects are located at the seed side of the dopant striations.

Swirl defects in quenched, dislocation-free FZ silicon crystals have been analyzed by Petroff and De Kock [96] using TEM. Silicon crystals were quenched from the melt to prevent impurity clustering and precipitation and to minimize their effects on swirl defect nucleation. The analysis shows that the A-swirl defects are perfect extrinsic dislocation loops elongated along the <100> directions. The smaller B-swirl defects remained undetectable by TEM because they cause very little lattice strain. From this, it was concluded that the silicon self-interstitials are the dominant point defects during the crystal growth process. An impurity interstitial clustering mechanism was proposed to explain the swirl defect formation during the growth of FZ single-crystal silicon. It is also reported that during the growth of FZ crystals, clusters of A-swirl and B-swirl defects form as a result of point defect condensation. The diffusion coefficient of these point defects in the temperature interval 1050°C to 1100°C is found to be 2.0×10^{-5} cm²/s. Section topographic analysis indicated that the A-swirl clusters rapidly increase in size as the crystal cooling rate decreases [97]. This is attributed to a combined climb-glide process.

From the studies of x-ray topographic and etching experiments, Chikawa and Shirai [98] found that dislocated crystals were found to melt homogeneously from their surfaces. Micro droplets of the silicon liquid, locally molten regions were observed inside dislocation-free crystals, simultaneously with melting from their surfaces. The formation process of the microdefects initiated from the droplet formation is explained by assuming some kind of impurities present are acting as the absorption centers for infrared radiation penetrating the crystals. As an origin of swirl defects in bulk crystals, the possibility of droplet formation near the growth interfaces during remelting periods is suggested by them to explain these formations.

New experimental data by de Kock and van de Wigert [99] strongly suggests that the formation of the different types of swirl defects is due to the parallel condensation process involving silicon interstitials and the vacancies. It is proposed that during CZ growth, three types of swirl defects, A, B, and C, can form. Doping with donor concentration species (particularly P, As, and Sb) more than 1.0×10^{17} cm^3 suppresses the formation of A-swirl defects. Doping with acceptors (B and Ga) with concentration values more than 1.0×10^{17} cm^3 eliminates the formation of B- and C-swirl defects. The observed doping effects are explained in terms of complex formation as a result of Coulomb attraction between dopants and charged thermal point defects.

9.8 Stacking Faults

These defects seriously affect device performance and yields, either alone or by causing further defects, particularly in epitaxial layers grown after their generation. These defects generally form in bulk silicon crystals during both crystal growth and cooling. Dyer and Voltmer [100] have reported that circular- and hexagonal-shaped stacking faults formed during the annealing treatment of melt-grown silicon crystals were found to contain radial line defects. The source of circular stacking faults has been attributed to the line defects and to precipitates. Figure 9.13 shows circular defects on one of the slices etched with Sirtl etchant. They appear as circles and grooves in the same vicinity lying in the <110> directions. The circular defects also have one, two, five, and six spokes. The spokes within the rings were identified as the dislocation dipoles, and the arcs of each ring have the same Burgers vector. The fact that the spokes are dipoles implies that they were formed by dislocations arriving on parallel but separate plates. Spherical precipitate formations are likely due to the circular faults on four intersecting {111} planes. Boundaries of stacking faults are concluded to be partial dislocations of the Shockley type [101].

Futagami [102] demonstrated that decorated faults could be distinguished from clean stacking faults using the Sirtl etching at any etching period. The decorated faults were delineated as hillocks, while the clean faults were delineated as flat and geometric pit-features by the Sirtl etch. The Sirtl etch was more effective in delineating the decorated stacking faults than were the Secco and Wright etches. Stacking faults free of precipitates are reported to be nearly inactive electrically. The influence of OISFs on the electrical characteristics of *p-n* junctions in silicon has been studied by Ravi et al. [103], who employed SEM and TEM in conjunction with electrical measurements. They found that the smaller faults, which are also more prone to impurity decoration, are electrically more active than larger faults, which are decorated to a lesser degree. IR light scattering tomography [104] is also used to analyze these oxidation-induced stacking faults.

FPDs often occur in vacancy-rich CZ silicon sections. Those defects are similar to the D-defects in FZ silicon caused by vacancy clustering. On the other hand, A-defects in FZ silicon are not observed in CZ silicon. However, at the boundary between the vacancy and interstitial-rich region in a CZ silicon wafer, an annular segment of high OISF density, commonly called an OISF ring or edge swirl, can occur [36]. The section inside the OISF ring is vacancy rich, so one finds a relatively high density of FPDs, as well as COPs and light point defects (LPDs). COPs are believed to have similar physical origins as FPDs. Enhanced precipitation of oxygen provides a further indication of excess vacancies inside the OISF ring. The outside area of the wafer, a region of interstitial dominance, has relatively few defects. There is a low density of dislocation loops in an area where it is somewhat more difficult to nucleate oxygen precipitation.

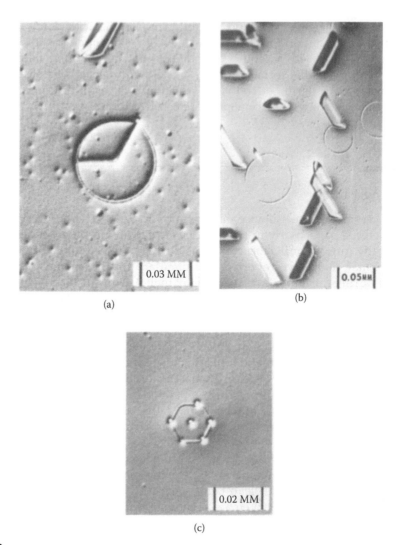

FIGURE 9.13
(a) and (b) Circular and (c) hexagonal stacking faults in <111> bulk silicon crystals. (From L. D. Dyer and F. W. Voltmer, *Journal of the Electrochemical Society*, **120**, 812–817, 1973 [100].)

For many years, OISFs were a problem in single-crystal silicon grown by the CZ method. An OISF is a type of lattice defect that can act as a sink/source for impurities. Fundamental causes of OISFs originate in the kinetics of crystal growth. The most dominant factor responsible for OISFs is the temperature of the molten silicon. Especially in the case of VLSI circuits, it is essential to produce OISF-free oxidized wafers. Bawa et al. [105] studied the dependence of OISF and oxygen incorporation upon the melt temperature. They discussed the perceived relationship between the two in detail.

Studies on the mechanism of OISF formation have been restricted almost exclusively to CZ crystals. It has been shown how the presence of a high concentration of oxygen species, which is typical of most CZ material, has complicated establishing the main causes for the formation of OISFs and their behavior in subsequent processing steps. However, it was observed that FZ material, which contains much less oxygen, did not show these complications. Because of this difficulty with CZ material, Dieleman and Martens [106] studied the formation of OISFs during oxidation of FZ silicon material. The crystal orientation of FZ silicon has a huge effect on the number of OISFs formed at the surface. Silicon <100> can be oxidized without any stacking faults forming; the same oxidations introduce numerous stacking faults in <111> silicon. The most striking result is that <100> silicon does not show any OISF after H_2O and HCl oxidations, and only a few after the oxygen oxidation. In contrast, <111> silicon has numerous OISFs for the same oxidation steps. Wet oxidation of <111> silicon induces about 1.0×10^7 OISF/cm^2; their size spread averages about 0.7 μm. Oxygen and HCl oxidation of <111> silicon induce about 1.0×10^4 OISF/cm^2 with a size spread averaging about 2.0 μm.

The nuclei for the formation of OISFs are formed during the cooling of the as-grown silicon crystal. The nature of stacking faults was identified some time ago, but their typical ring-like distribution has remained unexplained. In later experiments, Dornberger and Ammon [107] reported that the ring diameter is a sensitive function of the crystal pull rate and temperature gradients of the crystal. Further experiments showed that the OISF ring separates a region of high defect density (flow pattern/D-defects) and poor gate oxide quality inside the ring from a region of low defect density (large pits) and excellent gate oxide quality outside the ring. Experimental and theoretical investigations of this phenomenon have considerable technological relevance, as the gate oxide quality of silicon wafers has an impact on the final yield of devices fabricated using these silicon wafers. An important fact is that the predominant defect type changes across the OISF ring—from vacancy to interstitial defects. The suggestion is strongly supported by the enhanced oxygen precipitation inside the OISF ring and the sudden change across the ring to a suppressed precipitation outside the ring. More research is required in these areas. None of the theoretical models are able to quantitatively predict the variation of the OISF ring diameter with the crystal pull rate. A major obstacle has been the contraction and, finally, the disappearance of the OISF ring in the wafer center, which cannot be explained by radial out-diffusion of point defects. The radial variation of the OISF ring was investigated in a matrix experiment, where the crystal diameter, the thermal environment of the growing crystal, and the pull rate V were varied. The details are shown in Figure 9.14. It is seen that the OISF ring disappears in the crystal center, if $V/G = C_{crit} = 1.3 \times 10^{-3}$ cm^2 min^{-1} K^{-1}. Vacancy-related defects (D-defects) are formed if $V/G > C_{crit}$, and interstitial-related defects (large pits) are formed if $V/G < C_{crit}$. This behavior is reported for CZ crystals with a boron concentration of less than 5.0×10^{16} atoms/cm^3 and diameters between 4 and 8 inches grown with different heat shields.

According to Nakamura et al. [37], it is believed that the outside region of the OISF ring is self-interstitial dominant, its inside region is vacancy dominant, and D-defects are generated by the vacancies. It is well known that the position of the OISF ring can move from the crystal center to the crystal surface, depending upon the growth rate. The faster the crystal growth rate, the closer to the crystal surface the OISF ring approaches. Conversely, the slower the crystal growth rate, the closer the OISF ring moves to the center of the crystal. When the crystal growth rate is further slowed down, the OISF ring contracts toward the crystal center and disappears at a critical growth rate. The radius of the OISF ring also depends on temperature gradient. The relationship between the radius of the OISF ring and the V/G ratio is shown in Figure 9.15.

Choi et al. [108] studied the mechanical damage on structural and electrical properties of silicon wafers. It was found that as the degree of mechanical damage increases, the

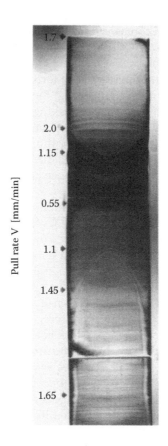

FIGURE 9.14
Axially cut slice of a 4-inch silicon crystal grown with heat shield and varying pull rate. The sample was heat treated for 4 h at 750°C and 16 h at 1050°C, and Secco etched for 8 min. (From E. Dornberger and W. von Ammon, *Journal of the Electrochemical Society*, **143**, 1648–1653, 1996, and the references therein [107].)

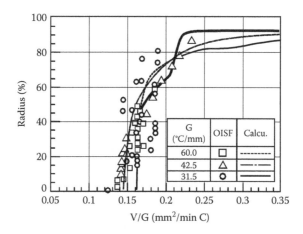

FIGURE 9.15
Radius of OISF ring and the calculated vacancy cluster region of 6-inch crystal as a function of V/G. The plots are the radius of the OISF ring, and the lines are the radius of the calculated vacancy cluster region. V, growth rate; G, axial temperature gradient. (From K. Nakamura, et al., *Journal of Crystal Growth*, **180**, 61–72, 1997, and the references therein [37].)

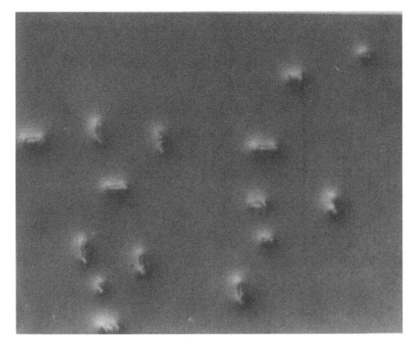

FIGURE 9.16
Stacking faults in silicon at 1000X. (From S. P. Murarka, T. E. Seidel, J. V. Dalton, J. M. Dishman, and M. H. Read, *Journal of the Electrochemical Society*, **127**, 716–724, 1980 [109].)

density of OISFs is increased, while the depth of OISF generation is almost independent of damage grade. The OISF density measurement is coincident with the lifetime result, which supports the suggestion that dislocations and/or stacking faults generated by mechanical damage play a role in carrier recombination near the surface of the wafers.

Murarka et al. [109] conducted systematic analysis of the stacking faults during CMOS device processing, covering the origin, elimination, and contribution to device leakage current. They investigated the formation of stacking faults in silicon at the oxide-silicon boundary. According to them, it is apparent that the large density ($\geq 10^4$ per cm^2) of OISFs was common during the processing, but not always associated with the wet oxidation following the high-temperature boron drive-in step to create deep junction steps. Figure 9.16 presents high-magnification photomicrographs showing the stacking faults in the ungettered regions of the process-bound silicon wafer. At a lower-temperature wet oxidation step, such as 1050°C, the density of stacking faults is reported to be less than 1000/cm^2.

Winkler and Sano [14] reported on the silicon dioxide gate integrity with different pull rates using various hot zone modifications to improve the oxide integrity. Quality of wafers was characterized by crystal defect density, as well as the GOI, and it was found to be strongly related to the crystal pulling conditions. Crystal pulling rate and subsequent thermal history of the crystal cooling process has strong effect on bulk parameters, such as grown-in defects, OISFs, and oxygen precipitates. Reducing the growth rate introduces a ring-like region, which appears at the center of the wafer, where a significantly higher stacking fault density is observed on the surface, as well as in the bulk after the wafers have undergone a test for oxidation. This OISF ring region is separating two areas on the wafers with different gate oxide quality, as shown in Figure 9.17 and Figure 9.18.

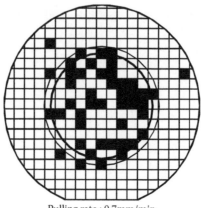

Pulling rate : 0.7mm/min

FIGURE 9.17
Wafer map showing all capacitors that failed in the gate oxide breakdown test. Outside the OISF ring, no break-down was recorded. (From R. Winkler and M. Sano, *Journal of the Electrochemical Society*, **141**, 1398–1401, 1994 [14].)

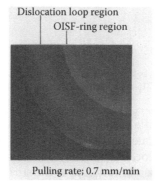

Pulling rate; 0.7 mm/min

FIGURE 9.18
X-ray topograph of silicon wafer after copper decoration showing two concentration areas inside and outside the OISF ring with the high-dislocation loop density separated by the OISF ring. (From R. Winkler and M. Sano, *Journal of the Electrochemical Society*, **141**, 1398–1401, 1994 [14].)

Inside the OISF ring region (center area), the GOI is poor when compared to the region that is outside the ring portion. At the peripheral area, the devices showed an almost-perfect GOI. These results prove that the bulk parameters strongly influence the GOI and are mainly related to the failures of DRAM. Point defect dynamics and an OISF ring in CZ crystals were also reported by Sinno et al. [110], who also proposed a model on the dynamics of intrinsic defects.

The parameter V/G ratio plays a dominant role in determining the prevalent defect type in a growing crystal. Ammon et al. [111] reported on the dependence of bulk defects on the axial temperature gradient of CZ silicon crystals. For this 4-, 6-, and 8-inch length crystals were grown with different heat shields to protect the growing crystals against radiation emitted from the melt surface and hot graphite parts. It was observed that the critical pull rate, at which the OISF ring vanishes from the wafer center, varies with the crystal diameter and the type of heat shield provided. A calculation of the axial temperature gradient at the solid–liquid interface for each crystal diameter to heat shield combination revealed that the critical pull rate is proportional to this axial temperature gradient G, which, in turn, is a function of the crystal diameter and heat shield. Here, the critical pull rate is entirely determined by the axial thermal gradient G. The C_{crit} where the OISF ring collapses in the wafer center and the excellent crystal quality extends over the entire wafer area. This was reported to be ~0.5 mm/min and independent of crystal diameter. This growth rate is, by no means, economically attractive, however, as the crystal output decreases not only by the low pull rate, but also due to a lower crystal yield. It is further reported that the OISF ring appears in the wafer center when the V/G ratio is equal to $C_{crit} = 1.3 \times 10^{-3}$ cm^2 min^{-1} K^{-1} condition holds. For ratios greater than C_{crit}, flow pattern/ D-defects are observed, whereas a ratio less than C_{crit} describes the condition for the growth of large-pit defects in the CZ-grown crystals. The enhanced O$_i$ precipitation inside the OISF ring indicates an excess of vacancies and, therefore, supports the vacancy model. On the other hand, the large-pit defects, which were identified as large dislocation loops, and the suppressed O$_i$ precipitation outside the OISF ring are an indication for an interstitial instead of a vacancy excess in the crystal.

Influence of the boron doping level in the range of 1.0×10^{15} to 2.0×10^{19} cm^{-3} on the position of the OISF ring in silicon crystals has been investigated by Dornberger et al. [112] through experiments as well as numerical simulation studies. For low boron-doped crystals, the position of the ring is described by a critical value C_{crit} defined by the ratio of the pull rate and the temperature gradient in the crystal at the solid-liquid interface. Boron concentrations higher than 10^{17} cm^{-3} shift the position of the ring toward the wafer center without changing the growth parameters. The critical value C_{crit} converts into a function C_{crit} (C_B), depending linearly on the boron concentration C_B. For zero boron concentration, the value matches that of low concentrations. The data scatter of approximately ±8% is in the range of computational and experimental errors. A linear and logarithmic representation of the data points are shown in Figure 9.19. This figure shows that the influence of C_B on the ring is negligibly small for (p^-) wafers with boron doping levels of less than 10^{17} cm^{-3}. The influence of boron becomes dominant for (p^+) and (p^{++}) wafers with doping levels above 10^{19} cm^{-3} as shown in Figure 9.19b. This experimental data shows that sufficiently high boron concentration values modify the thermodynamical properties of self-interstitials and vacancies present in the crystal.

The elimination of process-induced stacking faults is generally carried out using different methods. Petroff et al. [113] came out with a simple technique of depositing an Si$_3$N$_4$ film on the back of silicon wafers, followed by a preoxidation annealing approach, that effectively eliminated the formation of contamination-induced stacking faults on the device side of the wafers. This gettering of the Si$_3$N$_4$ layer has been found to effectively reduce the contamination of the wafers by impurities such as copper and gold. The defect gettering action of the Si$_3$N$_4$ film is related to the stress induced in the silicon wafers by the Si$_3$N$_4$ film. Rozgonyi and Kushner [114] reported that this Si$_3$N$_4$ presence deliberately introduces misfit dislocations, and the results show reduced leakage current and increased device yield. Further, it was reported that the occurrence of bulk stacking faults and the completeness of the misfit dislocation gettering process appears to be related to the density of native defects in the starting material.

At standard oxidation process temperatures above 1200°C, steam oxidation at 1300°C, and HCl oxidation at 1200°C, the stacking faults vanish from the silicon surface. Hattori [115] used a trichloroethylene (TCE) oxidation process to remove the stacking faults in silicon. Figure 9.20a shows the dependence of stacking fault length on the oxidation conditions. The physical size of the stacking fault was plotted as a function of nitrogen through a TCE bubbler. No OISFs appeared at a flow rate of more than 15 and 200 cc/min for 1200°C and 1100°C, respectively. Figure 9.20b shows the existing stacking faults shrinking and

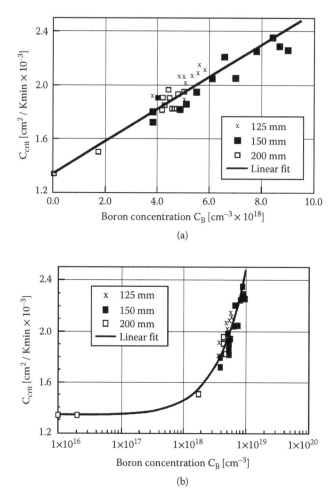

FIGURE 9.19
Dependence of the critical parameter on the boron concentration: (a) linear, (b) logarithmic representation. (From E. Dornberger, et al., *Journal of Crystal Growth*, **180**, 343–352, 1997, and the references therein [112].)

becoming totally elimated within 2 h of oxidation at 1200°C. Figure 9.20c shows the relationship between the temperatures and the minimum C_2HCl_3/O_2 ratio at which stacking faults could be completely eliminated. In this important experimental observation, the exact mechanism by which the TCE vapor acts to eliminate or shrink the stacking faults is not known. Annihilation of stacking faults is also reported in a nitrogen annealing environment by Hashimoto et al. [116]. The activation energy for this fault shrinkage was reported to be 5.2 eV, which is very close to the value of 5.13 eV for silicon self-diffusion.

9.9 Precipitations

Precipitation of nonsilicon atoms accumulating in one or more locations is often observed in the grown silicon crystals and subsequently in the process-bound wafers. Many precipitations have been identified using different methodologies, including oxygen [117–133],

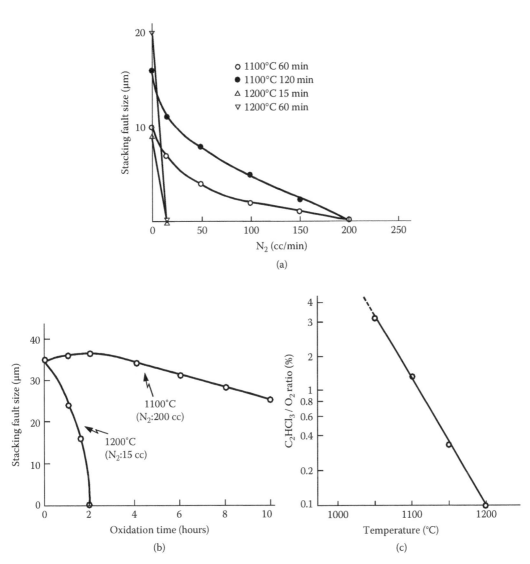

FIGURE 9.20
(a) Stacking fault size vs. C_2HCl_3-N_2 flow during the oxidation process; (b) shrinkage of existing stacking faults during additional oxidation in the presence of C_2HCl_3; and (c) oxidation temperature vs. C_2HCl_3/O_2 ratio whereby stacking faults are completely eliminated. (From T. Hattori, *Journal of the Electrochemical Society*, **123**, 945–946, 1976, and the references therein [115].)

copper [134], nickel [135], carbon [136–139], and many other metallic species [120]. Sometimes, these precipitations take place in (intentional) dopant species such as phosphorus, arsenic [140], and antimony [141] as well. These precipitations extend to more lattice locations, and their presence is a serious threat to the functionality of the device and ultimately determines the reliability of the device and the circuit. These precipitations may be physically small during the initial stages of VLSI and ULSI circuit fabrication, but may grow after each processing step. Among those listed earlier, oxide precipitation has drawn more attention because of crucible contamination in CZ-grown crystals, and its presence is difficult to avoid.

Chiou [117, 118] reported on the effects of preheating on the axial oxygen precipitation uniformity in CZ crystals. This step is carried out to reduce the thermal history effect by applying thermal annealing steps using the S-curve concept. Density spectra of grown-in oxide precipitate nuclei were measured by Kissinger et al. [119]. The oxygen precipitation behavior in 300 mm CZ silicon wafers with initial oxygen concentrations of 25–36 ppma has been studied by Ono et al. [121] following a two-step heat treatment. Oxygen precipitation, retardation, and recovery phenomena occurred in 300 mm silicon wafers with a low-carbon content. TEM observations show that the extended defect formation changes dramatically, from punched-out dislocations in precipitate retardation samples, to stacking faults in precipitated samples. Precipitate nucleation and growth during the initial low-temperature annealing likely plays a key role in oxygen precipitation recovery. With boron-doped CZ silicon wafers, the growth rate of oxide platelet precipitates is proportional to the square root of time, and the precipitate morphology changes from plate to polyhedral [122]. Strain around the precipitate decreases during the annealing cycle. These results indicate that changes in precipitate morphology occur because the effects of oxygen precipitation and boron atom size are enhanced as the boron concentration in silicon increases. Through selective ion etching [123, 128], it is possible to visualize these oxide precipitates.

According to Fujimori [142], who experimentally observed thermally induced oxygen precipitates in CZ silicon crystals, the morphology of oxygen precipitates depends not only upon the annealing temperature, but also on oxygen supersaturation. On the other hand, the oxygen precipitates near the denuded zone have an octahedral shape that is independent of the annealing temperature used, which demonstrates the important role oxygen concentration plays in morphology. The results stress the dependence of oxygen precipitate morphology upon oxygen supersaturation, which is in turn correlated with oxygen concentration. In bulk, the shape of the precipitates is strongly dependent upon the annealing temperature. This is because the supersaturation is small in this area due to out-diffusion. The precipitates formed in the bulk have different shapes, depending on the supersaturation of oxygen concentration. As an example, when $[Oi] = 1.44 \times 10^{18}$ atoms/cm^3, the precipitates are octahedral in shape. They become polyhedral when $[Oi] = 1.60 \times 10^{18}$ atom/cm^3. Figure 9.21a shows the annealing temperature and the oxygen supersaturation results obtained by the author. Morphology details are shown in Figure 9.21b. One such polyhedral precipitate shape is shown in Figure 9.22. The facet formation follows certain original silicon planes, as shown in the figure.

Precipitation of copper in silicon was reported by Schwuttke [134]. Oxygen present in the crystal forms a copper-oxygen complex in the shape of a needle and/or a plate. These precipitates have a definite preferred orientation with respect to the crystal lattice. Sueoka [135] has reported that both nickel and copper precipitations are reaction-limited in the crystal lattice, but are fast diffusants in silicon crystal. Precipitates of phosphorus and arsenic were reported on by Joshi [140] using TEM in silicon. In the case of phosphorus, the precipitates are believed to be complexes of various shapes—squares, rods, and hexagons—with no definite orientation in the silicon matrix. Silicon wafers diffused with a high concentration of arsenic also showed elongated precipitate structures with no definite orientations. Kodera [141] has reported on the antimony precipitations in silicon when the concentration exceeds a critical value. Corrugation appears on the silicon surface due to precipitation, and it is reported that that part of the crystal has become polycrystalline.

Refresh times for high-density DRAMs are, however, degraded by oxide (SiO$_2$) polyhedral precipitates. The problem is especially severe for wafers cut from the tail-end section of the crystal due to anomalous oxygen precipitation (AOP) [36]. The dwell time of the tail-end section at high temperatures below the freezing point of silicon (1412°C) is much

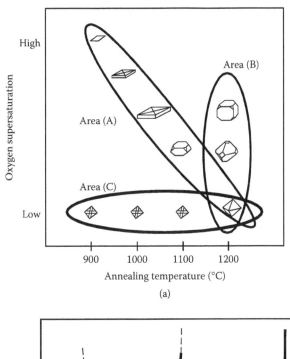

(a)

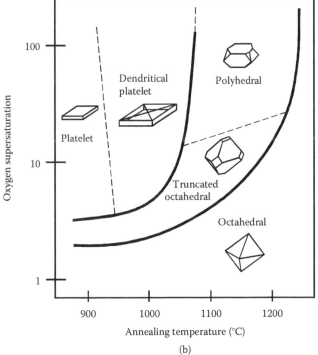

(b)

FIGURE 9.21

(a) Oxygen precipitate morphology summarized from the experimental results. (b) Diagram of oxygen precipitate morphology between annealing temperature and oxygen supersaturation. (From H. Fujimori, *Journal of the Electrochemical Society*, **144**, 3180–3184, 1997 [142].)

(a)

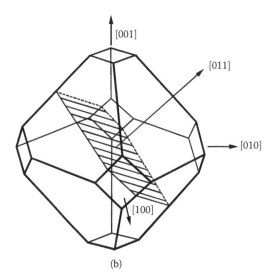

(b)

FIGURE 9.22
The variety of shapes of oxygen precipitates. (a) Polyhedral precipitate after 1170°C anneals. (b) Facets follow {111} and {200} silicon planes. (From R. Falster and V. V. Voronkov, "Lattice defects in silicon," 123-5.5&5.6.ppt, from Monsanto Electronic Materials Company (MEMC) website [143].)

shorter than that for the bulk of the crystal, due to fast cooling after completion of the CZ crystal growth process.

9.10 Surface Pits/Crystal-Originated Particles

With shrinking design rules, defect sizes less than 0.3 μm and less are becoming major issues for VLSI and ULSI circuits. Single-crystal wafers from conventional CZ silicon usually suffer from defects known as crystal-originated particles (COPs), or surface pits. Experimental results have shown that they are actually pits rather than particles. Often, they also referred to as localized light scatters (LLS), and they are becoming increasingly problematic. These COPs are grown-in agglomerations of vacancies and silicon self-interstitials that may cause degradation of GOI, isolation leakage, and junction leakage, resulting in lower device yield [4]. The origin, characterization, and effects were studied by Toktosunov and Bergholz [144], and their importance for devices, particularly for gate oxide defects and GOI, were examined. These COPs are identified as either single or octahedral multiple voids. The formation mechanism is not clear and is an active field of study at present. Vacancy agglomeration during the crystal cooling in the range of 1100°C to 1070°C has been reported to be the point where they merge to form these octahedral pits. The physical dimensions of these voids are around 0.1 to 0.15 μm, but the density is in the range of 10^5 to 10^6 cm^{-3}. Crystals without these grown-in defects, because of the agglomeration of silicon self-interstitials, tend to form at lower crystal growth rates, where agglomeration of vacancies are less likely to form. In order to prevent the effect of COPs on devices, polished wafers are sometimes replaced with epitaxial wafers or hydrogen-annealed wafers, thus increasing the cost of the starting material. These COPs can be mapped with a particle counter [30]. Problematic COPs are due to voids that intersect the wafer surface or that are present in the immediate subsurface. Octahedral voids are bounded by the lowest-energy {111} planes.

Iizuka and Kikuchi [145] reported that anomalous etch pits in highly doped silicon formed at the intersections of one-dimensional rods with the silicon crystal surface. Booker and Stickler [146] reported the presence of small particles in silicon crystals responsible for low yield in *p-n* junction devices. These defects were approximately circular in shape with gray area up to 1000 Å across containing sharply defined, black centers. The black centers are thought to be small particles, and the surrounding gray areas to be regions of either high strain or high segregation. Abe et al. [147] and Yukimoto et al. [148] also reported etch pits that are shallow in depth and rounded off when compared with the ordinary ones due to dislocations. Kashiwagi et al. [149] reported a good micro-pit delineation etchant that can delineate defects with extremely small defect energy and at the same time shows good preferential etching ability.

Due to the high pull rate of silicon ingot, also known as fast pulled silicon (FPS), the crystal passes quickly through the temperature range of 1150–1080°C—the COP forming range—so only COPs of smaller sizes and higher density can be formed. Furthermore, such small voids near the wafer surface can be easily removed by post-growth wafer annealing. Doping with nitrogen contributes to minimizing void formation in FPS. FPS development concerns include the effect of relatively high vacancy concentrations remaining in the wafer, which could cause AOP that degrades device performance and yield—for instance, in terms of DRAM refresh time [30]. The necessity of proper denuded zone formation in the wafer through high-temperature annealing steps is a way to minimize COPs.

Takano et al. [6] reported on the relationship between grown-in defects and thermal history during CZ silicon crystal growth. According to their studies, the grown-in defects in single crystals of silicon comprise FPDs, COPs, and LSTDs. They correlate with the C-mode yield of the time zero dielectric breakdown (TZDB) in MOS capacitors. They have reported that these defects are introduced during the solidification process and annihilated during the cooling process, from the melting point of silicon to about 1200°C. The heater power is automatically adjusted so that the temperature at the triple junction is the melting point of silicon. The densities were measured for FPDs, LSTDs, and COPs and the defects by using an oxygen precipitate profiler (OPP). The FPD density was measured under an optical microscope after Secco etching for 30 min that was carried out without any agitation. The LSTD density was measured using an MO-411 (Mitsui make). The densities of COPs whose detectable size was larger than 0.10 μm were measured on the polished wafers after repeated SC-1 cleaning (10 times) to check and support the experimental results. Figure 9.23

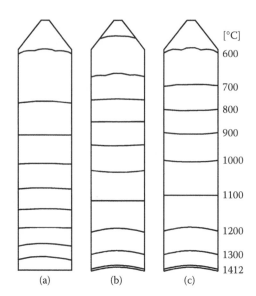

FIGURE 9.23
Temperature distribution in the crystals in three hot-zone configurations: (a) fast-cooling hot zone; (b) slow-cooling hot zone; and (c) very-slow-cooling hot zone. (From K. Takano, M. Iida, E. Iino, M. Kimura, and H. Yamagishi, *Journal of Crystal Growth*, **180**, 363–371, 1997 [6].)

shows the crystal temperature distributions simulated by using FEMAG, an axisymmetrical numerical simulation code, in a fast-cooling hot zone, a slow-cooling hot zone, and a very slow-cooling hot zone. The heater power was automatically adjusted to make the temperature at the triple junction the melting point of silicon. It was concluded that the cooling rate from 1412°C, the melting point of silicon, to 1150°C does not seem to be as important for the generation or annihilation of the defects listed earlier. Although the defect densities and the positions where the defect densities begin to change are different among the three different hot-zone configurations, the axial profiles are reported to be similar.

Suhren et al. [150] reported that high boron impurity concentration—on the order of 2.0×10^{19} cm^{-3}—influences the overall COP density of crystal defects on silicon wafers. They investigated the impact of the boron concentration, and determined it is an important additional parameter influencing radial defect distribution. A very low, homogeneous COP density was observed on the wafers with a resistivity below a diameter specific threshold. Above this threshold of 13.2 mΩ.cm (corresponding to a boron concentration of 4.8×10^{18} cm^{-3}) for 200-mm wafers and 11.0 mΩ.cm (6.3×10^{18} cm^{-3}) for 125 mm wafers, a pronounced increase of the COP counts is observed. These details are shown in Figure 9.24. In the resistivity range between 13.2 and 16.9 mΩ.cm for the 200 mm wafers and between 11.0 and 14.7 mΩ.cm for the 125 mm wafers, the COPs are no longer distributed homogeneously, but an area with increased COP density appears in the center of the wafer. A model was proposed attributing this effect to a decrease of the total atomic volume in the lattice at high boron doping levels.

Nishimura et al. [151] observed the TEM images, shown in Figure 9.25, of the COPs and the elemental analysis of them. The similarities in the shape and the precipitate content were analyzed. Both the octahedron and tetrahedron defects were 80 to 100 nm. The plane of the side-walls of the structures was identified as the {111} plane. The possible reason for this was that the voids in the crystalline lattice formed by agglomeration of vacancies during growth and/or cooling of CZ silicon crystal. The typical density of these defects is reported to be in the range of 2.0×10^6 cm^{-3}. The size and density of these COPs depend

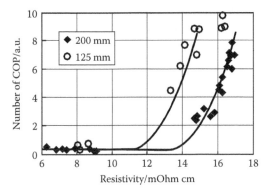

Number of COPs greater than 0.12 μm LSE on the entire wafer as a function of the electrical resistivity. (From M. Suhren, D. Gräf, U. Lambert, and P. Wagner, *Journal of the Electrochemical Society*, **144**, 4041–4044, 1997 [150].)

FIGURE 9.25
TEM image of a void in a silicon crystal. D-defects: voids (clusters of vacancies). Also known as crystal-originated particles (COPs). Typical density: about 2.0×10^6 cm^{-3}. Typical radius is about 80 nm. (From R. Falster and V. V. Voronkov, "Lattice Defects in Silicon"—Lecture notes from 123-5.5&5.6.ppt presentation. Source unknown. Downloaded through Internet. Probably from Monsanto Electronic Materials Company (MEMC) website. M. Nishimura, S. Yoshino, H. Motoura, S. Shimura, T. Mchedlidze, and T. Hikone, *Journal of the Electrochemical Society*, **143**, L243–L246, 1996 [143, 151].)

very strongly on the crystal pulling conditions and on the doping concentration levels. As crystal pulling increases, their total number increases but the size decreases. Different characterization tools are being used to study these important defects, and a few methods have been suggested to eliminate them [152–158]. Hydrogen annealing and nitrogen annealing are a couple of specific suggestions but more experimental data is needed to confirm they can totally eliminate these COPs.

9.11   Grown Vacancies and Defects

Almost all crystal silicon wafers for microelectronics device fabrication are produced by the CZ method or growth from the melt. It is well established that microdefects, called grown-in defects, exist in CZ-grown silicon crystals. Since these grown-in defects cause the deterioration of ULSI devices, it is increasingly important to reduce the number of defects and obtain the required crystal quality. These defects are considered to be agglomerates of intrinsic point defects, such as single vacancies and self-interstitials, which suggests that there is a strong relationship between the distribution of grown-in defects and the amount of excess point defects. It is empirically known that the type of grown-in defect can be controlled by the V/G ratio, where V is the growth rate and G is the temperature gradient at the solid-liquid interface. When the V/G value is higher than $(V/G)_{crit}$, vacancy grown-in defects or voids are formed under the condition of $V/G < (V/G)_{crit}$. Although many phenomenological models have been reported to interpret the physical meanings of the V/G rule to better predict defect distribution, the atomistic mechanism is not well understood. Since a large amount of intrinsic point defects are introduced at the interface, it is important to investigate the formation processes of point defects from a microscopic point of view. Motooka [159] proposed that the defect type is controlled by the structural difference in a ≈ 1.0 nm thick transition layer at the interface, while the crystallization mechanism is the same as that of solid-phase epitaxy in the high-temperature region. The nonequilibrium atomic diffusion in the transition layer is faster for a larger temperature gradient, and thus, the lower density region can be quickly filled by silicon atoms from the denser liquid side, which results in interstitial formation. On the other hand, the atomic diffusion constant becomes smaller as the gradient decreases, and space filling is not effective, which increases the opportunity for vacancies to form in the growing crystal.

Vacancy concentration wafer mapping has been suggested as a means to determine the quality of silicon wafers [160] for device processing. The concentration of vacancies is determined via measuring the concentration of fast-diffusing elements, especially the platinum or gold metals, which occupy these silicon lattice vacancies. These metals introduce deep energy levels in the bandgap, which act as recombination centers for excess carriers, presence of excess electrons or holes, and the associated excess carrier lifetime.

Grown-in void defects form due to the aggregation of excess vacancies during CZ crystal growth, and the grown-in defects have an octahedral shape surrounded by (111) surfaces at which a few nanometer-size oxide films grow. In order to decrease the density of grown-in defects and improve GOI characteristics, slow cooling in a temperature range around 1100°C has been used during CZ crystal growth. Adachi et al. [161] suggest that grown-in defects can be totally removed by annealing the wafers at 1300°C in argon ambient. It was further added that annealing in oxygen could not reduce the grown-in defects due to its inability to remove the inner oxide. However, wafers exposed to such large temperatures may develop plastic deformation and gravitational effects may arise.

The model proposed by Voronkov and Falster [162] accounts for complicated (strongly banded) precipitation patterns, particularly those observed in halted and quenched silicon crystals. The incorporated point defects agglomerate into microdefects upon lowering the temperature—in particular, the vacancies are agglomerated into voids. Oxygen in CZ crystals plays an important role by assisting with void formation, producing joint vacancy-oxygen agglomerates (oxide particles), and trapping vacancies into VO_2 species [163].

The results of Fujimori et al. [164] on the morphology of grown-in defects in nitrogen-doped crystals indicate that there is one kind of nitride precipitate in these crystals. The growth of whisker-like defects is identified after hydrogen annealing, and nitrogen and oxygen were detected from them. These defects contain both a distorted diamond structure and an amorphous structure. The formation behavior of grown-in voids for nitrogen-doped CZ crystals was also investigated by Umeno et al. [165] by means of a new quantitative defect-evaluation method using a bright-field IR laser interferometer. Quenching techniques were employed to study void formation, and in crystals grown without nitrogen doping, the total amount of vacancies composing the voids did not change during the crystal growth halt. However, the total amount of vacancies increased during the crystal growth halt. These results indicate that residual excess vacancies exist after void formation in crystals with nitrogen doping.

IR light scattering tomography [166], optical shallow defect analyzer [167], and bright field IR laser interferometer [168] are some of the other techniques used for analyzing these grown-in defects in silicon. The OPP and atomic force microscope, coupled with a laser particle counter [169], were also used to analyze these defects. However, the formation mechanism of grown-in defects is still unclear.

9.12   Thermal Donors

Thermal donors remain one of the much awaited and unsolved problems in defect physics. First reported nearly 50 years ago, their microscopic structure still remains a topic of much discussion. Many thermal donors have been reported, but only two major ones have been identified. The primary thermal donors, which are commonly observed in as-grown silicon, are related to oxygen. Referred to as oxygen thermal donors, these form in the temperature range between 400°C and 550°C. These thermal donors are formed by the polymerization of Si and O into complexes such as SiO_4 in interstitial sites in this temperature range. Careful quenching of the crystal annihilates these donors. Wijaranakula and Matlock [170] reported that annihilation of the oxygen thermal donors occurs after a short thermal anneal at 650°C. Although the exact microscopic structure of the oxygen thermal donors is still being debated, it is well established that this type of defect gives rise to photoluminescence at the photon energy of 0.767 eV.

The second type of thermal donor that forms in the temperature range between 500°C and 800°C is hypothesized to originate from an agglomerate consisting of interstitial carbon atoms. The thermal donor concentrations in the high-carbon and low-carbon samples are plotted as a function of isochronal annealing temperature, as shown in Figure 9.26. After an isochronal anneal, a peak of the maximum donor formation is observed at 460°C. This corresponds to the peak of the maximum oxygen thermal donor formation. In high-carbon samples, the oxygen thermal donor formation is suppressed by a factor of approximately one half. These carbons dissolve during the subsequent thermal annealing at 1050°C. The dissolution of carbon donors and, therefore, interstitial carbons causes the carbon atoms to return to their substitutional sites, while excess silicon interstitials are released and later condensed into interstitial-type dislocation loops.

Maddalon-Vinante et al. [171] studied the influence of rapid thermal annealing on thermal donor formation and hydrogen on the precipitation of oxygen. They reported that the oxygen loss during the subsequent precipitation steps is not directly correlated with the quantity of thermal donors formed. Two precipitation paths were proposed. It was reported that aside from the creation of thermal donors, oxygen atoms can create local oxygen agglomerates [172]. High-pressure argon treatment accelerates the kinetics of thermal donor introduction and suppresses the oxygen accumulation in agglomerates. The experimental results of Voronkova et al. [173] are considered a definite indication of the crucial role of silicon self-interstitials in the production of thermal donors. The generation rate is essentially enhanced at higher silicon self-interstitials. Neimash et al. [174] reported that tin doping leads to a considerable decrease of the oxygen-containing thermal donor

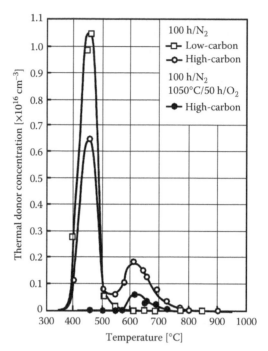

FIGURE 9.26
The thermal donor concentration as a function of isochronal annealing temperature. (From W. Wijaranakula and J. H. Matlock, *Journal of the Electrochemical Society*, **137**, 1964–1969, 1990, and the references therein [170].)

generation rate and its steady-state concentration values. A qualitative model has been proposed to explain the experimental observations. Rafi et al. [175, 176] reported that the use of direct plasma hydrogenation enhances the introduction rate of thermal donors in 450°C annealed *n*-type CZ silicon for shorter annealing times. The effect is more pronounced for longer hydrogenations and reaches a maximum at about 5 h of annealing at 450°C. With longer anneals, the hydrogen effect diminishes, resulting in a free carrier introduction rate of ~1.7 to 1.8 × 10^{14} cm^{-3}/h, irrespective of the hydrogen content. It is believed that in the hydrogenated samples, ultra-shallow donors, which are different from oxygen-containing thermal donors, are responsible for the increase in donor concentration. In further studies on this hydrogen plasma and its impact, Simoen et al. [177] observed that the formation kinetics of the thermal donors is a strong function of the hydrogen and oxygen concentration and the annealing temperature, but most of all, the doping type and doping concentration. The latter factor has a major impact on the charge state and, hence, the diffusion of hydrogen, dictating the formation kinetics of hydrogen-related shallow thermal donors.

9.13   Slips, Cracks, and Shape Irregularities

Conventionally prepared silicon wafers do exhibit curvatures from end to end, ranging from a few micrometers to several mils. This curvature arises from both the sawing operation and the subsequent mechanical and chemical polishing steps. The contribution of each to the slice curvature is a function of the particular processing technique being used. Although curve-free slices can be prepared by utilizing extreme care in each processing step, conventionally prepared slices do exhibit some curvature [178]. In practice, these curvatures create nonuniform contacts on the susceptor during chemical vapor deposition (CVD) operations, and can exhibit a substantial temperature variation with the slice curvature effects. This temperature difference may readily activate residual surface dislocation sources.

The heating and cooling of silicon wafers introduce many process-induced slips where a visible defect spreads, leading to wafer breakage and subsequent rejection from the batch of wafers. Presence of the slip and its associated edge dislocations degrade electrical parameters, especially if precipitation of metals or dopant atoms accumulate in those sites. Thus, minimizing these large defects in wafers is highly desirable [179]. Yoriume [180] conducted experimental measurements on silicon wafers by thermal oxidation and proved that the wafer warpage and physical sizes are not affected by the thermal oxidation process and will remain as they are in the beginning. Micro-indentation for fracture and stress-corrosion cracking studies in single crystals of silicon have been reported by Wong and Holbrook [181].

With the increase in a wafer diameter toward 300 mm and larger, thermal and gravitational stresses have greatly increased on the process-bound wafers. These stresses will cause slip dislocations in the wafers. These dislocations are harmful, since they cause warping of the wafers during thermal processes, or they degrade the VLSI and ULSI circuit performance by increasing current leakage. Therefore, it is important to decrease the thermal and gravitational stresses on these wafers. To minimize stress-related defect generation, Akatsuka et al. [182] developed a simulation model to calculate slip length during high-temperature thermal processes based on dislocation kinetics. The gravitational and thermal stresses were sheared to the slip systems of the silicon crystal. The dislocation velocity was calculated from the resolved shear stress. From this, the slip length was calculated by integrating the dislocation velocity with the duration time of the thermal process cycle. Accordingly, the slip occurrence in 300 mm wafers was predicted and compared with the experimental results. It was found that the model can predict the ramping rate, which causes the slip, and can calculate the slip length for various thermal processes with a high degree of accuracy. This model should aid in the establishment of slip-free sequences for 300 mm wafers. As an example, Figure 9.27a and b shows a schematic diagram of conventional four-point and ring-like jigs. The wafers were set on each jig with 9.5 mm spacing. The ring width is approximately 50 mm. After insertion into the furnace at 700°C, the wafers were ramped up from 700°C to 1000°C at a rate of 8°C/min, and from 1000°C to 1200°C at a rate of 1°C/min. After isothermal annealing at 1200°C for 60 min, the wafers were ramped down from 1200°C to 1000°C at a rate of 1°C/min, and from 1000°C to 700°C at a rate of 2.5°C/min. The wafers were withdrawn from the furnace at 700°C. The slip length during the thermal process was measured using x-ray topography. Figure 9.27c shows the results when wafers are supported using four-point and ring-like jigs. In the case of the four-point jig, the slip, with a length of approximately 30–50 mm, was caused by contact with the jig during the thermal process. In contrast, in the case of the ring-like jig, no slip was observed. These results mean the ring-like jig is an effective method for achieving slip-free processes in 300 mm wafers. These calculations and the model proposed by the team [182] strongly supports the analysis of the wafer condition to its jig locations and the possible slip formations before being exposed to different thermal cycles.

Kulshreshtha et al. [183] reported on oxygen precipitation–related, stress-modified crack propagation in high-growth-rate CZ silicon wafers using microscopic Raman and Fourier transform infrared spectroscopy (FTIR) techniques. The influence of stress gradients, due to oxygen precipitate distribution, and their deviation were reported in 200 mm diameter silicon wafers. The stress-modified crack behavior deviated considerably from the initial energetically favorable silicon [110]/(111) system, enabling scenarios in which a ductile fracture follows a radial path. Preferential etch-pit measurements confirmed the presence

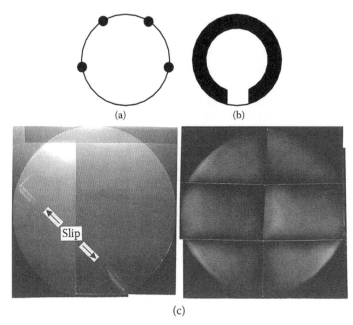

FIGURE 9.27
Support methods for 300 mm wafers: (a) conventional four-point jig, and (b) ring-like jig. (c) Results of slip observation using x-ray topography in both cases. (From M. Akatsuka, K. Sueoka, H. Katahama, and N. Adachi, *Journal of the Electrochemical Society*, **146**, 2683–2688, 1999, and the references therein [182].)

of oxygen precipitate rings, which influence stress-dependent energy transfer processes at the propagating crack tip, thereby dramatically modifying the crack path. These local changes may be primarily due to the presence of oxygen species.

9.14 Stress, Bowing, and Warpage

For process-bound wafers, particularly during the photolithographic step, a good planarity of the wafer surface is required. Tolerances on the order of 10 μm or less in vertical displacement is generally preferred, but the trend toward smaller device structures and larger wafer diameters will certainly aggravate this problem, and tight tolerances are mandatory [184]. Warpage (i.e., deviations from planarity from the center of the wafer to the edge) is a key acceptance parameter for process-bound wafers. Usually, the initial warping of the freshly cut and polished wafer is rather small, but can increase drastically after the wafer undergoes thermal treatments during subsequent high-temperature cycles. The warpage of silicon wafers has been an area of concern in VLSI and ULSI circuit manufacturing for many years. Film stresses and the mechanical strength of the wafer can both influence warpage. Warped wafers are difficult to handle, particularly in automatic processing equipment. The generation of complex circuit patterns by photolithography on the warped or bowed wafer is difficult to perform because of varying focal planes. This problem is especially serious with projection printing. In addition to wafer handling and photolithographic problems, warpage can affect the electrical characteristics of the fabricated devices and circuits. Dislocations introduced into the processed wafers as a result of plastic deformation are known to cause excessive leakage currents in the active device volume [185]. Due to a change of lateral dimensions on the wafer by warpage, which causes misalignment, exact device geometry cannot be realized, and this causes degradation of device characteristics.

A quantitative method for evaluating stress in crystal was established by photoelasticity. It consists of three procedures: measurement by plane polarized light, measurement by circularly polarized light, and piezooptical coefficient calculation. The last procedure is indispensable for stress evaluation in crystals. Kotake and Takasu [186] have reported that the piezooptical coefficients for silicon in different crystal orientations can be applied to crystal characterization. A practical approximation was made to correlate each isochromatic line directly to the difference between two principal stresses. This method was applied to evaluate the stress in silicon semiconductor pellets caused by the mounting and molding processes. Observation of isochromatic lines, according to temperature, was helpful in determining the zeroth order of retardation. These stress measurements are important from the device point of view.

High-temperature processing of silicon wafers often produces sufficient thermal stresses to generate slip and dislocations. The stresses arise during temperature transition as the wafers are stacked during high-temperature processes. During loading and unloading sequences, a temperature gradient arises between the center and edges of the wafer, resulting in stresses sufficient to generate dislocations and slip bands. Temperature, wafer dimensions, process history, initial bow, and the concentration levels of oxygen present in silicon wafers can cause slip damage to occur either at the periphery of the wafer or at the center of it, depending on the wafer size. Wafers that have a larger diameter over thickness ratio are more sensitive to warpage, and warped wafers show a much larger area affected by slip on the concave side than on the convex side, which leads to bending. Leroy and Plougonven [187] presented a model to predict the critical temperature above which warpage will occur as a function of temperature gradient, wafer curvature, and amount of oxygen precipitated in the silicon wafer.

Measurement of wafer warpage and film stresses relies mainly on techniques such as the surface profiling method, Newton-ring method, x-ray method, and the optical interference methods. The first three methods can only produce a single scan of a surface profile or can measure warpage at two reference points. Yang [188] adapted a simple optical imaging method for quick evaluation of wafer surface quality and warpage. Surface irregularities of 50–100 Å can be detected by this method. The detection limit for the radius of the curvature is about 270 m. The agreement between this method and the optical interference method is reported to be within 30%. A comparison of this method with the reflection x-ray method also shows a reasonable agreement of 20%. However, a substantial discrepancy exists between this method and the transmission x-ray method. The method is schematically shown in Figure 9.28. The figure shows that two beams are used for this purpose. By using a recorded picture and image planes, the diameter of wafer warpage can be calculated using simple geometry. The values are determined depending on whether a convex or a concave surface is being produced with a bright contrast. The film stress is evaluated based on the radius of curvature. This approach is simple and straightforward and is good for the nascent wafers. For process-bound wafers, however, this method may not work as well.

Lee and Tobin [192] experimentally verified the effect of CMOS processing on oxygen precipitation, warpage, and flatness in process-bound wafers. According to their observations, the process-induced wafer warpage was found to be independent of the initial warpage. This may be due to the relatively small warpage of the initial wafers selected for these studies, which was measured as 10.5 µm maximum. On the other hand, warpage was found to depend critically on the initial oxygen concentration. For oxygen concentrations

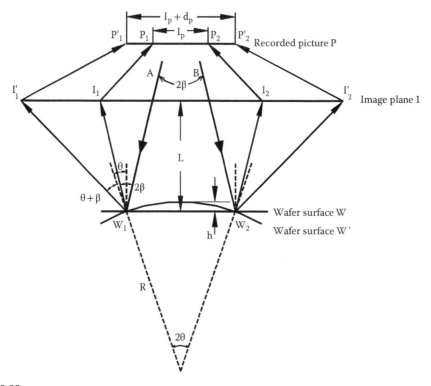

FIGURE 9.28

The principle of the optical imaging method. (From K. H. Yang, *Journal of the Electrochemical Society*, **132**, 1214–1218, 1985 [188].)

lower than 32 ppm, no systematic variation of warpage with processing was noted. In addition, wafer flatness changed by less than 1.0 μm through the entire CMOS process. Thermally induced warpage occurred in wafers with an initial oxygen concentration greater than 32 ppm during a 900°C push/pull ion implant annealing step. Lowering the push/pull temperatures to 650°C showed somewhat different results.

Elwell and Hahn [185] have further added that the warpage of laser-damaged silicon wafers was found to be less than that produced by mechanical abrasion damage. It is further reported that the silicon crystals from the seed ends are generally more resistant to warpage than those from tail end of the grown crystal. The oxygen present in the interstitial positions may be a possible reason for this. Epitaxial deposition can influence silicon wafer warpage as well. This is partly due to the lattice mismatch, the thermal expansion of dissimilar materials, and intrinsic stresses within the epitaxial layer [190]. These inherent forces will differ, depending on the particular substrate and the epitaxial layer being used. The presence of oxygen precipitation was shown not to influence warpage with this process. Chiou et al. [191] further found that the mechanical strength of wafers is a function of interstitial oxygen content, [Oi], amount of precipitation, precipitate morphology, type and dopant concentration, and other impurity content (mainly nitrogen and carbon). The mechanical superiority of CZ silicon has been known to occur because these crystals lock the available oxygen atoms to segregate at dislocations. The findings of Yonenaga [192] verify that interstitially dissolved oxygen atoms have no influence on the deformation behavior and mechanical strength of silicon crystals at temperatures close to the melting point when they are free from dislocations prior to deformation.

When growing single crystals of silicon with a large diameter, typically up to 300 mm, the thermal stresses and possible dislocation generation in them becomes a serious problem for both FZ- and CZ-grown crystals. A two-dimensional, problem-oriented code for the finite element method (FEM)-package ANSYS has been developed by Muižnieks et al. [193] to calculate the temperature field in the growing crystal, considering the radiation exchange with reflectors and environmental and thermal stresses. By comparing the calculated stresses with the critical stresses, the dislocated zone can be determined. The team developed a qualitative concept using the metastable state to explain the occurrence of dislocations. In a parametric study of different thermal boundary conditions and crystal geometries, the thermal stresses were also reported. The influence of crucible constraint on stress levels and dislocations was investigated by Chen et al. [194]. They reported that the crucible constraint had a significant influence on the thermal stresses and dislocations in the crystal ingot. These results indicate that it is important to reduce the crucible constraint in order to relax thermal stresses and reduce dislocations in a silicon ingot during the solidification process. Thermal stress distribution in the silicon ingot was solved by Chen et al. [195] using the displacement-based thermoelastic stress model. Several different melt-solid interface shapes were obtained by using different growth velocities, and the thermal stresses for different solidification times were compared. The simulation results suggested that the crucible constraint should be reduced and a longer solidification time should be used for growing silicon ingots with low thermal stress and low dislocation density.

9.15 Summary

In this chapter, we discussed the defects found in a regular crystal lattice. Anything that deviates from the regular crystal structure is considered a defect. This may be an isolated single defect or a group of defects joining together to form a cluster, or it can be a continuous defect extending to the physical dimensions of the crystal. We discussed various types of defects, including point, void, dislocation, line, screw dislocations, stacking faults, bulk, precipitations, surface pits and COPs, grown vacancies, and thermal donors. Silicon crystal slips, cracks, shape irregularities, built-in stress, wafer bowing, and warpage were also discussed.

References

1. W. C. Dash, "Silicon crystals free of dislocations," *Journal of Applied Physics*, **29**, 736, 1958.
2. M. S. Kulkarni, V. V. Voronkov, and R. Falster, "Quantification of defect dynamics in unsteady-state and steady-state Czochralski growth of monocrystalline silicon," *Journal of the Electrochemical Society*, **151**, G663–G678, 2004; "Dynamics of point defects and formation of microdefects in Czochralski crystal growth: Modeling, simulation and experiments," *Future Fab International*, Issue 14, Section 7, 2003, and the references therein.
3. E. Dornberger, E. Tomzig, A. Seidl, S. Schmitt, H.-J. Leister, Ch. Schmitt, and G. Müller, "Thermal simulation of the Czochralski silicon growth process by three different models and comparison with experimental results," *Journal of Crystal Growth*, **180**, 461–467, 1997, and the references therein.

4. H. Furuya, K. Harada, and J.-G. Park, "Defect reduction and improved gettering in CZ single crystal silicon," *Solid State Technology*, **44**, 109, 2001, and the references therein.
5. R. Falster and V. V. Voronkov, "Intrinsic point defects and their control in silicon crystal growth and wafer processing," *MRS Bulletin*, 28–32, June 2000, and the references therein.
6. K. Takano, M. Iida, E. Iino, M. Kimura, and H. Yamagishi, "Relationship between grown-in defects and thermal history during CZ Si crystal growth," *Journal of Crystal Growth*, **180**, 363–371, 1997.
7. J. E. Lawrence, "Behavior of dislocations in silicon semiconductor devices: Diffusion, electrical," *Journal of the Electrochemical Society*, **115**, 860–865, 1968.
8. L. E. Katz, "Relationship between process-induced defects and soft *p-n* junctions in silicon devices," *Journal of the Electrochemical Society*, **121**, 969–972, 1974.
9. C. M. Osburn and D. W. Ormond, "The effect of silicon wafer imperfections on minority carrier generation and dielectric breakdown in MOS structures," *Journal of the Electrochemical Society*, **121**, 1229–1233, 1974.
10. G. Kissinger, T. Grabolla, G. Morgenstern, H. Richter, D. Gräf, J. Vanhellemont, and W. von Ammon, "Grown-in oxide precipitate nuclei in Czochralski silicon substrates and their role in device processing," *Journal of the Electrochemical Society*, **146**, 1971–1976, 1999.
11. R. Falster, "Defect control in silicon crystal growth and wafer processing," MEMC Electronic Materials SpA, Novara 28100, Italy, and the references therein.
12. H. Abe, I. Suzuki, and H. Koya, "The effect of hydrogen annealing on oxygen precipitation behavior and gate oxide integrity in Czochralski Si wafers," *Journal of the Electrochemical Society*, **144**, 306–311, 1997, and the references therein.
13. Y. Ogita, K. Kobayashi, and H. Daio, "Photoconductivity characterization of silicon wafer mirror-polishing subsurface damage related to gate oxide integrity," *Journal of Crystal Growth*, **210**, 36–39, 2000.
14. R. Winkler and M. Sano, "Improvement of the gate oxide integrity by modifying crystal pulling and its impact on device failures," *Journal of the Electrochemical Society*, **141**, 1398–1401, 1994.
15. L. Jastrzebski and J. Lagowski, "Deep levels study in float zone Si used for fabrication of CCD imagers," *Journal of the Electrochemical Society*, **128**, 1957–1963, 1981.
16. L. Jastrzebski, R. Soydan, G. W. Cullen, W. N. Henry, and S. Vecrumba, "Silicon wafers for CCD images," *Journal of the Electrochemical Society*, **134**, 212–221, 1987.
17. L. Jastrzebski, R. Soydan, H. Elabd, W. Henry, and E. Savoye, "The effect of heavy metal contamination on defects in CCD imagers: Contamination monitoring by surface photovoltage," *Journal of the Electrochemical Society*, **137**, 242–249, 1990.
18. S. S. Kim and W. Wijaranakula, "The effect of the crystal grown-in defects on the pause tail characteristics of megabit dynamic random access memory devices," *Journal of the Electrochemical Society*, **141**, 1872–1878, 1994, and the references therein.
19. S. S. Kim and W. Wijaranakula, "The effect of the thermal history of Czochralski silicon crystals on the defect generation and refresh time degradation in high density memory devices," *Journal of the Electrochemical Society*, **142**, 553–559, 1995.
20. E. Dornberger, D. Temmler, and W. von Ammon, "Defects in silicon crystals and their impact on DRAM device characteristics," *Journal of the Electrochemical Society*, **149**, G226–G231, 2002, and the references therein.
21. M. Ruprecht, G. Benstetter, and D. Hunt, "A review of ULSI failure analysis techniques for DRAMs. Part II: Defect isolation and visualization," *Microelectronics Reliability*, **43**, 17–41, 2003.
22. Y. Hirano, K. Yamazaki, F. Inoue, K. Imaoka, K. Tanahashi, and H. Yamada-Kaneta, "Impact of defects in silicon substrate on flash memory characteristics," *Journal of the Electrochemical Society*, **154**, H1027–H1030, 2007.
23. N. I. Puzanov, A. M. Eidenzon, and D. N. Puzanov, "Modeling microdefect distribution in dislocation-free Si crystals grown from the melt," *Journal of Crystal Growth*, **178**, 468–478, 1997, and the references therein.
24. A. J. R. de Kock, P. J. Roksnoer, and P. G. T. Boonen, "Effect of growth parameters on formation and elimination of vacancy clusters in dislocation-free silicon crystals," *Journal of Crystal Growth*, **22**, 311–320, 1974.

25. A. G. Cullis, N. G. Chew, H. C. Webber, and D. J. Smith, "Orientation dependence of high speed silicon crystal growth from the melt," *Journal of Crystal Growth*, **68**, 624–638, 1984.
26. A. Miin-Ron, R. W. Dutton, D. A. Antoniadis, and W. A. Tiller, "The growth of oxidation stacking faults and the point defect generation at Si-SiO interface during thermal oxidation of silicon," *Journal of the Electrochemical Society*, **128**, 1121–1130, 1981.
27. V. V. Voronkov and R. Falster, "Intrinsic point defects and impurities in silicon crystal growth," *Journal of the Electrochemical Society*, **149**, G167–G174, 2002, and the references therein.
28. M. Budil, M. Heinrich, M. Schrems, and H. Pötzl, "Determination of the physical properties of point defects in silicon from back-side oxidation experiments," *Journal of the Electrochemical Society*, **137**, 3931–3934, 1990.
29. R. A. Brown, D. Maroudas, and T. Sinno, "Modeling point defect dynamics in the crystal growth of silicon," *Journal of Crystal Growth*, **137**, 12–25, 1994, and the references therein.
30. K.-M. Kim, "Materials: Silicon-pulling technology for 2000+," *Solid State Technology*, **43**, 69, January 2000, and the references therein.
31. Y. Matsushita, "Thermally induced microdefects in Czochralski-grown silicon crystals," *Journal of Crystal Growth*, **56**, 516–525, 1982.
32. H. Nakanishi, H. Kohda, and K. Hoshikawa, "Influence of annealing during growth on defect formation in Czochralski silicon," *Journal of Crystal Growth*, **61**, 80–84, 1983.
33. P. J. Roksnoer, "The mechanism of formation of microdefects in silicon," *Journal of Crystal Growth*, **68**, 596–612, 1984.
34. H. Lemke and K. Irmscher, "Proof of interstitial cobalt defects in silicon float zone crystals doped during crystal growth," *210 ECE Meeting Cancun*, Mexico, Oct. 29–Nov. 3, 2006.
35. J. H. Bleka, E. V. Monakhov, B. S. Avset, and B. G. Svensson, "DLTS study of room-temperature defect annealing in *n*-type high-purity FZ Si," *210 ECE Meeting Cancun*, Mexico, Oct. 29–Nov. 3, 2006, and the references therein.
36. K.-M. Kim, "Growing improved silicon crystals for VLSI/ULSI," *Solid State Technology*, **39**, 70, November, 1996, and the references therein.
37. K. Nakamura, T. Saishoji, T. Kubota, T. Iida, Y. Shimanuki, T. Kotooka, and J. Tomioka, "Formation process of grown-in defects in Czochralski grown silicon crystals," *Journal of Crystal Growth*, **180**, 61–72, 1997, and the references therein.
38. W. Wijaranakula, "Numerical modeling of the point defect aggregation during the Czochralski silicon crystal growth," *Journal of the Electrochemical Society*, **139**, 604–616, 1992.
39. W. A. Tiller, M. Friedman, R. Shaw, N. Cuendet, and T. Halicioglu, "Grown-in point defects and microscopic defect formation in CZ silicon: I. The one-dimensional, steady-state approximation," *Journal of Crystal Growth*, **186**, 113–127, 1998.
40. A. Muiznieks, I. Madzulis, K. Dadzis, K. Lacis, and Th. Wetzel, "Simplified Monte Carlo simulations of point defects during industrial silicon crystal growth," *Journal of Crystal Growth*, **266**, 117–125, 2004.
41. E. Dornberger, W. von Ammon, J. Virbulis, B. Hanna, and T. Sinno, "Modeling of transient point defect dynamics in Czochralski silicon crystals," *Journal of Crystal Growth*, **230**, 291–299, 2001.
42. T. L. Larsen, L. Jensen, A. Lüdge, H. Riemann, and H. Lemke, "Numerical simulation of point defect transport in floating-zone silicon single crystal growth," *Journal of Crystal Growth*, **230**, 300–304, 2001.
43. T. A. Frewen, T. Sinno, E. Dornberger, R. Hoelzl, W. von Ammon, and H. Bracht, "Global parameterization of multiple point-defect dynamics models in silicon," *Journal of the Electrochemical Society*, **150**, G673–G682, 2003.
44. K. Tanahashi, M. Kikuchi, T. Higashino, N. Inoue, and Y. Mizokawa, "Concentration of point defects changed by thermal stress in growing CZ silicon crystal: Effect of the growth rate," *Journal of Crystal Growth*, **210**, 45–48, 2000.
45. H. S. Woo, J. H. Jeong, and I. S. Kang, "Optimization of surface temperature distribution for control of point defects in the silicon single crystal," *Journal of Crystal Growth*, **247**, 320–332, 2003.
46. T. Ebe, "Effects of isotherm shapes on the point-defect behavior in growing silicon crystals," *Journal of Crystal Growth*, **244**, 142–156, 2002.

47. J.-S. Kim and T.-y. Lee, "Numerical study on the effect of operating parameters on point defects in a silicon crystal during Czochralski growth. I. Rotation effect," *Journal of Crystal Growth*, **219**, 205–217, 2000.

48. J.-S. Kim and T.-y. Lee, "Numerical study on the effect of operating parameters on point defects in a silicon crystal during Czochralski growth. II. Magnetic effect," *Journal of Crystal Growth*, **219**, 218–227, 2000.

49. S. Sama, M. Porrini, F. Fogale, and M. Servidori, "Investigation of Czochralski silicon grown with different interstitial oxygen concentrations and point defect populations," *Journal of the Electrochemical Society*, **148**, G517–G523, 2001.

50. D.-H. Hwang, S.-M. Hur, and K.-H. Lee, "The influence of point defect on the behavior of oxygen precipitation in CZ-Si wafers," *Journal of Crystal Growth*, **249**, 37–43, 2003.

51. T. Furuoya, "Dislocations in silicon single crystals," *Japanese Journal of Applied Physics*, **1**, 135–143, 1962.

52. G. H. Schwuttke, "X-ray diffraction microscopy study of imperfections in silicon single crystals," *Journal of the Electrochemical Society*, **109**, 27–32, 1962.

53. H. J. Queisser, "Properties of twin boundaries in silicon," *Journal of the Electrochemical Society*, **110**, 52–56, 1963.

54. Y. Sugita, "X-ray observations of defect structures in silicon crystals," *Japanese Journal of Applied Physics*, **4**, 962–972, 1965.

55. T. Abe and S. Maruyama, "Observations of defects in silicon single crystals," *Japanese Journal of Applied Physics*, **5**, 979–980. 1966.

56. Y. Yukimoto, "The origin and structures of the anomalous defects observed in silicon crystals," *Japanese Journal of Applied Physics*, **7**, 348–357, 1968.

57. J. M. Assour, "Identification of chemical constituents of defects in silicon," *Journal of the Electrochemical Society*, **119**, 1270–1272, 1972.

58. A. Rohatgi and P. Rai-Choudhury, "Process-induced effects on carrier lifetime and defects in float zone silicon," *Journal of the Electrochemical Society*, **127**, 1136–1139, 1980.

59. J. W. Cleland, "Heat-treatment studies of oxygen-defect-impurity interactions in silicon," *Journal of the Electrochemical Society*, **129**, 2127–2131, 1982.

60. J. T. McGinn, L. Jastrzebski, and J. F. Corboy, "Defect characterization in monocrystalline silicon grown over SiO_2," *Journal of the Electrochemical Society*, **131**, 398–403, 1984.

61. B. Lägel, I. D. Baikie, and U. Petermann, "A novel detection system for defects and chemical contamination in semiconductors based upon the scanning Kelvin probe," *Surface Science*, **433–435**, 622–626, 1999.

62. M. Ma, T. Ogawa, M. Watanabe, and M. Eguchi, "Study on defects in CZ-Si crystals grown under three different cusp magnetic fields by infrared light scattering tomography," *Journal of Crystal Growth*, **205**, 50–58, 1999.

63. S. Yang, Y. Li, Q. Ma, L. Liu, X. Xu, P. Niu, Y. Li, S. Niu, and H. Li, "Infrared absorption spectrum studies of the VO defect in fast-neutron-irradiated Czochralski silicon," *Journal of Crystal Growth*, **280**, 60–65, 2005.

64. M. Ma, T. Irisawa, T. Tsuru, T. Ogawa, M. Watanabe, and M. Eguchi, "Study on defects in CZ-Si crystals grown by normal, cusp magnetic field and electromagnetic field techniques using multi-chroic infrared light scattering tomography," *Journal of Crystal Growth*, **218**, 232–238, 2000.

65. L. Fedina, A. Gutakovskii, and A. Aseev, "FZ-Si crystal growth and HREM study of new types of extended defects during *in situ* electron irradiation," *Journal of Crystal Growth*, **229**, 1–5, 2001.

66. T. Taishi, T. Hoshikawa, M. Yamatani, K. Shirasawa, X. Huang, S. Uda, and K. Hoshikawa, "Influence of crystalline defects in Czochralski-grown Si multicrystal on minority carrier lifetime," *Journal of Crystal Growth*, **306**, 452–457, 2007.

67. T. Yamamoto and K. Nishihara, "Characterization of defects in annealed Czochralski-grown silicon wafers by photoluminescence method," *Journal of Crystal Growth*, **210**, 69–73, 2000.

68. S. Maruyama and O. Okada, "Crow track formed by mechanical force on silicon crystal wafer," *Japanese Journal of Applied Physics*, **3**, 300–301, 1964.

69. M. Y. Ben-Sira and S. Bukshpan, "Spontaneous generation of dislocations during growth of silicon single crystals," *Journal of Crystal Growth*, **2**, 248–250, 1968.

70. V. V. Kalaev, D. P. Lukanin, V. A. Zabelin, Yu. N. Makarov, J. Virbulis, E. Dornberger, and W. von Ammon, "Calculation of bulk defects in CZ Si growth: Impact of melt turbulent fluctuations," *Journal of Crystal Growth*, **250**, 203–208, 2003.

71. D. Yang, X. Yu, X. Ma, J. Xu, L. Li, and D. Que, "Germanium effect on void defects in Czochralski silicon," *Journal of Crystal Growth*, **243**, 371–374, 2002.

72. V. V. Voronkov and R. Falster, "The effect of nitrogen on void formation in Czochralski silicon crystals," *Journal of Crystal Growth*, **273**, 412–423, 2005.

73. M. Akatsuka, M. Okui, S. Umeno, and K. Sueoka, "Calculation of size distribution of void defects in CZ silicon," *Journal of the Electrochemical Society*, **150**, G587–G590, 2003.

74. M. Itsumi, "Octahedral void defects in Czochralski silicon," *Journal of Crystal Growth*, **237–239**, 1773–1778, 2002.

75. M. L. Joshi, "Effect of fast cooling on diffusion-induced imperfections in silicon," *Journal of the Electrochemical Society*, **112**, 912–916, 1965.

76. A. J. R. de Kock, "The elimination of vacancy-cluster formation in dislocation-free silicon crystals," *Journal of the Electrochemical Society*, **118**, 1851–1856, 1971.

77. F. Secco d'Aragona, "Dislocation etch for (100) planes in silicon," *Journal of the Electrochemical Society*, **119**, 948–951, 1972.

78. T. C. Chandler, "MEMC Etch—A chromium trioxide-free etchant for delineating dislocations and slip in silicon," *Journal of the Electrochemical Society*, **137**, 944–948, 1990.

79. S. Kawado, T. Taishi, S. Iida, Y. Suzuki, Y. Chikaura, and K. Kajiwara, "Three-dimensional structure of dislocations in silicon determined by synchrotron white x-ray topography combined with a topotomographic technique," *Journal of Physics D: Applied Physics*, **38** (10A), A17–A22, 2005.

80. M. C. Collet, "Recombination-generation centers caused by 60°-dislocations in silicon," *Journal of the Electrochemical Society*, **117**, 259–261, 1970.

81. I. Yonenaga, T. Taishi, X. Huang, and K. Hoshikawa, "Dislocation-impurity interaction in Czochralski-grown Si heavily doped with B and Ge," *Journal of Crystal Growth*, **275**, e501–e505, 2005.

82. T. Taishi, X. Huang, T. Wang, I. Yonenaga, and K. Hoshikawa, "Behavior of dislocations due to thermal shock in B-doped Si seed in Czochralski Si crystal growth," *Journal of Crystal Growth*, **241**, 277–282, 2002.

83. Y. Wang and K. Kakimoto, "Dislocation effect on crystal-melt interface: An *in situ* observation of the melting of silicon," *Journal of Crystal Growth*, **208**, 303–312, 2000.

84. H. Shimizu, S. Isomae, K. Minowa, T. Satoh, and T. Suzuki, "Gravitational stress-induced dislocations in large-diameter silicon wafers studied by x-ray topography and computer simulation," *Journal of the Electrochemical Society*, **145**, 2523–2529, 1998.

85. H. Hashimoto, S. Kozaki, and T. Ohkawa, "Observations of Pendellösung fringes and images of dislocations by x-ray shadow micrographs of Si crystals," *Applied Physics Letters*, **6**, 16–17, 1965.

86. H. Chen, "Observation of Pendellösung fringes in reflection-section topographs of bent silicon crystals," *Materials Letters*, **4**, 65–70, 1986.

87. T. F. Ciszek, "Characteristics of [115] dislocation-free float-zoned silicon crystals," *Journal of the Electrochemical Society*, **120**, 799–802, 1973.

88. www.tf.uni-kiel.de/matwis/amat/def_en/kap_1/illustr/t1_3_5.html. Prof. Helmut Föll, 1976, University of Kiel, Faculty of Engineering, Germany.

89. T. W. Fan, J. J. Qian, J. Wu, L. Y. Lin, and J. Yuan, "Tentative analysis of swirl defects in silicon crystals," *Journal of Crystal Growth*, **213**, 276–282, 2000.

90. H. Föll and B. O. Bolbesen, "Formation and nature of swirl defects in silicon," *Applied Physics*, **8**, 319–331, 1975.

91. P. J. Roksnoer, W. J. Bartels, and C. W. T. Bulle, "Effect of low cooling rates on swirls and striations in dislocation-free silicon crystals," *Journal of Crystal Growth*, **35**, 245–248, 1976.

92. V. V. Voronkov, "The mechanism of swirl defects formation in silicon," *Journal of Crystal Growth*, **59**, 625–643, 1982.

93. S. Yasuami, M. Ogino, and S. Takasu, "The swirl formation of defects in Czochralski-grown silicon crystals," *Journal of Crystal Growth*, **39**, 227–230, 1977.

94. A. Eyer, B. O. Kolbesen, and R. Nitsche, "Floating zone growth of silicon single crystals in a double-ellipsoid mirror furnace," *Journal of Crystal Growth*, **57**, 145–154, 1982.

95. A. Murgai, H. C. Gatos, and W. A. Westdorp, "Effect of microscopic growth rate on oxygen microsegregation and swirl defect distribution in Czochralski-grown silicon," *Journal of the Electrochemical Society*, **126**, 2240–2245, 1979.

96. P. M. Petroff and A. J. R. De Kock, "The formation of interstitial swirl defects in dislocation-free floating-zone silicon crystals," *Journal of Crystal Growth*, **36**, 4–10, 1976.

97. A. J. R. de Kock, P. J. Roksnoer, and P. G. T. Boonen, "The introduction of dislocations during the growth of floating-zone silicon crystals as a result of point defect condensation," *Journal of Crystal Growth*, **30**, 279–294, 1975.

98. J.-i. Chikawa and S. Shirai, "Melting of silicon crystals and a possible origin of swirl defects," *Journal of Crystal Growth*, **39**, 328–340, 1977.

99. A. J. R. de Kock and W. M. van de Wijgert, "The effect of doping on the formation of swirl defects in dislocation-free Czochralski-grown silicon crystals," *Journal of Crystal Growth*, **49**, 718–734, 1980.

100. L. D. Dyer and F. W. Voltmer, "Circular and hexagonal stacking faults in bulk silicon crystals," *Journal of the Electrochemical Society*, **120**, 812–817, 1973.

101. M. Yoshimatsu, "X-ray topographic study on stacking faults in silicon single crystals," *Japanese Journal of Applied Physics*, **3**, 94–103, 1964.

102. M. Futagami, "Distinction between clean and decorated oxidation-induced stacking faults by chemical etching," *Journal of the Electrochemical Society*, **127**, 1172–1177, 1980.

103. K. V. Ravi, C. J. Varker, and C. E. Volk, "Electrically active stacking faults in silicon," *Journal of the Electrochemical Society*, **120**, 533–541, 1973.

104. K. Marsden, S. Sadamitsu, M. Hourai, S. Sumita, and T. Shigematsu, "Observation of ring-OISF nuclei in CZ-Si using short-time annealing and infrared light scattering tomography," *Journal of the Electrochemical Society*, **142**, 996–1001, 1995.

105. M. S. Bawa, W. J. Bell, H. M. Grimes, and T. J. Shaffner, "Temperature dependence of oxidation induced stacking faults and oxygen incorporation in dislocation-free Czochralski silicon," *Journal of Crystal Growth*, **94**, 803–806, 1989.

106. J. Dieleman and T. H. G. Martens, "Strong orientation dependence of the formation of surface stacking faults during oxidation of float-zone silicon," *Applied Physics Letters*, **40**, 340–341, 1982.

107. E. Dornberger and W. von Ammon, "The dependence of ring-like distributed stacking faults on the axial temperature gradient of growing Czochralski silicon crystals," *Journal of the Electrochemical Society*, **143**, 1648–1653, 1996, and the references therein.

108. C.-Y. Choi, J.-H. Lee, and S.-H. Cho, "Characterization of mechanical damage on structural and electrical properties of silicon wafers," *Solid-State Electronics*, **43**, 2011–2020, 1999, and the references therein.

109. S. P. Murarka, T. E. Seidel, J. V. Dalton, J. M. Dishman, and M. H. Read, "A study of stacking faults during CMOS processing: Origin, elimination and contribution to leakage," *Journal of the Electrochemical Society*, **127**, 716–724, 1980.

110. T. Sinno, R. A. Brown, W. von Ammon, and E. Dornberger, "Point defect dynamics and the oxidation-induced stacking-fault ring in Czochralski-grown silicon crystals," *Journal of the Electrochemical Society*, **145**, 302–318, 1998.

111. W. von Ammon, E. Dornberger, H. Oelkrug, and H. Weidner, "The dependence of bulk defects on the axial temperature gradient of silicon crystals during Czochralski growth," *Journal of Crystal Growth*, **151**, 273–277, 1995, and the references therein.

112. E. Dornberger, D. Gräf, M. Suhren, U. Lambert, P. Wagner, F. Dupret, and W. von Ammon, "Influence of boron concentration on the oxidation-induced stacking fault ring in Czochralski silicon crystals," *Journal of Crystal Growth*, **180**, 343–352, 1997, and the references therein.

113. P. M. Petroff, G. A. Rozgonyi, and T. T. Sheng, "Elimination of process-induced stacking faults by preoxidation gettering of Si wafers: II. Si_3N_4 process," *Journal of the Electrochemical Society*, **123**, 565–570, 1976.

114. G. A. Rozgonyi and R. A. Kushner, "The elimination of stacking faults by preoxidation getter-ing of silicon wafers: III. Defect etch pit correlation with p-n junction leakage," *Journal of the Electrochemical Society*, **123**, 570–576, 1976.

115. T. Hattori, "Elimination of stacking faults in silicon by trichloroethylene oxidation," *Journal of the Electrochemical Society*, **123**, 945–946, 1976, and the references therein.

116. H. Hashimoto, H. Shibayama, H. Masaki, and H. Ishikawa, "Annihilation of stacking faults in silicon by impurity diffusion," *Journal of the Electrochemical Society*, **123**, 1899–1902, 1976.

117. H.-D. Chiou, "The effects of preheatings on axial oxygen precipitation uniformity in Czochralski silicon crystals," *Journal of the Electrochemical Society*, **139**, 1680–1684, 1992.

118. H.-D. Chiou, "Improving axial oxygen precipitation uniformity in CZ silicon crystals using the S-curve concept," *Journal of the Electrochemical Society*, **141**, 173–178, 1994.

119. G. Kissinger, J. Vanhellemont, U. Lambert, D. Gräf, E. Dornberger, and H. Richer, "Influence of residual point defect supersaturation on the formation of grown-in oxide precipitate nuclei in CZ-Si," *Journal of the Electrochemical Society*, **145**, L75–L78, 1998.

120. S. A. McHugo, E. R. Weber, S. M. Myers, and G. A. Petersen, "Gettering of iron to implantation induced cavities and oxygen precipitates in silicon," *Journal of the Electrochemical Society*, **145**, 1400–1405, 1998.

121. T. Ono, G. A. Rozgonyi, C. Au, T. Messina, R. K. Goodall, and H. R. Huff, "Oxygen precipitation behavior in 300-mm polished Czochralski silicon wafers," *Journal of the Electrochemical Society*, **146**, 3807–3811, 1999.

122. T. Ono, E. Asayama, H. Horie, M. Hourai, K. Sueoka, H. Tsuya, and G. A. Rozgonyi, "Effect of heavy boron doping on oxide precipitate growth in Czochralski silicon," *Journal of the Electrochemical Society*, **146**, 2239–2244, 1999.

123. K. Nakashima, Y. Watanabe, T. Yoshida, and Y. Mitsushima, "A method to detect oxygen pre-cipitates in silicon wafers by highly selective reactive ion etching," *Journal of the Electrochemical Society*, **147**, 4294–4296, 2000.

124. D.-H. Hwang, B.-Y. Lee, H.-D. Yoo, and O.-J. Kwon, "Anomalous oxygen precipitation near the vacancy and interstitial boundary in CZ-Si wafers," *Journal of Crystal Growth*, **213**, 57–62, 2000.

125. C. Claeys, E. Simoen, V. B. Neimash, A. Kraitchinskii, M. Kras'ko, O. Puzenko, A. Blondeel, and P. Clauws, "Tin doping of silicon for controlling oxygen precipitation and radiation hardness," *Journal of the Electrochemical Society*, **148**, G738–G745, 2001.

126. K. Sueoka, M. Akatsuka, M. Okui, and H. Katahama, "Computer simulation for morphology, size, and density of oxide precipitates in CZ silicon," *Journal of the Electrochemical Society*, **150**, G469–G475, 2003.

127. E. Leoni, L. Martinelli, S. Binetti, G. Borionetti, and S. Pizzini, "The origin of photolumines-cence from oxygen precipitates nucleated at low temperature in semiconductor," *Journal of the Electrochemical Society*, **151**, G866–G869, 2004.

128. K. Nakashima, T. Yoshida, and Y. Mitsushima, "Measurements of size, morphology, and spatial distribution of oxygen precipitates in Si wafers using RIE," *Journal of the Electrochemical Society*, **152**, G339–G344, 2005.

129. G. Kissinger, J. Dabrowski, A. Sattler, C. Seuring, T. Müller, H. Richer, and W. von Ammon, "Analytical modeling of the interaction of vacancies and oxygen for oxide precipitation in RTA treated silicon wafers," *Journal of the Electrochemical Society*, **154**, H454–H459, 2007.

130. Z. Zeng, X. Ma, J. Chen, D. Yang, I. Ratschinski, F. Hevroth, and H. S. Leipner, "Effect of oxy-gen precipitates on dislocation motion in Czochralski silicon," *Journal of Crystal Growth*, **312**, 169–173, 2010.

131. M. Arivanandhan, R. Gotoh, K. Fujiwara, T. Ozawa, Y. Hayakawa, and S. Uda, "The impact of Ge co-doping on grown-in O precipitates in Ga-doped Czochralski-silicon," *Journal of Crystal Growth*, **321**, 24–28, 2011.

132. Z. Zeng, J. Chen, Y. Zeng, X. Ma, and D. Yang, "Immobilization of dislocations by oxygen pre-cipitates in Czochralski silicon: Feasibility of precipitation strengthening mechanism," *Journal of Crystal Growth*, **324**, 93–97, 2011.

133. A. Sarikov, V. Litovchenko, I. Lisovskyy, M. Voitovich, S. Zlobin, V. Kladko, N. Slobodyan, V. Machulin, and C. Claeys, "Mechanisms of oxygen precipitation in CZ-Si wafers subjected to rapid thermal anneals," *Journal of the Electrochemical Society*, **158**, H772–H777, 2011.

134. G. H. Schwuttke, "Study of copper precipitation behavior in silicon single crystals," *Journal of the Electrochemical Society*, **108**, 163–167, 1961.

135. K. Sueoka, "Modeling of internal gettering of nickel and copper by oxide precipitates in Czochralski-Si wafers," *Journal of the Electrochemical Society*, **152**, G731–G735, 2005.

136. M. Yonemura, K. Sueoka, and K. Kamei, "Lattice strain around platelet oxide precipitates in C- and N-doped silicon epitaxial wafers," *Journal of the Electrochemical Society*, **148**, G630–G635, 2001.

137. J. Lu and G. Rozgonyi, "Oxygen and carbon precipitation in crystalline sheet silicon: Depth profiling by infrared spectroscopy, and preferential defect etching," *Journal of the Electrochemical Society*, **153**, G986–G991, 2006.

138. C. Reimann, M. Trempa, J. Friedrich, and G. Müller, "About the formation and avoidance of C and N related precipitates during directional solidification of multi-crystalline silicon from contaminated feedstock," *Journal of Crystal Growth*, **312**, 1510–1516, 2010.

139. M. Trempa, C. Reimann, J. Friedrich, and G. Müller, "The influence of growth rate on the formation and avoidance of C and N related precipitates during directional solidification of multi-crystalline silicon," *Journal of Crystal Growth*, **312**, 1517–1524, 2010.

140. M. L. Joshi, "Precipitates of phosphorus and of arsenic in silicon," *Journal of the Electrochemical Society*, **113**, 45–48, 1966.

141. H. Kodera, "Precipitation of antimony in heavily doped silicon identified by x-ray microanalyzer," *Japanese Journal of Applied Physics*, **2**, 193–194, 1963.

142. H. Fujimori, "Dependence on morphology of oxygen precipitates upon oxygen supersaturation in Czochralski silicon crystals," *Journal of the Electrochemical Society*, **144**, 3180–3184, 1997.

143. R. Falster and V. V. Voronkov, "Lattice defects in silicon", 123-5.5&5.6.ppt, from Monsanto Electronic Materials Company (MEMC) website, 2006.

144. A. Toktosunov and W. Bergholz, "Origin, characterization, and effect of crystal-originated particles (COPs) on device performance", Siemens AG, Semiconductor Division, Munich, Germany, 1990.

145. T. Iizuka and M. Kikuchi, "Anomalous etch patterns in heavily doped silicon," *Japanese Journal of Applied Physics*, **2**, 196, 1963.

146. G. R. Booker and R. Stickler, "Small particles in silicon," *Journal of the Electrochemical Society*, **111**, 1011–1012, 1964.

147. T. Abe, T. Samizo, and S. Maruyama, "Etch pits observed in dislocation-free silicon crystals," *Japanese Journal of Applied Physics*, **5**, 458–459, 1966.

148. Y. Yukimoto, A. Hirano, and Y. Sugioka, "X-ray observations of anomalous etch patterns in silicon crystals," *Japanese Journal of Applied Physics*, **6**, 420–421, 1967.

149. Y. Kashiwagi, R. Shimokawa, and M. Yamanaka, "Highly sensitive etchants for delineation of defects in single- and polycrystalline silicon materials," *Journal of the Electrochemical Society*, **143**, 4079–4087, 1996.

150. M. Suhren, D. Gräf, U. Lambert, and P. Wagner, "Crystal defects in highly boron doped silicon," *Journal of the Electrochemical Society*, **144**, 4041–4044, 1997.

151. M. Nishimura, S. Yoshino, H. Motoura, S. Shimura, T. Mchedlidze, and T. Hikone, "The direct observation of grown-in laser scattering tomography defects in Czochralski silicon," *Journal of the Electrochemical Society*, **143**, L243–L246, 1996.

152. W. Wijaranakula, "Characterization of crystal-originated defects in Czochralski silicon using nonagitated Secco etching," *Journal of the Electrochemical Society*, **141**, 3273–3277, 1994.

153. D. Gräf, M. Suhren, U. Lambert, R. Schmolke, A. Ehlert, W. von Ammon, and P. Wagner, "Characterization of crystal quality by crystal originated particle delineation and the impact on the silicon wafer surface," *Journal of the Electrochemical Society*, **145**, 275–284, 1998.

154. N. Shimoi, M. Kurokawa, A. Tanabe, N. Koizumi, and Y. Matsushita, "Accuracy of differential method to distinguish crystal originated particles from light point defects in Czochralski-grown silicon wafers," *Journal of Crystal Growth*, **210**, 31–35, 2000.

155. Q. Xiao and H. Tu, "New approach to remove crystal-originated pits in Czochralski-grown silicon: Combination of germanium ion implantation with solid-phase epitaxy," *Journal of Crystal Growth*, **271**, 368–375, 2004.
156. W. P. Lee, H. K. Yow, and T. Y. Tou, "Characterization of crystal-originated particles in silicon nitride doped, CZ-grown silicon wafers," *Journal of the Electrochemical Society*, **153**, G248–G252, 2006.
157. J. Chen, D. Yang, H. Li, X. Ma, D. Tian, L. Li, and D. Que, "Crystal-originated particles in germanium-doped Czochralski silicon crystal," *Journal of Crystal Growth*, **306**, 262–268, 2007.
158. M. Akatsuka, K. Sueoka, and T. Yamamoto, "Classification of etch pits at silicon wafer surface using image-processing instrument," *Journal of Crystal Growth*, **210**, 366–369, 2000.
159. T. Motooka, "Molecular-dynamics simulations of recrystallization processes in silicon: Nucleation and crystal growth in the solid-phase and melt," Department of Materials Science and Engineering, Kyushu University, Motooka 744, Fukuoka 819-0395, Japan.
160. H. Zimmermann, U. Gösele, M. Seilenthal, and P. Eichinger, "Vacancy concentration wafer mapping in silicon," *Journal of Crystal Growth*, **129**, 582–592, 1993.
161. N. Adachi, T. Hisatomi, M. Sano, and H. Tsuya, "Reduction of grown-in defects by high temperature annealing," *Journal of the Electrochemical Society*, **147**, 350–353, 2000.
162. V. V. Voronkov and R. Falster, "Grown-in microdefects, residual vacancies and oxygen precipitation bands in Czochralski silicon," *Journal of Crystal Growth*, **204**, 462–474, 1999.
163. V. V. Voronkov, "Grown-in defects in silicon produced by agglomeration of vacancies and self-interstitials," *Journal of Crystal Growth*, **310**, 1307–1314, 2008.
164. H. Fujimori, H. Fujisawa, Y. Hirano, and T. Okabe, "The morphology of grown-in defects in nitrogen-doped silicon crystals," *Journal of Crystal Growth*, **237–239**, 338–344, 2002.
165. S. Umeno, T. Ono, T. Tanaka, E. Asayama, H. Nishikawa, M. Hourai, H. Katahama, and M. Sano, "Nitrogen effect on grown-in defects in Czochralski silicon crystals," *Journal of Crystal Growth*, **236**, 46–50, 2002.
166. G. Kissinger, D. Gräf, U. Lambert, and H. Richter, "A method for studying the grown-in defect density spectra in Czochralski silicon wafers," *Journal of the Electrochemical Society*, **144**, 1447–1456, 1997.
167. K. Minowa, K. Takeda, S. Tomimatsu, and K. Umemura, "TEM observation of grown-in defects in CZ and epitaxial silicon wafers detected with optical shallow defect analyzer," *Journal of Crystal Growth*, **210**, 15–19, 2000.
168. K. Nakai, M. Hasebe, K. Ohta, and W. Ohashi, "Characterization of grown-in stacking faults and dislocations in CZ-Si crystals by bright field IR laser interferometer," *Journal of Crystal Growth*, **210**, 20–25, 2000.
169. H. Nishikawa, T. Tanaka, Y. Yanase, M. Hourai, M. Sano, and H. Ysuya, "Formation of grown-in defects during Czochralski silicon crystal growth," *Japanese Journal of Applied Physics*, **36**, 6595–6600, 1997.
170. W. Wijaranakula and J. H. Matlock, "A formation mechanism of the thermal donors related to carbon in silicon after an extended isochronal anneal," *Journal of the Electrochemical Society*, **137**, 1964–1969, 1990, and the references therein.
171. C. Maddalon-Vinante, J. P. Vallard, and D. Barbier, "Infrared study of the effect of rapid thermal annealing, thermal donor formation, and hydrogen on the precipitation of oxygen," *Journal of the Electrochemical Society*, **142**, 2071–2076, 1995.
172. I. V. Antonova, V. P. Popov, A. E. Plotnikov, and A. Misiuk, "Thermal donor and oxygen precipitate formation in silicon during 450°C treatments under atmospheric and enhanced pressure," *Journal of the Electrochemical Society*, **146**, 1575–1578, 1999.
173. V. V. Voronkov, G. I. Voronkova, A. V. Batunina, R. Falster, V. N. Golovina, A. S. Guliaeva, N. B. Tiurina, and M. G. Milvidski, "The sensitivity of thermal donor generation in silicon to self-interstitial sinks," *Journal of the Electrochemical Society*, **147**, 3899–3906, 2000.
174. V. B. Neimash, A. Kraitchinskii, M. Kras'ko, O. Puzenko, C. Claeys, E. Simoen, B. Svensson, and A. Kuznetsov, "Influence of tin impurities on the generation and annealing of thermal oxygen donors in Czochralski silicon at 450°C," *Journal of the Electrochemical Society*, **147**, 2727–2733, 2000.

175. J. M. Rafí, E. Simoen, C. Claeys, A. G. Ulyashin, R. Job, W. R. Fahrner, J. Versluys, P. Clauws, M. Lozano, and F. Campabadal, "Impact of direct plasma hydrogenation on thermal donor formation in n-type CZ silicon," *Journal of the Electrochemical Society*, **152**, G16–G24, 2005.

176. J. M. Rafí, E. Simoen, C. Claeys, A. G. Ulyashin, R. Job, W. R. Fahrner, J. Versluys, P. Clauws, M. Lozano, and F. Campabadal, "Analysis of oxygen thermal donor formation in *n*-type Cz silicon," *Analytical and Diagnostic Techniques for Semiconductor Materials, Devices and Processes*, **3**, 96–105, 2003.

177. E. Simoen, Y. L. Huang, Y. Ma, J. Lauwaert, P. Clauws, J. M. Rafi, A. Ulyashin, and C. Claeys, "What do we know about hydrogen-induced thermal donors in silicon?," *Journal of the Electrochemical Society*, **156**, H434–H442, 2009.

178. H. R. Huff, R. C. Bracken, and S. N. Rea, "Influence of silicon slice curvature on thermally induced stresses," *Journal of the Electrochemical Society*, **118**, 143–145, 1971.

179. A. W. Fisher and G. L. Schnable, "Minimizing process-induced slip in silicon wafers by slow heating and cooling," *Journal of the Electrochemical Society*, **123**, 434–435, 1976.

180. Y. Yoriume, "Deformation of silicon wafers by thermal oxidation," *Journal of the Electrochemical Society*, **129**, 2076–2081, 1982.

181. B. Wong and R. J. Holbrook, "Microindentation for fracture and stress-corrosion cracking studies in single-crystal silicon," *Journal of the Electrochemical Society*, **134**, 2254–2256, 1987.

182. M. Akatsuka, K. Sueoka, H. Katahama, and N. Adachi, "Calculation of slip length in 300-mm silicon wafers during thermal processes," *Journal of the Electrochemical Society*, **146**, 2683–2688, 1999, and the references therein.

183. P. K. Kulshreshtha, Y. Yoon, K. M. Youssef, E. A. Good, and G. Rozgonyi, "Oxygen precipitation related stress-modified crack propagation in high growth rate Czochralski silicon wafers," *Journal of the Electrochemical Society*, **159**, H125–H129, 2012.

184. A. E. Widmer and W. Rehwald, "Thermoplastic deformation of silicon wafers," *Journal of the Electrochemical Society*, **133**, 2403–2409, 1986.

185. D. Elwell and S. Hahn, "Effects of laser back-side damage upon mechanical properties of silicon," *Journal of the Electrochemical Society*, **131**, 1395–1400, 1984.

186. H. Kotake and S. Takasu, "Quantitative measurement of stress in silicon by photoelasticity and its application," *Journal of the Electrochemical Society*, **127**, 179–184, 1980.

187. B. Leroy and C. Plougonven, "Warpage of silicon wafers," *Journal of the Electrochemical Society*, **127**, 961–970, 1980, and the references therein.

188. K. H. Yang, "An optical imaging method for wafer warpage measurements," *Journal of the Electrochemical Society*, **132**, 1214–1218, 1985.

189. C.-O. Lee and P. J. Tobin, "The effect of CMOS processing on oxygen precipitation, wafer warpage, and flatness," *Journal of the Electrochemical Society*, **133**, 2147–2152, 1986.

190. D. Beauchaine, W. Wijaranakula, H. Mollenkopt, and J. Matlock, "Effect of thin film stress and oxygen precipitation on warpage behavior of large diameter p/p+ epitaxial wafers," *Journal of the Electrochemical Society*, **136**, 1787–1793, 1989.

191. H.-D. Chiou, Y. Chen, R. W. Carpenter, and J. Jeong, "Warpage and oxide precipitate distributions in CZ silicon wafers," *Journal of the Electrochemical Society*, **141**, 1856–1862, 1994.

192. I. Yonenaga, "Upper yield stress of Si crystals at high temperatures," *Journal of the Electrochemical Society*, **143**, L176–L178, 1996, and the references therein.

193. A. Muižnieks, G. Raming, A. Mühlbauer, J. Virbulis, B. Hanna, and W. v Ammon, "Stress induced dislocation generation in large FZ- and CZ-silicon single crystals – numerical model and qualitative considerations," *Journal of Crystal Growth*, **230**, 305–313, 2001.

194. X. Chen, S. Nakano, and K. Kakimoto, "Three-dimensional global analysis of thermal stress and dislocations in a silicon ingot during a unidirectional solidification process with a square crucible," *Journal of Crystal Growth*, **312**, 3261–3266, 2010.

195. X. J. Chen, S. Nakano, L. J. Liu, and K. Kakimoto, "Study on thermal stress in a silicon ingot during a unidirectional solidification process," *Journal of Crystal Growth*, **310**, 4330–4335, 2008.

10

Silicon Wafer Preparation for VLSI and ULSI Processing

10.1 Introduction

Metallic contamination in silicon wafers has harmful effects on device performance and circuits fabricated with them. It is well established that minor amounts of metals can degrade thin oxide film quality, leakage current densities, and minority carrier lifetimes. Fast-diffusing metal impurities, such as copper, nickel, and palladium, easily precipitate at the wafer surfaces, causing small structural defects called S-pits, or, in a higher density, resulting in haze [1]. The formation of the S-pits and surface haze has been correlated with the generation of oxidation-induced stacking faults (OISFs) and with the formation of epitaxial stacking faults. As the degree of circuit integration increases, the influence of metallic and particle impurities on wafer yields has led to higher chemical-purity requirements and advanced cleaning techniques, as well as to the development of different gettering schemes to reduce their effects.

10.2 Purity of Chemicals Used for Silicon Processing

A variety of organic and inorganic contaminations are often present on the surface of the silicon wafers, and these need to be removed. They may be due to the polishing chemicals or slurry components used to polish the wafers, or from some unknown sources. Sometimes, airborne dust particles, chemical residues, and other unexpected oil contamination may also be present on these surfaces. Prior to using the wafers, it is necessary to clean these contaminates from the silicon surface. Of course, the chemicals used to clean the surfaces should be highly pure and free from suspended particles. Particles in acids and solvents are typically either inherent in as-received containers, sinks, and process equipment, or they are generated during etching or cleaning processes from piping, filters, etc. The purity of the chemicals is determined by the trace materials present in them. It may become necessary to avoid certain chemicals that have trace metallic impurities, despite their high purity. Table 10.1 lists the purity of chemicals used for silicon integrated circuit fabrication, depending on the circuit complexity.

TABLE 10.1

Purity of Chemicals Used for Silicon Integrated Circuit Processing

Grade	SEMI Standard	Device Geometry
XLSI	Relates to SEMI Grade 4	Recommended for 0.09–0.20-μm feature size technologies Trace metals <0.1 ppb
SLSI	Relates to SEMI Grade 3	Recommended for 0.20–0.80-μm feature size technologies Trace metals <1 ppb
ULSI	Relates to SEMI Grade 2	Recommended for 0.80–1.20-μm feature size technologies Trace metals <10 ppb
VLSI	Relates to SEMI VLSI	Recommended for 0.80–0.20-μm feature size technologies Trace metals <50 ppb for most elements (chemistry dependent)
SEMI/EG	Relates to SEMI Grade 1	Recommended for >1.20-μm feature size technologies For standard applications and manufacturing of discrete devices

Source: Honeywell Electronic Materials & Electronic Chemicals, https://www.honeywell-pmt.com/sm/products-applications/electro-chem.html [2].

Notes: XLSI, a graphical user interface for a conceptual retrieval system; SLSI, super large scale integration; ULSI, ultra large scale integration; VLSI, very large scale integration; SEMI, Semiconductor Equipment and Materials International.

10.3   Degreasing of Silicon Wafers

Degreasing is generally done with a hot trichloroethylene solution. Very little ultrasonic agitation is recommended at this stage. Most of the adhesion chemicals are removed in this step. The wafers are rinsed in acetone solution before they are cleaned with deionized (DI) water.

10.4   Removal of Metallic and Other Impurities

It is not possible to have 100% pure chemicals to clean the silicon wafers. The electronic- and semiconductor-grade acids and chemical solvents most often used today for this purpose, have low concentrations of unknown and unspecified cations and anions. These can strongly degrade device performance. Transition metallic impurities diffuse quickly moving into the silicon wafer lattice, often through the interstitial paths. Once in the lattice, they replace silicon ions and change the bandgap of the material. If the parent lattice has defects, or if the metallic impurities have any inherently favorable properties, they get into the regular lattice sites. With the bandgap altered, the changed lattice sites act as a generation-recombination center. The result is both a decrease in minority carrier lifetimes and the creation of leakage currents at *p-n* junctions.

Wet chemical cleaning is often used to remove metallic impurities present on silicon surfaces. The most common method is piranha cleaning, which is often implemented in the beginning of the cleaning process. This is a mixture of two solutions: sulfuric acid (H_2SO_4) and hydrogen peroxide (H_2O_2). This mixture is also known as an SPM solution. The two solutions are combined slowly and continuously to avoid a vigorous reaction. This mixture creates an exothermic reaction and liberates a lot of heat, thus raising the solution temperature; hence, no external heating is required. The solution removes most of the heavy organic contaminants that arise because of photoresist (PR) residues, as well as tiny plastic materials attached to the silicon surface. In this process, a thin chemical oxide (SiO_x) on the order of few nanometers forms on the silicon surface, trapping all metallic contaminants present. Organic materials dissociate into smaller fragments and leave the silicon surface. Thorough rinsing of the wafers is immediately carried out using DI water to clear the surface from the chemicals used. At this stage, the silicon wafer surface will be hydrophilic and water will stick to it. Removing the thin oxide using diluted hydrofluoric acid (DHF) leaves a fresh silicon surface, free from surface-adsorbed particles and other residues.

After the piranha cleaning and removal of the thin oxide layer, the silicon wafers will undergo standard RCA cleaning. This method is widely used in the semiconductor industry to clean and free the silicon surface, especially from metallic impurities. This step is carried out in two stages. In the first stage, the wafers undergo a cleaning process identified as "RCA Standard Solution 1," also called the "RCA SC-1" or "APM—ammonium hydroxide-hydrogen peroxide-water mixture" method. The wet chemistry consists of an alkaline mixture of ammonium hydroxide (NH_4OH), hydrogen peroxide (H_2O_2), and water (H_2O). This mixture is not exothermic in nature and does not produce any heat; hence, the solution will be heated to a temperature of the order of 80°C. In this process, the solution removes organic surface residues and several metallic impurities. This SC-1 solution also effectively removes many inorganic particles sticking to the surface [1]. During this step, a thin chemical oxide, SiO_x, forms on the silicon surface. Most of the metals present on the silicon surface convert to their respective metal hydroxides and are dissolved in the solution. After thorough rinsing with DI water, the wafer is treated with DHF to strip off the oxide layer, where metal impurities are trapped inside.

The wafers then undergo the second process step, identified as "RCA Standard Solution 2" or the "RCA SC-2" or "HPM—hydrochloric acid-hydrogen peroxide-water mixture" method. In this step, an acidic mixture of hydrochloric acid (HCl), hydrogen peroxide (H_2O_2), and water (H_2O) is heated to a temperature on the order of 80°C. This method effectively removes most of the alkali ions present on the surface by converting them to their respective chlorides and causing them to leave the silicon surface. Most of these metal chlorides dissolve into the solution. This SC-2 solution can remove not only the alkali ions, but also the cations such as Al^{+3}, Fe^{+3}, and Mg^{+3} that form insoluble hydroxides that do not respond to the SC-1 solution.

Anttila et al. [1] conducted a systematic analysis on the metal impurity concentration on silicon surfaces using wet chemical cleaning routes and on the resident metal impurity levels using total reflection x-ray fluorescence analysis (TXRFA) techniques. Standard chemical solutions were used for this purpose as supplied by different vendors. The wafers were cleaned in RCA standard solutions, after which particle counts and metallic contamination were measured. Hot APM, an efficient particle remover, does deposit iron on silicon. The amount of deposited iron is strongly dependent on the quality of the peroxide used in this mixture; however, it can be reduced by using a shorter cleaning time. According to them, the deposition rate of iron on a wafer is likely diffusion-limited for shorter times before equilibrium is reached between the metallic concentrations on the wafer and in the cleaning solution. The iron concentration after APM is not sensitive to the age of the chemical mixture. There is, however, more zinc deposited in an aged chemical mixture. The same dependence of iron concentration on chemical supplier observed after APM was also found after hot HPM, but the iron concentrations were much lower here. The HPM mixture causes a considerable increase in the particle counts. These particles are loosely bound to the wafer surface and can, therefore, be easily removed through simple ultrasonic agitation. The copper concentration on silicon remained essentially constant, independent of chemical supplier, cleaning solution, or the time used for the cleaning. If APM is used as the last cleaning solution, the choice of peroxide supplier is critical here. However, if HPM is used last, all the tested chemical grades give essentially the same metal contamination levels. Figures 10.1 and 10.2 show the metallic concentration levels for iron, copper, and zinc impurities for the sequence adopted. Since the chemicals used for the wet cleaning process play a crucial role in the surface properties of silicon, it is always recommended to study the listed possible impurities present in those solutions.

The removal of metallic contaminants, particularly iron, from silicon wafers using very dilute acidic solutions has been reported by Anttila and Tilli [3]. The resulting surface concentrations for iron were measured and found to be well below the 1.0×10^{10} atoms/cm^2

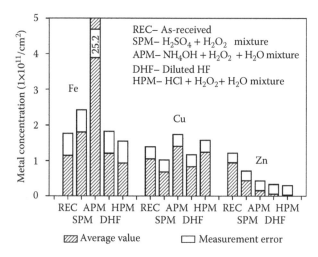

FIGURE 10.1
Metal concentrations in an SPM-APM-DHF-HPM clean. (From O. J. Anttila, M. V. Tilli, M. Schaekers, and C. L. Claeys, *Journal of the Electrochemical Society*, **139**, 1180–1185, 1992, and the references therein [1].)

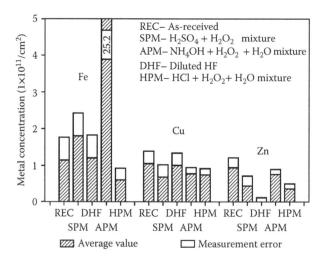

FIGURE 10.2
Metal concentrations in an SPM-DHF-APM-HPM clean. (From O. J. Anttila, M. V. Tilli, M. Schaekers, and C. L. Claeys, *Journal of the Electrochemical Society*, **139**, 1180–1185, 1992, and the references therein [1].)

level for most of the diluted solutions. The particle levels were also reported to be low, comparable to the standard RCA SC-1 clean. The hydrofluoric acid (HF) solutions behaved in a different way compared to the other tested acids, such as hydrochloric acid (HCl), nitric acid (HNO_3), and acetic acid (CH_3COOH), leaving a higher concentration of iron on the wafers, but a considerably lower surface recombination rate of the minority carriers in oxidized silicon wafers. The minority carrier bulk lifetimes were in excess of 100 µs for the oxidized acid–cleaned wafers, with the exception of the HF cleans. Iron was found to be the dominant lifetime-shortening contaminant for concentrations larger than 10^{11} atoms/cm^3. The exact reaction mechanisms involved in these observations need to be investigated.

A less critical cleaning procedure was also reported by Filho et al. [4] using a DHF dip followed by boiling in isopropyl alcohol as a final step. This occurs immediately after completing the RCA-based cleaning process. This step is effective in reducing particulate levels on the silicon surfaces without increasing the metallic contamination.

Monitoring of heavy metal contamination is an importance issue when considering the chemical solutions to be used for wafer cleaning. Typical traditional approaches include TXRF and atomic absorption spectroscopy (AAS), which have proven their usefulness in monitoring the wet chemical applications for silicon wafer cleaning. A new approach using surface photovoltage (SPV) was carried out by Jastrzebski et al. [5]. It is a fast, contactless method, and suitable for patterned product wafers as well. In this SPV method, the monitoring of chemical cleaning and the purity of the chemicals is achieved by mapping minority carrier diffusion length, iron concentration in the bulk, and surface contamination. This technique is uniquely suited for heavy metal monitoring. It was also used to monitor copper contamination in buffer HF by measuring its effect on surface recombination; iron contamination was monitored by measuring its effect on bulk recombination. Iron surface contamination was measured down to the 1.0×10^9 cm^{-2} level, while the detection limit of this approach was determined to be 2×10^8 cm^{-2}.

10.5 Gettering of Metallic Impurities

Silicon crystal wafers usually have a doped *p*- or *n*-type impurity, depending on wafer specifications. Only a thin layer of the wafer, approximately 1.0 µm to 15.0 µm deep, is used to define most of the very large scale integration (VLSI) and ultra large scale integration (ULSI) devices. In most cases, the remaining thickness of the silicon wafer acts as a mechanical support to the devices/circuits defined on the wafer. In typical device processing, when a silicon wafer is heated to ≈1000°C in a nitrogen or hydrogen chloride environment, most of the oxygen species present near the surface of the wafer are removed by the ambient atmosphere. Deep in the bulk of the silicon crystal, however, oxygen atoms remain at a high concentration and tend to precipitate as complexes. The surface of the wafer, where most of the active devices are fabricated, becomes depleted of oxygen species, while a region in the wafer several microns away from the surface becomes rich in unwanted defects. The defective region acts as a sink that attracts impurities present in the silicon lattice, such as heavy metals. As the wafer is heated, the defects are attracted to this sink and thus, they are removed from the wafer surface. This segregation mechanism is called intrinsic gettering [6]. Under certain annealing cycles, oxygen atoms in the bulk of the crystal can be precipitated as SiO_x clusters that act as trapping sites for impurities. This process is called internal gettering, and is one of the most effective means to remove unintentional impurities from the near-surface region where devices are fabricated. In addition to the intrinsic gettering, other treatments can be applied before or during device fabrication to remove defects and unwanted metallic impurities from regions where devices are fabricated. Gettering can be achieved by intentionally damaging the back of the silicon wafer and then subjecting it to high-temperature annealing treatments. Again, the damaged region in the wafer acts as a sink for unwanted impurities, such as heavy metals, which then diffuse from the surface to the back side and collect there. A common method to create strain in the back side of the wafer is to mechanically damage it or dope it heavily with specific impurities.

The term "gettering," as used in VLSI and ULSI process technology, refers to the removal of unwanted metallic impurities or process-induced contaminants from the active region of a semiconductor substrate. Although gettering has been commonly practiced, the mechanism is not well understood in most cases. It is well accepted that most of the metal atoms precipitate heterogeneously at crystal dislocations that are associated with oxide (SiO) precipitates. The metallic impurities act as carrier recombination and generation sites, and may even result in shorting of the device. These facts have forced the VLSI and ULSI community to impose strict specifications for metallic impurity contamination as low as 2.5×10^9 atoms/cm^2 [7]. Traditionally, metallic impurities have been removed, or gettered, from the near-surface region of the silicon wafer by creating metal-silicide precipitation sites in the form of oxygen precipitates and their related defects in the bulk of the wafer. This method is known as internal gettering or simply IG.

In single-crystal silicon wafers, intrinsic gettering involves mainly extracting the metallic impurities present from the surface of a silicon wafer and trapping them in the bulk. Since most of the devices are usually built within a short distance of several microns from the surface of the wafer, reducing surface impurities directly improves device performance. This gettering process involves the precipitation of oxygen in the bulk of the silicon lattice. The oxygen precipitates typically strain the silicon lattice, and thereby generate many dislocations in it. These dislocations attract metallic impurities, and they tend to accumulate near these crystal lattice sites. Since metals are fast-diffusing species, they respond quickly to the thermal treatments and will migrate from the top surface regions to the sites where dislocations are present. It is, therefore, of utmost importance to leave the silicon surface free of oxygen-precipitate-induced defects so that most of the metals present near the surface will be displaced from this critical area. An oxygen-free denuded zone [8] forms near the surface by means of a high-temperature anneal, which permits the out-diffusion of oxygen species. This thermal anneal is further followed by other treatments to enhance the nucleation and growth of oxygen precipitates in the bulk of the silicon.

Figure 10.3 shows a typical thermal cycle, which produces the desired oxygen profile for intrinsic gettering in silicon wafers. The first step is a high-temperature treatment at 1100°C to establish a denuded zone near the surface of the wafer. The second step is a low-temperature anneal step in the range of 600°C to 800°C to enhance nucleation sites so oxygen precipitation can take place. The third and final step involves another round of high-temperature thermal annealing [8]. Here, the temperature is raised to approximately 1000°C to allow for sites nucleated in the second stage to grow. It is postulated that under normal conditions, oxygen occupies an interstitial position in the lattice. In order to form and grow a stable nucleation site, a defect or small oxygen precipitate is necessary. In the first step, a denuded zone forms, and at the second step, they all nucleate.

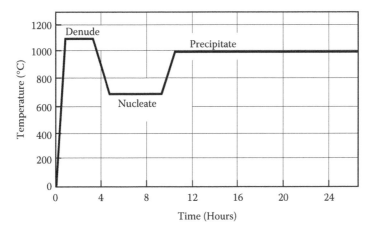

FIGURE 10.3
Typical thermal cycle to produce oxygen profile for intrinsic gettering. (From T. A. Baginski and J. R. Monkowski, *Journal of the Electrochemical Society*, **133**, 762–769, 1986, and the references therein [8].)

In the third step, they precipitate. Wafers that have received these prescribed thermal treatments have shown improvement in both leakage-limited yield and in minority carrier lifetime. These improvements have been theorized to result from the movement of metallic impurities to explain the gettering phenomenon. The details and the time spans are listed in the figure. In ULSI device processing, an IG of metallic contaminants is considered to be an important step. Metallic contaminants originating from contaminated chemicals, gaseous sources, or defective processing equipment either can cause structural defects in the fabricated device or will influence performance during operation; thus, the IG step improves device quality.

The control of carbon and oxygen in Czochralski (CZ) silicon is the subject of a great number of studies because of its importance in defect formation and gettering of impurities. Intrinsic gettering has been found to retard material slip, reduce S-pit formation, control epitaxial stacking fault formation, reduce *p-n* junction leakage currents, and improve metal-oxide semiconductor (MOS) generation lifetimes. Carbon is reported to enhance the formation of several types of crystallographic defects, including B-swirl defects. The amount of substitutional carbon that is considered to be nondetectable by the Fourier transform infrared spectroscopy (FTIR) is still higher than the concentration of an intentionally doped impurity such as boron [9]. Bailey et al. [10] used three heat cycles—1100°C, nitrogen ambient, 120 min; 650°C, nitrogen, 1200 min; and 1000°C, oxygen, 1200 min—with controlled ramp-up and ramp-down cycles to get denuded substrates for VLSI processing. In the higher carbon samples, a new defect, which we call a c-pit, also appears. Its axis of symmetry is oriented 45° to the <110> direction defined by the bulk stacking faults. Several such c-pits are shown in relationship to bulk stacking faults in Figure 10.4. Figure 10.5 is another transmission electron microscopy (TEM) image of a high-carbon-content wafer showing a higher density of these c-pits—in this case, four pairs.

CZ-grown single crystals of silicon are the main source for bipolar and MOS devices. However, as a result of the precipitation of supersaturated oxygen, a variety of defects are produced when the crystals are heat-treated at higher temperatures. The infrared measurements of Shimura et al. [11] are shown in Figure 10.6. During the first annealing step,

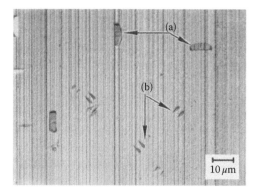

FIGURE 10.4
Optical photograph of (a) normal (100) stacking faults, and (b) c-pits as found in silicon with a high carbon concentration. (From W. E. Bailey, R. A. Bowling, and K. E. Bean, *Journal of the Electrochemical Society,* **132**, 1721–1725, 1985, and the references therein [10].)

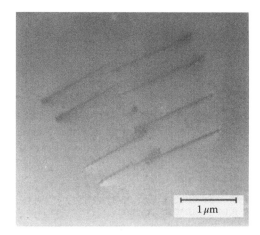

FIGURE 10.5
TEM micrograph of four pairs of c-pits in carbon silicon. (From W. E. Bailey, R. A. Bowling, and K. E. Bean, *Journal of the Electrochemical Society,* **132**, 1721–1725, 1985, and the references therein [10].)

the wafers were subjected to heat treatment at each temperature (600°C, 750°C, 950°C, and 1100°C) for 16 h in a dry oxygen environment, whereas the second annealing was carried out at 1230°C for 2 h in the same conditions. The heat-treated specimens were quenched by means of rapid withdrawal from a furnace and then were examined by infrared absorption, modified-Dash etching, and TEM techniques. A drastic change in the oxygen concentration was observed after the first annealing, and it is mainly due to the precipitation of the Si-O complex present in the crystal. The changes after the second annealing are mainly due to the synergism of Si-O precipitate dissolving and some of the oxygen out-diffusing from the silicon surface.

For complementary metal-oxide semiconductor (CMOS) technologies, intrinsic gettering treatments based on different temperature cycles are used to achieve a denuded zone of defects [12] in a shallow region below the silicon surface, where the performance of the devices is determined.

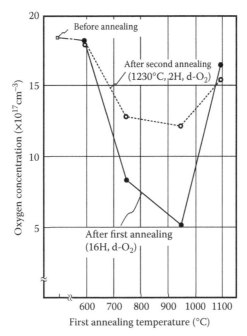

FIGURE 10.6
Interstitial oxygen concentration change before and after first and second annealing. (From F. Shimura, H. Tsuya, and T. Kawamura, *Journal of the Electrochemical Society*, **128**, 1579–1583, 1981 [11].)

Oxygen precipitation plays an important role in the gettering of CZ crystals. By using both low and high oxygen contents, Rivaud et al. [13] made a systematic study of the oxygen species through a four-step annealing process to measure the nucleation, precipitation, coalescence, and final formation defect colonies. The wafers were first heated at 950°C to form a pad oxide. After that, they were ramped to 1200°C at a rate of 10°C/min for the denuding step. They were kept at this temperature for 800 min in an ambient atmosphere consisting of oxygen and 0.5% HCl. Then, they were ramped down to 950°C in nitrogen ambient at a cooling rate of 5°C/min. In the nucleation cycle, the wafers were heat-treated isothermally at 700°C in nitrogen ambient for 960 min. In the third step, isothermal heat treatment was done at 950°C for 600 min for field oxide and for the other oxidation and annealing steps needed for device fabrication. The final annealing was done at 1050°C for 85 min in nitrogen ambient. Figure 10.7 shows the TEM image of an octahedral precipitate formation that was observed at step 2 of the sequence. The precipitate shape changes from octahedral to polyhedral and from polyhedral to octahedral were reported by Sakai et al. [14] for silicon wafers grown by the thermal annealing process.

As mentioned, controlling oxygen precipitation in CZ-grown single crystals of silicon is one of the major issues in the processing of integrated circuits. If not controlled, oxygen precipitation can be seriously detrimental to device performance and to circuit yield. Excess oxygen precipitation can also reduce silicon shear stress and hence, its resistance to plastic deformation from the large thermomechanical stresses involved in the processing cycles [15]. If the oxygen precipitates are located away from circuit elements and other defects associated with them, offer convenient gettering sites for fast-diffusing metal contaminants, thereby improving the circuit yields and the performance of the fabricated CMOS circuits.

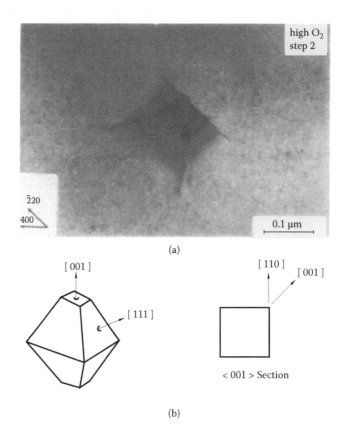

(a)

(b)

FIGURE 10.7
(a) High-magnification TEM micrograph of an octahedral precipitate. (b) Schematic illustration of the geometry of an octahedral precipitate. (From L. Rivaud, C. N. Anagnostopoulos, and G. R. Erikson, *Journal of the Electrochemical Society*, **135**, 437–442, 1988 [13].)

The behavior of oxygen precipitates in silicon at high temperatures was evaluated by Huber and Pagani [16]. The experimental results indicate that for extended high-temperature cycles, the temperature strongly influences the dynamics of oxygen precipitation. It was reported that for short anneals at 700°C, the clusters were rapidly dissolved at higher temperatures. An increase of the anneal time at 700°C and raising the temperature value to 800°C multiplies the number of stable oxygen precipitates. In Figure 10.8, the change in the interstitial oxygen is plotted as functions of the high-temperature anneal steps. The bottom curve shows the amount of precipitated oxygen in samples that were pre-annealed at 700°C for 1 h, while the upper two curves represent the data for a prolonged 700°C and 1 h 800°C pre-anneal. The plotted values clearly show that the amount of precipitated oxygen diminishes upon as the anneal temperature increases from 950°C to 1100°C. Although the growth of precipitates should proceed faster at higher temperatures, the drop-off in oxygen precipitation with temperature indicates a significant dissolution of oxygen clusters at higher temperatures. It was further recorded that the density for a given process temperature reaches a maximum value that is independent of the nucleation conditions. This maximum decreases exponentially with increasing temperature, with activation energy of 3.15 eV, and the data fits an empirical relationship with temperature, as shown in Figure 10.9.

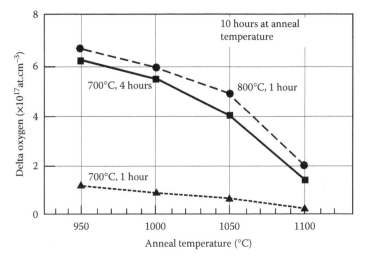

FIGURE 10.8
The change in interstitial oxygen as a function of the anneal temperature for different nucleation treatments. (From W. Huber and M. Pagani, *Journal of the Electrochemical Society*, **137**, 3210–3213, 1990 [16].)

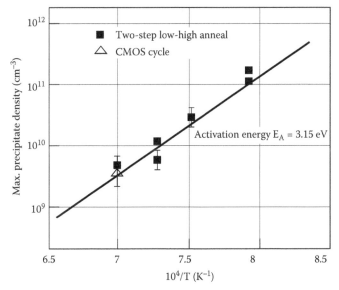

FIGURE 10.9
Maximum precipitate density plotted against the inverse of the high-temperature anneals. (From W. Huber and M. Pagani, *Journal of the Electrochemical Society*, **137**, 3210–3213, 1990 [16].)

Using thin polycrystalline silicon (on the order of 300 Å with the chemical vapor deposition [CVD] method) and diffusing this film by using a $POCl_3$ impurity source at 1080°C for phosphorus diffusion, followed by segregation anneal in the temperature range of 700°C to 800°C using nitrogen ambient for 60 min improves leakage current substantially [17]. Phosphorus gettering is known to be very effective at removing transition metals from the active regions of silicon devices, so it seems reasonable to expect that wafers from the top of an ingot would have higher lifetimes after gettering [18]. This approach is effective at reducing the gold metal present on the silicon surfaces.

Metallic impurities can be gettered from the bulk of the silicon by a variety of means, such as volatilization as chlorides during heat treatment in chlorine-containing ambient, diffusion into regions of enhanced metal solubility, or precipitation at deliberately introduced crystal defects. A number of techniques have been used for the controlled introduction of such crystal defects [19]. High-temperature hydrogen chloride gettering in the presence of a very thin oxide film has been shown to be a much more effective means of removing transition metal groups than is the conventional "dry oxidation with 1% HCl" treatment.

Katz et al. [19] described a pre-process gettering treatment for CZ-grown silicon substrates that removes essentially most of the transition-group metal contaminants present in the wafer and continues to getter contaminants during subsequent device processing as well. Laser-induced damage (LID) on the back surface of the wafer, a modified high-temperature hydrogen chloride treatment, and the formation of oxygen precipitates in the bulk of the wafer comprise the three-step gettering process. Neutron activation analysis was used to monitor the gettering behavior of each individual contaminant species, and it was proved that the combined treatment is indeed nearly 100% effective in removing unwanted metallic impurities, both from starting substrates and from processed wafers. Removing impurities from the starting substrates is important in order to prevent the formation of OISFs, which are nucleated by impurities and then become decorated with impurities, making them much harder to getter subsequently. The individual steps of the treatment suggested are as follows: (1) produce LID on the back side of the wafer surface by means of powerful, partially overlapping laser pulses; (2) ramp up the temperature from 950°C to 1250°C in a 0.5% oxygen, 0.5% HCl balance with argon gas; and (3) oxidize at 1250°C for 8 h in dry oxygen along with 0.5% HCl and then ramp down to 950°C. Keeping the oxygen concentration to 0.5% during step 2 is important to prevent the formation of a thick oxide film. The long oxidation at 1250°C both depletes interstitial oxygen at the silicon surface, thus preventing formation of OISFs, and causes the formation of oxygen precipitates in bulk if the initial intrinsic oxygen species concentration is high enough. Oxygen precipitates in the bulk are a valuable means of gettering the metallic impurities.

Copper is one of the heavy metals that has a detrimental effect on integrated circuit performance and the yield. Recently, its behavior in silicon has begun to attract notice with respect to copper interconnections used to increase performance speed. If copper contamination occurs during the integrated circuit fabrication process, such as during high-temperature heat cycles, it can diffuse throughout the wafer because of the high-diffusivity coefficient associated with it. As the wafer cools down to room temperature, copper precipitates or agglomerates near the wafer surface, unless gettering sites like oxygen-related defects exist in the wafer bulk. The size, density, and spatial distribution of copper precipitates depend on the copper contamination concentration and preexisting lattice defects, as well as on the cooling rate of the wafer from high temperatures down to room temperature. Isomae et al. [20] studied the gettering efficiency of this copper contamination relative to oxygen precipitates. Figure 10.10 shows the gettering efficiency as a function of copper contamination concentration—there is a clear increase. However, in the region above 5.0×10^{13} atoms/cm^2 of copper contamination, the gettering efficiency begins to saturate at a level above $\eta = 95\%$, which also is apparent from the figure.

Baginski and Monkowski [21] came out with a simpler method for gold gettering. This technique consists of depositing germanium approximately 75 μm thick on the back surface of a silicon wafer and then applying thermal annealing. The technique has been reported to be effective for removing gold from the front surface of a silicon wafer. In addition, a germanium-gettered wafer was compared with a control wafer, revealing that the germanium gettering is effective in preventing the formation of OISFs during

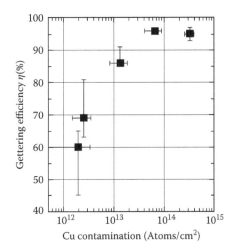

FIGURE 10.10
Gettering efficiency η as a function of the Cu contamination concentration in the intrinsic gettering–treated wafers, which were subjected to a two-step thermal treatment consisting of annealing for 16 h at 800°C and then for 4 h at 1000°C. (From S. Isomae, H. Ishida, T. Itoga, and K. Hozawa, *Journal of the Electrochemical Society*, **149**, G343–G347, 2002 [20].)

the high-temperature oxidation step. This approach appears to be simpler, except the germanium used is rather thick when compared to the actual silicon wafers.

The capability of oxide precipitate–related defects to capture impurities is widely utilized in the integrated circuit industry to remove harmful metallic contaminants from the device area. Iron is one of the most common and one of the most studied transition metallic contaminants in silicon, and is electrically not very active. The internal gettering occurs when iron becomes supersaturated and tends to precipitate heterogeneously. However, the experimental results show that at low iron concentrations, below 1.0×10^{13} cm^{-3}, the gettering process is much slower than predicted. Haarahiltunen et al. [22] discussed the modeling and optimization of iron gettering in silicon. Based on the simulation results, it was concluded that only the high initial iron concentrations, greater than 1.0×10^{13} cm^{-3}, can be gettered just by pulling the wafers out of the furnace at a relatively high cooling rate. With low concentrations, less than 1.0×10^{12} cm^{-3}, it is impossible to getter at any realistic cooling rate. Fast cooling rates may create stress-related issues.

Experimental investigations of McHugo et al. [7] indicated that iron gettering at both implantation-induced cavities and oxygen precipitates in silicon strongly interact with the iron such that the cavities effectively getter them. It was further shown that the cavities rapidly decrease the iron concentration values below the solubility concentrations at the gettering temperatures of 700°C and 850°C. Figure 10.11 shows a scanning electron microscope (SEM) micrograph of a preferentially etched silicon sample after a ramped-up internal gettering formation. Only shallow etch pits are visible in this figure. Lack of stacking faults or dislocation loops is most likely due to the low-temperature operation. The critical range of iron concentration reported by Hackl et al. [23] in CZ single-crystal silicon that changes internal gettering is smaller than 1.0×10^{11} atoms/cm^3. The lowest concentration of iron in the bulk with any detectable precipitation behavior for oxygen was found to be 6.5×10^{10} atoms/cm^3. During any implantation process step, one must take precautions to avoid extra iron ion implantation, even in the lowest concentration values, that may arise from unknown sources. Table 10.2 lists some of the metal impurity species and different methods adapted to getter them.

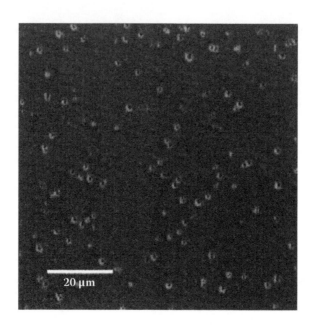

FIGURE 10.11
SEM micrograph of a preferentially etched silicon sample after a ramped internal gettering site-formation sequence. The shallow etch pits are characteristic of oxygen precipitates. (From S. A. McHugo, E. R. Weber, S. M. Myers, and G. A. Petersen, *Journal of the Electrochemical Society*, **145**, 1400–1405, 1998 [7].)

TABLE 10.2

Metal Impurity Species and Gettering Methods

Metal/Impurity Species	Method/Approach/Technique	Specific Device/Application	Ref.
Metallic impurities	Laser-induced damage	LSI devices	[24]
Transition metal impurities	Phosphorus gettering using $POCl_3$	Solar cells	[18]
Carbon	Three cycle heat treatment	VLSI devices	[10]
Carbon	Three-step internal gettering anneal	ULSI devices	[9]
Chromium	RTA lamp annealing and thermal treatment	ULSI devices	[25, 26]
Cobalt	Oxygen-related defects	CMOS circuits	[15]
Copper	Oxygen-related defects	CMOS circuits	[15]
Copper	Glass gettering (P_2O_5, B_2O_3)/by radioactive tracer method	Silicon devices	[27]
Copper	Chlorinated dry oxidation processes	MOS devices	[28]
Copper	Two-step thermal treatment	Silicon devices and circuits	[20]
Copper	Heat treatment/by radioactive tracer method	MOS devices	[29]
Copper	Two-step anneal cycle (low and high)	ULSI technology	[30, 31]
Copper	Oxide precipitates and thermal annealing (computer modeling)	300 mm double-side polished wafers	[32]
Copper	Phosphorus diffusion from PBr_3, $POCl_3$ sources	Silicon devices	[33]

Metal/Impurity Species	Method/Approach/Technique	Specific Device/Application	Ref.
Gold	Poly-Si deposition and predeposition of P impurity using $POCl_3$ source followed by segregation annealing	LSI devices (static RAM, ROM, microprocessors)	[17]
Gold	Glass gettering (P_2O_5, B_2O_3)/by radioactive tracer	Silicon devices	[27]
Gold	Germanium application on back side and thermal annealing	Silicon devices	[21]
Gold	Germanium ion implantation on back side and diode fabrication	Silicon diodes	[34]
Gold	Neutron activation	Silicon devices	[35]
Gold	Phosphorus gettering by use of neutron activation	Silicon devices	[36]
Gold	Phosphorus diffusion from PBr_3, $POCl_3$ sources	Silicon devices	[33]
Gold	Phosphoro- and boro-silica glasses	Silicon diodes	[37]
Iron	Oxygen-related defects	CMOS circuits	[15]
Iron	Glass gettering (P_2O_5, B_2O_3)/by radioactive tracer	Silicon devices	[27]
Iron	Two-step anneal cycle (low and high)	ULSI technology	[30]
Iron	Implantation-induced cavities and oxygen precipitates	CZ silicon wafers	[7]
Iron	Oxygen precipitation–related defects	CZ silicon wafers	[23]
Iron	Gettering simulations using oxygen precipitates	CZ silicon wafers	[38]
Iron	Heat treatment and annealing	Boron-doped silicon wafers	[39]
Iron	Boron implantation	Boron-doped silicon wafers	[40]
Nickel	Oxygen-related defects	CMOS circuits	[15]
Nickel	Oxide precipitates and thermal annealing (computer modeling)	300 mm double-side polished wafers	[32]
Nickel	Two-step anneal cycle (low and high)	ULSI technology	[31]
Nickel	Two-step isothermal annealing	CZ silicon wafers	[41]
Oxygen	Four-step annealing process at different temperatures	Silicon devices	[13]
Oxygen (interstitials)	Three-stage thermal annealing	Advanced memory circuits	[42]
Oxygen (interstitials)	Thermal-annealing	Bipolar and MOS devices	[11]
Oxygen (precipitation)	Low-temperature anneal	*n*-channel MOS dynamic RAM	[43]
Oxygen (precipitation)	Preannealing at 750°C in nitrogen ambient	n/n^+ epitaxial wafers	[44]
Oxygen (precipitation)	Three-step technique and ramped nucleation	n^+/p^- processed wafers	[45]
Palladium	Oxygen-related defects	CMOS circuits	[15]
Platinum	Phosphoro- and boro-silica glasses	Silicon diodes	[37]
Sodium	0.5–1.0% gaseous hydrogen chloride/chlorine by volume	MOS transistors	[46, 47]

10.6 Denuding of Silicon Wafers

Microdefects existing in the active region, close to the CZ silicon wafer surface, have many detrimental effects on the performance of semiconductor devices. Hence, it is desirable to eliminate them by providing a nearly defect-free zone, which is commonly referred to as a "denuded zone," in the active region of silicon wafers where most of the devices are fabricated. These are the subsurface regions where no significant microdefects are detected with an optical microscope. These denuded zones can form near the silicon surfaces if the wafers are subjected to a high-temperature treatment before being used for actual fabrication of integrated circuits. This proper pre-annealing treatment of wafers produces a denuded zone that is dependent on the annealing temperature, gas ambient conditions, time duration, and the sequence of events [48]. Two major mechanisms are suggested to explain this zone formation. The first is the oxygen out-diffusion mechanism, and the second is retardation of oxygen precipitation at the wafer surface by excess silicon interstitials.

Wang et al. [48] provided a systematic experimental proof to realize these almost-defect-free denuded zones in silicon wafers. This was achieved through various heat-treatment cycles with different thermal histories. The principal results are as follows: (1) a high-low, two-step annealing cycle (first at 1050°C and then at 800°C) results in a clear denuded zone at the wafer surface; (2) annealing in an inert or gaseous hydrogen chloride–containing inert ambient results in a much wider denuded zone (on the order of 20–30 µm); and (3) simple annealing of wafers for 16 h at 1050°C. These formations support the view that oxygen out-diffusion is the dominant mechanism by which denuded zones form in CZ silicon crystal wafers. Interstitial oxygen and carbon atoms change the length of the denuded zones. Figure 10.12 shows an optical photomicrograph of a silicon wafer that was annealed at 1050°C for 16 h in an argon environment followed by another heat anneal at 800°C in dry oxygen ambient. The wafer was treated with Wright etch to highlight the defects present. In all these methods, one has to be careful to execute the thermal cycle steps in a specific sequence only. Reversal may not yield the expected results in most of the cases.

In general, the distribution of oxygen precipitation achieved in an ensemble of silicon wafers depends strongly on the close coupling of the oxygen content associated with it. Many approaches have been developed over the years to manage these coupled complications and achieve the desired gettering effect without side effects. These include such wafer pre-treatments as the well-known "high-low-high" treatments [49] in which oxygen is first out-diffused at high temperatures, followed by a generally long low-temperature treatment to renucleate oxygen clusters, followed by another high-temperature treatment to grow them into the precipitates. Other approaches have included attempts to narrowly specify the oxygen concentration and crystal growth process, or even the segments of the crystal from which wafers should be taken for the specific application.

An illustration of the huge effect that vacancies have on oxygen precipitation behavior is shown in Figure 10.13 [49]. The steep, switch-like dependence of oxygen precipitate density on vacancy concentration means that a profile of vacancy concentration rising from the surface and going through the threshold value produces a rather sharply layered structure with a highly precipitating bulk underneath a nonprecipitating surface layer. The threshold for this layered design lies at a vacancy concentration of about 10^{12} cm^{-3}. Figure 10.14 schematically illustrates the design of such a wafer and compares it to the conventional oxygen out-diffusion approach to the problem of forming a denuded zone.

According to Falster [49], the simplest procedure for installing a useful profile of vacancies in a silicon wafer relies solely on Frenkel pair generation and the close proximity of

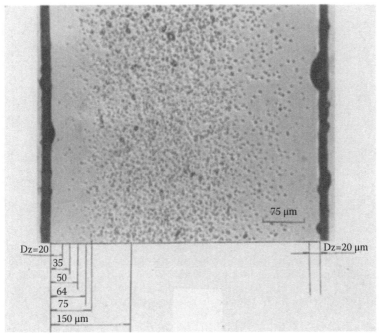

FIGURE 10.12
An optical photomicrograph of the cross-sectional (110) surface of a seed-end wafer annealed at 1050°C for 16 h in argon and at 800°C for 4 h in dry oxygen, after 8 min in Wright etch, showing clear denuded zones of ~20 μm at the subsurface regions with a high density of etch pits in the bulk. (From P. Wang, L. Chang, L. J. Demer, and C. J. Varker, *Journal of the Electrochemical Society*, **131**, 1948–1952, 1984 [48].)

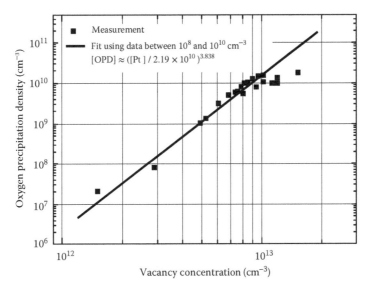

FIGURE 10.13
Oxygen precipitate densities produced following test heat treatments (800°C 4 h + 1000°C 16 h) as a function of wafer vacancy concentration. Vacancy concentration was determined by platinum diffusion experiments. (From R. Falster, "Defect control in silicon crystal growth and wafer processing," MEMC Electronic Materials SpA, Novara 28100, Italy, and the references therein [49].)

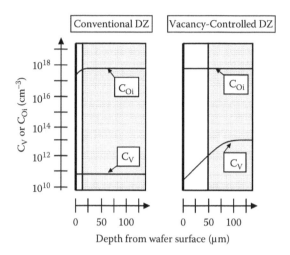

FIGURE 10.14

A schematic illustration of the difference between conventional methods of installing denuded zones in silicon wafers via oxygen out-diffusion and renucleation and a new method based on the installation of tailored vacancy concentration profiles. (From R. Falster, "Defect control in silicon crystal growth and wafer processing," MEMC Electronic Materials SpA, Novara 28100, Italy, and the references therein [49].)

the two wafer surfaces. Heating a thin silicon wafer to high temperature results in the rapid equilibration of vacancy and interstitials within the system. First, Frenkel pairs, vacancies, and self-interstitials are produced in equal amounts. This very fast reaction leads to recombination-generation equilibrium. The product of vacancy-interstitials of the two concentrations acquires the equilibrium value $C_i^* C_v^*$, with the concentration of both equal to $(C_i^* C_v^*)^{1/2}$. In the sample to be cooled at this point, under the condition of equal concentration, the vacancies and interstitials would mutually annihilate each other in the reverse process of their generation, resulting in no vacancy concentration enhancement by the time the samples reach room temperature. This coupled process is controlled mainly by diffusion of self-interstitials, which are the faster diffusers, with the two concentrations being comparable. An example of depth distribution of oxygen precipitates produced by vacancy concentration control is shown in Figure 10.15.

The formation mechanism and a model of these defect-free denuded zones in antimony-doped epitaxial substrate wafers are provided by Wijaranakula and Matlock [50]. This model considers that "grown-in" micro-precipitate nuclei play a significant role in the formation of the denuded zones. This model explained the formation mechanism and denuded zone in Sb-doped epitaxial wafers.

Intrinsic gettering of germanium-doped CZ (GCZ) silicon with different concentrations of germanium was investigated by Yu et al. [51]. They analyzed both conventional CZ and GCZ silicon samples that were annealed using a one-step high-temperature process followed by a sequence of low- and high-temperature annealing cycles. The good defect-free denude zones in the near surface of the GCZ silicon could be achieved simply by using a one-step high-temperature annealing process. Furthermore, the density of the bulk micro-defects as intrinsic gettering sites was reported to be higher than in the CZ silicon as a result of germanium-enhancing oxygen precipitation during a three-step annealing process. These experimental results showed that germanium in silicon enhanced the out-diffusion of oxygen species. Furthermore, it is believed that germanium doping can increase the intrinsic gettering ability in CZ silicon wafers. Chen et al. [52] also reported on germanium's effect on as-grown oxygen precipitation in GCZ silicon crystals. The as-grown

(a)

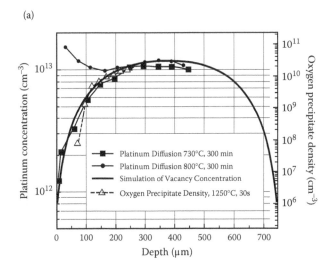

(b)

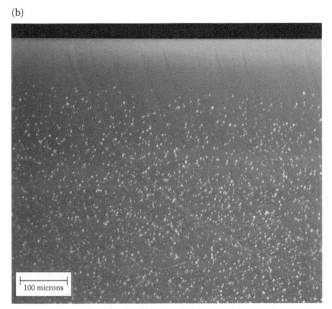

FIGURE 10.15
(a) Depth profiles of Pt diffusion profiles (~vacancy concentration) measured at 730° and 800°C, calculated vacancy concentrations, and measured oxygen precipitate densities in an RTP-treated sample processed at 1250°C. The oxygen precipitate density axis is scaled to correspond to the vacancy-precipitate density calibration. (b) An etched cross-section of a silicon wafer containing such a profile following a precipitation heat treatment (800°C 4 h + 1000°C 16 h) showing the depth profile of oxygen precipitates resulting from an RTP-installed vacancy concentration profile. The bulk density of precipitates in this example is 1×10^{10} cm^{-3}. (From R. Falster, "Defect control in silicon crystal growth and wafer processing," MEMC Electronic Materials SpA, Novara 28100, Italy, and the references therein [49].)

oxygen precipitation can be enhanced in comparison to the conventional CZ crystals at high temperatures, even above the formation temperature of void. This has been ascribed to the effect of heterogeneous nucleus sites supplied by the Ge-related complexes generated in these GCZ crystals. This phenomenon is considered to be associated with the reduction in the critical radius of oxygen precipitates at the elevated temperatures. This approach appears to be a simpler way to achieve defect-free denuded zones, which is a basic requirement for VLSI and ULSI chips.

Available thermal processes can establish a precipitate-free surface region for active devices. Also, denuded zone separation from bulk silicon that contains precipitates or bulk microdefects is effective in gettering metallic impurities from the active devices. The higher the oxygen level in silicon wafers, the greater the possibility of detrimental oxygen precipitates occurring in the active device region. Control of oxygen level, uniformity, and precipitation behavior in CZ silicon has been and will continue to be a challenge in the face of ever-increasing crystal diameters from 200 mm to 300–450 mm or larger in the near future [53]. Optimization of crystal growth parameters using such approaches as the magnetically-grown-CZ (MCZ) technique has been largely effective in meeting the challenge so far, but with the larger crystal growth systems (with 150–600 kg melts in 24–36-inch or larger crucibles) required for larger wafers, the problems with growing perfect crystals will persist.

10.7 Neutron Irradiation

Li et al. [54] reported on the effect of neutron irradiation on oxygen precipitation behavior in CZ silicon. They found that a heavy dose of neutron irradiation accelerated the oxygen precipitation in CZ-grown silicon under a higher annealing temperature, in the range of 1070°C to 1130°C. They attributed this precipitation effect to the defect cluster formation induced by the neutron irradiation. This may be due to the fact that the core for the oxygen precipitation for heterogeneous nuclei and to shorten the nucleation time and subsequently accelerate the oxygen precipitates in the bulk of the silicon wafer. In addition, oxygen diffusion is reported to enhance the presence of a large quantity of vacancies near the surface of the silicon wafer.

Pietila and Masson [35] have reported on the intrinsic gettering of gold in CZ single-crystal silicon, which was evaluated using neutron activation analysis and electron beam microprobe analysis. Silicon wafers were subjected to oxygen denuding and heat-treatment cycles that produced a defect structure adjacent to a defect-free region, oxide particles and stacking faults at mixed ratios, or stacking faults alone. Gold distributed homogeneously throughout the wafer at the level of the solubility limit and preferential precipitation of gold at oxide particles or stacking faults did not occur. Aside from the surface effects, the concentration of gold in the wafer was not significantly enhanced by the intrinsic defect structures produced. Murarka [36] used neutron activation analysis to determine the concentration of gold in silicon wafers before and after phosphorus gettering.

10.8 Argon Annealing of Wafers

Argon annealing is generally carried out on silicon wafers to control intrinsic point defects, such as vacancies and interstitial silicon atoms. Prostomolotov et al. [55] suggested a rapid thermal annealing approach be carried out quickly on the wafers—20 to 60 s duration—to a temperature in the range of 1200°C to 1250°C in the argon atmosphere. This technique creates the possibility of increasing the vacancy concentration in a silicon wafer higher than the threshold value (10^{12} cm^{-3}); this leads to the sharp decomposition of the supersaturated solid solution of oxygen that occurs during multistep thermal annealing of silicon wafers.

Carbon is considered to be one of the more harmful impurities in CZ silicon; it not only affects the behavior of oxygen present in silicon, but also generates swirl defects. The silicon wafers with a high carbon content are not generally recommended for VLSI and ULSI circuits. Chen et al. [56] reported on the effect of carbon on oxygen precipitation behavior in internal gettering processing for CZ silicon. Through high-low-high, high-temperature anneals in argon ambient, it has been shown that both good-quality denuded zone and high-density bulk microdefects could be formed in CZ silicon wafers with a high carbon content. Figure 10.16 shows cross-sectional microphotographs of

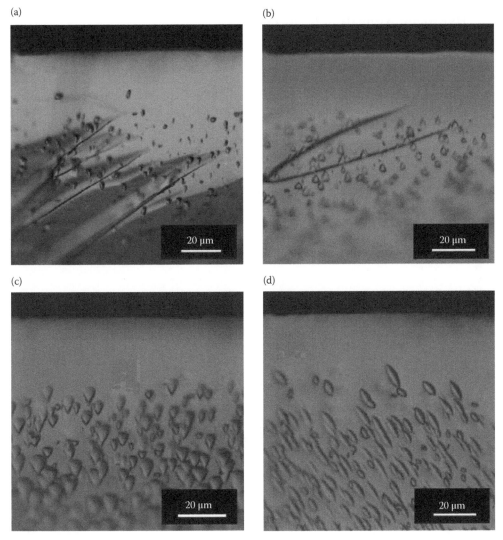

FIGURE 10.16
Cross-sectional optical microphotograph of (a) reference wafer, (b) low-carbon content, (c) medium-carbon content, and (d) high-carbon content CZ wafers subjected to intrinsic gettering thermal cycles of 1200°C (2 h), 750°C (16 h), and 1050°C (16 h) in argon ambient. (From J. Chen, D. Yang, X. Ma, and D. Que, *Journal of Crystal Growth*, **290**, 61–66, 2006, and the references therein [56].)

low-carbon content L[C], medium carbon content M[C], and high carbon content H[C] wafers selected from the lot for the argon anneal experiments. The samples were subjected to the internal gettering process performed at 1200°C (for 2 h), 750°C (for 16 h), and 1050°C (for 16 h). Defect-free denuded zones approximately 15 μm thick were formed in all these wafers. It was further reported that the concentration of substitutional carbon remained for almost the entire thermal cycle. This carbon-doped CZ silicon remains a complicated issue, and more experiments are needed to confirm its role in the silicon crystal lattice site.

10.9 Hydrogen Annealing of Wafers

A hydrogen-annealed internal gettering wafer (also known as a Hi-WAFER) provides a high-quality substrate for ULSI because it has a higher reliability of the gate oxide [57] for the integrated circuits fabricated in them. With severe restrictions on the starting silicon wafers, particularly CZ crystals, the demand is high for properly getter-treated wafers, such as intrinsic gettered wafers or extrinsic ones. The relationship between gate oxide integrity and grown-in defects has been evaluated by many researchers, and will likely continue to be so.

The silicon (100) wafer surface after the hydrogen anneal is predominantly hydride covered. The experimental observations of Gräf et al. [58] indicate that only silicon hydride bending and stretching vibrations are observed. High gate oxide integrity yield is obtained for annealing in hydrogen or argon at sufficiently high annealing temperatures, around 1200°C, and annealing times of 1 to 2 h. The improvement in gate oxide integrity is the result of a reduction in near-surface defect densities. In addition, the surface roughness increased after annealing when compared to the polished wafer surfaces. The annealing

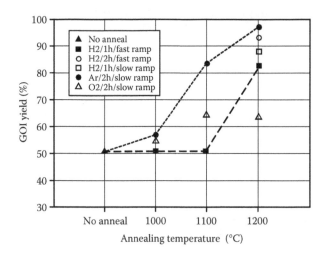

FIGURE 10.17
Gate oxide integrity yield as a function of annealing temperature and various annealing conditions as indicated in the insert ambient/temperature/ramping conditions. (From D. Gräf, U. Lambert, M. Brohl, A. Ehlert, R. Wahlich, and P. Wagner, *Journal of the Electrochemical Society*, **142**, 3189–3192, 1995 [58].)

time of several hours required in order to obtain a sufficient improvement in wafer performance aggravates this roughening in hydrogen and argon ambient, when compared to epitaxial wafers with significantly shorter processing times. For hydrogen-terminated surfaces, the oxygen uptake proceeds rather slowly, as shown in Figure 10.17. However, in comparison to a polished and HF-treated wafer, the hydrogen-annealed silicon (100) surface oxidizes faster with an earlier onset of SiO_{2-x} growth, as shown in Figure 10.18, indicating a reduced stability of the reconstructed hydrogen-terminated surface.

Abe et al. [59] investigated the effect of hydrogen annealing in the temperature range of 850°C to 1200°C on oxygen precipitation and the gate oxide integrity for CZ-grown wafers. The gate oxide integrity of the silicon wafers improved and OISFs were suppressed by this high-temperature annealing in hydrogen atmosphere. Furthermore, these wafers have denuded zone layers and bulk microdefects regions, so they are expected to act as an intrinsic getter. The studies of Abe et al. confirm that these wafers showed improved dielectric breakdown characteristics of the gate oxide grown on them. To confirm this breakdown phenomenon, time dependent dielectric breakdown (TDDB) characteristics of 10 μm polished (or thickness removed) wafers after hydrogen annealing degraded when they are compared with nonpolished silicon wafers. This experimental evaluation showed that the effect of hydrogen annealing is limited to just the surface or near-surface regions.

Sueoka et al. [60] reported on the mechanical strength of hydrogen-annealed CZ silicon wafers, with an emphasis on the generation of slip dislocations by oxide precipitations. Six-inch *p*-type wafers were used for the experiments. Hydrogen annealing was performed on these wafers at 1200°C for 1 h. It was found that slip generation by oxide precipitates did not occur in the low-temperature simulation, while many slip dislocations were generated in the high-temperature simulation. The precipitate morphology, size, and density of the simulated wafers were investigated by TEM analysis. The experimental results were explained by the effect of precipitate size on slip generation. The investigators concluded

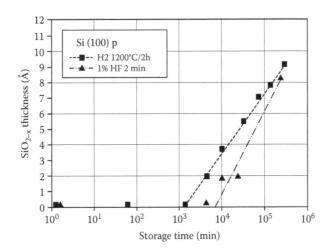

FIGURE 10.18
Native oxide growth after storage of silicon (100) wafers after hydrogen anneal and after HF treatment, respectively. The earlier onset of the oxygen uptake of the hydrogen annealed surface indicates reduced surface stability. (From D. Gräf, U. Lambert, M. Brohl, A. Ehlert, R. Wahlich, and P. Wagner, *Journal of the Electrochemical Society*, **142**, 3189–3192, 1995 [58].)

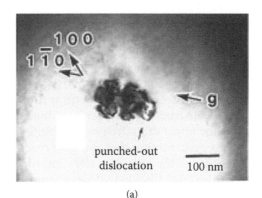

(a)

(b)

FIGURE 10.19
(a) Two-beam bright-field TEM image of a polyhedral precipitate after hydrogen annealing and (b) definition of the polyhedral precipitate size S. (From K. Sueoka, M. Akatsuka, H. Katahama, and N. Adachi, *Journal of the Electrochemical Society*, **146**, 364–366, 1999 [60].)

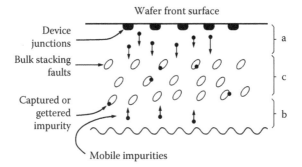

FIGURE 10.20
Schematic of a denuded zone in a wafer cross-section and gettering sites. "a" and "b" are zones denuded of defects, and "c" represents the region of intrinsic gettering. (From C. W. Pearce, "Crystal growth and wafer preparation" in *VLSI Technology*, edited by S. M. Sze, McGraw-Hill, New York, 1988, and the references therein [61].)

that hydrogen-annealed wafers have no mechanical strength problems during actual low-temperature processes. Figure 10.19 shows a two-beam bright-field TEM image of a typical polyhedral precipitate and its size after hydrogen annealing.

The ability to capture harmful impurities can be utilized so as to separate or drive them away from the active device/circuit area. Figure 10.20 shows the wafer front area where the devices are fabricated. The area is free from metallic impurities and from their interference. Different zones are identified in the wafer cross-section for illustration purposes [61].

10.10   Final Cleaning, Rinsing, and Wafer Drying

In the final cleaning, a wet chemical process is still employed due to its remarkable characteristics: it causes no damage to the wafer, it is effective, and it is carried out at room temperature. The wafer is rinsed with 18.3 MΩ-cm DI water. The active bare silicon surface is exposed to dilute HF and finally with the ultrapure DI water to remove the traces of fluoride ions. The concentration of impurities at this point must be extremely low.

The segregation of metallic impurities at the interface between the silicon wafer and the liquid in the wet cleaning process was analyzed by Ohmi et al. [62]. They found that metals—particularly copper ions, which have higher electronegativity than silicon—are directly adsorbed on the silicon surface by an electron exchange mechanism. On the other hand, metals such as iron and potassium, which have lower electronegativity than silicon, are not adsorbed on the surface. In the normal cleaning process, when a native oxide is formed on the silicon surface, metals such as iron and potassium are oxidized more easily and are included into the growing native oxide. Trace metals are automatically included into the oxide, making it rich with metallic ions. The presence of metal in concentration values in the range of 1 ppb will get embedded into this nascent oxide of silicon. Hence, metallic impurities must be suppressed to at least below the 1-ppt (parts per trillion) level in ultrapure water and high-purity HF acid, which are employed in the final step of the cleaning process. In addition, it is important to maintain an inert atmosphere, such as nitrogen or argon, to suppress native oxide formation and to reduce metallic impurities in the ultrapure water used for rinsing. These studies indicate that chemicals with higher metallic ions will disturb the top surface of the silicon. Only select high-purity wet chemicals can be used in the final cleaning process.

Kim et al. [63] studied how effective various chemical solutions composed of noble metals such as copper, silver, and gold were at cleaning silicon wafer surfaces. Based on their results, it appears that the conventional RCA cleaning recipe used presently throughout the world is not as effective for cleaning as it was thought to be. The metal deposition behavior and removal mechanisms for copper, silver, and gold in a solution are similar, yet fundamentally different, even though these metals are categorized in the same noble group which depends on their redox potential, electronegativity, and formation enthalpy of metal-oxide for compound formation and removal from the surface.

Effective cleaning of the silicon surface is essential for obtaining high product yields and reliability for VLSI and ULSI devices. Peroxide-based solutions are capable of removing organic and metallic contaminants from silicon quite efficiently. The presence of a native oxide layer on the silicon surface is suspected to inhibit the surface cleaning, and proper planning is certainly required when carrying out the wafer cleaning steps.

10.11   Summary

In this chapter, we discussed the preparation of silicon wafers for VLSI and ULSI processing. This included the purity of the chemicals, degreasing of wafers, removing metallic and other impurities from the silicon surface, the gettering process for metallic impurities, and creating denuded zones in silicon wafers. Other methods discussed included neutron irradiation, argon annealing, and hydrogen annealing of wafers. Finally, we looked at how the wafers are given a final cleaning and properly dried for shipment.

References

1. O. J. Anttila, M. V. Tilli, M. Schaekers, and C. L. Claeys, "Effect of chemicals on metal contamination on silicon wafers," *Journal of the Electrochemical Society*, **139**, 1180–1185, 1992, and the references therein.
2. Honeywell Electronic Materials & Electronic Chemicals, https://www.honeywell-pmt.com/sm /products-applications/electro-chem.html.
3. O. J. Anttila and M. V. Tilli, "Metal contamination removal on silicon wafers using dilute acidic solutions," *Journal of the Electrochemical Society*, **139**, 1751–1756, 1992.
4. S. G.d. S. Filho, C. M. Hasenack, L. C. Salay, and P. Mertens, "A less critical cleaning procedure for silicon wafer using diluted HF dip and boiling in isopropyl alcohol as final steps," *Journal of the Electrochemical Society*, **142**, 902–907, 1995.
5. L. Jastrzebski, O. Milic, M. Dexter, J. Lagowski, D. DeBusk, K. Nauka, R. Witowski, M. Gordon, and E. Persson "Monitoring of heavy metal contamination during chemical cleaning with surface photovoltage," *Journal of the Electrochemical Society*, **140**, 1152–1159, 1993.
6. B. El-Kareh, *Fundamentals of Semiconductor Processing Technology*, Kluwer Academic Publishers, Boston, 1995.
7. S. A. McHugo, E. R. Weber, S. M. Myers, and G. A. Petersen, "Gettering of iron to implantation induced cavities and oxygen precipitates in silicon," *Journal of the Electrochemical Society*, **145**, 1400–1405, 1998.
8. T. A. Baginski and J. R. Monkowski, "Correlation of precipitation defects and gold profiles in intrinsically gettered silicon," *Journal of the Electrochemical Society*, **133**, 762–769, 1986, and the references therein.
9. W. Wijaranakula and J. H. Matlock, "A formation of crystal defects in carbon-doped Czochralski-grown silicon after a three-step internal gettering anneal," *Journal of the Electrochemical Society*, **138**, 2153–2159, 1991.
10. W. E. Bailey, R. A. Bowling, and K. E. Bean, "Gettering of carbon in silicon processing," *Journal of the Electrochemical Society*, **132**, 1721–1725, 1985, and the references therein.
11. F. Shimura, H. Tsuya, and T. Kawamura, "Thermally induced defect behavior and effective intrinsic gettering sink in silicon wafers," *Journal of the Electrochemical Society*, **128**, 1579–1583, 1981.
12. M. Lozano, M. Ullán, C. Martínez, L. Fonseca, J.M. Rafi, F. Campabadal, E. Cabruja, C. Fleta, M. Key, and S. Bermejo "Effect of combined oxygenation and gettering on minority carrier lifetime in high-resistivity FZ silicon," *Journal of the Electrochemical Society*, **151**, G652–G657, 2004.
13. L. Rivaud, C. N. Anagnostopoulos, and G. R. Erikson, "A transmission electron microscopy (TEM) study of oxygen precipitation induced by internal gettering in low and high oxygen wafers," *Journal of the Electrochemical Society*, **135**, 437–442, 1988.
14. K. Sakai, T. Yamagami, and K. Ojima, "Change in shape of oxygen precipitate grown by thermal annealing," *Journal of Crystal Growth*, **210**, 65–68, 2000.
15. R. Falster and W. Bergholz, "The gettering of transition metals by oxygen-related defects in silicon," *Journal of the Electrochemical Society*, **137**, 1548–1559, 1990, and the references therein.
16. W. Huber and M. Pagani, "The behavior of oxygen precipitates in silicon at high process temperature," *Journal of the Electrochemical Society*, **137**, 3210–3213, 1990.
17. L. Baldi, G. Cerofolini, and G. Ferla, "Heavy metal gettering in silicon-device processing," *Journal of the Electrochemical Society*, **127**, 164–169, 1980.
18. D. Macdonald, A. Cuevas, and F. Ferrazza, "Response to phosphorus gettering of different regions of cast multicrystalline silicon ingots," *Solid-State Electronics*, **43**, 575–581, 1999.
19. L. E. Katz, P. F. Schmidt, and C. W. Pearce, "Neutron activation study of a gettering treatment for Czochralski silicon substrates," *Journal of the Electrochemical Society*, **128**, 620–624, 1981.
20. S. Isomae, H. Ishida, T. Itoga, and K. Hozawa, "Intrinsic gettering of copper in silicon wafers," *Journal of the Electrochemical Society*, **149**, G343–G347, 2002.

21. T. A. Baginski and J. R. Monkowski, "Germanium back-side gettering of gold in silicon," *Journal of the Electrochemical Society*, **133**, 142–147, 1986.
22. A. Haarahiltunen, H. Väinölä, M. Yli-Koski, J. Sinkkonen, and O. Anttila, "Modeling and optimization of internal gettering of iron in silicon," *210 ECE Meeting Cancun*, Mexico, Oct. 29–Nov. 3, 2006.
23. B. Hackl, K.-J. Range, H. J. Gores, L. Fabry, and P. Stallhofer, "Iron and its detrimental impact on the nucleation and growth of oxygen precipitates during internal gettering processes," *Journal of the Electrochemical Society*, **139**, 3250–3254, 1992.
24. C. W. Pearce and V. J. Zaleckas, "A new approach to lattice damage gettering," *Journal of the Electrochemical Society*, **126**, 1436–1437, 1979.
25. S. Krieger-Kaddour, N-E. Chabane-Sari, and D. Barbier, "Transmission electron microscopic study of the morphology of oxygen precipitates and of chromium precipitation during intrinsic gettering in Czochralski-grown silicon: Influence of lamp pulse annealing," *Journal of the Electrochemical Society*, **140**, 495–500, 1993.
26. N-E. Chabane-Sari, S. Krieger-Kaddour, C. Vinante, M. Berenguer, and D. Barbier, "A deep level transient spectroscopy study of the internal gettering of Cr in Czochralski-grown silicon: Efficiency and reversibility upon lamp pulse annealing," *Journal of the Electrochemical Society*, **139**, 2900–2904, 1992.
27. M. Nakamura, T. Kato, and N. Oi, "A study of gettering effect of metallic impurities in silicon," *Japanese Journal of Applied Physics*, **7**, 512–519, 1968.
28. T. A. Baginski and J. R. Monkowski, "The role of chlorine in the gettering of metallic impurities from silicon," *Journal of the Electrochemical Society*, **132**, 2031–2033, 1985.
29. K.-S. Kim, S.-W. Lee, H.-B. Kang, B.-Y. Lee, and S.-M. Park, "Quantitative evaluation of gettering efficiencies below 1×10^{12} atoms/cm^3 in p-type silicon using a ^{65}Cu tracer," *Journal of the Electrochemical Society*, **155**, H912–H917, 2008.
30. B. Hacki, K.-J. Range, P. Stallhofer, and L. Fabry, "Correlation between DLTS and TRXFA measurements of copper and iron contaminations in FZ and CZ silicon wafers; Application to gettering efficiencies," *Journal of the Electrochemical Society*, **139**, 1495–1498, 1992.
31. R. Hoelzl, D. Huber, K.-J. Range, L. Fabry, J. Hage, and R. Wahlich, "Gettering of copper and nickel in p/p+ epitaxial wafers," *Journal of the Electrochemical Society*, **147**, 2704–2710, 2000.
32. K. Sueoka, "Modeling of internal gettering of nickel and copper by oxide precipitates in Czochralski-Si wafers," *Journal of the Electrochemical Society*, **152**, G731–G735, 2005.
33. R. L. Meek, T. E. Seidel, and A. G. Cullis, "Diffusion gettering of Au and Cu in silicon," *Journal of the Electrochemical Society*, **122**, 786–796, 1975.
34. T. A. Baginski, "Back-side germanium ion implantation gettering of silicon," *Journal of the Electrochemical Society*, **135**, 1842–1843, 1988.
35. D. A. Pietila and D. B. Masson, "Evaluation of intrinsic gettering of gold by oxide precipitation in Czochralski silicon," *Journal of the Electrochemical Society*, **135**, 686–690, 1988.
36. S. P. Murarka, "A study of the phosphorus gettering of gold in silicon by use of neutron activation analysis," *Journal of the Electrochemical Society*, **123**, 765–767, 1976.
37. K. P. Lisiak and A. G. Milnes, "A comparison of the process-induced gettering of atomic platinum and gold in silicon," *Journal of the Electrochemical Society*, **123**, 305–308, 1976.
38. H. Hieslmair, A. A. Istratov, S. A. McHugo, C. Flink, and E. R. Weber, "Analysis of iron precipitation in silicon as a basis for gettering simulations," *Journal of the Electrochemical Society*, **145**, 4259–4264, 1998.
39. Y. Kamiura, F. Hashimoto, and M. Yoneta, "Effects of heat-treatments on electrical properties of boron-doped silicon crystals," *Journal of the Electrochemical Society*, **137**, 3642–3647, 1990.
40. J. L. Benton, P. A. Stolk, D. J. Eaglesham, D.C. Jacobson, C.Y. Cheng, J.M. Poate, S.M. Myers, and T.E. Haynes "The mechanisms of iron gettering in silicon by boron ion-implantation," *Journal of the Electrochemical Society*, **143**, 1406–1409, 1996.
41. K. Sueoka, S. Sadamitsu, Y. Koike, T. Kihara, and H. Katahama, "Internal gettering for Ni contamination in Czochralski silicon wafers," *Journal of the Electrochemical Society*, **147**, 3074–3077, 2000.

42. T. Tuomi, M. Tuominen, E. Prieur, J. Partanen, J. Lahtinen, and J. Laakkonen, "Synchrotron section topographic study of Czochralski-grown silicon wafers for advanced memory circuits," *Journal of the Electrochemical Society*, **142**, 1699–1701, 1995.

43. H. R. Huff, H. F. Schaake, J. T. Robinson, S. C. Baber, and D. Wong, "Some observations on oxygen precipitation/gettering in device processed Czochralski silicon," *Journal of the Electrochemical Society*, **130**, 1551–1555, 1983.

44. W. Wijaranakula, J. H. Matlock, and H. Mollenkopf, "Effect of postannealing on the oxygen precipitation and internal gettering process in n/n+ (100) epitaxial wafers," *Journal of the Electrochemical Society*, **135**, 3113–3119, 1988.

45. A. R. Comeau, "Improvement to bulk oxygen precipitate density necessary for internal gettering in antimony doped n/n+ epitaxial wafers using the 3-step technique and ramped nucleation," *Journal of the Electrochemical Society*, **139**, 1455–1463, 1992.

46. P. H. Robinson and F. P. Heiman, "Use of HCl gettering in silicon device processing," *Journal of the Electrochemical Society*, **118**, 141–143, 1971.

47. R. S. Ronen and P. H. Robinson, "Hydrogen chloride and chlorine gettering: An effective technique for improving performance of silicon devices," *Journal of the Electrochemical Society*, **119**, 747–752, 1972.

48. P. Wang, L. Chang, L. J. Demer, and C. J. Varker, "Denuded zones in Czochralski silicon wafers," *Journal of the Electrochemical Society*, **131**, 1948–1952, 1984.

49. R. Falster, "Defect control in silicon crystal growth and wafer processing," MEMC Electronic Materials SpA, Novara 28100, Italy, www.suneditionsemi.com/assests/file/technology/papers, and the references therein.

50. W. Wijaranakula and J. H. Matlock, "A formation mechanism of the defect-free denuded zone in antimony-doped epitaxial substrate wafers," *Journal of the Electrochemical Society*, **137**, 1262–1270, 1990.

51. X. Yu, D. Yang, X. Ma, H. Li, Y. Shen, D. Tian, L. Li, and D. Que "Intrinsic gettering in germanium-doped Czochralski crystal silicon crystals," *Journal of Crystal Growth*, **250**, 359–363, 2003.

52. J. Chen, D. Yang, H. Li, X. Ma, and D. Que, "Germanium effect on as-grown oxygen precipitation in Czochralski silicon," *Journal of Crystal Growth*, **291**, 66–71, 2006, and the references therein.

53. K.-M. Kim, "Growing improved silicon crystals for VLSI/ULSI," *Solid State Technology*, **39**, 70, November 1996, and the references therein.

54. Y. X. Li, H. Y. Guo, B. D. Liu, T.J. Liu, Q.Y. Hao, C.C. Liu, D.R. Yang, and D.L. Que "The effects of neutron irradiation on the oxygen precipitation in Czochralski-silicon," *Journal of Crystal Growth*, **253**, 6–9, 2003.

55. A. I. Prostomolotov, N. A. Verezub, M. V. Mezhennii, and V. Ya. Reznik, "Thermal optimization of CZ bulk growth and wafer annealing for crystalline dislocation-free silicon," *Journal of Crystal Growth*, **318**, 187–192, 2011.

56. J. Chen, D. Yang, X. Ma, and D. Que, "Effect of carbon doping on oxygen precipitation behavior in internal gettering processing for Czochralski silicon," *Journal of Crystal Growth*, **290**, 61–66, 2006, and the references therein.

57. R. Takeda, P. Xin, J. Yoshikawa, Y. Kirino, Y. Matsushita, Y. Hosoki, N. Tsuchiya, and O. Fujii "300-mm diameter hydrogen annealed silicon wafers," *Journal of the Electrochemical Society*, **144**, L280–L282, 1997.

58. D. Gräf, U. Lambert, M. Brohl, A. Ehlert, R. Wahlich, and P. Wagner, "Improvement of Czochralski silicon wafers by high-temperature annealing," *Journal of the Electrochemical Society*, **142**, 3189–3192, 1995.

59. H. Abe, I. Suzuki, and H. Koya, "The effect of hydrogen annealing on oxygen precipitation behavior and gate oxide integrity in Czochralski Si wafers," *Journal of the Electrochemical Society*, **144**, 306–311, 1997.

60. K. Sueoka, M. Akatsuka, H. Katahama, and N. Adachi, "Investigation of the mechanical strength of hydrogen-annealed Czochralski silicon wafers," *Journal of the Electrochemical Society*, **146**, 364–366, 1999.

61. C. W. Pearce, "Crystal growth and wafer preparation" in *VLSI Technology*, edited by S. M. Sze, McGraw-Hill, New York, 1988, and the references therein.

62. T. Ohmi, T. Imaoka, I. Sugiyama, and T. Kezuka, "Metallic impurities segregation at the interface between Si wafer and liquid during wet cleaning," *Journal of the Electrochemical Society*, **139**, 3317–3335, 1992.

63. J.-S. Kim, H. Morita, G.-M. Choi, and T. Ohmid, "Cleaning efficiencies of various chemical solutions for noble metals such as Cu, Ag, and Au on Si wafer surfaces," *Journal of the Electrochemical Society*, **146**, 4281–4289, 1999.

11

Packing of Silicon Wafers

11.1   Packing of Fully Processed Blank Silicon Wafers

A wide variety of wafer containers, holders, storage cases, and shipping boxes for wafers are presently available, and the packing technology has changed over the years in response to changes in wafer size. There is no single industry-standard method of wafer-level packaging at present. Most small wafers are positioned vertically with the top opening, and these boxes are shipped in a sealed container. Larger wafers are positioned horizontally with the front opening. All of the wafers are placed separately without touching each other. Enough gaps are provided inside to handle the individual silicon wafers. Cleaned wafers are packed in these shipping boxes, and the number of wafers in each box varies. Packaging technology has improved in terms of materials and design that add fewer particles during shipping and in storage. Figure 11.1 shows some of the wafer containers [1], and Figure 11.2 outlines the different generations of wafer storage and transportation methods followed by the wafer suppliers [2–4]. Packing materials vary from different plastics to Teflon.

The plastic containers are a source of micro-plastics. They are produced in the environment as a consequence of the breakdown of larger plastic material into smaller and smaller fragments, called secondary micro-plastics. The breakdown is caused by mechanical forces and/or photochemical processes triggered by sunlight or by ultraviolet radiation inside the clean rooms. The abundance and wide distribution of micro-plastics in the environment has steadily increased as plastic consumption has increased. Other sources of contamination are suspended air particles in the ambient—particularly those that float in the air—and other organic compounds.

When 8-inch silicon wafers were primarily being produced, they often were contaminated with organic compounds in the clean room because open cassettes were used for storage and transportation. Since the front-opening unified pod (FOUP) system was adopted in the fabrication of 300 mm wafers, the wafer surface has no longer been exposed in a clean room environment, but has almost been exposed under the controlled atmosphere in semiconductor equipment environment only [4]. Therefore, the amount of organic contaminants on the silicon wafer is reduced. However, in the future, organic contaminants will need to be reduced even further. The FOUP is a specialized plastic enclosure designed to hold silicon wafers securely and safely in a controlled environment, and to allow the wafers to be removed for processing or measurement using tools equipped with appropriate load ports and robotic handling systems. The schematic drawing of a FOUP is shown in Figure 11.3 with the associated gas flow connections.

(a)

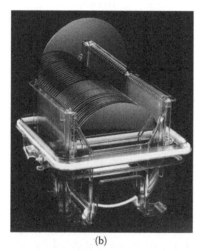

(b)

(c)

FIGURE 11.1
(a) Silicon wafer shipping container circa 1970; (b) 1990s 200 mm polycarbonate package; and (c) current 300 mm front opening shipping box (FOSB) package, which can be opened and closed by robotic handlers. (From G. Fisher, M. R. Seacrist, and R. W. Standley, *Proceedings of the IEEE*, **100**, 1454–1474, 2012, and the references therein [1].)

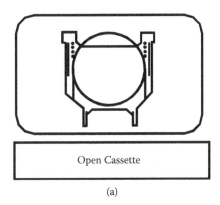

Open Cassette

(a)

(b)

FIGURE 11.2
Different generations of wafer storage and transportation methods. (a) Schematic diagram (b) Packages in use. (From O. Anttila, "Challenges of silicon materials research: manufacture of high resistivity, low oxygen Czochralski silicon," Okmetic Fellow, Okmetic Oyj; www.engineering-ed.org/Semiconductor/documents/ Unit%203%20crystal%20Growth%20and%20wafer%20Prep.ppt; *Unit 3 Crystal Growth and Wafer Preparation*; M. Saito, K. Anbai, and T. Hayashi, *IEEE Transactions on Semiconductor Manufacturing*, **18**, 575–583, 2005 [2–4].)

The formation of dust particles containing chemical aggressive compounds in the range of 0.2 µm from humid air and with inorganic reactive pollutants is a relatively well-investigated process. Sophisticated filtration techniques are currently being developed to reduce the level of contamination in clean room environments. Another approach to solving the contamination problem is storing the silicon wafers in a dry, clean atmosphere during both transportation and pauses in processing steps. The fabrication design for the processing of 300 mm wafers makes that idea easy to implement as wafers are transported and stored in FOUPs. In order to remove airborne molecular contamination (AMC), these pods can be purged with pure nitrogen. As shown in Figure 11.3, the shell and door of

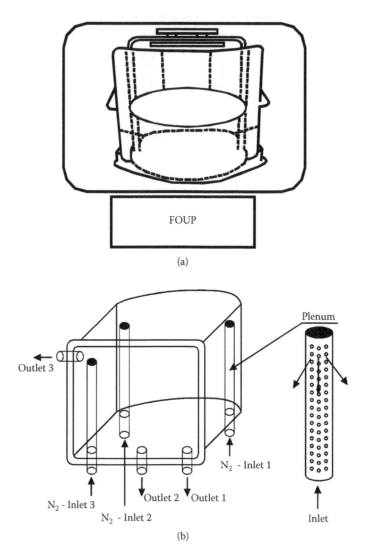

FIGURE 11.3
(a) Schematic drawing of the FOUP, (b) with the inlet and outlets for purge experiments. (From M. Saito, K. Anbai, and T. Hayashi, *IEEE Transactions on Semiconductor Manufacturing*, **18**, 575–583, 2005; J. Frickinger et al., *IEEE Transactions on Semiconductor Manufacturing*, **13**, 427–433, 2000 [4, 5].)

the FOUP is made of polycarbonate (PC), and the wafer support structure is made of polyether etherketone (PEEK) [5]. Typically, these FOUPs will accommodate 25 wafers. Nitrogen gas with a purity on the order of 99.999% is used for purging. The nitrogen flow rate is typically maintained at 6.6–18 liter/min. The FOUP will be equipped with three positions each for nitrogen inlet and outlet through the bottom plate and at the left side. A distributor column (plenum) is attached to each inlet to distribute the incoming gas flow over the whole height of the FOUP parallel to the wafers. Frickinger et al. [5] showed that the FOUP can be purged effectively as far as oxygen is concerned. Oxygen does not absorb into or permeate through the plastic material of the FOUP to a significant extent. Hu and Wu [6] studied the air flow and particle characteristics of a 300 mm FOUP load port unit and evaluated how wafers may be contaminated by potential particle sources during docking, and they proposed how particle contamination could be mitigated. Each FOUP has various coupling plates, pins, and holes to allow the FOUP to be located on a load port and to be manipulated by the automated material handling system (AMHS). FOUPs may also contain radio frequency (RF) tags that allow them to be identified by readers on tools, in the AMHS, etc. Figure 11.4 shows a typical FOUP with a 25-capacity wafer holder [7].

FOUPs began to appear along with the first 300 mm wafer processing tools in the mid-1990s. The size of the wafers and their comparative lack of rigidity meant that Standard Mechanical Interface (SMIF) was not a viable technology. FOUPs were designed with the constraints of 300 mm in mind, with the removable cassette being replaced by fins in the FOUP that hold the wafers in place, and the bottom opening door being replaced by a front-opening door to allow robotic handling mechanisms to access the wafers directly from the FOUP. The weight of a fully loaded 25-wafer FOUP, at around 9 kg, means that AMHSs are essential for all but the smallest of fabrication plants. A new factor in the current change in wafer size is the effect it has on packaging processes. With wafer-level packaging emerging, many of the assembly and test equipment makers now have to take wafer size into account [8]. Another factor is that devices are driving the move to 300 mm wafers. The technology of packaging or bumping integrated circuits (ICs) in wafer form is applicable to a wide range of devices. Low input/output (I/O) devices such as integrated passive devices or electrically erasable programmable read-only memory (EEPROMs) rely on wafer-level chip-scale package (CSP) technology for a small form factor, low-cost

FIGURE 11.4
Typical 25-wafer-capacity FOUP used for holding large-diameter silicon wafers. (From www.entegris.com. Entegris, Inc., 3500 Lyman Boulevard, Cheska, Minnesota, 55318, USA, 2002 Entegris, Inc [7].)

packaging solution. High-performance, high I/O devices such as microprocessors and peripheral logic use flip-chip mounting, not only for its superior electrical performance and low parasitics, but also for its ability to enable pad-limited designs that cannot be wire bonded. Fabrication facilities are being built around the world to produce a number of different devices on 300 mm wafers, including dynamic random access memory (DRAM), logic, and microprocessors. These high-performance devices will be the first products to be bumped on 300 mm wafers. As the technology and availability of 300 mm processing broadens, more cost-sensitive devices for consumer products will begin to be produced on the larger wafers.

In fact, with 450 mm silicon wafers around the corner, many industries are offering FOUPs with additional features [9] that offer special protection for silicon wafers. This includes a fire-safe plastic body, airtight door structures, purge facility, wafer support, and proper interface mechanisms to avoid contamination from ambient. It also includes optimized wafer support locations to minimize wafer sag by using a noncontact sensor. The complete package and its interiors are shown in Figure 11.5.

When finished silicon wafers are exposed to a clean room atmosphere for processing, gaseous organic molecules in the air rapidly adsorb onto the surface of the wafers. The arrival rate of these gaseous species is several orders of magnitude higher than that of the dust particles. To avoid the adsorption of airborne organic and inorganic species on to silicon surfaces, the wafers are typically stored and handled in wafer storage boxes. These boxes are normally made of polypropylene and/or polycarbonate materials. However, it has been shown recently that while wafers can be stored in boxes to protect against airborne contaminants, organic additives can vaporize from the plastic materials from which the boxes are constructed and the outgassed organics adsorb onto the silicon wafer surfaces. Since silicon wafer vendors ship their wafers to their customers by packing them in plastic boxes of their own design, the wafer surfaces will likely be contaminated by this outgassing.

Trace levels of organic compounds may arise within the clean room environments, wafer storage boxes, and process chambers. The organic contaminants comprise an extensive list of polar and nonpolar compounds, such as various hydrocarbons, amines, organosilicons, and organophosphate compounds. Acetone and xylene are two major organics often used in the silicon processing technology, and their presence in the clean room environment is quite possible. Adsorption and desorption rate constants of these compounds on silicon wafers were measured by Tlili et al. [10] as a function of the relative humidity (RH) present in the clean rooms. Acetone exhibited stronger linear dependences than RH dependences than did xylene. There are strong dipole–dipole interactions between the molecules of acetone and the outermost layer of adsorbed water on the silicon surface. It was proposed that such electrostatic interactions may lead to the formation of a hydrogen-bonded complex between these two items. However, xylene exhibits weak dispersion forces with the top layer of adsorbed water molecules on the silicon wafer. The latter observation is consistent with the physical-chemical properties of both organic compounds.

It is well known that organic contaminants adsorbed on a silicon wafer surface cause a number of detrimental effects, such as electrical property degradation, yield losses in the fabricated ICs, and so on. The adsorption rate on silicon wafers varies from one type of organic contaminant to another. In addition, the ease with which organic contaminants are absorbed on wafers depends on the boiling point. Hence, reducing organic contamination is a very important issue while storing the wafers in a container or when they are in line for various unit processes [4]. In fact, it is increasingly considered a necessary and effective element. From a practical point of view, the effect of organic materials in packaging is an issue.

(a)

(b)

(c)

FIGURE 11.5
(a) and (b) The 450 mm silicon wafer FOUP packing used for shipment, and (c) for storage. (From P. Lee, "450 mm FOUP/FOSB Development Status in Taiwan," www.gudeng.com.tw. Gudeng Precision Industrial Co., Ltd., 428 Bade St., Shulin City, Taipei County 238, Taiwan, Republic of China [9].)

Sugimoto and Okamura [11] analyzed the adsorption behavior of organic contamination on silicon surfaces. When the silicon wafers were exposed to typical clean room atmosphere, various gaseous organic molecules in the air were easily adsorbed onto their surfaces. Figure 11.6 shows a typical gas chromatogram of the organic contaminants on a silicon wafer surface that was exposed for a period of 24 h. This chromatogram reveals a number of peaks, each of which corresponds to a different organic contaminant. Abundance represents the total ion intensity of the m/z (atomic or molecular mass number to charge ratio) monitored for each peak in mass spectroscopy (MS) analysis. The peaks themselves represent the detected amount, and the retention times are associated with the boiling points of the detected organic contaminants in this case. Because not all of the detected organic contaminants have a definitive boiling point, the boiling points of the contaminants listed here are based on the retention time. It was further reported that the amount of organic contamination increased with the time the wafers were exposed. The type of contaminants adsorbed on the wafer for longer times was different from that on wafers exposed for a shorter duration. The organic vapors with a low boiling point tended to adsorb quickly than with higher boiling points. When the new plastic storage box was used, the organic contaminants were mainly plastic additives contained in the box. When a used plastic storage box that was left open in the clean room was used, the organic contaminants were airborne ones that originated in the clean room. Contaminants containing extreme polar groups were adsorbed immediately to the silicon surfaces.

Jackson et al. [12] attempted to explain the interaction and adsorption behavior of different organic molecules using benzene (C_6H_6). The charge effects of the organic materials on actual silicon surfaces were evaluated by noting the change in conductance of surface-controlled field-effect transistors prepared without gate electrodes. Cavalcoli et al. [13] reported on the possibility of detecting surface contamination on silicon wafers using fast, simple, nondestructive, and noncontacting methods using a scanning Kelvin probe technique to map surface contaminants on silicon wafers.

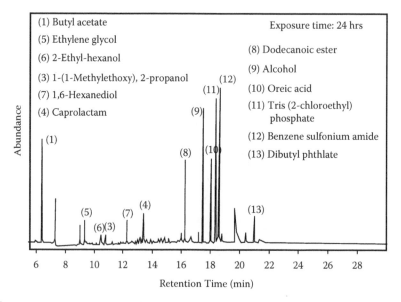

FIGURE 11.6
A typical gas chromatogram of organic contaminants on a silicon wafer surface exposed to a clean room environment for 24 h. Some of them are listed along with the retention time in minutes. (From F. Sugimoto and S. Okamura, *Journal of the Electrochemical Society*, **146**, 2725–2729, 1999 [11].)

According to Saga and Hattori [14], when the silicon wafers are stored in plastic boxes, volatile organics from the polymeric construction material adsorb onto the wafer surfaces. A very small quantity of additives in the plastic material is apt to adsorb onto the wafers more easily than the unpolymerized monomers and oligomers outgassing in large quantities. Various wet cleaning solutions were evaluated, and dilute hydrofluoric acid (HF) as well as ozonized ultrapure water have been found to completely remove these organic contaminants from the silicon surfaces. After wet cleaning, organic contaminants adsorb more easily on the ozonized water-treated silicon surface than on the dilute HF-treated surface. Adsorption of the organic additives on the silicon surfaces can be inhibited by preventing the native oxide growth in a nitrogen atmosphere after dilute HF cleaning. Further analysis using thermal desorption gas chromatography–mass spectrometry (TDGC–MS) methods for organic contaminants on the surface of silicon wafers stored in various types of plastic boxes. These boxes were supplied by major silicon wafer vendors, as well as volatiles outgassing from the materials that compose these plastic boxes in order to identify the source of the organic contaminants on the wafers. They also evaluated various wet cleaning solutions in terms of their ability to remove trace organic contaminants from the silicon wafer surfaces. Table 11.1 lists the major organics adsorbing on the silicon wafer stored in plastic boxes and the gases that outgas from the boxes.

Figure 11.7 shows details of the gas chromatogram of the organic contaminants adsorbed on the surface of the silicon wafers stored in a plastic box. These observations were made on cleaned silicon wafers, which were stored again in the boxes for one month for later surface analysis. The chemical structures of the aromatic organic compounds identified from the mass spectra are also shown in this figure. A large amount of 2,6-di-t-butyl-2,5-cyclohexadiene-1,4-dione and 2,6-di-t-butyl-4-methylene-2,5-cyclohexadiene-1-one, corresponding to the two adjacent peaks in the middle of the chromatogram, as well as a large amount of dibutyl phthalate (DBP), were detected on this wafer. It was confirmed that the aromatic hydrocarbons actually adsorb on the silicon surfaces in larger quantities than do the aliphatic hydrocarbons. Typical gas chromatograms of gases from the cassette and lid are shown in Figure 11.8. Details are summarized in Table 11.1.

TABLE 11.1

Volatile Organic Gaseous Species Adsorbed on the Wafers Stored in Plastic Boxes and Those Outgassing from the Container Boxes. Data Compiled from Different Sources and Vendors.

Major Organics Adsorbing on the Wafers (from Different Companies)	Organics Outgassing from Different Box Parts
2,6-Di-t-butyl-2,5-cyclohexadiene-1,4-dione	BHT (cassette)
2,6-Di-t-butyl-4-methylene-2,5-cyclohexadiene-1-one	
Dibutyl phthalate (DBP)	DBP (lid)
	Aliphatic hydrocarbons (all)
2,6-Di-t-butyl-2,5-cyclohexadiene-1,4-dione	BHT (box)
2,6-Di-t-butyl-4-methylene-2,5-cyclohexadiene-1-one	
Tetradecane nitrile	Aliphatic hydrocarbons (all)
2,6-Di-t-butyl-2,5-cyclohexadiene-1,4-dione	BHT (packing)
2,6-Di-t-butyl-4-methylene-2,5-cyclohexadiene-1-one	DBP (box)
Dibutyl phthalate (DBP)	Aliphatic hydrocarbons (all)
2,6-Di-t-butyl-2,5-cyclohexadiene-1,4-dione	BHT (box, packing)
2,6-Di-t-butyl-4-methylene-2,5-cyclohexadiene-1-one	DBP (lid)
Dibutyl phthalate (DBP)	Diacetyl benzene (lid)
Diacetyl benzene	Aliphatic hydrocarbons (all)

Source: K. Saga and T. Hattori, *Journal of the Electrochemical Society*, **143**, 3279–3284, 1996 [14].

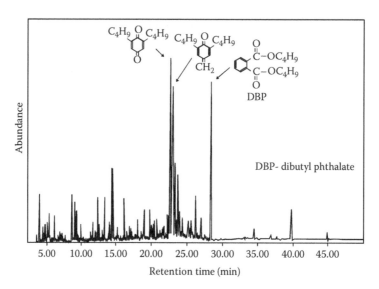

FIGURE 11.7
Typical gas chromatogram for organic contaminants absorbing on the surface of a silicon wafer stored in a plastic box. (From K. Saga and T. Hattori, *Journal of the Electrochemical Society*, **143**, 3279–3284, 1996 [14].)

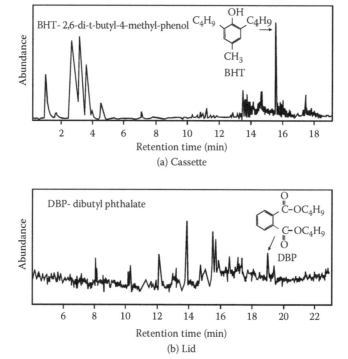

FIGURE 11.8
Typical chromatograms of gases from the (a) cassette and (b) lid of a plastic box. (From K. Saga and T. Hattori, *Journal of the Electrochemical Society*, **143**, 3279–3284, 1996 [14].)

Habuka et al. [15] modeled the rate theory of multicomponent adsorption of organic species on the silicon wafer surface. It deals mainly with the mechanism of the time-dependent change in the surface concentration of organic species. For this study the team has used numerical calculations to model the phenomenon. An equation composed of the adsorption rate from the gas phase to the silicon wafer surface and the desorption rate from the silicon wafer surface was developed that accounts for competitive processes in a multicomponent system. This equation can describe and predict the actual increase and decrease in the surface concentrations of propionic acid ester, siloxane (D9), and di(2-ethylhexyl)phthalate (DOP). Organic species with a large adsorption rate and a small desorption rate remain in significant abundance on the silicon wafer surface for a very long period after cleaning. Based on the concept of fruit basket phenomenon a computer model, multicomponent organic species adsorption-induced contamination, MOSAIC, was proposed to demonstrate the adsorption rate and desorption rates. These are the basic dominant mechanisms observed for silicon surfaces. The MOSAIC model was validated for the increase and the decrease in the measured concentrations of the nine organic species [16]. The results were reproduced by accounting for the adsorption and desorption properties on silicon wafer surfaces. Suppressing the increase in the surface concentration of DOP due to the coexisting organic compounds was observed by the silicon plate sampling (SPS) method and was explained by the MOSAIC model [17]. Therefore, an integration of the MOSAIC model and the SPS method has great capability for evaluating the nature of airborne organic contamination in detail on silicon surfaces.

Ishiwari et al. [18] developed a new method, called the silicon plate method, to evaluate organic contamination on a silicon wafer surface. Using this method, the concentration of bis(2-ethylhexyl)phthalate on the silicon wafer surface was experimentally shown to be in a steady state that has a correlation with its concentration in the clean room air. The details are shown in Figure 11.9. The figure shows the total organic species in the clean room air, which were collected using the Teanx adsorbent solid trap and measured using thermal desorption gas chromatograph mass spectrometry (TDGC-MS). The existence of various organic species, including DOP and DBP, is indicated in Figure 11.9a. Figure 11.9b through d shows the organic species adsorbed on the surface of the silicon plate exposed for different time intervals of 0.5, 2, and 24 h, respectively. The organic species adsorbed on the silicon plate in the clean room in this study are mainly DOP and DBP, although various organic species do exist in the environment. Time-dependent variations are highlighted in the rest of the figure. With time, the presence of DOP dominates on the surfaces, whereas DBP shows an initial rising trend but subsequently diminishes from the silicon surface.

Tsai et al. [19] used a simple white-light illumination method to clean the silicon wafer surfaces using rapid optical surface treatment (ROST). They demonstrated that a 60 s long ROST using a halogen lamp in ambient air is effective in removing light hydrocarbons from silicon surfaces that were adsorbed during wafer shipping and storage. Their experiment indicates that the cleaning action during lamp exposure is predominantly a thermally driven oxidation step, although the effect of optical interactions cannot be entirely excluded. It was demonstrated further that the ROST cleaning action is not as effective on wafers stored for a prolonged period in frequently opened boxes, whereas it is effective on silicon wafers freshly removed from the sealed shipping boxes.

Most of the organic compounds adsorbing on the wafer surface are desorbed when the wafers are heated to higher temperatures. Ozone-based cleaning recipes [20], applied in a wet bench setup and a spray-processing tool, are also effective at removing organic contamination from the silicon surfaces. The vapor drying technique, which uses isopropyl alcohol (IPA) [21], is excellent for obtaining particle-free silicon wafer surfaces. The other options involve using $NH_4OH-H_2O_2-H_2O$ solutions (RCA clean SC-1) [22] with NH_4OH content around the ratio of 0.05:1:5 of the aforementioned combination. If the wafers have undergone any spin operation, for the purpose of wafer drying or a coating for surface

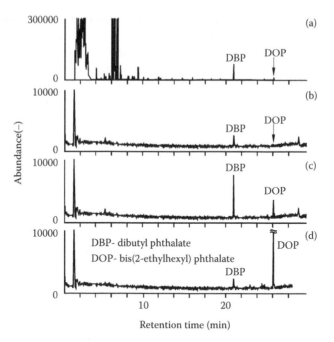

FIGURE 11.9
Thermal desorption mass spectrometry spectra measured for dibutyl phthalate (DBP) and bis (2-ethylhexyl) phthalate (DOP): (a) Organic species in the clean room air; (b), (c), and (d) organic species on the silicon surface after exposure for 1800 s, 7200 s, and 86,400 s, respectively. (From S. Ishiwari, H. Kato, and H. Habuka, *Journal of the Electrochemical Society*, **148**, G644–G648, 2001 [18].)

adhesion promoters, that implements a vacuum chuck holder. This way backside contamination, due to the presence of hard polymers, may get onto the wafer surface. These materials may mitigate [23] with the front of the silicon surface even though all the process steps are carried out only on the front and polished sides of the wafer.

11.2 Storage of Wafers and Control of Particulate Contamination

Münter et al. [24] analyzed the effect of time-dependent haze on silicon wafers. They demonstrated that after storing silicon wafers in polymer boxes, there is an increase in the number of localized light scatters (LLS) on the silicon surfaces. This phenomenon is caused by chemical processes on the surface called time-dependent haze (TDH), which the authors proved by intentionally generating it. Increases in LLS are generally observed after the contaminated wafers have been stored for a certain period. Preparations with inorganic compounds exhibited the formation of crystallites on the silicon surfaces. High vapor pressure and good solubility in water of organic compounds favored the generation of organic TDH. In both experiments, the presence of copper promoted the formation of TDH. Preparations with mixtures of contaminants revealed the interaction of chemical compounds on the surface. Ammonium sulfate had a strong impact on the formation of TDH. Pulsed-force-mode atomic-force microscopy measurements were used to identify the different chemical species constituting TDH by adhesion force mapping. The details are shown in Figure 11.10a and b. Figure 11.10a shows the scanning electron microscope (SEM) image of a wafer exposed to SO_2. The energy-dispersive x-ray spectroscopy (EDX) spectra, as shown in Figure 11.10b, of particles indicate the presence of sulfur, nitrogen, and oxygen. Figure 11.11 shows the increase in the number of LLSs, which was significant, was a criterion for TDH when they were intentionally exposed to volatile organics and stored for a longer time in the normal clean room environment.

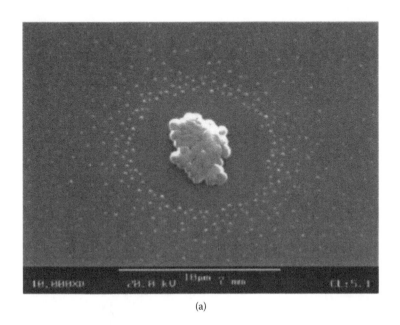

(a)

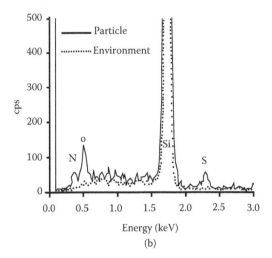

(b)

FIGURE 11.10
(a) SEM image of a wafer region exposed to SO_2. An agglomeration of large particles is visible, surrounded by smaller particles. At the immediate environment of the agglomeration, no particles exit. The small particles recrystallize to large ones. (b) In comparison to the surrounding surface, EDX spectra of particles indicate the elements sulfur, nitrogen, and oxygen. (From M. Münter, B. O. Kolbesen, W. Storm, and T. Müller, *Journal of the Electrochemical Society*, **150**, G192–G197, 2003 [24].)

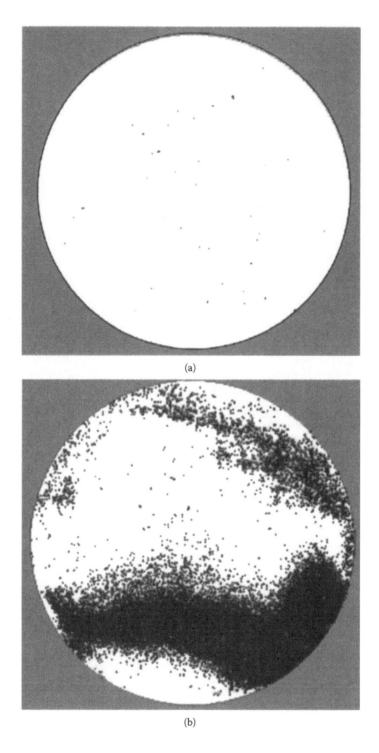

(a)

(b)

FIGURE 11.11
(a) Laser light scattering image of an SC-1 cleaned wafer. (b) Laser light scattering image of the same wafer after four months of storage in a gas atmosphere containing acetone at room temperature. The increase in localized light scatter (LLS) count in the range from 0.12 to 0.24 μm is about 15,000. (From M. Münter, B. O. Kolbesen, W. Storm, and T. Müller, *Journal of the Electrochemical Society*, **150**, G192–G197, 2003 [24].)

Particles are known to have a dramatic effect on the manufacture of ICs and are a major yield loss factor. They become more important as the size of individual semiconductor devices are scaled down, and their presence has become a point of concern. Particles are present all around us in the atmosphere in great abundance, and they are difficult to avoid. They include pollens, dusts, fibers, metals, metal oxides, hydrocarbons, and various other organic materials. These particles are generated by several means, including mechanical abrasion, chemical reactions, and combustion processes. People are the main source of dust particles in any semiconductor production facility; the particles come from their clothes, their skin and hair, and from their breath, particularly if they smoke.

The detection of particulate contamination on the silicon surfaces used for wafer processing is a very important aspect of the continued improvement of the yield and reliability of manufactured devices. The current methods for measuring particulate contamination fall into three general types: manual particle counting (by unaided eye or with a microscope) of the light scattered under the illumination of an intense white-light source; automatic particle counting with a commercially available instrument that uses a helium/neon or helium/cadmium laser; and automatic counting with a mercury or mercury-xenon arc lamp source and a vidicon display [25]. Commercial systems, such as the Aeronca wafer inspection system (WIS-150), can examine the various thicknesses of silicon dioxide layers on the wafer surface. Particle sizes from 0.215 to 6.40 μm, nominal oxide thickness values of 0.1, 0.2, 0.3, 0.4, and 1.0 μm, and polystyrene latex (PSL) particles/spheres of sizes between 0.215 and 2.35 μm can be evaluated.

As mentioned, controlling particulate contamination of wafer surfaces is important, both during storage and when the wafer is being considered for fabrication. A scanner is usually used to inspect incoming bare silicon wafers to determine their cleanliness level and to evaluate which ones will be selected for processing. Commercially available systems, such as Tencor Surfscan 4000, Estek, PMS, Censor, and Hitachi products [26], are typically used. Because of the high refractive index of silicon, the listed systems can detect particles as small as 0.1 μm in size with 90% efficiency.

R. A. Bowling [27] analyzed the particle adhesion properties on semiconductor surfaces and estimated the driving forces that are responsible for their sticking. Small particles on a dry surface are primarily held in place by strong van der Waals forces, which can increase in magnitude over time due to deformation of the particle surface increasing the total contact area with the semiconductor. Immersion generally reduces the total force of adhesion. Upon removal from the liquid, however, the predominating capillary forces of adhesion due to the formation of liquid bridges between the particle and the surface then have an effect. The same phenomenon of capillary action may also occur in high humidity conditions. The total forces of adhesion for small particles are so large that they exceed the gravitational force on those particles by many orders of magnitude and permanently stay with the surface. Hence, emphasis should be placed on preventing particle deposition on the surfaces rather than counting on their subsequent removal. Slight ultrasonic agitation may be recommended to release the adhered particles from the surface, but may not remove all of them.

11.3 Storage of Wafers and Control of Particulate Contamination with Process-Bound Wafers

A systematic study by Raider et al. [28] using electron spectroscopy for chemical analysis (ESCA) and ellipsometry methods on the oxide growth on silicon provided interesting results. According to them, on any fresh silicon surface, impurities rapidly adsorb on any moderately doped silicon substrates. This is particularly after the chemical etching step is carried out to remove the existing silicon dioxide layer from the surface using HF-based solutions. The initial oxide growth rates on the silicon surface in air or water at room temperature are lower than on cleaved silicon surfaces. HF etching enhances the rate of impurity adsorption on a silicon surface relative to the rate of actual surface oxidation. An induction period is sometimes observed during which oxide growth does not occur on the etched silicon surfaces, but it is not observed once oxidation is initiated. The thickness of the impurity layer on etched silicon decreases after oxidation is initiated, and its effect on the rate of subsequent oxidation of silicon also decreases. A number of factors affect the rate of oxide growth on etched silicon at room temperature. An increased oxide growth rate is observed if the substrate dopant (*n*- or *p*-type) concentration is greater than or equal to 10^{19} cm^{-3}. The oxide growth rate in air is greater on <111> oriented substrates than on <100> surfaces. Ultraviolet (UV) radiation (2537 Å) also increases the oxide growth rate in air. However, the major effect of UV is to increase the nature and film thickness of the carbon impurities at the silicon surface, presumably by a mechanism different from that of adsorption in the dark. The equilibrium oxide thickness in air at room temperature is less than 14 Å, unless the substrate dopant concentration is greater than 10^{19} cm^{-3} or the substrate is exposed to UV radiation during the oxidation process.

Antistatic technology is one of the key technologies affecting the device yield of silicon wafers. Open manufacturing systems are mainly employed when the wafers would normally be exposed to air while they are transported between different unit processes. In order to overcome problems such as wafer surface contamination due to impurities in the air and native oxide formation, closed manufacturing systems were introduced, and are preferred to handle wafers in high-purity nitrogen ambient conditions. The wafers usually are charged to several kV when they come into contact with insulating materials or when the gas from a nozzle blows the wafer. When the wafer is charged, the airborne particles adhere on it due to the electrostatic attraction, which affects process yield. Inaba et al. [29] reported that the charge-up of the wafers in nitrogen and argon gas ambient can be prevented by ionizing the gas molecule with UV light. This method features excellent neutralization capability without accompanying particle generation. In addition, it hardly depends on the gas flow rate or the moisture concentration present inside the container. Further, the residual potential after neutralization is always kept at 0 V, and this method can be applied to the reduced-pressure ambient conditions.

11.4 Summary

In this chapter, we discussed wafer packaging. Fully processed and cleaned silicon wafers are shipped to users located at different locations in packing boxes that should not cause any damage to the wafers. A broken wafer is useless and cannot be used for very large scale integration (VLSI) and ultra large scale integration (ULSI) processing. Early packages were simple, but as wafers increase, shipment and handling have become major issues. New designs have been developed to control particulate contamination, both during processing and while in storage. Proper care of the environment is necessary; otherwise, the wafer surface may attract dust and chemical vapors and may develop haze on the surface, which can make the wafer unfit for circuit fabrication.

References

1. G. Fisher, M. R. Seacrist, and R. W. Standley, "Silicon crystal growth and wafer technologies," *Proceedings of the IEEE*, **100**, 1454–1474, 2012, and the references therein.
2. O. Anttila, "Challenges of silicon materials research: Manufacture of high resistivity, low oxygen Czochralski silicon," Okmetic Fellow, Okmetic Oyj. Technical presentation, CERN_RD-50_06_2005_Okmetic_OA.ppt.
3. www.powershow.com/view/f1659-NjRhN/Unit_3_Crystal_Growth_and_Wafer_preparation_ powerpoint.ppt. *Unit 3 Crystal Growth and Wafer Preparation.*
4. M. Saito, K. Anbai, and T. Hayashi, "Organic outgassing behavior of plastic material and reduction of organic contamination in semiconductor equipment," *IEEE Transactions on Semiconductor Manufacturing*, **18**, 575–583, 2005.
5. J. Frickinger, J. Bügler, G. Zielonka, L. Pfitzner, H. Ryssel, S. Hollemann, and H. Schneider, "Reducing airborne molecular contamination by efficient purging of FOUPs for 300-mm wafers– The influence of materials properties," *IEEE Transactions on Semiconductor Manufacturing*, **13**, 427–433, 2000.
6. S.-C. Hu and T.-M. Wu, "Experimental study of airflow and particle characteristics of a 300-mm FOUP/LPU mini-environment system," *IEEE Transactions on Semiconductor Manufacturing*, **16**, 660–667, 2003.
7. www.entegris.com. Entegris, Inc., 3500 Lyman Boulevard, Cheska, Minnesota, 55318, USA, 2002 Entegris, Inc.
8. EBSCO Host Trade Publication, "Suppliers' successes with 300 mm tools and materials," *Solid State Technology*, **44**, p.51, May 2001.
9. P. Lee, "450-mm FOUP/FOSB Development Status in Taiwan," Technical Presentation, www. gudeng.com.tw. Gudeng Precision Industrial Co., Ltd., 428 Bade St., Shulin City, Taipei County 238, Taiwan, Republic of China.
10. S. Tlili, L. I. Nieto-Gligorovski, B. Temime-Rousell, S. Gligorovski, and H. Wortham, "Humidity and temperature dependences of the adsorption and desorption rates for acetone and xylene on silicon wafer," *Journal of the Electrochemical Society*, **157**, P43–P48, 2010.
11. F. Sugimoto and S. Okamura, "Adsorption behavior of organic contaminants on a silicon wafer surface," *Journal of the Electrochemical Society*, **146**, 2725–2729, 1999.
12. J. A. Jackson, J. R. Szedon, and T. A. Temofonte, "An effect of organic electron donors and acceptors on a real silicon surface," *Journal of the Electrochemical Society*, **119**, 1424–1425, 1972.
13. D. Cavalcoli, A. Cavallini, M. Rossi, S. Binetti, F. Izzia, and S. Pizzini, "Surface contaminant detection in semiconductors using noncontacting techniques," *Journal of the Electrochemical Society*, **150**, G456–G460, 2003.
14. K. Saga and T. Hattori, "Identification and removal of trace organic contamination on silicon wafers stored in plastic boxes," *Journal of the Electrochemical Society*, **143**, 3279–3284, 1996.
15. H. Habuka, M. Shimada, and K. Okuyama, "Rate theory of multicomponent adsorption of organic species on silicon wafer surface," *Journal of the Electrochemical Society*, **147**, 2319–2323, 2000.

16. H. Habuka, M. Shimada, and K. Okuyama, "Adsorption and desorption rate of multicomponent organic species on silicon wafer surface," *Journal of the Electrochemical Society*, **148**, G365–G369, 2001.

17. H. Habuka, S. Ishiwari, H. Kato, M. Shimada, and K. Okuyama, "Airborne organic contamination behavior on silicon wafer surface," *Journal of the Electrochemical Society*, **150**, G148–G154, 2003.

18. S. Ishiwari, H. Kato, and H. Habuka, "Development of evaluation method for organic contamination on silicon wafer surfaces," *Journal of the Electrochemical Society*, **148**, G644–G648, 2001.

19. C.-L. Tsai, P. Roman, C.-T. Wu, C. Pantano, J. Berry, E. Kamieniecki, and J. Ruzyllo, "Control of organic contamination of silicon surfaces using white light illumination in ambient air," *Journal of the Electrochemical Society*, **150**, G39–G44, 2003.

20. M. Claes, S. De Gendt, C. Kenens, T. Conard, H. Bender, W. Storm, T. Bauer, P. Mertens, and M.M. Heyns, "Controlled deposition of organic contamination and removal with ozone-based cleanings," *Journal of the Electrochemical Society*, **148**, G118–G125, 2001.

21. H. Mishima, T. Ohmi, T. Mizuniwa, and M. Abe, "Desorption of isopropanol (IPA) and moisture from IPA vapor dried silicon wafers," *IEEE Transactions on Semiconductor Manufacturing*, **2**, 121–129, 1989.

22. M. Itano, F. W. Kern, Jr., R. W. Rosenberg, M. Miyashita, I. Kawanabe, and T. Ohmi, "Particle deposition and removal in wet cleaning processes for ULSI manufacturing," *IEEE Transactions on Semiconductor Manufacturing*, **5**, 114–120, 1992.

23. N. Iyer, N. Saka, and J.-H. Chun, "Contamination of silicon surface due to contact with solid polymers," *IEEE Transactions on Semiconductor Manufacturing*, **14**, 85–96, 2001.

24. M. Münter, B. O. Kolbesen, W. Storm, and T. Müller, "Analysis of time-dependent haze on silicon surfaces," *Journal of the Electrochemical Society*, **150**, G192–G197, 2003.

25. B. R. Locke and R. P. Donovan, "Particle sizing uncertainties in laser scanning of silicon wafers: Calibration/evaluation of the Aeronca Wafer Inspection System 150," *Journal of the Electrochemical Society*, **134**, 1763–1771, 1987.

26. B. Y. H. Liu, S.-K. Chae, and G.-N. Bae, "Sizing accuracy, counting efficiency, lower detection limit and repeatability of a wafer surface scanner for ideal and real-world particles," *Journal of the Electrochemical Society*, **140**, 1403–1409, 1993.

27. R. A. Bowling, "An analysis of particle adhesion on semiconductor surfaces," *Journal of the Electrochemical Society*, **132**, 2208–2214, 1985.

28. S. I. Raider, R. Flitsch, and M. J. Palmer, "Oxide growth on etched silicon in air at room temperature," *Journal of the Electrochemical Society*, **122**, 413–418, 1975.

29. H. Inaba, T. Ohmi, M. Morita, M. Nakamura, T. Yoshida, and T. Okada, "Neutralization of wafer charging in nitrogen gas," *IEEE Transactions on Semiconductor Manufacturing*, **5**, 359–367, 1992, and the references therein.

Index

A

Acceptors, 8
Accommodative property of silicon traps, 247
Accurate boundary conditions, 71
Acetic acid, 167
Acetone, 382
Acid leaching, 254
Acoustic laser probing technique, 175–178
Actual silicon concentration, 255
Adverse thermal gradients, 63
Aeronca wafer inspection system, 391
AES, *see* Auger electron spectroscopy
Airborne molecular contamination (AMC), 379
Aluminum, 250
Aluminum oxide, 64, 67
Aluminum-silicon liquid alloy cathode, 29
AMC, *see* Airborne molecular contamination
American Society for Testing and Materials
        (ASTM), 204, 279
AMHS, *see* Automated material handling
        system
Ammonium sulfate, 388
Amorphous silicon, 46
    source of, 31
Anisotropic etching techniques, 46
Annihilation of stacking faults, 322
Anomalous oxygen precipitation (AOP), 324
Antimony, 97–98, 250–251
Antimony-doped crystals, 226
Antimony-doped silicon crystals, 251
Antistatic technology, 392
AOP, *see* Anomalous oxygen precipitation
APF, *see* Atomic packing factors
Application-specific integrated circuits
        (ASICs), 18
Aqua regia, 33
Argon annealing of wafers, 366–368
Argon gas, 82
Aromatic organic compounds, chemical
        structures of, 385
Arrhenius equation, 65
Arrhenius plot, 258
    of oxygen concentration in molten silicon, 66
Arsenic, 96, 251–252
Arsenic-doped crystals, 226
ASICs, *see* Application-specific integrated
        circuits

ASTM, *see* American Society for Testing and
        Materials
A-swirls, 312
Asymmetric fields of velocity, 113
Asymmetric radial thermal field, 108
Atomic absorption spectroscopy (AAS), 351
Atomic configuration, 259
Atomic-force microscopes, 178, 331
Atomic force microscopy observations, analysis
        of silicon wafer using, 179
Atomic hydrogen, 262
Atomic nitrogen interstitials, 267
Atomic packing factors (APF), 48
Auger electron, 31
Auger electron spectroscopy (AES), 178
    analysis, 247
    of metallurgical-grade silicon, 180
    studies, 178–180
Automated material handling system
        (AMHS), 381
Axial carbon distributions, 258
Axial distribution of oxygen
        concentration, 89, 90
Axially symmetric cusp magnetic field, 274, 275
Axial magnetic field, 111, 114, 131
Axial oxygen distributions, 79, 193, 275
Axial resistivity distribution of boron-doped
        and phosphorus-doped silicon
        crystal, 227
Axial-symmetric heat field, 92
Axial temperature gradient, 300, 301
Axial thermal gradient, 321
Axial thermal profiles of silicon crystals, 75, 185
Axisymmetric magnetic field, 106, 111, 131
Axisymmetric spoke pattern, 71

B

Back-reflection section topographs, 273
Ball-and-stick models of seed-SiO, 272
Band edge photoluminescence intensity, 200
Band-to-band emission, 200
Band-to-band intensity, 200
Baroclinic instability, 105, 211
B-defects, 258
Benard instability, 112
Binding energy, 267

Bis(2-ethylhexyl)phthalate, 387
Boolean applications, silicon devices for, 11–12
Boron, 94–95, 252–253
    diffusion, 262
Boron-doped crystals, 252
Boron-doped CZ silicon wafers, 324
Boron-doped silicon
    crystal, axial resistivity distribution of, 227
    dopant concentration density to resistivity
        conversion for, 236–238
    resistivity to dopant concentration density
        conversion for, 230–232
Boron doping level, influence of, 321
Boron impurity atom, 58
Bottom-masked crucible, 88
Boule formation, 158–162
Boussinesq approximation, 116
Bragg diffraction method, 210
Bridgman crystal growth technique, 60
Bright-field infrared laser interferometer, 177, 330
Broad probe tip, use of, 190
B-swirls, 312
Bulk microdefects, density of, 364
Bulk nickel impurities, detection and
        removal of, 265
Bulk silicon crystal lattice, defects in, 295
Buoyancy convection, 85
Buoyancy-driven convection, 63
Buoyancy-driven flow, 114
Buoyancy force, 62, 103
Buoyant convection, 67, 114
Burgers vectors, 310, 312, 315

C

CACRT, *see* Czochralski accelerated crucible
        rotation technique
Capability of oxide precipitate–related
        defects, 359
Carbon, 255–259, 367
    principal allotropic forms of, 7
Carbon-containing globular clusters, 258
Carbon contamination, 178
Carbon content of semiconductor-grade
        silicon, 257
Carbon-doped CZ silicon, 368
Carbon impurity, 258
Carbon monoxide (CO), 29, 64
Carbon striations, 211
Carrier mobility, 195
CCDs, *see* Charge coupled devices
Centrifugal pumping, 67
    flows, 114

Chamber temperature profile, 72–77
Charge-coupled devices (CCDs), 186, 249
    sensitivity to defects in, 297
Charge particle activation analysis, 199
Chemical cleaning of silicon wafers, 171
Chemical etching, 167, 181
    and mechanical damage removal, 167–168
Chemical polishing of silicon wafers, 169
Chemical staining, 181–184
Chemical vapor deposition (CVD), 138, 254
    operations, 333
    reactor, 35
Chemimechanical polishing for planar wafers,
        168–169
Chip-scale package (CSP) technology, 381
Chlorosilanes, MGS to, 34
Chromium, 259–260
Circuit complexity, devices of, 21
Circuit miniaturization, art of, 12–18
Circular movement of the heat wave, 92
Circular stacking faults, 316
Cluster-of-point defects, 312
CMOS, *see* Complementary metal-oxide
        semiconductor
CO, *see* Carbon monoxide
Cobalt-doped crystals, 187, 303
Cochran flow, 183, 278
Co-doping method, 226, 261
Commercial microcrystalline silicon, 255
Complementary metal-oxide
        semiconductor (CMOS)
    applications, diameter silicon wafer
        specifications for, 13–14
    devices, 3, 18, 296
    fabrication, 250
    technologies, 186, 354
Complementary metal-oxide semiconductor
        (CMOS) VLSI circuits
    300 mm diameter silicon wafer
        specifications for, 16–17
    200 mm diameter silicon wafer
        specifications for, 15–16
Computational algorithms, 84
Computer-controlled eddy current system,
        185
Computer simulation-based tools, 17
Concentration contours, 228–229
Conductive geometric singularity, formation
        of, 261
Contactless capacitance doping profiling
        measurement technique, 184
Contactless characterization, 184–185
Contact mode, 178

Contamination management schemes, 279
Continuous liquid feeding technique, 226
Conventional CZ method, 270
Conventional growth condition, 115
Conventional optical spectroscopic
    methods, 177
Conventional oxygen out-diffusion
    approach, 362
Conventional quantitative analyses, 248
Conventional RCA cleaning recipe, 371
Conventional silicon wafers, 278
Copper, 260–261, 358
Copper-oxygen complex, formation of, 260
Copper precipitate colonies, 187
COPs, *see* Crystal-oriented particles
Coriolis force, 71, 72, 103, 106
Coupled Navier-Stokes, 225
Coupled thermal-electromagnetic-
    hydrodynamic problem, 131
Covalent bond, 3
Covalent crystal, 3
Cross-sectional microphotographs, 367
Cross-sectional optical microphotograph, 367
Crucible constrain, influence of, 337
Crucible holding molten silicon, 255
Crucible-related contaminations, 124
Crucible rotation (CR), 62
Crystal chamber, environmental and ambient
    control in, 82–84
Crystal cooling, 124
Crystal defects
    decoration, 187
    in hypothetical silicon lattice, 294
Crystal evaluation, 183
Crystal growth parameters
    optimization of, 281, 366
    variation of, 115
Crystal growth processes, 72
    of liquid silicon, 259
Crystal growth striations, 107–108
Crystal growth technique, 120
Crystal homogeneity, 111
Crystalline defects in single-crystal silicon
    wafer, 294
Crystalline ingot, 294
Crystalline solids, 1
Crystalline state, salient feature of, 3
Crystallographic defects, 249
    types of, 353
Crystal-melt interface, 120
    shape of, 105
Crystal-melt phase boundary, 74
Crystal order, 48–50

Crystal orientations, 50–54
    of FZ silicon, 317
    identification of, 158–162
Crystal-oriented particles (COPs), 96, 252, 326,
    327, 329
Crystal perfection, 48–50
Crystal planes, 46
Crystal pulling
    rate, 84–93, 319
    seed selection for, 77–81
Crystal purification techniques, 263
Crystal rotation rates, 279
Crystals, 1
    systems, 47
Crystal shaping methods, 197
Crystal solidification process, 247
CSP technology, *see* Chip-scale package
    technology
Cusp magnetic fields, 107
Cusp-shaped magnetic fields, 112
CVD, *see* Chemical vapor deposition
Czochralski accelerated crucible rotation
    technique (CACRT), 63
Czochralski crystal
    inhomogeneities and defects in, 91
    pulling, parameters in, 85
Czochralski crystal growth/pulling technique,
    78, 303
    chamber temperature profile, 72–77
    crystal chamber, environmental and
        ambient control in, 82–84
    crystal growth striations, 107–108
    crystal pulling, seed selection for, 77–81
    crystal pull rate and seed/crucible rotation,
        84–93
    growing doped crystals, dopant addition for,
        94–100
    liquid and grown silicon crystals, impurity
        segregation between, 102–107
    methods for continuous, 100–102
    molten silicon, crucible choice for, 64–72
    post-growth thermal gradient and crystal
        cooling after pull-out, 122–124
    use of magnetic field in, 108–117
    VLSI And ULSI applications, large-area
        silicon crystals for, 117–122
Czochralski gas-controlled crystals, 83
Czochralski-grown crystals, 83, 187
Czochralski-grown single-crystal silicon, 82
Czochralski method, 59
Czochralski process, 293
Czochralski pulling method, 3, 138
Czochralski silicon, 353

Czochralski silicon crystals, 98, 177, 225, 293
Czochralski silicon furnace with heat shield, 75
Czochralski silicon wafers
     effect of artificial mechanical damage in, 197
     specifications of, 296
Czochralski single-crystal silicon, 248, 263
Czochralski techniques, 19, 92, 158, 299
Czochralski wafers, 18

D

Dash etch, 181
Dash necking method, 118, 310
Dash pedestal method, 136
Dash technique, 77, 81
DBP, *see* Dibutyl phthalate
D-defects, 297
Decay-time curve method, 200
Deep-level photoluminescence, 200
Deep-level transient spectroscopy (DLTS),
     185–187, 297
     analysis, 260
Defect decoration by metals, 187
Defect formation mechanism, 123
Defect-free denuded zones, 364, 368
Demountable molds, use of, 136
Density dynamic access memory (DRAM), 105
Density spectra of grown-in oxide precipitate
     nuclei, 324
Denude zone (DZ), 97, 362
     in silicon wafers, 364
Derivative surface photovoltage spectroscopy
     (DSPS), 186, 297
Desirable centrifugal pumping flow, 114
Device fabrication processes, 252, 254
Device miniaturization, art of, 4–6
Device performance, degradation of, 297
Device-processing technology, 1
DHF, *see* Diluted hydrofluoric acid
Diameter silicon wafer specifications
     for CMOS and MEMS applications, 13–14
     CMOS VLSI circuits, 15–17
Diamond crystal structure, 47–48
Dibutyl phthalate (DBP), 385, 388
Differential-infrared absorption (DIR), 194, 273
Digital applications, MOS and CMOS devices
     for, 18
Diluted hydrofluoric acid (DHF), 349
Di(2-ethylhexyl)phthalate (DOP), 387
DIR, *see* Differential-infrared absorption
Direct current (DC) power supply, 72
Discrete transistors, 2
Disilane, 31

Dislocation density, x-ray technique for, 210–213
Dislocation-free crystals, 253
Dislocation-free silicon crystals, 92, 124, 130, 299
Displacement-based thermoelastic stress
     model, 337
Distillation process, 34
DLTS, *see* Deep-level transient spectroscopy
Donor atoms, 8
Donor concentration species, doping with, 94,
     250, 314
Dopant concentration, 135, 228
     homogeneity, 111
Dopant distribution in growing crystals,
     128–129
Dopant impurities, 225
Dopant impurity distributions, 117
Doped crystals, dopant addition for growing
     antimony, 97–98
     arsenic, 96
     boron, 94–95
     germanium, 99–100
     nitrogen, 96–97
     phosphorus, 95–96
Doping silicon with Group IV isovalent
     impurity, 281
Double-beam x-ray radiography system, 105
Double-beam x-ray visualization technique,
     67, 211
Double-crystal topographs, 211
Double-layered CZ process, 183
Double-sided polishing technology, 97
DPW, *see* Gross die per wafer
DRAM, *see* Dynamic random access memory
DSPS, *see* Derivative surface photovoltage
     spectroscopy
Dual-type octahedral void defects, 308
Dust particles, formation of, 379
Dynamic random access memory (DRAM), 248
     reliability tests, 296

E

EBIC-mode SEM, 202
Eddy current sensor, 75
Eddy current testing, 185
Edge contour profiling of silicon wafers, 168
Edge profiling of slices, 167
Edge-supported pulling method, 137
EDX spectra, *see* Energy-dispersive x-ray
     spectroscopy spectra
EEPROMs, *see* Electrically erasable
     programmable read-only memory
Effective cleaning of silicon surface, 371

Effective melt replenishment technique, 101
Effective segregation coefficient, 226
Efficient Newton method, 121
Efficient oxygen doping, 276
8-inch silicon wafers, 377
Elastic anisotropy, 81
Elastic stress parameters, 164
Electrical conduction, 8, 9
Electrical conductivity of silicon, 8
Electrically erasable programmable read-only memory (EEPROMs), 381
Electrical parameters, 59
Electrical properties, determination of, 262
Electrical resistivity, 27
Electrochemical solid ionic sensor, 65
Electrolysis method, 33
Electromagnetic force (EMF) CZ method, 120
Electron beam, 188
Electron beam-induced current (EBIC)-mode scanning electron microscopy (SEM), 183
Electron beam microprobe analysis, 366
Electron exchange mechanism, 371
Electronic-grade polysilicon, 37
Electronic-grade silicon, 35, 124
    pieces, 60
    production of, 36, 37
    purifying metallurgical-grade silicon to, 32
Electronic power devices, 97
Electronic scattering, 9
Electron irradiation, 191
Electron microscopy, 213
    studies, 262
Electron mobility, 253
Electron spectroscopy for chemical analysis (ESCA), 392
Elemental silicon, 7
EMF CZ method, *see* Electromagnetic force CZ method
Energy-dispersive x-ray spectroscopy (EDX) spectra, 389
Epitaxial silicon, 206
    wafers, grown-in defects in, 209
Equilibrium concentrations of vacancy species, 274
Equilibrium segregation coefficient, 225
ESCA, *see* Electron spectroscopy for chemical analysis
Etching techniques, 181–184
Etch-pit-count technique, 182
Etch-pit techniques, 253

F

Fast-diffusing metal impurities, 347
    control of, 248
Fast pulled silicon (FPS), 327
FEM, *see* Finite element method
FEM-package, *see* Finite element method-package
Fibers, 137
Fickian-typed diffusion, 115
Finite-difference methods, 121
Finite-element analysis, 70
Finite element method (FEM), 63, 76, 80, 121, 311
Finite element method (FEM)-package, 337
Flame emission spectrometry, 188
Flash memory characteristics, defects in, 298
Float-zone crystal growth technique, 124–125
    crystal growth rate and seed rotation, 126–127
    dopant distribution in growing crystals, 128–129
    heating mechanisms and RF coil shape, 125–126
    impurity segregation between liquid and grown silicon crystals, 130
    large area silicon crystals and limitations of shape and size, 131–135
    thermal gradient and post-growth crystal cooling, 135
    use of magnetic fields for, 130–131
Float-zone crystals, 226
Float-zone–grown crystals, 181, 251
Float-zone method, 3, 59
Float-zone process, 139, 293
Float-zone silicon crystals, 130
Float-zone techniques, 276, 299
Flow-mode transition, 212
Flow pattern defects (FPDs), 303, 315
Flow velocity fields of molten silicon, 68
Flow visualization, 105
    using x-ray radiography, 87
Forced convection dominant procedure, 275
Fourier IR spectroscopy, 271
Fourier transform infrared (FTIR), 84
    measurements for impurity identification, 191–195
    spectroscopy, 191, 257, 333, 353
Four-point probe technique, 91, 189
    mapping system, resistivity contour map for 6-inch silicon wafer using, 228, 244
    resistivity mapping system, 242
    for resistivity measurement and mapping, 189–191
Fowler-Nordheim tunneling, 28

FPD, *see* Flow pattern defects
FPS, *see* Fast pulled silicon
Fractional melting process, 33
"Free carrier absorption," 197
Frenkel pair generation, 362
Frenkel reaction, 122, 293, 300
Front-opening unified pods (FOUPs), 380, 381
    system, 377
Fruit basket phenomenon, concept of, 387
FTIR, *see* Fourier transform infrared
Fully processed blank silicon wafers, packing
    of, 377–388
FZ-inverted pedestal techniques, 130

G

Ga-doped silicon crystals, 130
Gallium, 96, 253
Gallium-doped CZ silicon, 253
    wafers, 96
Gamma-ray diffractometry, 201
Gas chromatogram of organic contaminants, 384
Gate oxide integrity (GOI), 99
    for MOS devices, 296
    of silicon wafers, 369
    yield, 368
Gate oxides, 28
GCZ, *see* Germanium-doped CZ
Geometric configuration of an 8-inch-diameter
    continuous CZ system, 101
Germanium, 2, 99–100, 261
Germanium-doped CZ (GCZ)
    silicon wafers, 261
    wafer, 99
Gettering schemes, 278
Gold, 261–262
Grain boundaries, 45
Grain-boundary and bulk-diffusion
    mechanisms, 73
Granular electronic-grade polysilicon, 39, 40
Granular polysilicon, 39
Graphite heating elements, 60
Graphite materials, 74
Graphite, usage of, 64
Gross die per wafer (DPW), 59
Group V impurities, 94
Grown-in microdefects, 296
    in dislocation-free silicon crystals, 298–299
"Grown-in" micro-precipitate nuclei, 364
Grown-in point defects, 304
Grown-in void defects, 330
Grown silicon crystals, liquid silicon crystals
    vs., 102–107, 130

H

Half-loop dislocations, types of, 311
Hall effect, 195
    measurements, 253
Hall mobility, 195–196
Halogens, 32
Halting crystal growth methods,
    308
Hard sphere model, 55
Harmful impurities, 370
HCl, *see* Hydrochloric acid
Heat exchange method, 66
Heat-transfer fluids, 35
Heat transfer mechanisms, 69
Heat-treated specimens, 354
Heavy metal contamination
    on defect formation, 249, 297
    monitoring of, 351
Helium, 262
Helium-neon (He-Ne) laser, 175
Hemlock Semiconductor process, 34
Heterogeneous nucleation of amorphous SiO_2, 272
Hexagonal stacking faults, 316
High-energy electron diffraction studies,
    188
Higher tensile residual stresses, 209
High-frequency inductor, 126
High gate oxide integrity yield, 368
High-intensity light source, reflection of,
    162
High-intensity reflectograms, 158
    use of, 160
"High-low-high" treatments, 362
High-magnification TEM micrograph of an
    octahedral precipitate, 356
High-performance devices, 382
High-pressure plasma (HPP) deposition
    process, 136
High-quality material properties, 2
High-quality oxide insulator, 2
High-quality wafers, 297
High-resolution electron microscopy (HREM),
    304
High-resolution x-ray diffraction, 177
High-temperature vacancy trapping, evidence
    of, 273
Hillock formation, mechanism of, 209
Homogeneous COP density, 328
Hotoacoustic displacement, 197
HPP deposition process, *see* High-pressure
    plasma deposition process

HREM, *see* High-resolution electron microscopy
Hydrochloric acid (HCl), 349
Hydrodynamic equations, solution of, 108
Hydrofluoric acid (HF)-based solutions, 392
Hydrofluoric acid (HF) solutions, 351
Hydrogen, 262–263
Hydrogen-annealed CZ silicon wafers, 369
Hydrogen-annealed internal gettering wafer, 368
Hydrogen-annealed silicon, 369
Hydrogen-annealed wafers, 370
Hydrogen annealing, 329, 369
    effect of, 263
    on oxygen precipitation behavior, effect of, 262
    of wafers, 368–370
Hydrogen halides, 32
Hydrogen-terminated surfaces, 369

I

ICs, *see* Integrated circuits
ID, *see* Internal diameter
Identify wafer imperfections, 175
IHTCM, *see* Integrated hydrodynamic thermal-capillary model
Impurities, 92, 282
    concentrations, photoluminescence method for determining, 199–201
    distribution, 107
    electrically active and inactive, 228
    evaluation, optical methods for, 199
Impurity doping
    effect of, 183, 202, 250
    striations, 182
"Inclination fringes," 209, 311
Induction heating, 72
Industrial silicon production, 119
Infrared absorption spectrophotometry, 199, 255
Infrared absorption spectroscopy, 257
Infrared absorption technique, 183
Infrared light-scattering tomography, 304, 331
Infrared spectroscopy, 191, 255
Infrared temperature sensor output, 82
Ingot slicing, 162–164
Inhomogeneous distribution of oxygen, 212
Inhomogeneous oxygen distribution, 105
Inline array configurations, 190
Inorganic compounds, preparations with, 388
Input/output (I/O) devices, 381
In situ gamma-ray diffractometry technique, 201

Integrated circuit fabrication, 13, 358
    crystal order and perfection, 48–50
    crystal orientations and planes, 50–54
    crystal structures, 45–47
    diamond crystal structure, 47–48
    influence of dopants and impurities in silicon crystals, 54–58
Integrated circuit fabrication technology, 27
    revolution in, 4–6
Integrated circuit fabrication technology, key material for, 27–28
    metallurgical-grade silicon, 29–31
    polycrystalline silicon feed for crystal growth, 37–41
    raw silicon material, preparation of, 28–29
    ultra-high pure silicon for electronics applications, 37
Integrated circuits (ICs), packaging/ bumping, 381
Integrated hydrodynamic thermal-capillary model (IHTCM), 120
Integrated thermal-capillary model (ITCM), 80
Intensity-modulated energy source, 177
Intentional dopant impurities in silicon wafers, 247–253
Intentional dopant species, 107
Interface control algorithm, 100
Interface shape prediction, 104
Intermediate compounds, 37
Internal diameter (ID)
    slicing, 164
    wafering, 163
Internal gettering, 352
Interstitial-and vacancy-type microdefects, 296
Interstitial carbon diffusivity, 258
Interstitial defects, 50
Interstitial iron, 186
Interstitial oxygen, 269
    concentration, 355
    reduction, 200
    striations, 276
Intrinsic conductivity, 8
Intrinsic gettering, 353
    of germanium-doped CZ, 364
Intrinsic point defect, 296
Intrinsic silicon semiconductor, band diagram of, 7
Intrinsic stacking faults, formation of, 50
Ion-exchanged zeolites, 39
Ionized impurities, distribution of, 210
Iron, 263–265, 359
Iron concentration
    in single-crystal Czochralski silicon, 264
    of wafers, 73

Iron surface contamination, 351
Irregular-shaped wafers, van der Pauw
    resistivity measurement technique
    for, 210
Irvin's curves, 228
Isoelectronic impurities, 183, 202, 250, 282
ITCM, *see* Integrated thermal-capillary model

J

Joulean heat distribution, 126
Joulean heating, 125

K

Kelvin probe, 304

L

Laplace's model, 118
Large-diameter silicon growth system, 100
Large scale integration (LSI), 18
    circuits and applications, 18–20
Laser-induced damage (LID), 358
Laser scattering tomography (LSTDs),
    303, 306
Last cleaning solution, 350
Laue back-reflection x-ray technique, 158
Leakage current measurement technique, 185
LID, *see* Laser-induced damage
Light point defects (LPDs), 315
Linear four-point probe, 191
Liquid etchants, 181
Liquid feeding CZ crystal growth method,
    103
Liquid feeding method, 102
Liquid silicon crystals *vs.* grown silicon
    crystals, 102–107, 130
Liquid–solid interface, 79, 107
Liquid-solid monocomponent growth system,
    84
LLS, *see* Localized light scatters
Localized light scatters (LLS), 326
    on silicon surfaces, 388
Longitudinal oxygen distribution in silicon
    crystals, 88
Longitudinal temperature gradient, 127
Lorentz force, 103, 115
Low boron-doped crystals, 321
Low pressure CVD (LPCVD)-deposited
    amorphous silicon, 202
Low pressure CVD (LPCVD) technique, 249
Low-pressure CZ puller technique, 258

LPCVD technique, *see* Low pressure CVD
    technique
LPDs, *see* Light point defects
LSI, *see* Large scale integration
LSTDs, *see* Laser scattering tomography
Luminescence spectroscopy, 263

M

Macroscopic radial distribution of phosphorus
    dopant, 130, 253
Magnetically-grown-CZ (MCZ) technique, 366
Magnetic CZ (MCZ)–grown silicon wafers, 297
Magnetic CZ (MCZ) techniques, 281
Magnetic fields, 131, 275
Mapping system, command menu for, 228, 243
Marangoni convection, 114
Marangoni forces, 103
Mass-action law, 274
Mass spectra analysis, 196
Mass spectrography, 188
Mass spectroscopy (MS) analysis, 384
Matrix ion species, 204
Maxwell's equations, 112, 185
MCZ techniques, *see* Magnetic CZ techniques
Mechanical superiority of CZ silicon, 336
Megasonic cleaning, 170–171
Melt-grown silicon crystals, 315
Melt oxygen fraction, 65
Melt-quenching technique, 277
Melt-solid interface, 92
Melt temperature distribution, 115
Meridional flow field, 107
Metal concentration in solar-grade silicon,
    255
Metal impurity species, 359
    and gettering methods, 360–361
Metallic contaminations, 27, 177, 185
    precipitation of, 296
    removal of, 350
    in silicon wafers, 347
Metallic impurities, 28, 32, 248, 254, 282
    gettering of, 351–361
    in metallurgical-grade silicon, 135
    removal of, 348–351
    segregation of, 371
    in wafers, concentration of, 247
Metallic precipitates, presence of, 28
Metallurgical-grade polysilicon, 28
Metallurgical-grade silicon (MGS), 28–31, 124,
    247, 254, 255
    for elemental analysis, 179
    purification of, 31–37

Metallurgical silicon, 29
    ways to purify, 32, 33
Metal-oxide semiconductor (MOS), 225
    capacitors, 249
    devices for digital applications, 18
    generation, 353
    structures, 296
Metal-oxide semiconductor (MOS)-based
        circuits, 3
Metal-oxide semiconductor field-effect
        transistor (MOSFET), 261
    devices, 18, 96, 188, 306
    electronic properties of, 12
MGS, *see* Metallurgical-grade silicon
Microcrystalline polysilicon, 255
Microdefect agglomerates, 294
Microdefect decoration, quality of, 187
Microdefect density, 104
Microdefect distribution in quenched crystals,
        303
Microdefect dynamics, 293
Microdefect formation, 302
Microdefects, 294
    in CZ-grown silicon crystals, 194, 302
    types of, 257, 303
Micro droplets of silicon liquid, 314
Microelectromechanical systems
        (MEMS), 77
Microelectromechanical systems (MEMS)
        applications, 46, 168
    diameter silicon wafer specifications for, 13–14
    silicon for, 20–23
Microelectronic devices, ever-shrinking size
        of, 294
Micro-FTIR, 193
    absorption spectroscopy, 105
    measurements, 278
Micro-inhomogeneities, 183, 211, 314
Micro-plastics, abundance and wide
        distribution of, 377
Microprocessors, 18
Microroughness, 202
Micro-segregation traces, 136
Microstructures, forms of, 46
Minority carrier diffusion length, 197–198
Misfit factor, 55
Mobility, 195
Modern CZ crystals, 258
Modern FZ silicon crystals, 128
Modified circular plate theory, 207
Moisture-free argon, 125
Mold-shaping technique, 269
Molecular beam epitaxy method, 137

Molecular-dynamic simulations of crystal
        growth, 79
Molten salt electrolysis method, 31
Molten silicon
    convection of, 211
    crucible choice for, 64–72
    typical impurities in, 254
Molten silicon flow
    effects of, 211
    instability, 80
Monocrystalline silicon, 187
MOS, *see* Metal-oxide semiconductor
MOSAIC model, 387
MOS-based circuits, *see* Metal-oxide
        semiconductor-based circuits
MOSFET transistors, *see* Metal-oxide
        semiconductor field-effect transistor
MS analysis, *see* Mass spectroscopy analysis
Multilayer step-junction theory, 205
Multizone adaptive grid, 121

N

Nano-IC manufacturing, 19
Nanosecond laser radiation pulses, 81
Nascent oxygen, 64
Navier-Stokes, 112
    continuity, 191
    equations, 70
Needle-eye inductor, 126
Needle-eye technique, 131, 132, 135
Neutron activation, 188
    analysis, 358, 366
Neutron irradiation, 366
New donors (NDs), 99
Newton method, 63
Nickel, 265
    impurities in CZ single-crystal silicon, 265
Nitrogen, 96–97, 265–268
Nitrogen annealing, 329
Nitrogen-doped CZ (NCZ), 97, 117
    crystals, 330
    silicon, 276
Nitrogen doping, 97, 192, 204
Nitrogen gas, 380
Nitrogen monomers, 84
Nitrogen out-diffusion profiles, 266
Nitrogen-oxygen defects, 192
Nitrogen-vacancy-oxygen defects, 267
Noble metals, 371
Noncontact sheet resistance, 185
Noncrystalline materials, 7
    exhibit, 45

Nondestructive optical method, 202
Nondimensional parameters, estimation of, 79
Nonequilibrium atomic diffusion, 330
Nonideal crystal lattice, 211
Nonlinear optical characterization, 185
Nonlinear optical techniques, potential of, 207
Nonrotating crystals, 130
Nonsilicon atoms, 247
    precipitation of, 322
Nonuniform axial field, 114
Normal cleaning process, 371
No-self-interstitial-rich defects, 123
Novel unbalanced magnetic technique, 115
No-wafer-preparation detection system, 304
n-type dopants, 204
n-type silicon, 11, 12
    wafers, 160
Nucleation temperature of interstitial
    microdefects, 258
Numerical modeling, 126
Numerical simulation of complete crystal
    growth system, 76

O

Octet rule, 3
OISFs, *see* Oxidation-induced stacking faults
One-dimensional doping profile
    reconstruction, 185
1.1 eV band-edge emissions, 200
One-step high-temperature process, 364
Open manufacturing systems, 392
OPP, *see* Oxygen precipitate profiler
Optical band spectra, 199
Optical microscope photographs of flow
    pattern defects, 308
Optical photograph, 354
Optical photomicrograph of silicon wafer, 362
Optical projection lithography, requirements
    of, 168
Optical techniques, 185
Optimum rotation parameters, 127
Organic additives, adsorption of, 385
Organic compounds, trace levels of, 382
Organic contaminants, 382
    gas chromatogram of, 384–386
Organic materials, 349
Organic species, 387
    rate theory of multicomponent adsorption
        of, 386
Orthogonal components of the cusp magnetic
    field, 275
Oscillation amplitude, 104

Oscillatory flows of molten silicon, 91
Oscillatory melt flow fields, 93
Out-diffusion of boron impurity, 82
Out-of-plane deformation, 207
Ovalent crystal, 3
Oxidation-induced stacking faults (OISFs), 64,
    248, 297, 347
    density measurement, 319
    formation of, 317
    ring, 318, 320
Oxide precipitation, sequence of, 272
Oxidized acid–cleaned wafers, 351
Oxygen, 2
    in silicon wafers, 269
    unintentional dopant impurities in silicon
        wafers, 268–281
Oxygen concentration, 87
    in crystal grown, 88
    distribution in silicon, 276
Oxygen distribution in CZ single crystals, 277
    of silicon, 193
Oxygen equilibrium solubility, effects of
    antimony doping on, 251
Oxygen-free argon, 125
Oxygen impurity in silicon, 273
Oxygen micro-precipitation, 276
Oxygen microsegregation, 183, 192, 206
Oxygen out-diffusion mechanism, 362
Oxygen precipitate morphology, 325
Oxygen precipitate profiler (OPP), 327, 331
Oxygen precipitation, 276, 314, 355
    behavior, 252
    in CZ silicon, 261, 269
    distribution of, 362
    formation of, 271
    in hydrogen annealing, 296
    variety of shapes of, 326
Oxygen segregation coefficient, 98
Oxygen transportation mechanism, 269
Ozone-based cleaning recipes, 387

P

Pancake induction coil, 125, 133
Pancake-shaped inductor, 126
Particle activation analysis, 258
Particulate contamination, control of, 388–391
Parts per million (ppm), 31
PC, *see* Polycarbonate
Peclet number, 79
Pedestal pulling technique, 257, 303
PEEK, *see* Polyether etherketone
Pendellösung fringes, 209, 273, 311, 313

Perfect crystal, 55
Peroxide-based solutions, 371
Phase-difference scanning optical microscope, 202
Phosphorus, 95–96, 253
Phosphorus-diffused silicon, 135
Phosphorus-doped crystals, 226
Phosphorus-doped silicon
    crystal, axial resistivity distribution of, 227
    dopant concentration density to resistivity conversion for, 239–241
    resistivity to dopant concentration density conversion for, 233–235
Phosphorus gettering, 357
Phosphorus impurity in silicon, 253
Photoacoustic displacement, 178
Photoacoustic methods, 177
Photoemission electron microscope, 188
Photoexcitation, 199
Photoluminescence, 184, 185, 263
Photoluminescence-based commercial tool, 201
Photoluminescence measurements, 188
Photoluminescence spectroscopy, 199, 200
Photo-scanning measurements, 132
Photo-scanning technique, 132
Photosensitive memories, 297
Photovoltaic applications, casting of polycrystalline silicon for, 138
Photovoltaic devices, cost of, 137
Piranha cleaning, 348, 349
Plane-polarized light ray, 208
Planes, 50–54
Plasma treatment, 179
Plastic deformation of the CZ-grown crystals, 201
p-n junction, 12
    reverse currents, 296
Point defects
    aggregation, 304
    concentration gradients, 300
    dynamics, 320
    in FZ silicon, 303
    in silicon, types of, 293
Polycrystalline, 46
    materials, 45
    sheets, 255, 269
Polycrystalline ingot, 294
Polycrystalline silicon, 3, 82
    feed for crystal growth, 37–41
Polycrystalline silicon rods, 102

Polycrystalline silicon surface roughness, optical evaluation of, 202
Polyether etherketone (PEEK), 380
Polyhedral morphology, 260
Polysilicon, 28
Polysilicon feed, 138
Post-growth crystal cooling, 135
Post-growth thermal gradient, 122–124
Potassium, 188
ppm, *see* Parts per million
Pre-annealing treatment of wafers, 362
Primary thermal donors, 331
Problematic COPs, 327
Process-bound silicon wafers, 22
Process-bound wafers, 188, 334
    and control of particulate contamination with, 392
Process-induced crystal defects, 187
Process-induced crystallographic defects, 293
Process-induced stacking faults, elimination of, 321
Process-induced wafer warpage, 335
Processing single-crystal silicon using the CZ method, 72
Propagation mechanisms, 136
PR residues, *see* Photoresist residues
Pt diffusion profiles, depth profiles of, 365
p-type dopants, 204, 206
p-type silicon, 11, 12
    wafers, 262
Pulsed-force-mode atomic-force microscopy measurements, 388
Pulverized MGS, partial purification of, 33
Pure silicon semiconductor, band diagram of, 7

Q

Quadrant detector, 202
Qualitative model, 332
Qualitative transition model, 70
Quantitative evaluation of the micrograph, 312
Quartz crucible mechanical holder, 60
Quartz silica crucibles, 64, 66
Quasi-steady-state modeling, 80
    of heat transfer, 120
Quenched crystal, 122, 303
Quenching techniques, 330

R

Radial carbon, 258
Radial distribution of oxygen content, 273
Radial dopant distribution, 257, 269
    in silicon, 94

Radial dopant nonuniformities, 131
Radial impurity distribution, 226
Radial oxygen distribution, 278
Radial resistivity measurements, 157–158
Radial resistivity uniformity for n-doped
    material, 247
Radial resistivity variations, 226
    in boron-doped and phosphorus-doped
        silicon crystal, 227
Radial temperature gradient, 114
Radial variation of oxygen concentration, 90
Radio frequency (RF), 381
    coil shape, heating mechanisms and,
        125–126
    heating, 72
Raman scattering, 191
Raman spectroscopy, 184, 208
Random-access memory (RAM) circuits, 197
Rapid optical surface treatment (ROST), 387
Rapid thermal annealing, 266
    approach, 366
    influence of, 331
Rapid thermal processing (RTP), 252
Raw silicon materials, 28
    preparation of, 28–29
RCA-based cleaning process, 351
"RCA Standard Solution 1," 349
"RCA Standard Solution 2," 349
Reflection x-ray method, 335
Regular lattice structure, vacancy in, 49
Relative humidity (RH), 382
Residual stresses, 207, 208
Resistivity contour map for 6-inch silicon
    wafer, 228, 244
Resistivity distributions, 127
Resistivity variations, 188
Reynolds number, 70
    of crystal rotation, 91
Reynolds stress turbulence closure, 68
RF, *see* Radio frequency
RH, *see* Relative humidity
Robotic handling mechanisms, 381
Robust finite-volume/Newton method, 126
ROST, *see* Rapid optical surface treatment
Rotational flow distribution, 116
RTP, *see* Rapid thermal processing
Rutherford backscattering spectroscopy, 177

S

Sb-doped CZ silicon, 250
    crystals, 98
Scanning electron microscope (SEM), 389

for defect analysis, 201–202
    micrograph, 359, 360
Scanning optical microscope, 202–203
SC-1 cleaned wafer, laser light scattering image
    of, 390
Screw dislocations, 310–312
SCSC, *see* Single-crystalline silicon cylinders
S-curve concept, 324
Secco etching, 181, 303
"Secondary flat," 159
Secondary ion mass spectrometry (SIMS), 266
Secondary micro-plastics, 377
Seed/crucible rotation, crystal pull rate and,
    84–93
Seed crystals, 77
Seed rotation, crystal growth rate and, 126–127
Selective etching patterns, 182
Self-interstitial–dominant silicon, 123
Self-interstitials, 50, 293
Self-seeding process, 60
SEM, *see* Scanning electron microscope
SEM-energy dispersive x-ray
    spectroscopy, 194
Semiconductor, 2
    electronic properties of, 1
    properties, 45
    use of silicon as, 6–11
Semiconductor device applications, 7
Semiconductor device technology, 138
Semiconductor electronics, 2
Semiconductor-grade silicon, flowchart for
    obtaining, 39, 40
Semiconductor silicon, concentration of carbon
    in, 258
Sensitive surface analysis technique, 204
Sensitive synchrotron methods, 211
Shaped magnetic field, effect of, 111
Sharp-edged silicon wafer, 167
Sheet resistance, 189
Shockley-Read recombination-generation
    center, 310
Siemens process, 39
Siemens reactor, 38
Siemens technique, 37
Si-H-Cl gas phase system, 34
Silane
    pyrolysis of, 37
    use of, 136
Silane molecule, 39
Silica crucible, oxygen transport from, 87
Silicide formation, 187
Silicon, 2
    in circular and spherical shapes, 137

conversion between resistivity and dopant density in, 229
for MEMS applications, 20–23
as semiconductor, use of, 6–11
Silicon atoms, 4, 47
Silicon-based semiconductors, performance of, 175
Silicon-bulk single crystal, 81
Silicon carbide, 64, 66
    film, 74
Silicon carbide–coated graphite, 73, 74
Silicon chip, packaged, 22
Silicon complexity, devices of, 21
Silicon compounds, 29
Silicon crystalline structures, 136–138
Silicon crystals, 48, 55
    grown-in defect formation in, 297
    influence of dopants and impurities in, 54–58
    planes, 12
    structure, 47
    wafers, 351
Silicon device-processing technology, 4
Silicon devices
    effect of influence on, 247–250
    integration of, 12–18
Silicon diamond structure, 53
Silicon dioxide gate integrity, 319
Silicon electrical resistivity, 228
Silicon hollow tubes, 138
Silicon ingots to silicon wafers
    Boule formation, identification of crystal orientation, and flats, 158–162
    chemical etching and mechanical damage removal, 167–168
    chemimechanical polishing for planar wafers, 168–169
    edge profiling of slices, 167
    final cleaning and inspection, 171
    ingot slicing, 162–164
    megasonic cleaning, 170–171
    radial resistivity measurements, 157–158
    surface roughness and overall wafer topography, 170
    wafer slices, mechanical lapping of, 164–166
Silicon lattice
    misfit factors of typical impurities in, 11
    position of interstitial oxygen in, 271
Silicon multicrystals, 77
Silicon nitride, 64, 67
Silicon nitride–based crucibles, 67
Silicon-on-insulator (SOI)
    substrate wafers, 135
    wafers, 187

Silicon plate sampling (SPS) method, 387
Silicon processing, purity of chemicals for, 347–348
Silicon processing technology, 181
Silicon ribbons, 136–137
    technology, 136
Silicon self-interstitials, 130
Silicon sheets, 137
Silicon single crystals, 60
Silicon technology, 12
Silicon tetrachloride, 34
Silicon wafer preparation
    Argon annealing of wafers, 366–368
    degreasing of, 348
    denuding of, 362–366
    final cleaning, rinsing, and wafer drying, 371
    hydrogen annealing of wafers, 368–370
    metallic impurities, 348–361
    neutron irradiation, 366
    silicon processing, purity of chemicals for, 347–348
Silicon wafers, 22, 45, 188, 247, 252, 293, 324, 366, 367, 384
    acoustic laser probing technique, 175–178
    AES studies, 178–180
    atomic-force microscope studies on surfaces, 178
    bulk defects and voids, 306–310
    chemical staining and etching techniques, 181–184
    contactless characterization, 184–185
    defect decoration by metals, 187
    defects in, 296
    dislocations and screw dislocations, 310–312
    DLTS, 185–187
    edge contouring of, 167
    flame emission spectrometry, 188
    four-point probe technique for resistivity measurement and mapping, 189–191
    FTIR measurements for impurity identification, 191–195
    fully processed blank silicon wafers, packing of, 377–388
    gamma-ray diffractometry, 201
    GFA, 195
    grown vacancies and defects, 329–331
    hall mobility, 195–196
    healing and cooling of, 333
    impurity concentrations, photoluminescence method for determining, 199–201
    industry, 164, 175

line defects, 304–306
mass spectra analysis, 196
mechanical property of, 162
minority carrier diffusion length/lifetime/
    surface photovoltage, 197–198
notch on, 161
optical methods for impurity evaluation, 199
point defects and vacancies, 298–304
precipitations, 322–326
quality of, 175
resistivity measurements, 189
scanning optical microscope, 202–203
SEM for defect analysis, 201–202
shipping container, 378
in silicon devices and structures, 294–298
SIMS for impurity distribution, 203–205
slips, cracks, and shape irregularities,
    332–334
spreading resistance and two-point probe
    measurement technique, 205–207
stacking faults, 315–322
storage of wafers and control of particulate
    contamination, 388–391
stress, bowing, and warpage, 334–337
stress measurements, 207–209
surface pits/crystal-originated particles,
    326–329
swirl defects, 312–315
TEM, 209–210
thermal donors, 331–332
typical lapping and grinding machine
    for, 166
valuation techniques used to study, 176–177
van der Pauw resistivity measurement
    technique for irregular-shaped wafers,
    210
x-ray technique for crystal perfection and
    dislocation density, 210–213
Silicon wafers, impurities in
    carbon, 255–259
    chromium, 259–260
    copper, 260–261
    germanium, 261
    gold, 261–262
    helium, 262
    hydrogen, 262–263
    intentional and unintentional impurities
        and influence on silicon devices, effect
        of, 247–250
    intentional dopant, 250–253
    iron, 263–265
    metallic impurities, 282
    nickel, 265

    nitrogen, 265–268
    oxygen, 268–281
    tin, 281–282
    unintentional dopant, 254–255
Silicon wafer-specific material parameter,
    critical assessment of, 4
Silicon wafers, resistivity and impurity
    concentration mapping of, 225–227
    electrically active and inactive impurities,
        228
    surface mapping and concentration
        contours, 228–229
    surface roughness mapping on complete
        wafer, 229, 243
Silicon whiskers, 137
SIMS, *see* Secondary ion mass spectrometry
SIMS for impurity distribution, 203–205
Single-bulk crystals, growth process of, 85
Single-crystal CZ silicon, metal impurities in,
    248
Single-crystalline silicon cylinders (SCSC), 137
Single-crystal properties, 46
Single-crystal semiconductor, 1
Single-crystal silicon, 3, 52, 59, 293, 304
    Bridgman crystal growth technique, 60
    chamber temperature profile, 72–77
    crystal chamber, environmental and
        ambient control in, 82–84
    crystal growth striations, 107–108
    crystal pulling, seed selection for, 77–81
    crystal pull rate and seed/crucible rotation,
        84–93
    Czochralski crystal growth/pulling
        technique, 60–64
    float-zone crystal growth technique,
        124–135
    growing doped crystals, dopant addition for,
        94–100
    growth orientations, 77
    ingots, 157
    liquid and grown silicon crystals, impurity
        segregation between, 102–107
    methods for continuous, 100–102
    mobility, 195
    molten silicon, crucible choice for, 64–72
    post-growth thermal gradient and crystal
        cooling after pull-out, 122–124
    properties, 20
    silicon crystalline structures and growth
        techniques, 136–138
    use of magnetic field in, 108–117
    VLSI And ULSI applications, large-area
        silicon crystals for, 117–122

wafers, 352
  zone refining of single-crystal silicon, 135–136
Single-crystal substrates, 46
Single-crystal wafers, 326
Single silicon crystals, defects in, 18
Single-wafer chemical polishing, 169
SiO_2, *see* Oxygen precipitates
Si-O-Si network, 28
Sirtl etching, 181, 183, 315
6-inch silicon wafer, measurement of, 163
Slicing methods, types of, 164
Slight ultrasonic agitation, 170
Small-angle x-ray scattering, 251
SMIF, *see* Standard Mechanical Interface
Solar cell efficiencies, 198
Solar-grade silicon, 29, 255
  flowchart for obtaining, 39, 40
  metal content in, 255
Solenoids, 115
Solidification interface, 62
Solid–liquid interface, 68, 79, 130
  configuration, 63
Solid–liquid silicon interface, 79
Solid materials, 45
Solid-melt interface fluctuations, 202
Solid solubility of typical impurities in silicon, 55–56
Solid-source mass spectrometry, 196
Solid-state single crystals, 48
Sophisticated filtration techniques, 379
Spectrophotometer, 202
Spherical precipitate formations, 315
Spherical silicon multicrystals, 137
Spherical single crystals for solar cell substrates, 137
S-pits, formation of, 347
Spoke patterns, 72
Spray-processing tool, 387
SPS method, *see* Silicon plate sampling method
Sputtering process, 179
SPV, *see* Surface photovoltage
Square array configurations, 190
Stacking faults in silicon, 319
Standard cleaning techniques, 202
Standard linear four-point probe method, 157
Standard Mechanical Interface (SMIF), 381
Standard turbulence model, 121
Static magnetic field, 115
Steady-state condition, 76
Steady-state growth process, 100
Steady-state model, 275
Steady-state temperature distribution, 69

STHAMAS3D software package, 138
Stillinger-Weber interatomic potential, 300
Stress measurements, 207–209
Stress-related defect generation, 333
Stress relief mechanism, 280
Striation patterns, 109, 184
Strong dipole–dipole interactions, 382
Structural characterization method, 177
Structural perfection testing, 202
Structural properties of as-grown silicon, 299
Submerged electrode arc furnace, 31
Submicron-sized crystalline silicon oxide, 65
Substitutional impurities, 55
  concentration of, 54
Substitutional impurity atom, 57
Substrate wafers for microelectronic devices, 62, 78
Subtle selective etching method, 271
Super-cooling degree for the nucleation, 273
Supersaturated interstitial oxygen atoms, 99
Surface-controlled field-effect transistors, conductance of, 384
Surface defects, 1, 52
Surface grinding of ingot, 157
Surface mapping, 228–229
Surface photovoltage (SPV), 197–198, 351
Surface profiler, 164
Surface roughness mapping on wafer, 229, 243
Surface temperature model, 83
Surface-tension-driven flow, 108
Surface tension–induced flows, 91
Swirl defects, 296, 312–315
  distributions, 276, 314
  formation, 183, 192, 206, 282
  types of, 129
Swirl-like pattern, 313
Switch-like dependence of oxygen precipitate density, 362
Synchrotron techniques, 263
Synchrotron topographic methods, 273
Synchrotron white x-ray topography, 310

T

Tapping mode, 178
TDDB, *see* Time dependent dielectric breakdown
TDH, *see* Time-dependent haze
Teanx adsorbent solid trap, 387
TEM, *see* Transmission electron microscopy
Temperature distribution
  of 4-inch crystal growth furnace, 76
  of furnace, 74
Temperature–heat field pulsation, 92

Temperature oscillations, effect of, 104
Tetrahedral bonding directions, 2
Thermal asymmetry, 106
Thermal capillary model, 63
Thermal conductivity, degradation of, 178
Thermal convection behavior of silicon melt, 226
Thermal convection dominant procedure, 275
Thermal desorption gas chromatography–mass spectrometry (TDGC–MS) methods, 385, 387
Thermal desorption mass spectrometry spectra, 388
Thermal donors (TDs), 84, 331–332
   formation of, 261
   kinetics of, 332
Thermal fluctuations, effects of, 108
Thermal gradient, 135
Thermal mapping, 194
Thermal radiation energy, 70
Thermal simulations of CZ growth process, 74
Thermal stress distribution in the silicon ingot, 337
Thermal symmetry, 111
Thermal waves of nonaxisymmetric flow, 72, 80
Thermocapillary convection, 67
Thermocouples, 74, 75
Thermohydrodynamics, 72
"Thickness fringes," 311
Thin polycrystalline silicon, 357
Three-dimensional finite-element program, 81
Three-dimensional simulation approaches, 136
Three-dimensional time-dependent simulations, 81
Time dependent dielectric breakdown (TDDB), 369
Time-dependent 3D simulations, 103
Time-dependent haze (TDH), 388
Time-dependent model, 275
   for oxygen incorporation, 80
Time-dependent Navier-Stokes equations, 63
Time-dependent numerical simulation, 116
Time-dependent temperature distribution, 118
Time-dependent variations, 387
Time-of-flight SIMS (TOF-SIMS) approach, 204
Time-saving numerical scheme, 120
Time zero dielectric breakdown (TZDB), 327
Tin, 281–282
TOF-SIMS approach, *see* Time-of-flight SIMS approach
Total reflectance x-ray fluorescence spectroscopy (TXRF), 204

Total reflection x-ray fluorescence analysis (TXRFA) techniques, 349
Total reflection x-ray fluorescence (TXRF) spectroscopy, 250
Trace metals, 371
Traditional crystal-growing techniques, 3
Transition-group metal contaminants, 358
Transition-metal contamination, 27
Transition metallic impurities, 348
Transition metals, interstitial and substitutional defects of, 187, 303
Transmission electron microscopy (TEM), 209–210, 252, 353
   micrograph, 354
Transverse magnetic (TM) field, 83, 130
Trench-type memory devices, 297
Trichloroethylene (TCE) oxidation process, 321
Trichlorosilane, 34
   hydrogen reduction of, 35
   to polycrystalline silicon, thermal conversion of, 36
Turbulence, 121
Two-beam bright-field TEM image, 370
2D axisymmetric models, 103
Two-dimensional stress maps, 185
Two-point probe measurement technique, 205–207
Two-side lapping method, 164
Two-step annealing process, 273
TXRF, *see* Total reflectance x-ray fluorescence spectroscopy
TXRFA techniques, *see* Total reflection x-ray fluorescence analysis techniques
TXRF spectroscopy, *see* Total reflection x-ray fluorescence spectroscopy
TZDB, *see* Time zero dielectric breakdown

U

ULSI, *see* Ultra large scale integration
Ultra-high-purity silicon, 129
   for electronics applications, 37
Ultra large scale integration (ULSI)
   applications, 293
   circuits and applications, 18–20
   device processing, 353
   devices, 294, 351
Ultrasonic agitation, 183, 391
Ultraviolet (UV) radiation, 392
Unbalanced magnetic technique, 115
Uncut ingot, 165
Uniaxial stress, 208
Uniform magnetic field, 131

Unintended impurities, presence of, 12
Unintentional dopant impurities in silicon
        wafers, 247–250, 254–255
    carbon, 255–259
    chromium, 259–260
    copper, 260–261
    germanium, 261
    gold, 261–262
    helium, 262
    hydrogen, 262–263
    iron, 263–265
    nickel, 265
    nitrogen, 265–268
    oxygen, 268–281
    tin, 281–282
Unintentional dopant species, 107
Unintentional impurity, 259
Unintentional iron impurity in silicon, 263
Union Carbide process, 34
Unwanted metallic impurities, 352, 358

V

Vacancy clusters, formation of, 124
Vacancy concentration profiles, use of, 296
Vacancy concentration wafer mapping, 197, 330
Vacancy-dominant silicon, 123
Vacancy-interstitial annihilation mechanism,
        105
Vacancy-interstitials, product of, 364
Vacancy/nitrogen species, 268
Vacancy-type defects, 297
Valence electrons, 7
van der Pauw resistivity measurement
        technique for irregular-shaped wafers,
        210
van der Waals forces, 391
Vapor drying technique, 387
Vapor-liquid-solid (VLS) mechanism, 137
Vertical magnetic field, 112
Very large scale integration (VLSI), 294, 351
    circuits and applications, 18–20
    transistors, 225
    and ULSI applications, large-area silicon
        crystals for, 117–122
Void formation in silicon crystals, 306
Volatile organic gaseous species, 385
Voronkov's theory, 123, 297

W

Wafer map, 320
Wafer resistivity values, measuring, 228

Wafers, 18, 74, 297
    argon annealing of, 366–368
    breakage of, 160
    drying, 371
    hydrogen annealing of, 368–370
    iron concentrations of, 73
    quality of, 319
    storage of, 388–392
    surface roughness mapping on, 229, 243
Wafer several microns, 351
Wafer slices, mechanical lapping of, 164–166
Wafer-slicing methods, 164
Wafer storage, different generations of, 379
Wafer topography, 170
Wafer uniformity, demand for, 247
Wafer warpage, measurement of, 335
Warpage, 334
    of laser-damaged silicon wafers, 336
Warped wafers, 334
Water-model melt-flow simulation, 89
Water-model method, 90
Wet chemical cleaning, 348
Wet chemical etching, 167
Wet cleaning process, 350, 371
White-light illumination method, 387
Wire saw, 164
Wright etchant, 181, 182

X

XPS, *see* X-ray photoelectron spectroscopy
X-ray diffraction (XRD), 251
    analysis, 39
    topography, 211, 311
X-ray fluorescence spectra, 263
X-ray methods, 211
X-ray photoelectron spectroscopy (XPS), 251
X-ray radiography, 67, 85, 211
X-ray Raman scattering, 250
X-ray technique for crystal perfection, 210–213
X-ray topographic methods, 211
X-ray topography, 253, 313, 333
    of cone section, 305
    of silicon wafer, 320
XRD, *see* X-ray diffraction
Xylene, 382

Z

Zero boron concentration, 321
Zero-etch-pit-density (0-EPD), 130
Zone refining, 139